SCHAUM'S OUTLINE OF

THEORY AND PROBLEMS

OF

BEGINNING CALCULUS

•

ELLIOTT MENDELSON, Ph.D.
Professor of Mathematics
Queens College

•

SCHAUM'S OUTLINE SERIES
McGRAW-HILL, INC.
New York St. Louis San Francisco Auckland Bogotá
Caracas Lisbon London Madrid Mexico Milan
Montreal New Delhi Paris San Juan Singapore
Sydney Tokyo Toronto

ELLIOTT MENDELSON is Professor of Mathematics at Queens College of the City University of New York. He also has taught at the University of Chicago, Columbia University, and the University of Pennsylvania, and was a member of the Society of Fellows of Harvard University. He is the author of several books, including *Schaum's Outline of Boolean Algebra and Switching Circuits.* His principal area of research is mathematical logic and set theory.

This book is printed on recycled paper containing a minimum of 50% total recycled fiber with 10% postconsumer de-inked fiber.

Schaum's Outline of Theory and Problems of
BEGINNING CALCULUS

9 10 11 12 13 14 15 SH SH 9 8 7 6 5 4 3 2

ISBN 0-07-041465-3

Sponsoring Editor, David Beckwith
Editing Supervisor, Marthe Grice
Production Manager, Nick Monti
Cover design by Amy E. Becker.

Library of Congress Cataloging in Publication Data

Mendelson, Elliott.
Schaum's outline of theory and problems of beginning calculus.

(Schaum's outline series)
Includes index.
1. Calculus. 2. Calculus - - Problems, exercises, etc.
I. Title. II. Title: Beginning calculus.
QA303.M387 1985 515 84-12607
ISBN 0-07-041465-3

Preface

This Outline is limited to the essentials of calculus. It carefully develops, *giving all steps*, the principles of differentiation and integration on which the whole of calculus is built. The book is suitable for reviewing the subject, or as a self-contained text for the first part of an elementary calculus course.

Because the author has found that many of the difficulties students encounter in calculus are due to weakness in algebra and arithmetical computation, emphasis has been placed on reviewing algebraic and arithmetical techniques whenever they are used. Every effort has been made—especially in regard to the composition of the solved problems—to ease the beginner's entry into calculus. There are also some 1400 supplementary problems (with a complete set of answers at the end of the book).

High-school courses in calculus can readily use this Outline. Many of the problems are adapted from questions in the Advanced Placement Examination in Calculus AB, so that students will automatically receive preparation for that test.

The author wishes to thank his editor, David Beckwith, for encouragement and practical advice. He also wishes to thank his wife and family, and all his friends, for their patience and understanding while he became a hermit during the writing of this book.

ELLIOTT MENDELSON

Contents

CONTENTS

Chapter 1

Coordinate Systems on a Line

1.1 THE COORDINATE OF A POINT

Let $\mathscr{L}$ be a line. Choose a point O on the line, and call this point the *origin*.

Now, select a direction along $\mathscr{L}$; say, the direction from left to right on the diagram.

For every point P to the right of the origin O, let the *coordinate* of P be the distance between O and P.

(Of course, to specify such a distance, it is first necessary to establish a unit distance by arbitrarily picking two points and assigning the number 1 to the distance between these two points.)

In the diagram

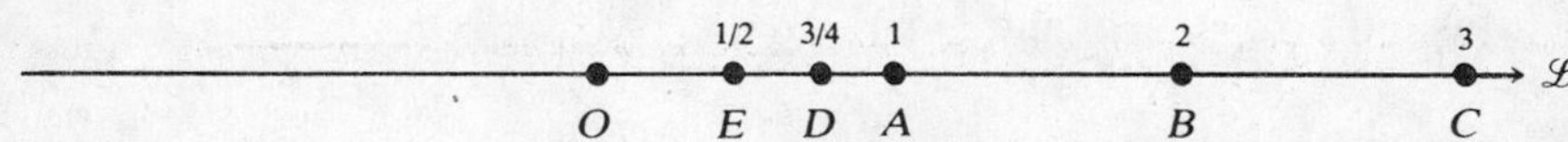

the distance $\overline{OA}$ is assumed to be 1, so that the coordinate of A is 1. The point B is two units away from O; therefore, B has coordinate 2. Every positive real number r is the coordinate of a unique point on $\mathscr{L}$ to the right of the origin O; namely, of that point to the right of O whose distance from O is r.

To every point Q on $\mathscr{L}$ to the left of the origin O,

we assign a negative real number as its coordinate; namely, the number $-\overline{QO}$, the negative of the distance between Q and O. For example, in the diagram

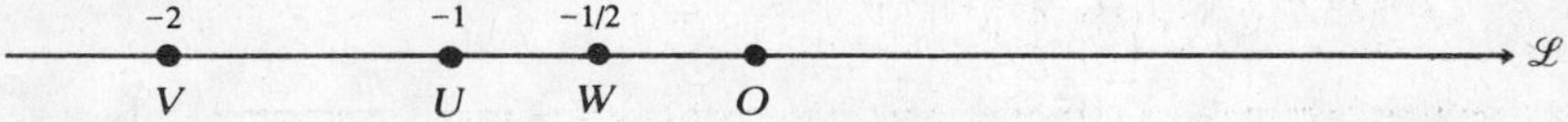

the point U is assumed to be a distance of one unit from the origin O; therefore, the coordinate of U is -1. The point W has coordinate $-\frac{1}{2}$, which means that the distance $\overline{WO}$ is $\frac{1}{2}$. Clearly, every negative real number is the coordinate of a unique point on $\mathscr{L}$ to the left of the origin.

The origin O is assigned the number 0 as its coordinate.

This assignment of real numbers to the points of the line $\mathscr{L}$ is called a *coordinate system* on $\mathscr{L}$.

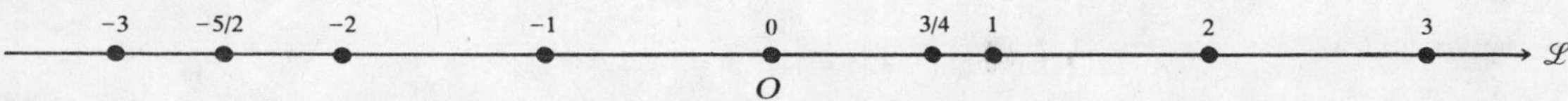

Choosing a different origin, a different direction along the line, or a different unit distance, would result in a different coordinate system.

1.2 ABSOLUTE VALUE

For any real number b, define the *absolute value* $|b|$ to be the magnitude of b; that is,

$$|b| = \begin{cases} b & \text{if } b \geq 0 \\ -b & \text{if } b < 0 \end{cases}$$

In other words, if b is a positive number or zero, its absolute value $|b|$ is b itself. But, if b is negative, its absolute value $|b|$ is the corresponding positive number $-b$.

EXAMPLES

$$|3| = 3 \qquad \left|\frac{5}{2}\right| = \frac{5}{2} \qquad |0| = 0 \qquad |-2| = 2 \qquad \left|-\frac{1}{3}\right| = \frac{1}{3}$$

Properties of the Absolute Value

Notice that any number r and its negative $-r$ have the same absolute value:

$$|r| = |-r| \tag{1.1}$$

An important special case of (*1.1*) results from choosing $r = u - v$:

$$|u - v| = |v - u| \tag{1.2}$$

If $|a| = |b|$, then either a and b are the same number or a and b are negatives of each other:

$$|a| = |b| \quad \text{implies} \quad a = \pm b \tag{1.3}$$

Moreover, since $|a|$ is either a or $-a$, and $(-a)^2 = a^2$,

$$|a|^2 = a^2 \tag{1.4}$$

Replacing a in (*1.4*) by ab gives

$$|ab|^2 = (ab)^2 = a^2b^2 = |a|^2\,|b|^2 = (|a|\,|b|)^2$$

whence, the absolute value being nonnegative,

$$|ab| = |a|\,|b| \tag{1.5}$$

Absolute Value and Distance

Consider a coordinate system on a line $\mathscr{L}$, and let A_1 and A_2 be points on $\mathscr{L}$ with coordinates a_1 and a_2. Then:

$$|a_1 - a_2| = \overline{A_1A_2} = \text{distance between } A_1 \text{ and } A_2 \tag{1.6}$$

EXAMPLES

(*i*)

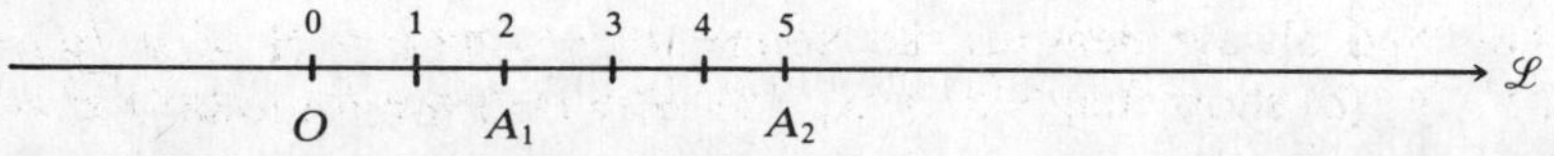

$$|a_1 - a_2| = |2 - 5| = |-3| = 3 = \overline{A_1A_2}$$

(*ii*)

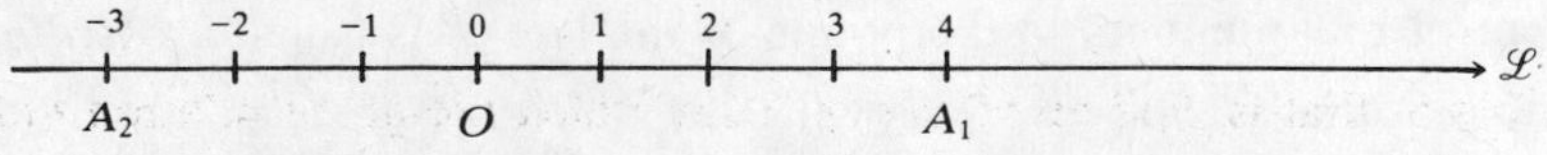

$$|a_1 - a_2| = |4 - (-3)| = |4 + 3| = |7| = 7 = \overline{A_1A_2}$$

A special case of (*1.6*) is very important: If a is the coordinate of A, then

$$|a| = \text{distance between } A \text{ and the origin} \tag{1.7}$$

The Triangle Inequality

For any positive number c,

$$|u| \leq c \quad \text{is equivalent to} \quad -c \leq u \leq c \tag{1.8}$$

EXAMPLE $|u| \leq 3$ if and only if $-3 \leq u \leq 3$.

Similarly,

$$|u| < c \quad \text{is equivalent to} \quad -c < u < c \tag{1.9}$$

EXAMPLE To find a simpler form for the condition $|x-3| < 5$, substitute $x-3$ for u in (*1.9*), obtaining $-5 < x-3 < 5$. Adding 3, we have: $-2 < x < 8$.

It follows immediately from the definition of the absolute value that, for any two numbers a and b,

$$-|a| \leq a \leq |a| \qquad \text{and} \qquad -|b| \leq b \leq |b|$$

Hence, by addition,

$$-(|a| + |b|) \leq a + b \leq |a| + |b|$$

and so, by (*1.8*), with $u = a + b$ and $c = |a| + |b|$,

$$|a+b| \leq |a| + |b| \quad \textbf{triangle inequality} \tag{1.10}$$

In (*1.10*), the sign $<$ applies when and only when a and b are of opposite signs.

EXAMPLE $|3 + (-2)| = |1| = 1$, but $|3| + |-2| = 3 + 2 = 5$.

Solved Problems

1.1 Recalling that $\sqrt{u}$ always denotes the *nonnegative* square root of u, (*a*) evaluate $\sqrt{3^2}$; (*b*) evaluate $\sqrt{(-3)^2}$; (*c*) show that $\sqrt{x^2} = |x|$. (*d*) Why isn't the formula $\sqrt{x^2} = x$ always true?

(*a*) $\sqrt{3^2} = \sqrt{9} = 3$. (*b*) $\sqrt{(-3)^2} = \sqrt{9} = 3$. (*c*) By (*1.4*), $x^2 = |x|^2$; hence, since $|x| \geq 0$, $\sqrt{x^2} = |x|$. (*d*) Because $|x| = x$ is false when $x < 0$; see part (*c*).

1.2 Solve $|x+3| \leq 5$; that is, find all values of x for which the given relation holds.

By (*1.8*), $|x+3| \leq 5$ if and only if $-5 \leq x + 3 \leq 5$. Subtracting 3, $-8 \leq x \leq 2$.

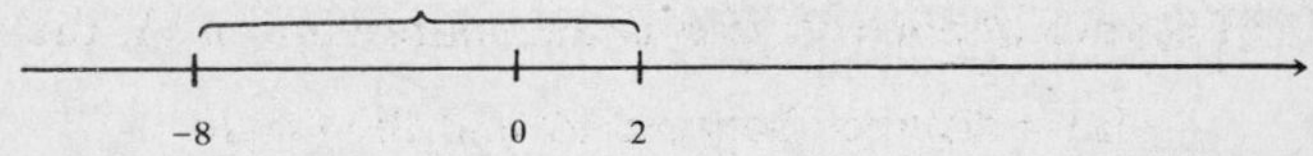

1.3 Solve $|3x+2|<1$.

By (*1.9*), $-1<3x+2<1$. Subtracting 2, $-3<3x<-1$. Dividing by 3,

$$-1<x<-\frac{1}{3}$$

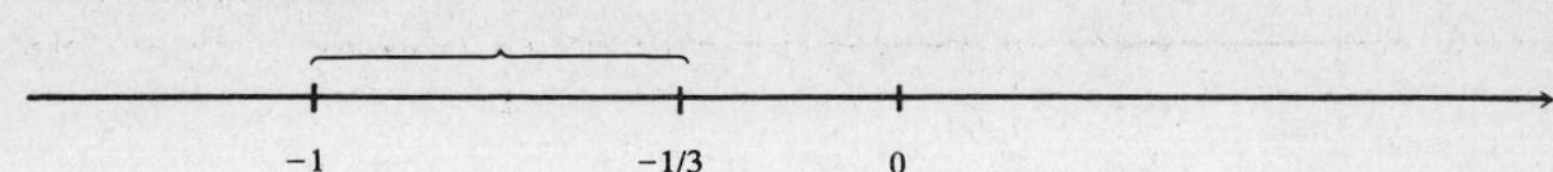

1.4 Solve $|5-3x|<2$.

By (*1.9*), $-2<5-3x<2$. Subtracting 5, $-7<-3x<-3$. Dividing by -3,

$$\frac{7}{3}>x>1$$

ALGEBRA REVIEW Multiplying or dividing both sides of an inequality by a negative number *reverses* the inequality: if $a<b$ and $c<0$, then $ac>bc$.

To see this, notice that $a<b$ implies $b-a>0$. Hence, $(b-a)c<0$, since the product of a positive number and a negative number is negative. So, $bc-ac<0$, or $bc<ac$.

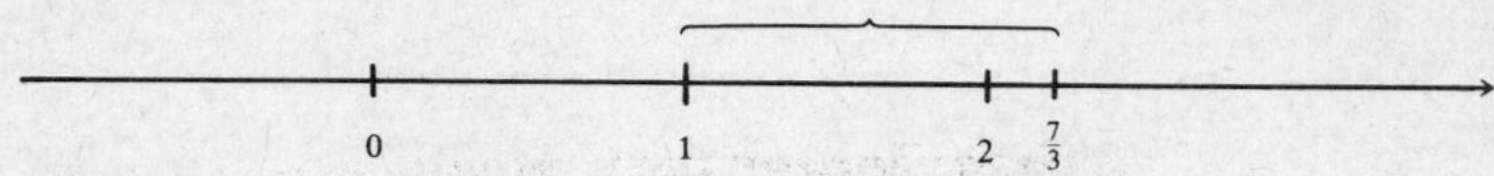

1.5 Solve

$$\frac{x+4}{x-3}<2 \qquad (1)$$

We cannot simply multiply both sides by $x-3$, because we do not know whether $x-3$ is positive or negative.

Case 1: $x-3>0$. Multiplying (*1*) by this positive quantity preserves the inequality:

$$x+4<2x-6$$

Subtract x: $4<x-6$
Add 6: $10<x$

Thus, when $x>3$, (*1*) holds when and only when $x>10$.

Case 2: $x-3<0$. Multiplying (*1*) by this negative quantity reverses the inequality:

$$x+4>2x-6$$

Subtract x: $4>x-6$
Add 6: $10>x$

Thus, if $x<3$, (*1*) holds if and only if $x<10$. But, $x<3$ implies that $x<10$. Hence, when $x<3$, (*1*) is true.

From cases 1 and 2, (*1*) holds for $x>10$ and for $x<3$.

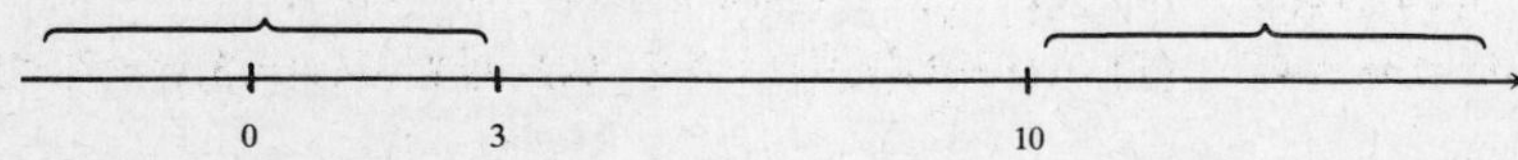

1.6 Solve $(x-2)(x+3)>0$.

A product is positive when and only when both factors are of like sign.

Case 1: $x-2>0$ and $x+3>0$. Then $x>2$ and $x>-3$. But these are equivalent to $x>2$ alone, since $x>2$ implies $x>-3$.

Case 2: $x-2<0$ and $x+3<0$. Then $x<2$ and $x<-3$, which are equivalent to $x<-3$, since $x<-3$ implies $x<2$.

Thus, $(x-2)(x+3)>0$ holds when either $x>2$ or $x<-3$.

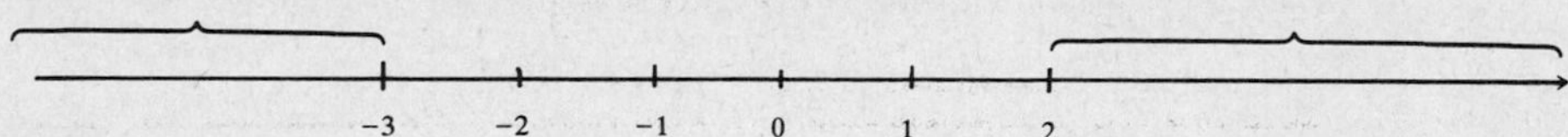

1.7 Solve $|3x-2| \geq 1$.

Let us solve the negation of the given relation, $|3x-2|<1$. By (*1.9*),

$$-1<3x-2<1$$

$$1<3x<3 \qquad \text{[add 2]}$$

$$\frac{1}{3}<x<1 \qquad \text{[divide by 3]}$$

Therefore, the solution of $|3x-2| \geq 1$ is $x \leq \frac{1}{3}$ or $x \geq 1$.

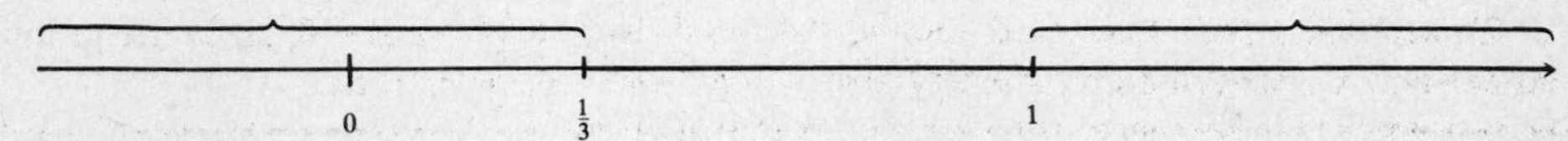

Supplementary Problems

1.8 (*a*) For what kind of number u is $|u|=-u$? (*b*) For what values of x does $|3-x|$ equal $x-3$? (*c*) For what values of x does $|3-x|$ equal $3-x$?

1.9 Show that u is equal to either $|u|$ or $-|u|$. (*Hint*: Consider two separate cases.)

1.10 Solve: (*a*) $|2x+3|=4$, (*b*) $|5x-7|=1$.

1.11 Solve:

(*a*) $|x-1|<1$ (*b*) $|3x+5| \leq 4$ (*c*) $|x+4|>2$

(*d*) $|2x-5| \geq 3$ (*e*) $|x^2-10| \leq 6$ (*f*) $\left|\frac{x}{2}+3\right|<1$

1.12 Solve:

(*a*) $\frac{x}{x+5}<1$ (*b*) $\frac{x-7}{x+3}>2$ (*c*) $\left|\frac{1}{x}-2\right|<4$

(*d*) $\left|1+\frac{3}{x}\right|>2$ (*e*) $1<3-2x<5$ (*f*) $3 \leq 2x+1<4$

1.13 Solve:

(*a*) $x(x+2)>0$ (*b*) $(x-1)(x+4)<0$ (*c*) $x^2-6x+5>0$ (*d*) $x^2+7x-8<0$

(*e*) $x^2<3x+4$ (*f*) $x(x-1)(x+1)>0$ (*g*) $(2x+1)(x-3)(x+7)<0$

(*Hints*: In (*c*), factor; in (*g*), there are four cases.)

1.14 Show that if $b \neq 0$,

$$\left|\frac{a}{b}\right|=\frac{|a|}{|b|}$$

(*Hint*: Use (*1.5*).)

1.15 Show that (*a*) $|a^2| = |a|^2$, (*b*) $|a^3| = |a|^3$. (*c*) Generalize the results (*a*) and (*b*).

1.16 Solve:

$$(a)\quad |2x-3| = |x+2| \qquad (b)\quad |7x-5| = |3x+4| \qquad (c)\quad 2x-1 = |x+7|$$

1.17 Solve: (*a*) $|2x-3| < |x+2|$, (*b*) $|3x-2| \le |x-1|$. (*Hint*: In (*a*), consider the three cases $x \ge \frac{3}{2}$, $-2 \le x < \frac{3}{2}$, and $x < -2$.)

1.18 (*a*) Prove: $|a-b| \ge ||a|-|b||$. (*Hint*: Use the Triangle Inequality to prove that $|a| \le |a-b| + |b|$ and $|b| \le |a-b| + |a|$.) (*b*) Prove: $|a-b| \le |a| + |b|$.

1.19 Determine whether $\sqrt{a^4} = a^2$ holds for all real numbers a.

1.20 Does $\sqrt{a^2} < \sqrt{b^2}$ always imply that $a < b$?

1.21 Let O, I, A, B, C, D be points on a line, with respective coordinates 0, 1, 4, -1, $\frac{3}{2}$, and $-\frac{1}{3}$. Draw a diagram showing these points and find: $\overline{IA}$, $\overline{AI}$, $\overline{OC}$, $\overline{BC}$, $\overline{IB} + \overline{BD}$, $\overline{ID}$, $\overline{IB} + \overline{BC}$, $\overline{IC}$.

1.22 Let A and B be points with coordinates a and b. Find b, if (*a*) $a = 7$, B is to the right of A, and $|b-a| = 3$; (*b*) $a = -1$, B is to the left of A, and $|b-a| = 4$; (*c*) $a = -2$, $b < 0$, and $|b-a| = 3$.

1.23 Prove: (*a*) $a < b$ is equivalent to $a + c < b + c$.

ALGEBRA $a < b$ means that $b - a$ is positive. The sum and product of two positive numbers are positive, the product of two negative numbers is positive, and the product of a positive and a negative number is negative.

(*b*) If $0 < c$, then $a < b$ is equivalent to $ac < bc$ and to

$$\frac{a}{c} < \frac{b}{c}$$

Chapter 2

Coordinate Systems in a Plane

2.1 THE COORDINATES OF A POINT

We shall establish a correspondence between the points of a plane and pairs of real numbers.

Choose two perpendicular lines in the plane, as in Fig. 2-1. Let us assume for the sake of simplicity that one of the lines is horizontal and the other vertical. The horizontal line will be called the *x-axis* and the vertical line will be called the *y-axis*.

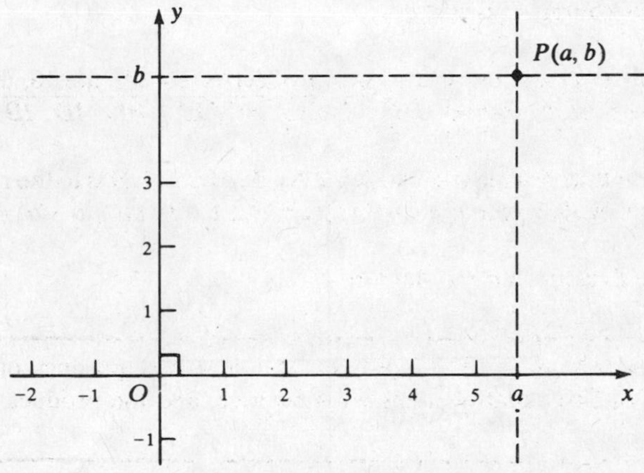

Fig. 2-1

Next, choose coordinate systems on the x-axis and the y-axis. The origin for both coordinate systems is taken to be the point O where the axes intersect. The x-axis is directed from left to right, and the y-axis from bottom to top. The part of the x-axis with positive coordinates is called the *positive x-axis*, and the part of the y-axis with positive coordinates the *positive y-axis*.

Consider any point P in the plane. Take the vertical line through the point P, and let a be the coordinate of the point where the line intersects the x-axis. This number a is called the *x-coordinate* of P (or the *abscissa* of P). Now, take the horizontal line through P, and let b be the coordinate of the point where the line intersects the y-axis. The number b is called the *y-coordinate* of P (or the *ordinate* of P). Every point has a unique pair (a, b) of coordinates associated with it.

EXAMPLES In Fig. 2-2, the coordinates of several points have been indicated. We have limited ourselves to integer coordinates only for simplicity.

Conversely, every pair (a, b) of real numbers is associated with a unique point in the plane.

EXAMPLES In the coordinate system of Fig. 2-3, to find the point having coordinates (3, 2), start at the origin O, move three units to the *right* and then two units *upward*. To find the point with coordinates $(-2, 4)$, start at the origin O, move two units to the *left* and then four units *upward*. To find the point with coordinates $(-1, -3)$, start from the origin, move one unit to the *left* and then three units *downward*.

Given a coordinate system, the entire plane except for the points on the coordinate axes can be divided into four equal parts, called *quadrants*. All points with both coordinates positive form the first quadrant, quadrant I, in the upper right-hand corner (see Fig. 2-4). Quadrant II consists of all points with negative x-coordinate and positive y-coordinate; quadrants II, III, and IV are also shown in Fig. 2-4.

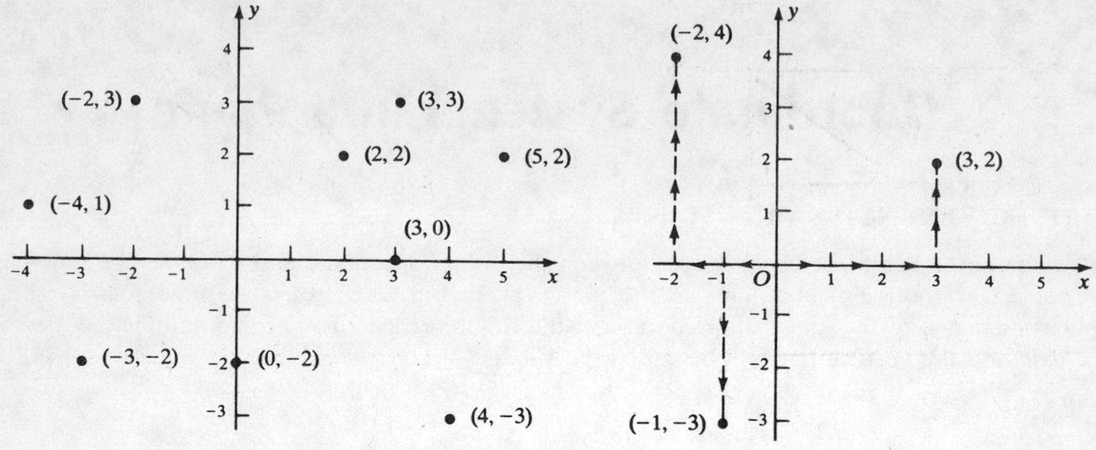

Fig. 2-2

Fig. 2-3

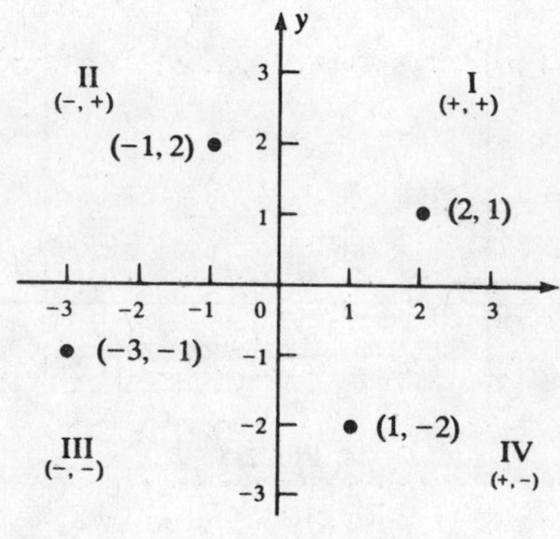

Fig. 2-4

The points having coordinates of the form $(0, b)$ are precisely the points on the y-axis. The points having coordinates $(a, 0)$ are the points on the x-axis.

If a coordinate system is given, it is customary to refer to the point with coordinates (a, b) simply as "the point (a, b)." Thus, one might say: "The point $(1, 0)$ lies on the x-axis."

2.2 THE DISTANCE FORMULA

Let P_1 and P_2 be points with coordinates (x_1, y_1) and (x_2, y_2) in a given coordinate system (Fig. 2-5). We wish to find a formula for the distance $\overline{P_1P_2}$.

Let R be the point where the vertical line through P_2 intersects the horizontal line through P_1. Clearly, the x-coordinate of R is x_2, the same as that of P_2; and the y-coordinate of R is y_1, the same as that of P_1. By the Pythagorean Theorem,

$$\overline{P_1P_2}^2 = \overline{P_1R}^2 + \overline{P_2R}^2$$

Now, if A_1 and A_2 are the projections of P_1 and P_2 on the x-axis, the segments P_1R and A_1A_2 are opposite sides of a rectangle. Hence, $\overline{P_1R} = \overline{A_1A_2}$. But $\overline{A_1A_2} = |x_1 - x_2|$, by (*1.6*). Thus, $\overline{P_1R} = |x_1 - x_2|$. Similarly, $\overline{P_2R} = |y_1 - y_2|$. Consequently,

$$\overline{P_1P_2}^2 = |x_1 - x_2|^2 + |y_1 - y_2|^2 = (x_1 - x_2)^2 + (y_1 - y_2)^2$$

whence

$$\overline{P_1P_2} = \sqrt{(x_1 - x_2)^2 + (y_1 - y_2)^2} \quad \textbf{distance formula} \tag{2.1}$$

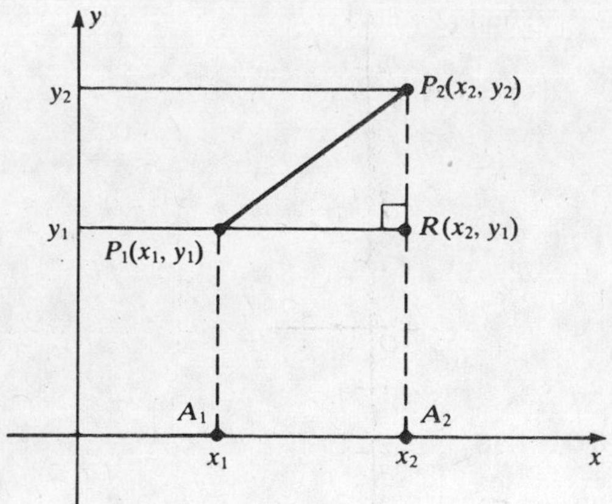

Fig. 2-5

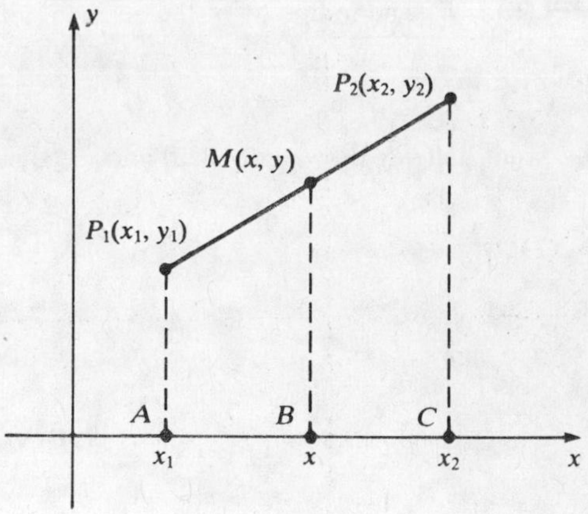

Fig. 2-6

The reader should check that this formula also holds when P_1 and P_2 lie on the same horizontal line or on the same vertical line.

EXAMPLES (*a*) The distance between (3, 8) and (7, 11) is:

$$\sqrt{(3-7)^2+(8-11)^2} = \sqrt{(-4)^2+(-3)^2} = \sqrt{16+9} = \sqrt{25} = 5$$

(*b*) The distance between (4, −3) and (2, 7) is:

$$\sqrt{(4-2)^2+(-3-7)^2} = \sqrt{2^2+(-10)^2} = \sqrt{4+100} = \sqrt{104}$$
$$= \sqrt{4 \cdot 26} = \sqrt{4} \cdot \sqrt{26} = 2\sqrt{26}$$

ALGEBRA For any positive numbers u and v, $\sqrt{uv} = \sqrt{u}\,\sqrt{v}$, since

$$(\sqrt{u}\,\sqrt{v})^2 = (\sqrt{u})^2\,(\sqrt{v})^2 = uv$$

(*c*) The distance between any point (a, b) and the origin $(0, 0)$ is $\sqrt{a^2+b^2}$.

2.3 THE MIDPOINT FORMULAS

Again considering two arbitrary points $P_1(x_1, y_1)$ and $P_2(x_2, y_2)$, we shall find the coordinates (x, y) of the midpoint M of the segment P_1P_2 (Fig. 2-6). Let A, B, C be the perpendicular projections of P_1, M, P_2 on the x-axis. The x-coordinates of A, B, C are x_1, x, x_2, respectively. Since the lines P_1A, MB, and P_2C are parallel, the ratios $\overline{P_1M}/\overline{MP_2}$ and $\overline{AB}/\overline{BC}$ are equal. But $\overline{P_1M} = \overline{MP_2}$; hence, $\overline{AB} = \overline{BC}$. Since $\overline{AB} = x - x_1$ and $\overline{BC} = x_2 - x$,

$$x - x_1 = x_2 - x$$
$$2x = x_1 + x_2$$
$$x = \frac{x_1 + x_2}{2}$$

(The same result is obtained when P_2 is to the left of P_1, in which case, $\overline{AB} = x_1 - x$ and $\overline{BC} = x - x_2$.) Similarly, $y = (y_1 + y_2)/2$. Thus, the coordinates of the midpoint M are determined by the **midpoint formulas**

$$x = \frac{x_1 + x_2}{2} \qquad y = \frac{y_1 + y_2}{2} \tag{2.2}$$

In words: the coordinates of the midpoint are the averages of the coordinates of the endpoints.

EXAMPLES (*a*) The midpoint of the segment connecting (1, 7) and (3, 5) is

$$\left(\frac{1+3}{2}, \frac{7+5}{2}\right) = (2, 6)$$

(*b*) The point halfway between (−2, 5) and (3, 3) is

$$\left(\frac{-2+3}{2}, \frac{5+3}{2}\right) = (\tfrac{1}{2}, 4)$$

Solved Problems

2.1 Determine whether the triangle with vertices $A(-1, 2)$, $B(4, 7)$, $C(-3, 6)$ is isosceles.

$$\overline{AB} = \sqrt{(-1-4)^2 + (2-7)^2} = \sqrt{(-5)^2 + (-5)^2} = \sqrt{25+25} = \sqrt{50}$$
$$\overline{AC} = \sqrt{[-1-(-3)]^2 + (2-6)^2} = \sqrt{(2)^2 + (-4)^2} = \sqrt{4+16} = \sqrt{20}$$
$$\overline{BC} = \sqrt{[4-(-3)]^2 + (7-6)^2} = \sqrt{7^2 + 1^2} = \sqrt{49+1} = \sqrt{50}$$

Since $\overline{AB} = \overline{BC}$, the triangle is isosceles.

2.2 Determine whether the triangle with vertices $A(-5, -3)$, $B(-7, 3)$, $C(2, 6)$ is a right triangle.

Use (*2.1*) to find the squares of the sides:

$$\overline{AB}^2 = (-5+7)^2 + (-3-3)^2 = 2^2 + (-6)^2 = 4 + 36 = 40$$
$$\overline{BC}^2 = (-7-2)^2 + (3-6)^2 = 81 + 9 = 90$$
$$\overline{AC}^2 = (-5-2)^2 + (-3-6)^2 = 49 + 81 = 130$$

Since $\overline{AB}^2 + \overline{BC}^2 = \overline{AC}^2$, $\triangle ABC$ is a right triangle with right angle at B.

GEOMETRY The converse of the Pythagorean Theorem is also true: If $\overline{AC}^2 = \overline{AB}^2 + \overline{BC}^2$ in $\triangle ABC$, then $\measuredangle ABC$ is a right angle.

2.3 Prove by use of coordinates that the midpoint of the hypotenuse of a right triangle is equidistant from the three vertices.

Let the origin of a coordinate system be located at the right angle C; let the positive x-axis contain leg CA, and the positive y-axis leg CB. (See Fig. 2-7(*a*).)

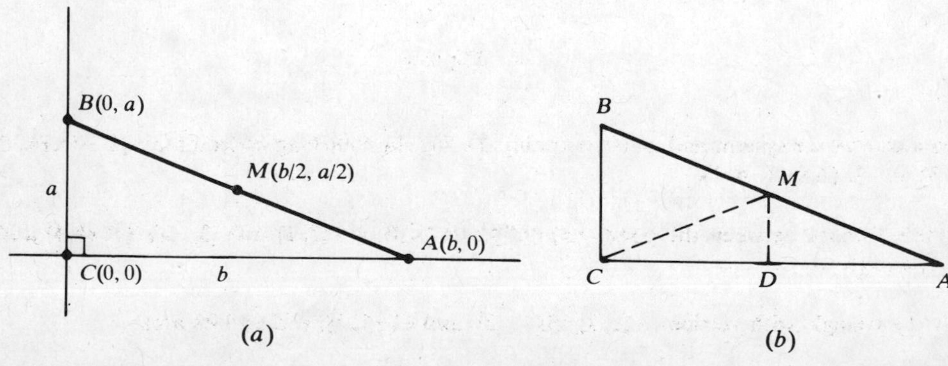

Fig. 2-7

Vertex A has coordinates $(b, 0)$, where $b = \overline{CA}$; and vertex B has coordinates $(0, a)$ where $a = \overline{BC}$. Let M be the midpoint of the hypotenuse. By the Midpoint Formulas, (*2.2*), the coordinates of M are $(b/2, a/2)$. Now, by the Pythagorean Theorem,

$$\overline{MA} = \overline{MB} = \frac{\overline{AB}}{2} = \frac{\sqrt{a^2 + b^2}}{2}$$

and, by the Distance Formula, (*2.1*),

$$\overline{MC} = \sqrt{\left(\frac{b}{2} - 0\right)^2 + \left(\frac{a}{2} - 0\right)^2} = \sqrt{\frac{b^2}{4} + \frac{a^2}{4}} = \sqrt{\frac{a^2 + b^2}{4}}$$

ALGEBRA For any positive numbers u, v,

$$\sqrt{\frac{u}{v}} \cdot \sqrt{v} = \sqrt{\frac{u}{v} \cdot v} = \sqrt{u} \qquad \text{and so} \qquad \sqrt{\frac{u}{v}} = \frac{\sqrt{u}}{\sqrt{v}}$$

$$= \frac{\sqrt{a^2 + b^2}}{\sqrt{4}} = \frac{\sqrt{a^2 + b^2}}{2}$$

Hence, $\overline{MA} = \overline{MC}$. [For a simpler, geometrical proof, see Fig. 2-7(*b*); MD and BC are parallel.]

Supplementary Problems

2.4 In Fig. 2-8, find the coordinates of points A, B, C, D, E, and F.

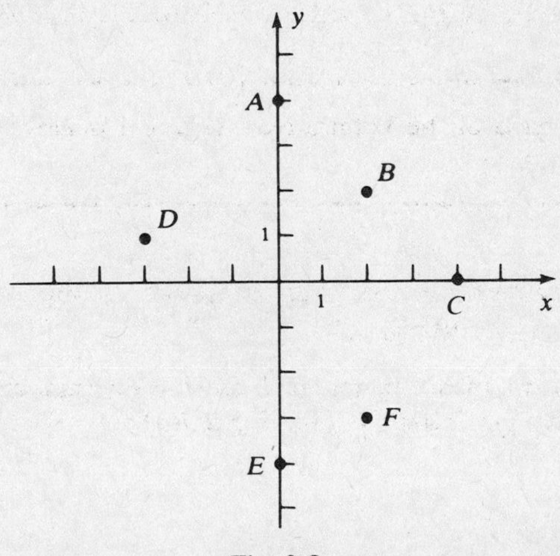

Fig. 2-8

2.5 Draw a coordinate system and mark the points having the following coordinates: $(1, -1)$, $(4, 4)$, $(-2, -2)$, $(3, -3)$, $(0, 2)$, $(2, 0)$, $(-4, 1)$.

2.6 Find the distance between the points (*a*) $(2, 3)$ and $(2, 8)$; (*b*) $(3, 1)$ and $(3, -4)$; (*c*) $(4, 1)$ and $(2, 1)$; (*d*) $(-3, 4)$ and $(5, 4)$.

2.7 Draw the triangle with vertices $A(4, 7)$, $B(4, -3)$, and $C(-1, 7)$, and find its area.

2.8 If $(-2, -2)$, $(-2, 4)$, and $(3, -2)$ are three vertices of a rectangle, find the fourth vertex.

2.9 If the points $(3, 1)$ and $(-1, 0)$ are opposite vertices of a rectangle whose sides are parallel to the coordinate axes, find the other two vertices.

2.10 If $(2, -1)$, $(5, -1)$, and $(3, 2)$ are three vertices of a parallelogram, what are the possible locations of the fourth vertex?

2.11 Give the coordinates of a point on the line passing through the point $(2, 4)$ and parallel to the y-axis.

2.12 Find the distance between the points (*a*) $(2, 6)$ and $(7, 3)$; (*b*) $(3, -1)$ and $(0, 2)$; (*c*) $(4, \frac{1}{2})$ and $(-\frac{1}{4}, 3)$.

2.13 Determine whether the three given points are vertices of an isosceles triangle or of a right triangle (or of both). Find the area of each right triangle.

(*a*) $(-1, 2), (3, -2), (7, 6)$ (*b*) $(4, 1), (1, 2), (3, 8)$ (*c*) $(4, 1), (1, -4), (-4, -1)$

2.14 Find the value of k such that $(3, k)$ is equidistant from $(1, 2)$ and $(6, 7)$.

2.15 (*a*) Are the three points $A(1, 0)$, $B(\frac{7}{2}, 4)$, and $C(7, 8)$ collinear (that is, all on the same line)? (*Hint*: If A, B, C form a triangle, the sum of two sides, $\overline{AB} + \overline{BC}$, must be greater than the third side, $\overline{AC}$. If B lies between A and C on a line, $\overline{AB} + \overline{BC} = \overline{AC}$.) (*b*) Are the three points $A(-5, -7)$, $B(0, -1)$, and $C(10, 11)$ collinear?

2.16 Find the midpoints of the line segments with endpoints (*a*) $(1, -1)$ and $(7, 5)$; (*b*) $(\frac{3}{2}, 4)$ and $(1, 0)$; (*c*) $(\sqrt{2}, 1)$ and $(5, 3)$.

2.17 Find the point (a, b) such that $(3, 5)$ is the midpoint of the line segment connecting (a, b) and $(1, 2)$.

2.18 Prove by use of coordinates that the line segment joining the midpoints of two sides of a triangle is one-half the length of the third side.

Chapter 3

Graphs of Equations

Consider the following equation involving the variables x and y:

$$2y - 3x = 6 \tag{i}$$

Notice that the point (2, 6) satisfies the equation; that is, when the x-coordinate 2 is substituted for x and the y-coordinate 6 is substituted for y, the left-hand side, $2y - 3x$, assumes the value of the right-hand side, 6. The *graph* of (i) consists of all points (a, b) that satisfy the equation when a is substituted for x and b is substituted for y. We tabulate some points that satisfy (i) in Fig. 3-1(a), and indicate these points in Fig. 3-1(b). It is apparent that these points all lie on a straight line. In fact, it will be shown later that the graph of (i) actually is a straight line.

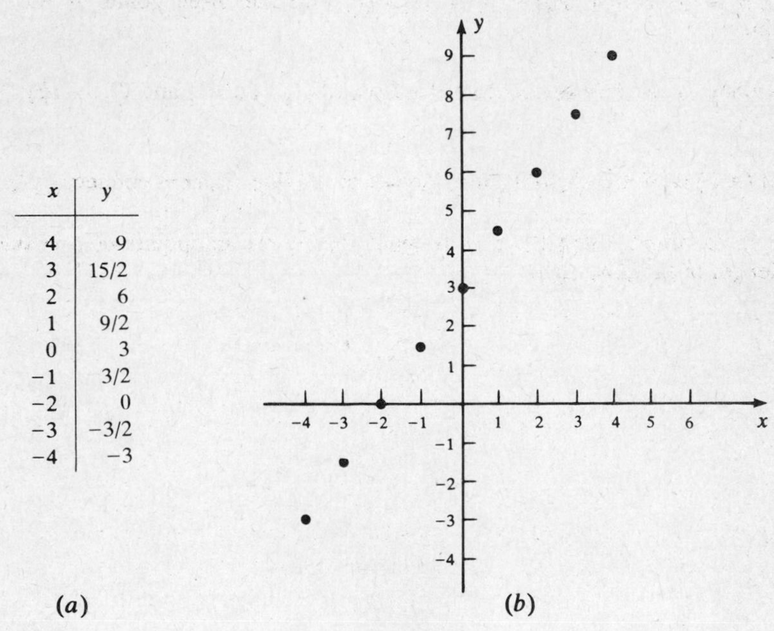

x	y
4	9
3	15/2
2	6
1	9/2
0	3
−1	3/2
−2	0
−3	−3/2
−4	−3

(a) (b)

Fig. 3-1

In general, the *graph* of an equation involving x and y as its only variables consists of all points (x, y) satisfying the equation.

EXAMPLES (a) Some points on the graph of $y = x^2$ are computed in Fig. 3-2(a) and shown in Fig. 3-2(b). These points suggest that the graph looks like the dashed curve—a type known as a *parabola*. (b) The graph of the equation $xy = 1$ is a type of curve called a *hyperbola*. As shown in Fig. 3-3(b), the graph splits into two separate pieces. The points on the hyperbola get closer and closer to the axes as they move farther and farther from the origin. (c) The graph of the equation

$$\frac{x^2}{9} + \frac{y^2}{4} = 1$$

is a closed curve of a type called an *ellipse*. See Fig. 3-4.

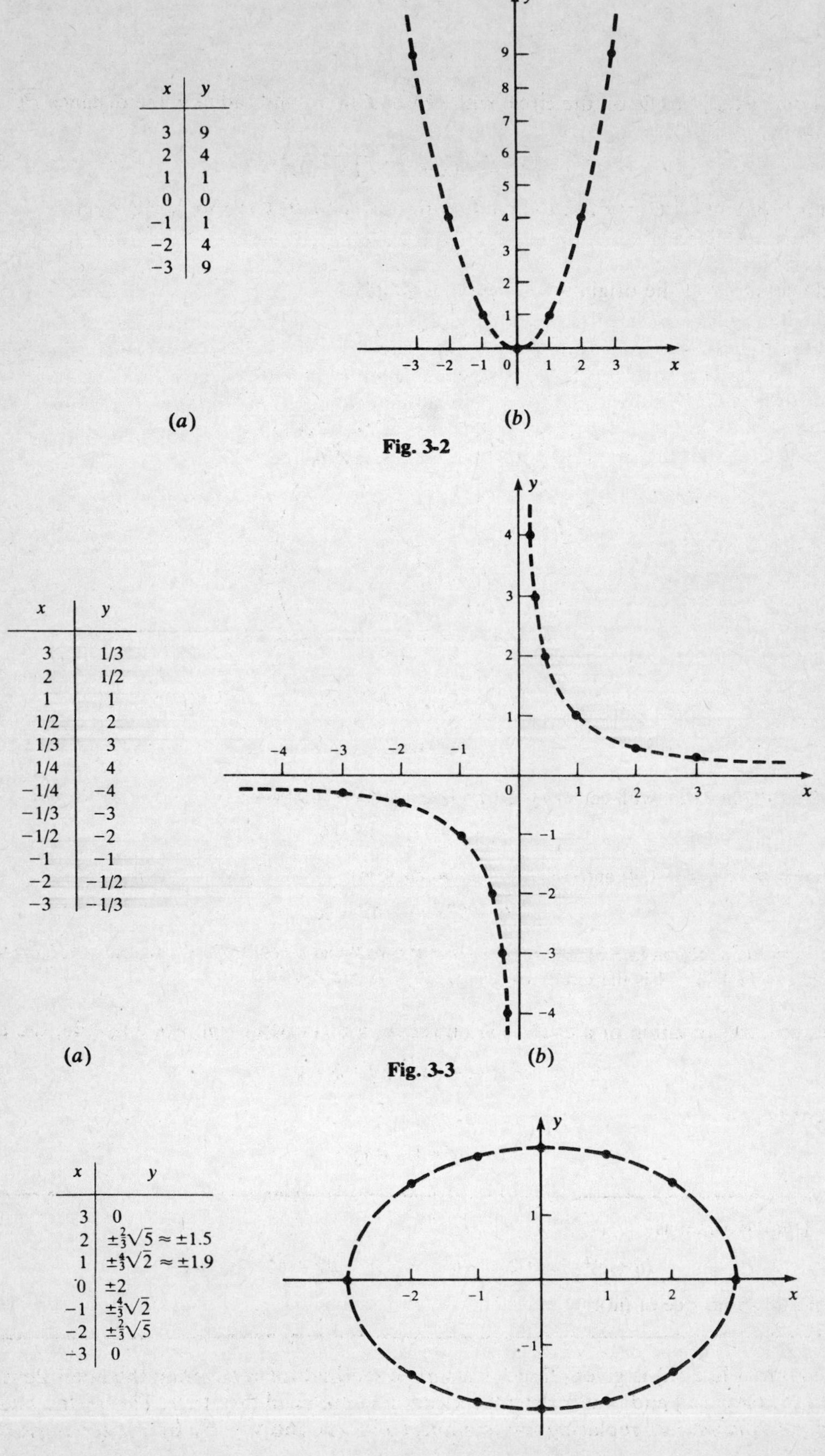

x	y
3	9
2	4
1	1
0	0
-1	1
-2	4
-3	9

(a) (b)

Fig. 3-2

x	y
3	1/3
2	1/2
1	1
1/2	2
1/3	3
1/4	4
$-1/4$	-4
$-1/3$	-3
$-1/2$	-2
-1	-1
-2	$-1/2$
-3	$-1/3$

(a) (b)

Fig. 3-3

x	y
3	0
2	$\pm\frac{2}{3}\sqrt{5} \approx \pm 1.5$
1	$\pm\frac{4}{3}\sqrt{2} \approx \pm 1.9$
0	± 2
-1	$\pm\frac{4}{3}\sqrt{2}$
-2	$\pm\frac{2}{3}\sqrt{5}$
-3	0

(a) (b)

Fig. 3-4

Circles

For a point $P(x, y)$ to lie on the circle with center $C(a, b)$ and radius r, the distance $\overline{PC}$ must be r (Fig. 3-5). Now, by (*2.1*),

$$\overline{PC} = \sqrt{(x-a)^2 + (y-b)^2}$$

The *standard equation*, $\overline{PC}^2 = r^2$, of the circle with center (a, b) and radius r is then

$$(x-a)^2 + (y-b)^2 = r^2 \tag{3.1}$$

For a circle centered at the origin, (*3.1*) becomes simply

$$x^2 + y^2 = r^2 \tag{3.2}$$

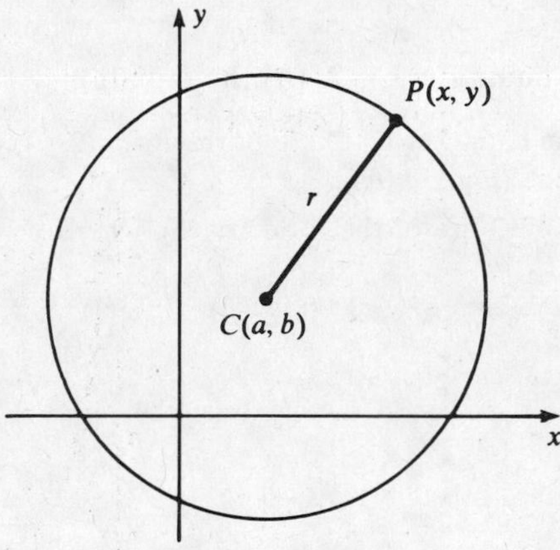

Fig. 3-5

EXAMPLES (*a*) The circle with center (1, 2) and radius 3 has the equation

$$(x-1)^2 + (y-2)^2 = 9$$

(*b*) The circle with center (−1, 4) and radius 6 has the equation

$$(x+1)^2 + (y-4)^2 = 36$$

(*c*) The graph of the equation $(x-3)^2 + (y-7)^2 = 16$ is the circle with center (3, 7) and radius 4. (*d*) The graph of the equation $x^2 + (y+2)^2 = 1$ is the circle with center (0, −2) and radius 1.

Sometimes, the equation of a circle will appear in a disguised form. For example, the equation

$$x^2 + y^2 - 6x + 2y + 6 = 0 \tag{ii}$$

is equivalent to

$$(x-3)^2 + (y+1)^2 = 4 \tag{iii}$$

ALGEBRA Use the formulas

$$(u+v)^2 = u^2 + 2uv + v^2 \qquad (u-v)^2 = u^2 - 2uv + v^2$$

to expand the left-hand side of (iii).

If an equation like (ii) is given, there is a simple method for recovering the equivalent standard equation of the form (iii) and thus finding the center and radius of the circle. This method depends on *completing the squares*; i.e., replacing the quantities $x^2 + Ax$ and $y^2 + By$ by the equal quantities

$$\left(x + \frac{A}{2}\right)^2 - \frac{A^2}{4} \qquad \text{and} \qquad \left(y + \frac{B}{2}\right)^2 - \frac{B^2}{4}$$

EXAMPLE Let us find the graph of the equation

$$x^2 + y^2 + 4x - 2y + 1 = 0$$

Completing the squares, replace $x^2 + 4x$ by $(x+2)^2 - 4$ and $y^2 - 2y$ by $(y-1)^2 - 1$:

$$(x+2)^2 - 4 + (y-1)^2 - 1 + 1 = 0 \qquad \text{or} \qquad (x+2)^2 + (y-1)^2 = 4$$

This is the equation of a circle with center $(-2, 1)$ and radius 2.

Solved Problems

3.1 Find the graph of (*a*) the equation $x = 2$, (*b*) the equation $y = -3$.

(*a*) The points satisfying the equation $x = 2$ are of the form $(2, y)$, where y can be any number. These points form a vertical line (Fig. 3-6(*a*)).

(*b*) The points satisfying $y = -3$ are of the form $(x, -3)$, where x is any number. These points form a horizontal line (Fig. 3-6(*b*)).

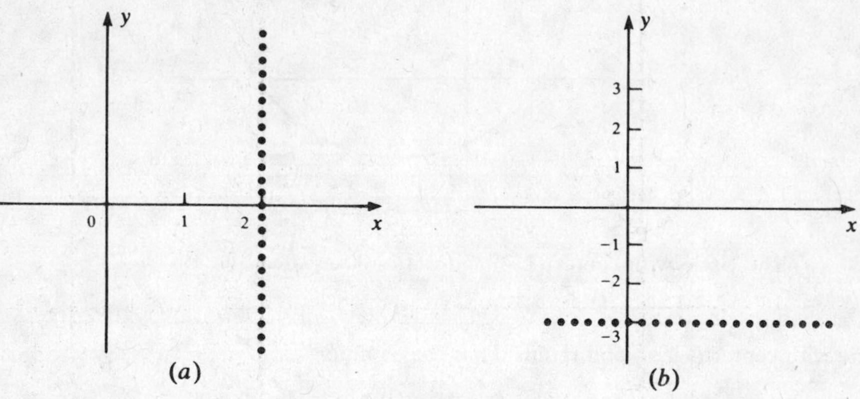

Fig. 3-6

3.2 Find the graph of the equation $x = y^2$.

Plotting several points suggests the curve shown in Fig. 3-7. This curve is a parabola, which may be obtained from the graph of $y = x^2$ (Fig. 3-2) by switching the x- and y-coordinates.

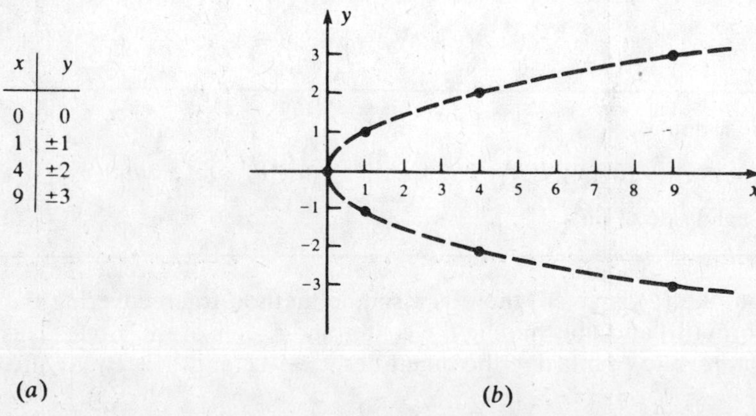

x	y
0	0
1	±1
4	±2
9	±3

Fig. 3-7

3.3 Find the graphs of

$$(a)\quad y = x^2 + 2 \qquad (b)\quad y = x^2 - 2 \qquad (c)\quad y = (x-2)^2 \qquad (d)\quad y = (x+2)^2$$

(*a*) The graph of $y = x^2 + 2$ is obtained from the graph of $y = x^2$ (Fig. 3-2) by raising each point two units in the vertical direction. See Fig. 3-8(*a*).

(*b*) The graph of $y = x^2 - 2$ is obtained from the graph of $y = x^2$ by lowering each point two units. See Fig. 3-8(*b*).

(*c*) The graph $\mathscr{G}^*$ of $y = (x-2)^2$ is obtained from the graph $\mathscr{G}$ of $y = x^2$ by moving every point of $\mathscr{G}$ two units to the right. To see this, assume (a, b) is on $\mathscr{G}$. Then $b = a^2$. Hence, $y = (x-2)^2$ is satisfied when $y = b$ and $x = a + 2$; that is, $(a + 2, b)$ lies on $\mathscr{G}^*$. See Fig. 3-8(*c*).

(*d*) The graph of $y = (x+2)^2$ is obtained from the graph of $y = x^2$ by moving every point two units to the left. The reasoning is as in (*c*); see Fig. 3-8(*d*).

Parts (*c*) and (*d*) can be generalized as follows: If c is a positive number, the graph of $y = f(x - c)$ is obtained from the graph of $y = f(x)$ by moving each point c units to the right. The graph of $y = f(x + c)$ is obtained from the graph of $y = f(x)$ by moving each point c units to the left.

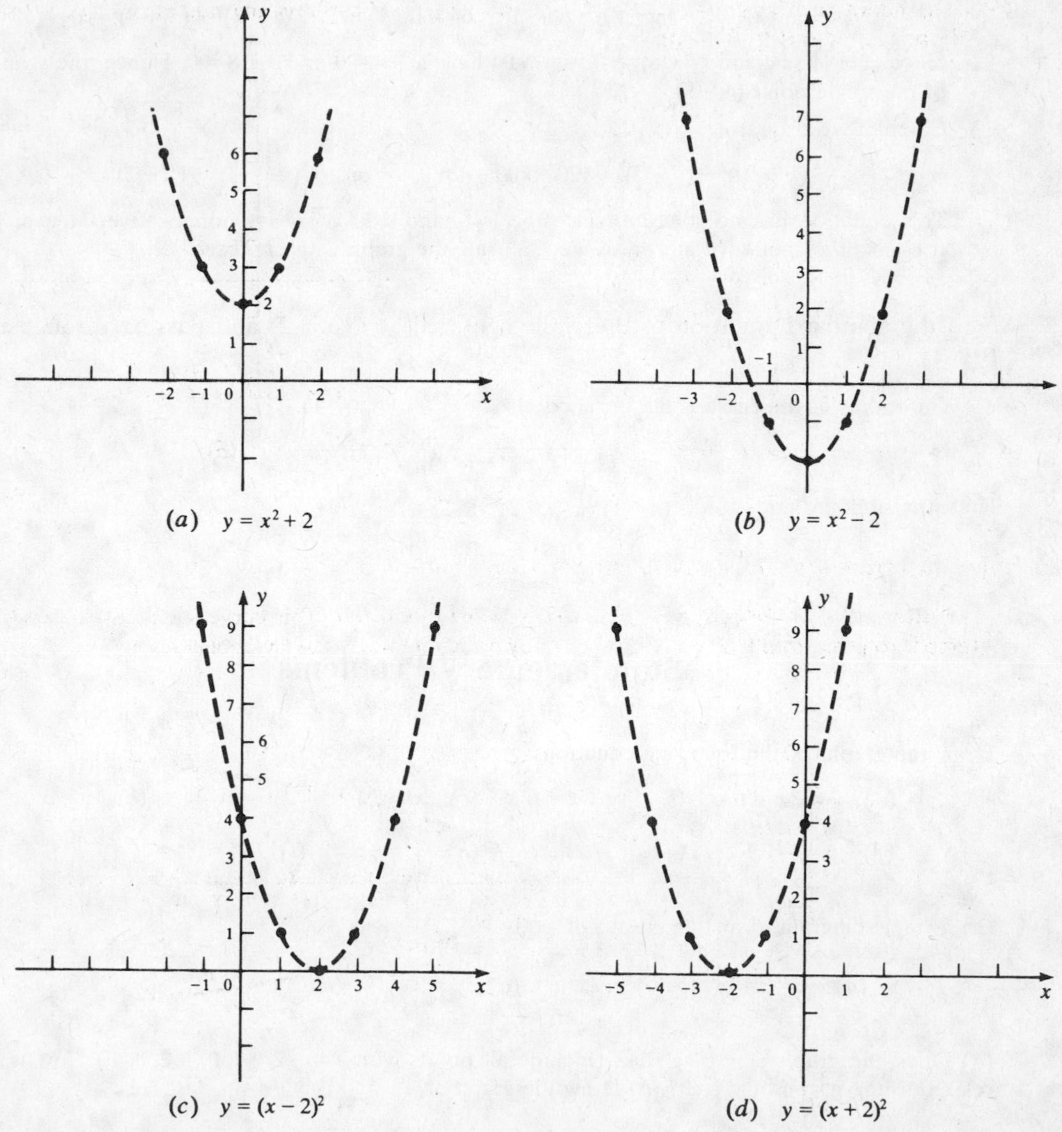

Fig. 3-8

3.4 Identify the graphs of

(a) $3x^2 + 3y^2 - 6x - y + 1 = 0$ (b) $x^2 + y^2 - 8x + 16y + 80 = 0$

(c) $x^2 + y^2 + 20x - 4y + 120 = 0$

(a) First, divide both sides by 3:

$$x^2 + y^2 - 2x - \frac{1}{3}y + \frac{1}{3} = 0$$

Complete the squares:

$$(x-1)^2 + \left(y - \frac{1}{6}\right)^2 + \frac{1}{3} - 1 - \frac{1}{36} = 0$$

or $$(x-1)^2 + \left(y - \frac{1}{6}\right)^2 = 1 + \frac{1}{36} - \frac{1}{3} = \frac{36}{36} + \frac{1}{36} - \frac{12}{36} = \frac{25}{36}$$

Hence, the graph is a circle with center $(1, \frac{1}{6})$ and radius $\frac{5}{6}$.

(b) Complete the squares:

$$(x-4)^2 + (y+8)^2 + 80 - 16 - 64 = 0 \quad \text{or} \quad (x-4)^2 + (y+8)^2 = 0$$

Since $(x-4)^2 \geq 0$ and $(y+8)^2 \geq 0$, we must have $x - 4 = 0$ and $y + 8 = 0$. Hence, the graph consists of the single point $(4, -8)$.

(c) Complete the squares:

$$(x+10)^2 + (y-2)^2 + 120 - 100 - 4 = 0 \quad \text{or} \quad (x+10)^2 + (y-2)^2 = -16$$

This equation has no solution, since the left-hand side is always nonnegative. Hence, the graph consists of no points at all; or, as we shall say, the graph is the *null set.*

3.5 Find the standard equation of the circle centered at $C(1, -2)$ and passing through the point $P(7, 4)$.

The radius of the circle is the distance

$$\overline{CP} = \sqrt{(7-1)^2 + [4-(-2)]^2} = \sqrt{36+36} = \sqrt{72}$$

Thus, the standard equation is $(x-1)^2 + (y+2)^2 = 72$.

Supplementary Problems

3.6 Draw the graphs of the following equations:

(a) $3y - x = 6$ (b) $3y + x = 6$ (c) $x = -1$ (d) $y = 4$ (e) $y = x^2 - 1$

(f) $y = \frac{1}{x} + 1$ (g) $y = x$ (h) $y = -x$ (i) $y^2 = x^2$

3.7 On a single diagram, draw the graphs of

(a) $y = x^2$ (b) $y = 2x^2$ (c) $y = 3x^2$ (d) $y = \frac{1}{2}x^2$ (e) $y = \frac{1}{3}x^2$

3.8 (a) Draw the graph of $y = (x-1)^2$. (Include all points with $x = -2, -1, 0, 1, 2, 3, 4$.) How is this graph related to the graph of $y = x^2$? (b) Draw the graph of

$$y = \frac{1}{x-1}$$

(*c*) Draw the graph of $y = (x + 1)^2$. How is this graph related to that of $y = x^2$? (*d*) Draw the graph of $y = 1/(x + 1)$.

3.9 Sketch the graphs of the following equations.

(*a*) $\dfrac{x^2}{4} + \dfrac{y^2}{9} = 1$ (*b*) $4x^2 + y^2 = 4$ (*c*) $x^2 - y^2 = 1$

(*d*) $y = x^3$ (*e*) $|x| + |y| = 1$ (*f*) $|x| - |y| = 1$

(*g*) $\dfrac{(x-1)^2}{4} + \dfrac{(y-2)^2}{9} = 1$ (*h*) $y = \dfrac{1}{2}(x + |x|)$

(*Hints*: (*c*) is a hyperbola; in (*e*), look at each quadrant separately; obtain (*g*) from (*a*).)

3.10 Find an equation whose graph consists of all points $P(x, y)$ whose distance from the point $F(0, p)$ is equal to its distance $\overline{PQ}$ from the horizontal line $y = -p$ (p is a fixed positive number). See Fig. 3-9.

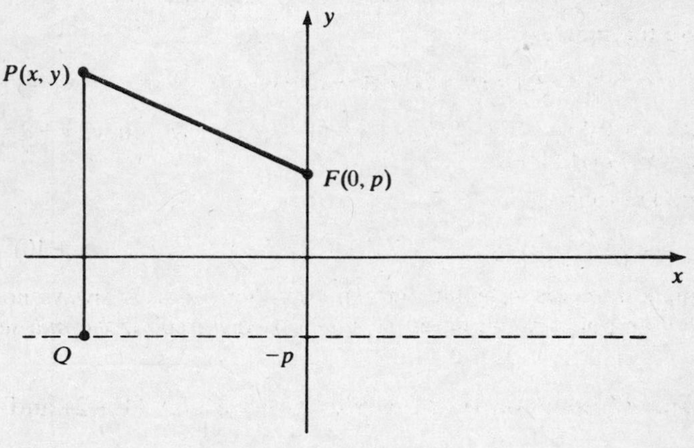

Fig. 3-9

3.11 Find the standard equations of the circles satisfying the given conditions: (*a*) center (4, 3), radius 1; (*b*) center (−1, 5), radius $\sqrt{2}$; (*c*) center (0, 2), radius 4; (*d*) center (3, 3), radius $3\sqrt{2}$; (*e*) center (4, −1) and passing through (2, 3); (*f*) center (1, 2) and passing through the origin.

3.12 Identify the graphs of the following equations:

(*a*) $x^2 + y^2 - 12x + 20y + 15 = 0$ (*b*) $x^2 + y^2 + 30y + 29 = 0$ (*c*) $x^2 + y^2 + 3x - 2y + 4 = 0$

(*d*) $2x^2 + 2y^2 - x = 0$ (*e*) $x^2 + y^2 + 2x - 2y + 2 = 0$ (*f*) $x^2 + y^2 + 6x + 4y = 36$

3.13 (*a*) Problem 3.4 suggests that the graph of an equation

$$x^2 + y^2 + Dx + Ey + F = 0$$

is either a circle, a point, or the null set. Prove this. (*b*) Find a condition on the numbers D, E, F which is equivalent to the graph's being a circle. (*Hint*: Complete the squares.)

3.14 Find the standard equation of the circle passing through the points (3, 8), (9, 6), and (13, −2). (*Hint*: Write the equation in nonstandard form,

$$x^2 + y^2 + Dx + Ey + F = 0$$

and then substitute the values of x and y given by the three points. Solve the three resulting equations for D, E, and F.)

3.15 For what value(s) of k does the circle $(x - k)^2 + (y - 2k)^2 = 10$ pass through the point (1, 1)?

3.16 Find the standard equations of the circles of radius 3 that are tangent to both the lines $x = 4$ and $y = 6$.

Chapter 4

Straight Lines

4.1 SLOPE

If $P_1(x_1, y_1)$ and $P_2(x_2, y_2)$ are two points on a line $\mathscr{L}$, the number m defined by the equation

$$m = \frac{y_2 - y_1}{x_2 - x_1}$$

is called the *slope* of $\mathscr{L}$. The slope measures the "steepness" of $\mathscr{L}$; it is the ratio of the change $y_2 - y_1$ in the y-coordinate to the change $x_2 - x_1$ in the x-coordinate (see Fig. 4-1(a)).

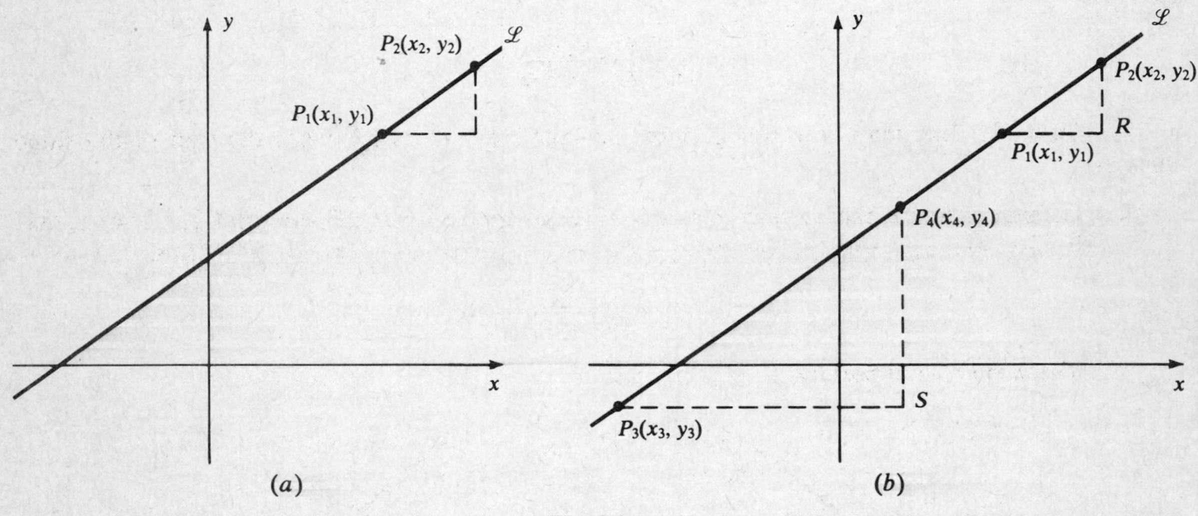

Fig. 4-1

Notice that the value m of the slope does not depend upon the pair of points P_1, P_2 selected: if another pair, $P_3(x_3, y_3)$ and $P_4(x_4, y_4)$, is chosen, the same value of m is obtained. In fact, in Fig. 4-1(b), $\triangle P_3P_4S$ is similar to $\triangle P_1P_2R$.

GEOMETRY The angles at R and S are both right angles, and the angles at P_1 and P_3 are equal because they are corresponding angles determined by the line $\mathscr{L}$ cutting the parallel lines P_1R and P_3S.

Consequently,

$$\frac{\overline{RP_2}}{\overline{P_1R}} = \frac{\overline{SP_4}}{\overline{P_3S}} \qquad \text{or} \qquad \frac{y_2 - y_1}{x_2 - x_1} = \frac{y_4 - y_3}{x_4 - x_3}$$

i.e., the slope determined from P_1 and P_2 is the same as the slope determined from P_3 and P_4.

EXAMPLE In Fig. 4-2, the slope of the line connecting the points (1, 3) and (3, 6) is

$$\frac{6-3}{3-1} = \frac{3}{2} = 1.5$$

Notice that, as a point on the line moves 2 units to the right, it moves 3 units upward. Observe also that the order in which the points are taken has no effect on the slope:

$$\frac{3-6}{1-3} = \frac{-3}{-2} = 1.5$$

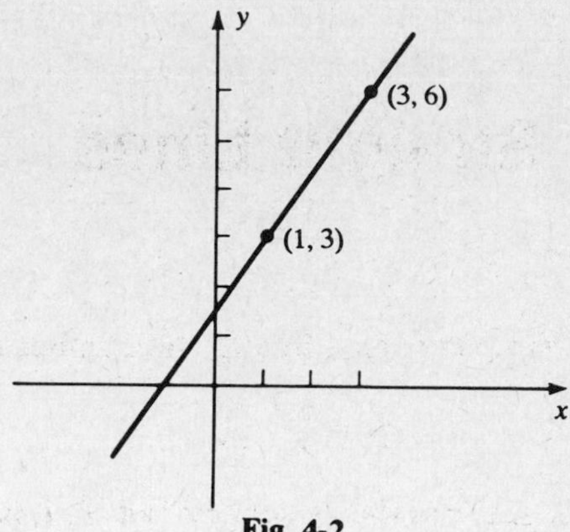

Fig. 4-2

In general,

$$\frac{y_2 - y_1}{x_2 - x_1} = \frac{y_1 - y_2}{x_1 - x_2}$$

The slope of a line may be positive, zero, or negative. Let us see what the sign of the slope indicates.

(i) Consider a line $\mathscr{L}$ that extends upward as it extends to the right. From Fig. 4-3(*a*), we see that $y_2 > y_1$; hence, $y_2 - y_1 > 0$. In addition, $x_2 > x_1$, and, therefore, $x_2 - x_1 > 0$. Thus,

$$m = \frac{y_2 - y_1}{x_2 - x_1} > 0$$

The slope of $\mathscr{L}$ is positive.

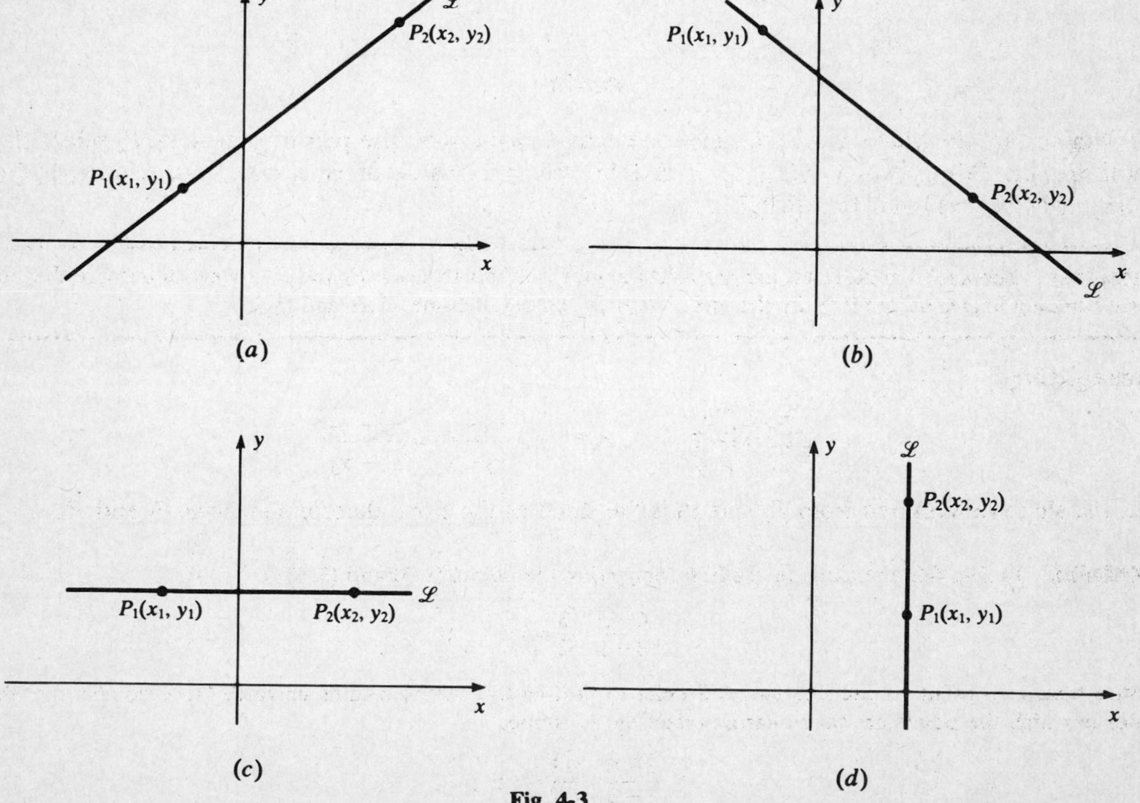

Fig. 4-3

(ii) Consider a line $\mathscr{L}$ that extends downward as it extends to the right. From Fig. 4-3(*b*), we see that $y_2 < y_1$; therefore, $y_2 - y_1 < 0$. But, $x_2 > x_1$; so $x_2 - x_2 > 0$. Hence,

$$m = \frac{y_2 - y_1}{x_2 - x_1} < 0$$

The slope of $\mathscr{L}$ is negative.

(iii) Consider a horizontal line $\mathscr{L}$. From Fig. 4-3(*c*), $y_1 = y_2$, or $y_2 - y_1 = 0$. Moreover, $x_2 > x_1$, and, therefore, $x_2 - x_1 \neq 0$. Hence,

$$m = \frac{y_2 - y_1}{x_2 - x_1} = \frac{0}{x_2 - x_1} = 0$$

The slope of $\mathscr{L}$ is zero.

(iv) Consider a vertical line $\mathscr{L}$. From Fig. 4-3(*d*), $y_2 > y_1$; so that $y_2 - y_1 \neq 0$. But $x_2 = x_1$, or $x_2 - x_1 = 0$. Hence, the expression

$$\frac{y_2 - y_1}{x_2 - x_1}$$

is undefined. *The concept of slope is not defined for $\mathscr{L}$.* (Often we express this situation by saying that the slope of $\mathscr{L}$ is "infinite.")

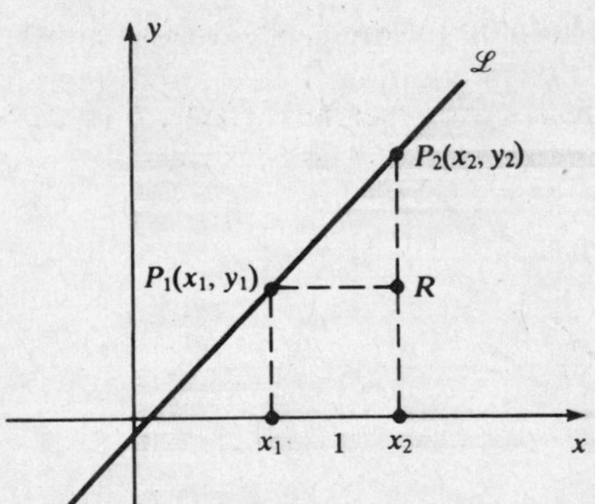

Fig. 4-4

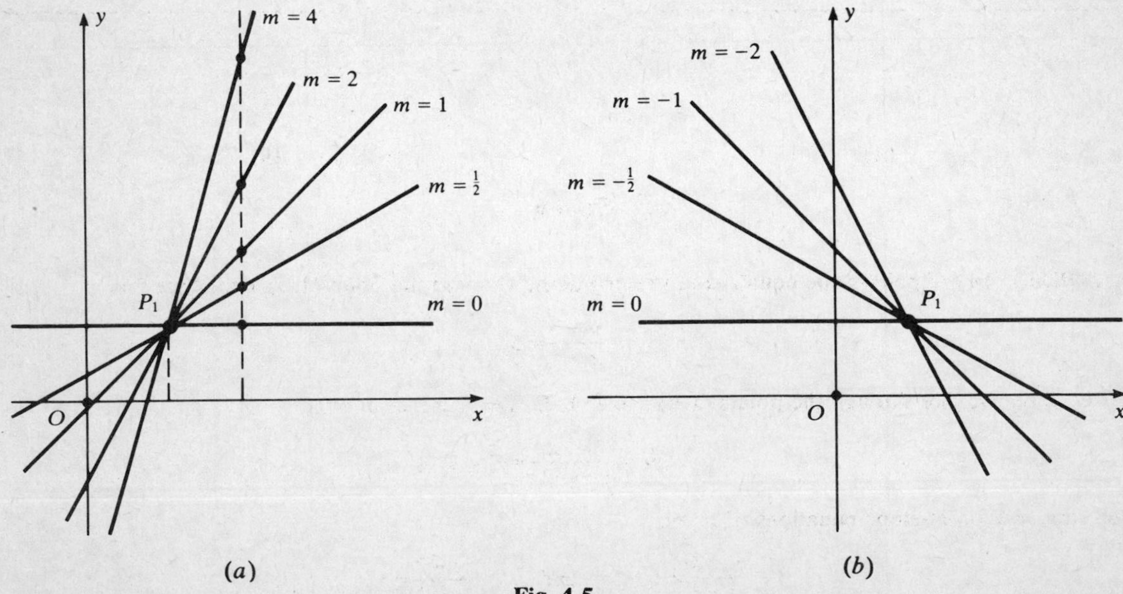

Fig. 4-5

Now let us see how the slope varies with the "steepness" of the line. First, let us consider lines with positive slopes, passing through a fixed point $P_1(x_1, y_1)$; one such line is shown in Fig. 4-4. Take another point $P_2(x_2, y_2)$ on $\mathscr{L}$ such that $x_2 - x_1 = 1$. Then, by definition, the slope m is equal to the distance $\overline{RP_2}$. Now, as the steepness of the line increases, $\overline{RP_2}$ increases without limit (see Fig. 4-5(*a*)). Thus, the slope of $\mathscr{L}$ increases from 0 (when $\mathscr{L}$ is horizontal) to $+\infty$ (when $\mathscr{L}$ is vertical).

By a similar construction, we show that as a negatively-sloped line becomes steeper and steeper, the slope steadily decreases from 0 (when the line is horizontal) to $-\infty$ (when the line is vertical). See Fig. 4-5(*b*).

4.2 EQUATIONS OF A LINE

Consider a line $\mathscr{L}$ that passes through the point $P_1(x_1, y_1)$ and has slope m (Fig. 4-6(*a*)). For any other point, $P(x, y)$, on the line, the slope m is, by definition, the ratio of $y - y_1$ to $x - x_1$. Hence,

$$\frac{y - y_1}{x - x_1} = m \tag{4.1}$$

On the other hand, if $P(x, y)$ is not on the line $\mathscr{L}$ (Fig. 4-6(*b*)), then the slope

$$\frac{y - y_1}{x - x_1}$$

of the line PP_1 is different from the slope m of $\mathscr{L}$, so that (*4.1*) does not hold. Thus, $\mathscr{L}$ is the graph of (*4.1*).

Equation (*4.1*) is called a *point-slope equation* of the line $\mathscr{L}$. Such an equation can be written when the slope m and some point (x_1, y_1) of $\mathscr{L}$ are known.

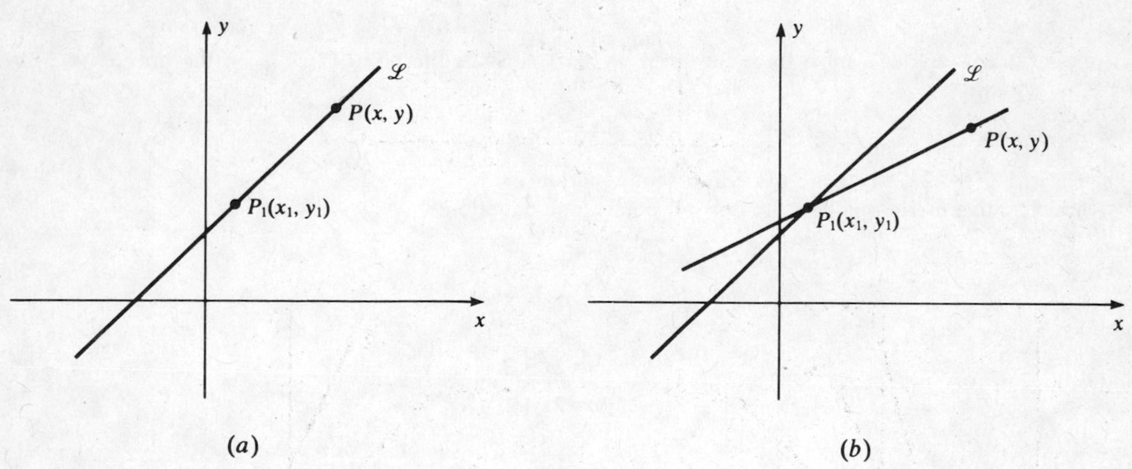

Fig. 4-6

EXAMPLES (*a*) A point-slope equation of the line going through the point (1, 3) with slope 5 is

$$\frac{y-3}{x-1} = 5$$

(*b*) Let $\mathscr{L}$ be the line through the points (1, 4) and (−1, 2). The slope of $\mathscr{L}$ is:

$$m = \frac{4-2}{1-(-1)} = \frac{2}{2} = 1$$

Therefore, two point-slope equations of $\mathscr{L}$ are

$$\frac{y-4}{x-1} = 1 \qquad \text{and} \qquad \frac{y-2}{x+1} = 1$$

Multiplication of (4.1) by $x - x_1$ yields $y - y_1 = m(x - x_1)$, which is equivalent to

$$y - y_1 = mx - mx_1 \qquad \text{or} \qquad y = mx + (y_1 - mx_1)$$

Let b stand for the number $y_1 - mx_1$; then the equation becomes:

$$y = mx + b \tag{4.2}$$

When $x = 0$, (4.2) yields the value $y = b$. Hence, the point $(0, b)$ lies on $\mathscr{L}$. Thus, b is the y-coordinate of the point where $\mathscr{L}$ intersects the y-axis (see Fig. 4-7). The number b is called the *y-intercept* of $\mathscr{L}$, and (4.2) is called the *slope-intercept equation* of $\mathscr{L}$.

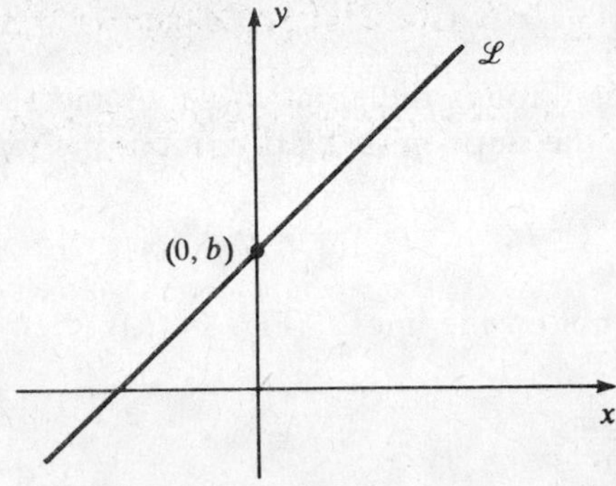

Fig. 4-7

EXAMPLE Let $\mathscr{L}$ be the line through points (1, 3) and (2, 5). Its slope m is

$$\frac{5-3}{2-1} = \frac{2}{1} = 2$$

Its slope-intercept equation must have the form $y = 2x + b$. Since the point (1, 3) is on the line $\mathscr{L}$, (1, 3) must satisfy the equation:

$$3 = 2(1) + b$$

So, $b = 1$, giving $y = 2x + 1$ as the slope-intercept equation.

An alternative method is to write down a point-slope equation:

$$\frac{y-3}{x-1} = 2$$

whence

$$y - 3 = 2(x - 1)$$
$$y - 3 = 2x - 2$$
$$y = 2x + 1$$

4.3 PARALLEL LINES

Assume that $\mathscr{L}_1$ and $\mathscr{L}_2$ are parallel, nonvertical lines, and let P_1 and P_2 be the points where $\mathscr{L}_1$ and $\mathscr{L}_2$ cut the y-axis (see Fig. 4-8(a)). Let R_1 be one unit to the right of P_1, and R_2 one unit to the right of P_2. Let Q_1 and Q_2 be the intersections of the vertical line through R_1 and R_2 with $\mathscr{L}_1$ and $\mathscr{L}_2$. Now, $\triangle P_1R_1Q_1$ is congruent to $\triangle P_2R_2Q_2$.

GEOMETRY Use the ASA (Angle-Side-Angle) Congruence Theorem. $\measuredangle R_1 = \measuredangle R_2$ since both are right angles;

$$\overline{P_1R_1} = \overline{P_2R_2} = 1$$

$\measuredangle P_1 = \measuredangle P_2$, since $\measuredangle P_1$ and $\measuredangle P_2$ are formed by pairs of parallel lines.

Hence, $\overline{R_1Q_1} = \overline{R_2Q_2}$, and

$$\text{slope of } \mathscr{L}_1 = \frac{\overline{R_1Q_1}}{1} = \frac{\overline{R_2Q_2}}{1} = \text{slope of } \mathscr{L}_2$$

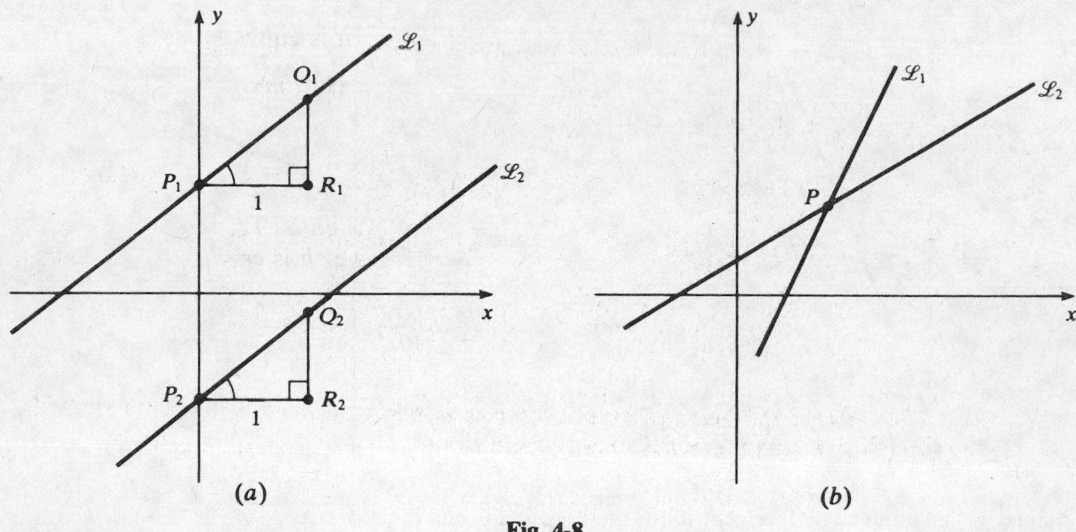

Fig. 4-8

Parallel lines have equal slopes.

Conversely, if different lines $\mathscr{L}_1$ and $\mathscr{L}_2$ are not parallel, then their slopes must be different. For, if $\mathscr{L}_1$ and $\mathscr{L}_2$ meet at the point P and if their slopes are the same, then $\mathscr{L}_1$ and $\mathscr{L}_2$ would have to be the same line. Thus, we have proved:

Theorem 4.1: Two distinct lines are parallel if and only if their slopes are equal.

EXAMPLE Let us find an equation of the line $\mathscr{L}$ through (3, 2) and parallel to the line $\mathscr{M}$ having the equation

$$3x - y = 2$$

The line $\mathscr{M}$ has slope-intercept equation $y = 3x - 2$; hence, the slope of $\mathscr{M}$ is 3, and the slope of the parallel line $\mathscr{L}$ also must be 3. The slope-intercept equation of $\mathscr{L}$ must then be of the form $y = 3x + b$. Since (3, 2) lies on $\mathscr{L}$,

$$2 = 3(3) + b \qquad \text{or} \qquad b = -7$$

Thus, the slope-intercept equation of $\mathscr{L}$ is $y = 3x - 7$.

4.4 PERPENDICULAR LINES

Theorem 4.2: Two nonvertical lines are perpendicular if and only if the product of their slopes is -1.

For a proof, see Problem 4.5.

Solved Problems

4.1 Find the slope of the line having the equation $5x - 2y = 4$. Draw the line, and determine whether the points (10, 23) and (6, 12) are on the line.

Solve the equation for y:

$$y = \frac{5}{2}x - 2 \qquad (1)$$

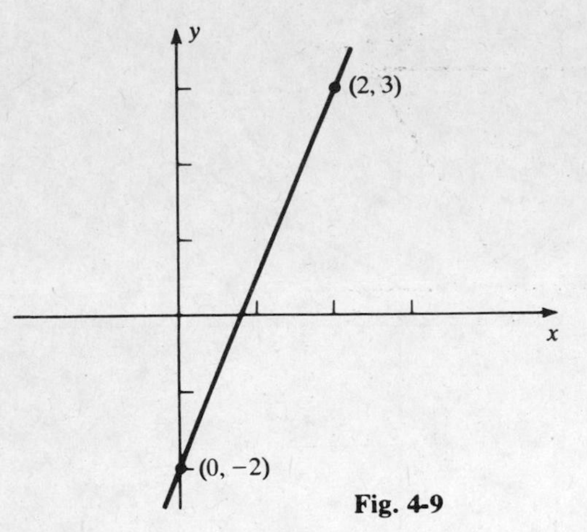

Fig. 4-9

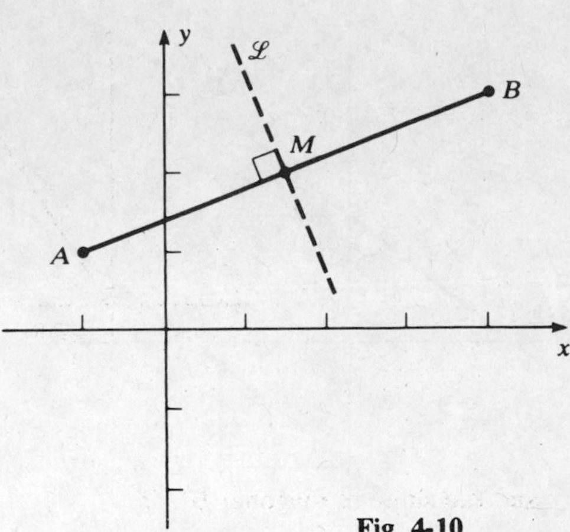

Fig. 4-10

Thus, we have the slope-intercept form; the slope is 5/2 and the y-intercept is -2. The line goes through the point $(0, -2)$. To draw the line, we need another point on the line. Substitute 2 for x in (*1*), giving $y = 3$. Hence, $(2, 3)$ is a point on the line; see Fig. 4-9. (We could have found other points on the line by substituting numbers other than 2 for x.) To test whether $(10, 23)$ is on the line, substitute 10 for x and 23 for y and see whether the equation $5x - 2y = 4$ holds. The two sides turn out to be equal; so $(10, 23)$ is on the line. A similar check shows that $(6, 12)$ is not on the line.

4.2 Find an equation of the line $\mathscr{L}$ which is the perpendicular bisector of the line segment connecting the points $A(-1, 1)$ and $B(4, 3)$. See Fig. 4-10.

$\mathscr{L}$ must pass through the midpoint M of segment AB. By the Midpoint Formulas, the coordinates of M are $(3/2, 2)$. The slope of the line through A and B is

$$\frac{3-1}{4-(-1)} = \frac{2}{5}$$

Hence, by Theorem 4.2, the slope of $\mathscr{L}$ is

$$\frac{-1}{2/5} = -\frac{5}{2}$$

A point-slope equation of $\mathscr{L}$ is

$$\frac{y-2}{x-\frac{3}{2}} = -\frac{5}{2}$$

4.3 Determine whether the points $A(-1, 6)$, $B(5, 9)$, and $C(7, 10)$ are collinear.

A, B, C will be collinear if and only if the line AB is the same as the line AC, which is equivalent to the slope of AB being equal to the slope of AC.

The slopes of AB and AC are

$$\frac{9-6}{5-(-1)} = \frac{3}{6} = \frac{1}{2} \qquad \text{and} \qquad \frac{10-6}{7-(-1)} = \frac{4}{8} = \frac{1}{2}$$

Hence, A, B, C are collinear.

4.4 Prove by use of coordinates that the diagonals of a rhombus (a parallelogram of which all sides are equal) are perpendicular to each other.

Represent the rhombus as in Fig. 4-11. (How do we know that the x-coordinate of D is $v + u$?) Then the slope of diagonal AD is

$$m_1 = \frac{w-0}{v+u-0} = \frac{w}{v+u}$$

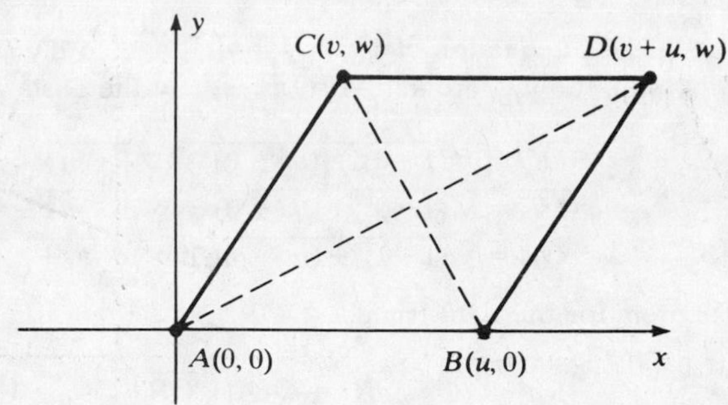

Fig. 4-11

and the slope of diagonal BC is

$$m_2 = \frac{w-0}{v-u} = \frac{w}{v-u}$$

Hence $$m_1 m_2 = \left(\frac{w}{v+u}\right)\left(\frac{w}{v-u}\right) = \frac{w^2}{v^2-u^2}$$

Since $ABDC$ is a rhombus, $\overline{AB} = \overline{AC}$. But, $\overline{AB} = u$ and $\overline{AC} = \sqrt{v^2+w^2}$; so,

$$\sqrt{v^2+w^2} = u \qquad \text{or} \qquad v^2+w^2 = u^2 \qquad \text{or} \qquad w^2 = u^2 - v^2$$

Consequently,

$$m_1 m_2 = \frac{w^2}{v^2-u^2} = \frac{u^2-v^2}{v^2-u^2} = -1$$

and, by Theorem 4.2, lines AD and BC are perpendicular.

4.5 Prove Theorem 4.2.

Assume that $\mathscr{L}_1$ and $\mathscr{L}_2$ are perpendicular, nonvertical lines, of respective slopes m_1 and m_2. We shall show that $m_1 m_2 = -1$.

Let $\mathscr{L}_1^*$ be the line through the origin O and parallel to $\mathscr{L}_1$, and let $\mathscr{L}_2^*$ be the line through the origin and parallel to $\mathscr{L}_2$ (Fig. 4-12(*a*)). Since $\mathscr{L}_1$ is parallel to $\mathscr{L}_1^*$ and $\mathscr{L}_2$ is parallel to $\mathscr{L}_2^*$, the slope of $\mathscr{L}_1^*$ is m_1 and the slope of $\mathscr{L}_2^*$ is m_2 (Theorem 4.1). Also, $\mathscr{L}_1^*$ is perpendicular to $\mathscr{L}_2^*$, since $\mathscr{L}_1$ is perpendicular to $\mathscr{L}_2$. Let R be the point on $\mathscr{L}_1^*$ with x-coordinate 1, and let Q be the point on $\mathscr{L}_2^*$ with x-coordinate 1

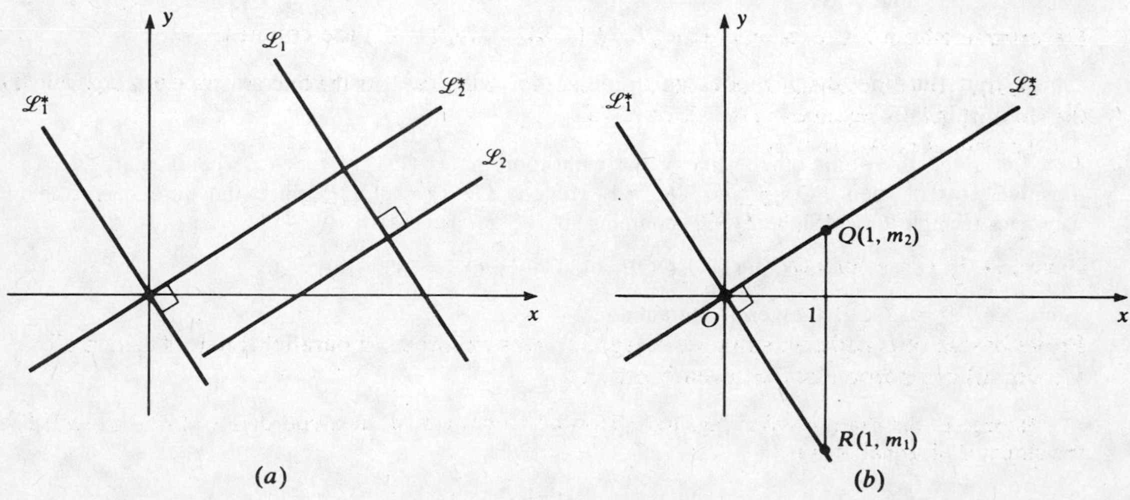

Fig. 4-12

(Fig. 4-12(*b*)). The slope-intercept equation of $\mathscr{L}_1^*$ is $y = m_1x$, and so the y-coordinate of R is m_1, since its x-coordinate is 1. Similarly, the y-coordinate of Q is m_2. By the Distance Formula,

$$\overline{OQ} = \sqrt{(1-0)^2 + (m_2 - 0)^2} = \sqrt{1 + m_2^2}$$
$$\overline{OR} = \sqrt{(1-0)^2 + (m_1 - 0)^2} = \sqrt{1 + m_1^2}$$
$$\overline{QR} = \sqrt{(1-1)^2 + (m_2 - m_1)^2} = \sqrt{(m_2 - m_1)^2}$$

By the Pythagorean Theorem for the right triangle QOR,

$$\overline{QR}^2 = \overline{OQ}^2 + \overline{OR}^2$$
$$(m_2 - m_1)^2 = (1 + m_2^2) + (1 + m_1^2)$$
$$m_2^2 - 2m_2m_1 + m_1^2 = 2 + m_1^2 + m_2^2$$
$$-2m_1m_2 = 2$$
$$m_1m_2 = -1$$

Conversely, assume that $m_1m_2 = -1$, where m_1 and m_2 are the slopes of lines $\mathscr{L}_1$ and $\mathscr{L}_2$. Then $\mathscr{L}_1$ is not parallel to $\mathscr{L}_2$. (Otherwise, by Theorem 4.1, $m_1^2 = -1$, which contradicts the fact that a square is never negative.) Let $\mathscr{L}_1$ intersect $\mathscr{L}_2$ at point P (Fig. 4-13). Let $\mathscr{L}_3$ be the line through P and perpendicular to $\mathscr{L}_1$. If m_3 is the slope of $\mathscr{L}_3$, then, by what we have just shown,

$$m_1m_3 = -1 = m_1m_2 \qquad \text{or} \qquad m_3 = m_2$$

Since $\mathscr{L}_2$ and $\mathscr{L}_3$ pass through P and have the same slope, they must coincide. Since $\mathscr{L}_1$ is perpendicular to $\mathscr{L}_3$, $\mathscr{L}_1$ is perpendicular to $\mathscr{L}_2$.

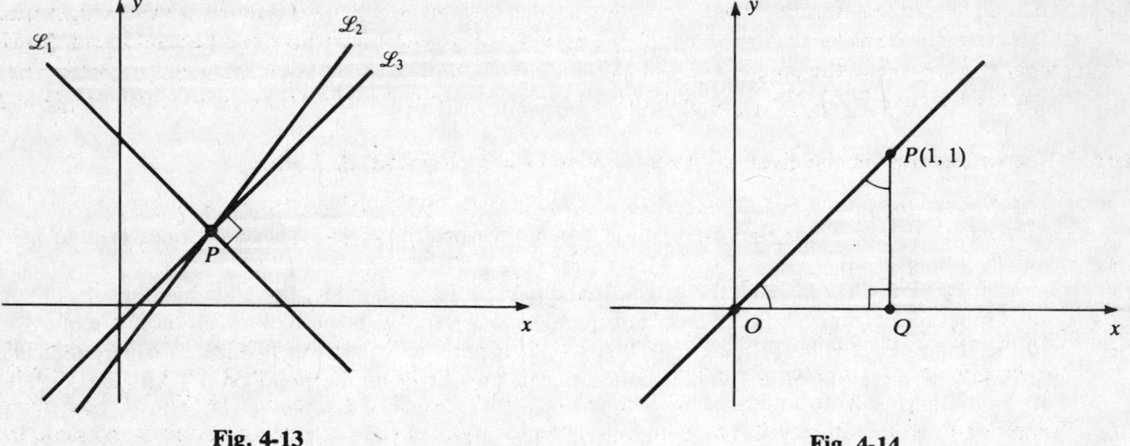

Fig. 4-13

Fig. 4-14

4.6 Show that the line $y = x$ makes an angle of 45° with the positive x-axis (i.e., $\measuredangle POQ$ in Fig. 4-14 contains 45°).

Let P be the point on the line $y = x$ with coordinates (1, 1). Drop a perpendicular, PQ, to the positive x-axis. Then $\overline{PQ} = 1$ and $\overline{OQ} = 1$. Hence, $\measuredangle OPQ = \measuredangle QOP$, since they are base angles of isosceles triangle QPO. Since $\measuredangle OQP$ contains 90°,

$$\measuredangle OPQ + \measuredangle QOP = 180° - \measuredangle OQP = 180° - 90° = 90°$$

Since $\measuredangle OPQ = \measuredangle QOP$, they each contain 45°.

Supplementary Problems

4.7 Find a point-slope equation of the line through (*a*) (2, 5) and (−1, 4), (*b*) (1, 4) and the origin, (*c*) (7, −1) and (−1, 7).

4.8 Find the slope-intercept equation of the line (*a*) through the points (−2, 3) and (4, 8); (*b*) having slope 2 and y-intercept −1; (*c*) through the points (0, 2) and (3, 0); (*d*) through (1, 4) and parallel to the x-axis; (*e*) through (1, 4) and rising 5 units for each unit increase in x; (*f*) through (5, 1) and falling 3 units for each unit increase in x; (*g*) through (−1, 4) and parallel to the line with the equation $3x + 4y = 2$; (*h*) through the origin and parallel to the line with the equation $y = 1$; (*i*) through (1, 4) and perpendicular to the line with the equation $2x - 6y = 5$; (*j*) through the origin and perpendicular to the line with the equation $5x + 2y = 1$; (*k*) through (4, 3) and perpendicular to the line with the equation $x = 1$; (*l*) through the origin and bisecting the angle between the positive x-axis and the positive y-axis.

4.9 Find the slopes and y-intercepts of the lines given by the following equations. Also find the coordinates of a point other than $(0, b)$ on each line.

(*a*) $y = 5x + 4$ (*b*) $7x - 4y = 8$ (*c*) $y = 2 - 4x$ (*d*) $y = 2$ (*e*) $\frac{y}{4} + \frac{x}{3} = 1$

4.10 If the point $(2, k)$ lies on the line with slope $m = 3$ passing through the point (1, 6), find k.

4.11 Does the point (−1, −2) lie on the line through the points (4, 7) and (5, 9)?

4.12 (*a*) Use slopes to determine whether the points (4, 1), (7, 3), and (3, 9) are the vertices of a right triangle. (*b*) Use slopes to show that (5, 4), (−4, 2), (−3, −3), and (6, −1) are vertices of a parallelogram. (*c*) Under what conditions are the points $(u, v + w)$, $(v, u + w)$, and $(w, u + v)$ on the same line? (*d*) Determine k so that the points $A(7, 5)$, $B(-1, 2)$, and $C(k, 0)$ are the vertices of a right triangle with right angle at B.

4.13 Determine whether the given lines are parallel, or perpendicular, or neither.

(*a*) $y = 5x - 2$ and $y = 5x + 3$ (*b*) $y = x + 3$ and $y = 2x + 3$
(*c*) $4x - 2y = 7$ and $10x - 5y = 1$ (*d*) $4x - 2y = 7$ and $2x + 4y = 1$
(*e*) $7x + 3y = 6$ and $3x + 7y = 14$

4.14 Temperature is usually measured either in Fahrenheit or in Celsius degrees. The relation between Fahrenheit and Celsius temperatures is given by a linear equation. The freezing point of water is 0° Celsius or 32° Fahrenheit, and the boiling point of water is 100° Celsius or 212° Fahrenheit. (*a*) Find an equation giving Fahrenheit temperature y in terms of Celsius temperature x. (*b*) What temperature is the same in both scales?

4.15 For what values of k will the line $kx + 5y = 2k$ (*a*) have y-intercept 4? (*b*) have slope 3? (*c*) pass through the point (6, 1)? (*d*) be perpendicular to the line $2x - 3y = 1$?

4.16 A triangle has vertices $A(1, 2)$, $B(8, 0)$, $C(5, 3)$. Find equations of (*a*) the median from A to the midpoint of the opposite side, (*b*) the altitude from B to the opposite side, (*c*) the perpendicular bisector of side AB.

4.17 Draw the line determined by the equation $4x - 3y = 15$. Find out whether the points (12, 9) and (6, 3) lie on this line.

4.18 (*a*) Prove that any linear equation $ax + by = c$ is the equation of a line, it being assumed that a and b are not both zero. (*Hint*: Consider separately the cases $b \neq 0$ and $b = 0$.) (*b*) Prove that any line is the graph of a linear equation. (*Hint*: Consider separately the cases where the line is vertical and where the line is not vertical.)

4.19 If the line $\mathscr{L}$ has the equation $3x + 2y - 4 = 0$, prove that a point $P(x, y)$ is above $\mathscr{L}$ if and only if $3x + 2y - 4 > 0$.

4.20 If the line $\mathcal{M}$ has the equation $3x - 2y - 4 = 0$, prove that a point $P(x, y)$ is below $\mathcal{M}$ if and only if $3x - 2y - 4 > 0$.

4.21 Use two inequalities to define the set of all points above the line

$$4x + 3y - 9 = 0$$

and to the right of the line $x = 1$. Draw a diagram indicating the set.

4.22 The leading car rental company, Heart, charges \$30 per day and 15¢ per mile for a car. The second-ranking company, Bird, charges \$32 per day and 12¢ per mile for the same kind of car. If you expect to drive x miles per day, for what values of x would it cost less to rent the car from Heart?

4.23 Prove the following geometric theorems by using coordinates: (*a*) The figure obtained by joining the midpoints of consecutive sides of any quadrilateral is a parallelogram. (*b*) The altitudes of any triangle meet at a common point. (*c*) A parallelogram with perpendicular diagonals is equilateral (a rhombus). (*d*) If two medians of a triangle are equal, the triangle is isosceles. (*e*) An angle inscribed in a semicircle is a right angle. (*Hints*: For (*a*), (*b*), and (*c*), choose coordinate systems as in Fig. 4-15.)

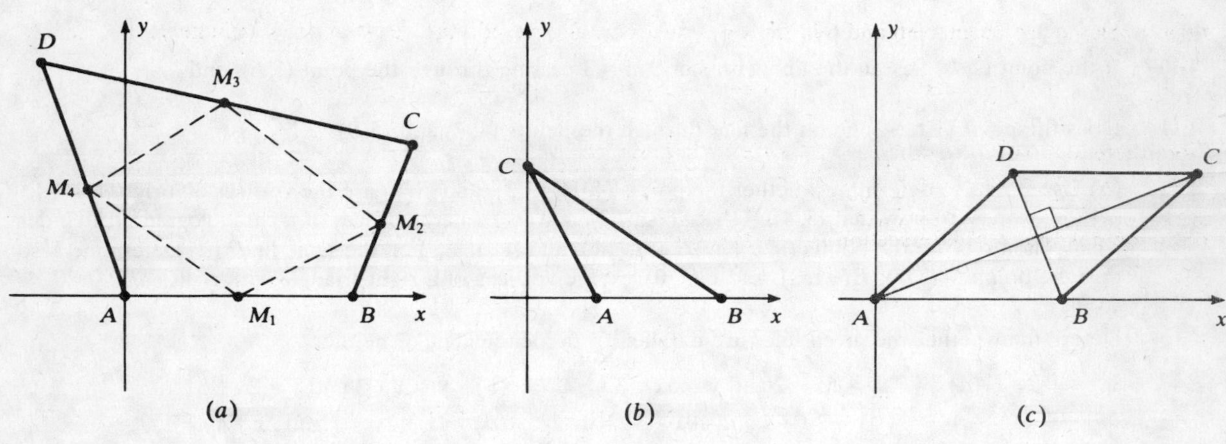

Fig. 4-15

4.24 Describe geometrically the families of lines

$$(a)\quad y = mx + 2 \qquad (b)\quad y = 3x + b$$

where m and b are any real numbers.

4.25 The *x-intercept* of a line $\mathcal{L}$ is defined to be the x-coordinate of the unique point where $\mathcal{L}$ intersects the x-axis. Thus, it is the number a for which $(a, 0)$ lies on $\mathcal{L}$. (*a*) Which lines do not have x-intercepts? (*b*) Find the x-intercepts of (i) $2x - 3y = 4$, (ii) $x + 7y = 14$, (iii) $5x - 13y = 1$, (iv) $x = 3$, (v) $y = 1$. (*c*) If a and b are the x-intercept and y-intercept of a line, show that

$$\frac{x}{a} + \frac{y}{b} = 1 \tag{1}$$

is an equation of the line. (*d*) If (*1*) is an equation of a line, show that a is the x-intercept and b is the y-intercept.

4.26 In the triangle with vertices $A(3, 1)$, $B(2, 7)$, and $C(4, 10)$, find the slope-intercept equation of (*a*) the altitude from A to side BC; (*b*) the median from B to side AC; (*c*) the perpendicular bisector of side AB.

Chapter 5

Intersections of Graphs

The *intersection* of two graphs consists of the points that the graphs have in common. These points can be found by solving simultaneously the equations that determine the graphs.

EXAMPLES (*a*) To find the intersection of the lines $\mathscr{L}$ and $\mathscr{M}$ determined by

$$\mathscr{L}: \quad 4x - 3y = 15$$
$$\mathscr{M}: \quad 3x + 2y = 7$$

multiply the first equation by 2 and the second equation by 3:

$$8x - 6y = 30$$
$$9x + 6y = 21$$

Now y has the coefficients -6 and 6 in the two equations, and we add the equations to eliminate y:

$$17x = 51 \qquad \text{or} \qquad x = \frac{51}{17} = 3$$

From the equation for $\mathscr{M}$, when $x = 3$, $3(3) + 2y = 7$. Hence,

$$9 + 2y = 7 \qquad \text{or} \qquad 2y = -2 \qquad \text{or} \qquad y = -1 .$$

Thus, the only point of intersection is $(3, -1)$ (Fig. 5-1).

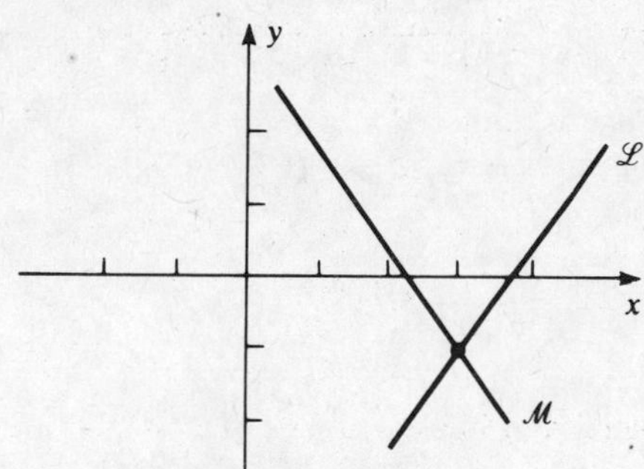

Fig. 5-1

(*b*) Let us find the intersection of the line $\mathscr{L}$: $y = 2x + 1$, and the circle

$$(x - 1)^2 + (y + 1)^2 = 16$$

We must solve the system

$$y = 2x + 1 \tag{1}$$
$$(x - 1)^2 + (y + 1)^2 = 16 \tag{2}$$

By (*1*), substitute $2x + 1$ for y in (*2*):

$$(x - 1)^2 + (2x + 2)^2 = 16$$
$$(x^2 - 2x + 1) + (4x^2 + 8x + 4) = 16$$
$$5x^2 + 6x + 5 = 16$$
$$5x^2 + 6x - 11 = 0$$
$$(5x + 11)(x - 1) = 0$$

Hence, either $5x + 11 = 0$ or $x - 1 = 0$; that is, either $x = -11/5 = -2.2$ or $x = 1$. By (*1*), when $x = 1$, $y = 3$; and,

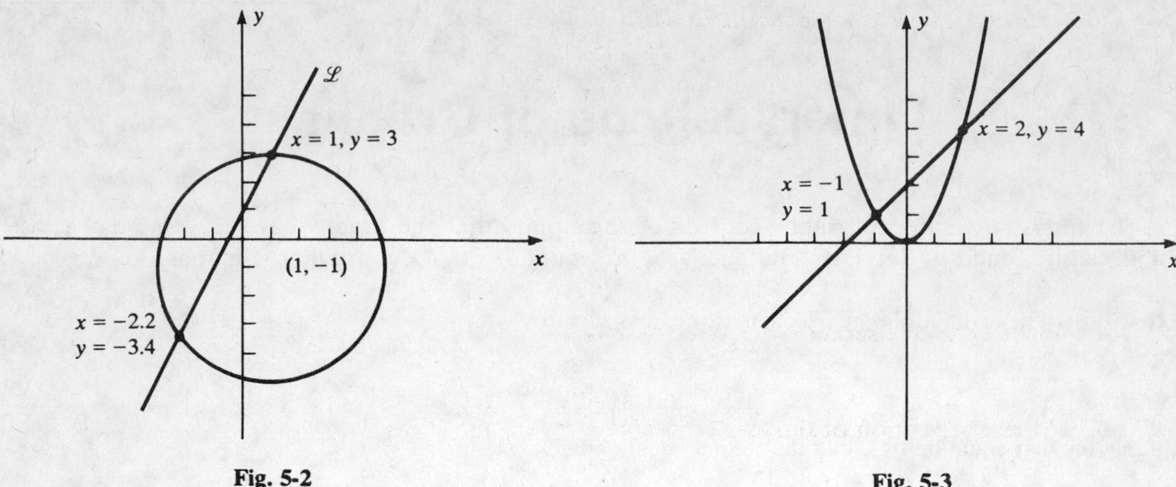

Fig. 5-2

Fig. 5-3

when $x = -2.2$, $y = -3.4$. Thus, there are two intersection points, (1, 3) and (−2.2, −3.4), as indicated in Fig. 5-2.
(*c*) To find the intersection of the line $y = x + 2$ and the parabola $y = x^2$, we solve the system:

$$y = x + 2 \tag{3}$$

$$y = x^2 \tag{4}$$

By (*4*), substitute x^2 for y in (*3*):

$$x^2 = x + 2$$
$$x^2 - x - 2 = 0$$
$$(x - 2)(x + 1) = 0$$

Hence, either $x - 2 = 0$ or $x + 1 = 0$; that is, either $x = 2$ or $x = -1$. By (*3*) or (*4*), when $x = 2$, $y = 4$; and, when $x = -1$, $y = 1$. Thus, the intersection points are (2, 4) and (−1, 1) (Fig. 5-3).

Solved Problems

5.1 Find the intersection of the lines

$$\mathscr{L}:\ 3x - 3y = 1 \qquad \mathscr{M}:\ 4x + 2y = 3$$

We must solve the system

$$3x - 3y = 1$$
$$4x + 2y = 3$$

In order to eliminate y, multiply the first equation by 2 and the second equation by 3:

$$6x - 6y = 2$$
$$12x + 6y = 9$$

Add the two equations:

$$18x = 11 \qquad \text{or} \qquad x = 11/18$$

Substitute 11/18 for x in the first equation:

$$3\left(\frac{11}{18}\right) - 3y = 1$$
$$\frac{11}{6} - 1 = 3y$$
$$\frac{5}{6} = 3y$$
$$\frac{5}{18} = y$$

So, the point of intersection is (11/18, 5/18).

5.2 Find the intersection of the line $\mathscr{L}$: $y = x + 3$ and the ellipse

$$\frac{x^2}{9} + \frac{y^2}{4} = 1$$

To solve the system of two equations, substitute $x + 3$ (as given by the equation of $\mathscr{L}$) for y in the equation of the ellipse:

$$\frac{x^2}{9} + \frac{(x+3)^2}{4} = 1$$

Multiply by 36 to clear the denominators:

$$4x^2 + 9(x+3)^2 = 36$$
$$4x^2 + 9(x^2 + 6x + 9) = 36$$
$$4x^2 + 9x^2 + 54x + 81 = 36$$
$$13x^2 + 54x + 45 = 0$$

ALGEBRA The solutions of the quadratic equation $ax^2 + bx + c = 0$ are given by the *quadratic formula*:

$$x = \frac{-b \pm \sqrt{b^2 - 4ac}}{2a}$$

By the quadratic formula,

$$x = \frac{-54 \pm \sqrt{(54)^2 - 4(13)(45)}}{2(13)} = \frac{-54 \pm \sqrt{2916 - 2340}}{26}$$
$$= \frac{-54 \pm \sqrt{576}}{26} = \frac{-54 \pm 24}{26}$$

Hence, either

$$x = \frac{-54 + 24}{26} = -\frac{30}{26} = -\frac{15}{13} \quad \text{and} \quad y = x + 3 = -\frac{15}{13} + \frac{39}{13} = \frac{24}{13}$$

or

$$x = \frac{-54 - 24}{26} = \frac{-78}{26} = -3 \quad \text{and} \quad y = x + 3 = -3 + 3 = 0$$

The two intersection points are shown in Fig. 5-4.

Notice that we could have solved $13x^2 + 54x + 45 = 0$ by factoring the left side:

$$(13x + 15)(x + 3) = 0$$

However, such a factorization is sometimes difficult to discover, whereas the quadratic formula can be applied automatically.

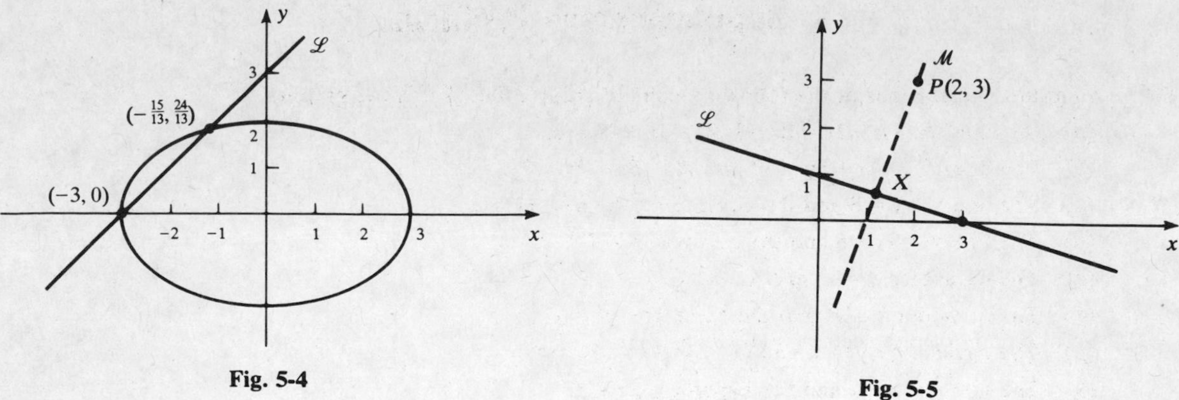

Fig. 5-4

Fig. 5-5

5.3 Find the perpendicular distance from the point $P(2, 3)$ to the line

$$\mathcal{L}: \quad 3y + x = 3$$

Let the perpendicular from P to $\mathcal{L}$ hit $\mathcal{L}$ at the point X (Fig. 5-5). If we can find the coordinates of X, then the distance $\overline{PX}$ can be computed by the Distance Formula, (*2.1*). But, X is the intersection of line $\mathcal{L}$ with the line $\mathcal{M}$ through P perpendicular to $\mathcal{L}$. The slope-intercept equation of $\mathcal{L}$ is

$$y = -\frac{1}{3}x + 1$$

which shows that the slope of $\mathcal{L}$ is $-1/3$. Therefore, by Theorem 4.2, the slope of $\mathcal{M}$ is

$$\frac{-1}{-1/3} = 3$$

so that a point-slope equation of $\mathcal{M}$ is

$$\frac{y-3}{x-2} = 3$$

Solving for y, we obtain the slope-intercept equation of $\mathcal{M}$:

$$y - 3 = 3(x-2) \qquad \text{or} \qquad y - 3 = 3x - 6 \qquad \text{or} \qquad y = 3x - 3$$

Now solve the system

$$\mathcal{L}: \quad 3y + x = 3$$
$$\mathcal{M}: \quad y = 3x - 3$$

By the second equation, substitute $3x - 3$ for y in the first equation:

$$3(3x-3) + x = 3$$
$$9x - 9 + x = 3$$
$$10x = 12$$
$$x = \frac{12}{10} = \frac{6}{5}$$

By the equation for $\mathcal{M}$, when $x = 6/5$,

$$y = 3\left(\frac{6}{5}\right) - 3 = \frac{18}{5} - 3 = \frac{3}{5}$$

Thus, point X has coordinates $(6/5, 3/5)$, and

$$\overline{PX} = \sqrt{\left(2-\frac{6}{5}\right)^2 + \left(3-\frac{3}{5}\right)^2} = \sqrt{\left(\frac{4}{5}\right)^2 + \left(\frac{12}{5}\right)^2} = \sqrt{\frac{16}{25} + \frac{144}{25}} = \sqrt{\frac{160}{25}}$$
$$= \frac{\sqrt{160}}{\sqrt{25}} = \frac{\sqrt{16}\sqrt{10}}{5} = \frac{4\sqrt{10}}{5} = .8\sqrt{10}$$

Use the approximation $\sqrt{10} \approx 3.16$ (the symbol $\approx$ means "is approximately equal to") to obtain

$$\overline{PX} \approx (.8)(3.16) \approx 2.53$$

Supplementary Problems

5.4 Find the intersections of the following pairs of graphs, and sketch the graphs:

(*a*) the lines $\mathscr{L}$: $x - 2y = 2$ and $\mathscr{M}$: $3x + 4y = 6$.

(*b*) The lines $\mathscr{L}$: $4x + 5y = 10$ and $\mathscr{M}$: $5x + 4y = 8$.

(*c*) The line $x + y = 8$ and the circle $(x-2)^2 + (y-1)^2 = 25$.

(*d*) The line $y = 8x - 6$ and the parabola $y = 2x^2$.

(*e*) The parabolas $y = x^2$ and $x = y^2$.

(*f*) The parabola $x = y^2$ and the circle $x^2 + y^2 = 6$.

(*g*) The circles $x^2 + y^2 = 1$ and $(x-1)^2 + y^2 = 1$.

(*h*) The line $y = x - 3$ and the hyperbola $xy = 4$.

(*i*) The circle of radius 3 centered at the origin and the line through the origin with slope $\frac{2}{3}$.

5.5 Find the distance from the point $(0, 3)$ to the line $x - 2y = 2$.

5.6 By using the method employed in Problem 5.3, show that a formula for the distance from the point $P(x_1, y_1)$ to the line $\mathscr{L}$: $Ax + By + C = 0$ is

$$\frac{|Ax_1 + By_1 + C|}{\sqrt{A^2 + B^2}}$$

5.7 Let x represent the number of million pounds of mutton that farmers offer for sale per week. Let y represent the number of dollars per pound that buyers are willing to pay for mutton. Assume that the *supply equation* for mutton is

$$y = .02\,x + .25$$

that is, $.02\,x + .25$ is the price per pound at which farmers are willing to sell x million pounds of mutton per week. Assume also that the *demand equation* for mutton is

$$y = -.025\,x + 2.5$$

that is, $-.025\,x + 2.5$ is the price per pound at which buyers are willing to buy x million pounds of mutton. Find the intersection of the graphs of the supply and demand equations. This point (x, y) indicates the supply x at which the seller's price is equal to what the buyer is willing to pay.

5.8 Find the center and radius of the circle passing through the points $A(3, 0)$, $B(0, 3)$, and $C(6, 0)$. (*Hint*: The center is the intersection of the perpendicular bisectors of any two sides of $\triangle ABC$.)

5.9 Find the equations of the lines through the origin that are tangent to the circle with center at $(3, 1)$ and radius 3. (*Hint*: A tangent to a circle is perpendicular to the radius at the point of contact. Therefore, the Pythagorean Theorem may be used to give a second equation for the coordinates of the point of contact.)

5.10 Find the coordinates of the point on the line $y = 2x + 1$ that is equidistant from $(0, 0)$ and $(5, -2)$.

Chapter 6

Symmetry

6.1 SYMMETRY ABOUT A LINE

Two points P and Q are said to be *symmetric with respect to a line* $\mathcal{L}$ if P and Q are mirror images in $\mathcal{L}$; more precisely, the segment PQ is perpendicular to $\mathcal{L}$ at a point A such that $\overline{PA} = \overline{QA}$ (see Fig. 6-1).

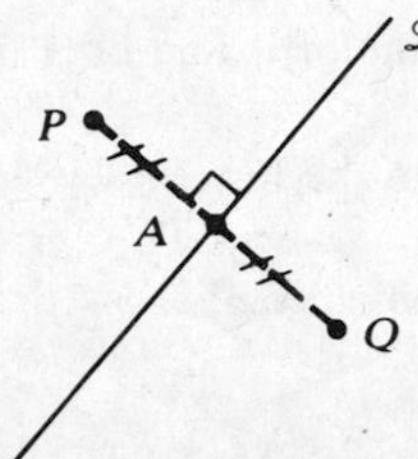

Fig. 6-1

(i) If $Q(x, y)$ is symmetric to the point P with respect to the y-axis, then P is $(-x, y)$. See Fig. 6-2(a).

(ii) If $Q(x, y)$ is symmetric to the point P with respect to the x-axis, then P is $(x, -y)$. See Fig. 6-2(b).

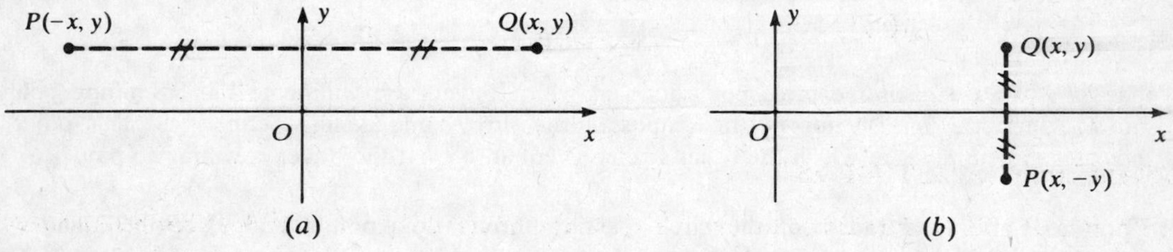

Fig. 6-2

A graph is said to be *symmetric with respect to a line* $\mathcal{L}$ if, for any point P on the graph, the point Q that is symmetric to P with respect to $\mathcal{L}$ is also on the graph. $\mathcal{L}$ is then called an *axis of symmetry* of the graph.

Fig. 6-3

Consider the graph of an equation $f(x, y) = 0$. Then, by (i) above, the graph is symmetric with respect to the y-axis if and only if $f(x, y) = 0$ implies $f(-x, y) = 0$. And, by (ii) above, the graph is symmetric with respect to the x-axis if and only if $f(x, y) = 0$ implies $f(x, -y) = 0$.

EXAMPLES (*a*) The y-axis is an axis of symmetry of the parabola $y = x^2$ (Fig. 6-4(*a*)). For, if $y = x^2$, then $y = (-x)^2$. The x-axis is not an axis of symmetry of this parabola; although (1, 1) is on the parabola, (1, −1) is not on the parabola. (*b*) The ellipse

$$\frac{x^2}{4} + y^2 = 1$$

(Fig. 6-4(*b*)) has both the y-axis and the x-axis as axes of symmetry. For, if

$$\frac{x^2}{4} + y^2 = 1$$

then

$$\frac{(-x)^2}{4} + y^2 = 1 \qquad \text{and} \qquad \frac{x^2}{4} + (-y)^2 = 1$$

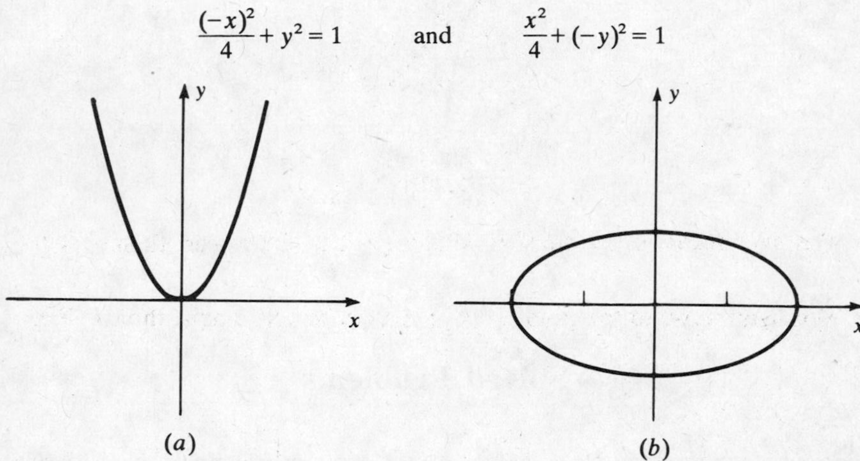

Fig. 6-4

6.2 SYMMETRY ABOUT A POINT

Two points P and Q are said to be *symmetric with respect to a point* A if A is the midpoint of the line segment PQ (Fig. 6-5(*a*)).

The point Q symmetric to the point $P(x, y)$ with respect to the origin has coordinates $(-x, -y)$. (In Fig. 6-5(*b*), $\triangle POR$ is congruent to $\triangle QOS$. Hence, $\overline{OR} = \overline{OS}$ and $\overline{RP} = \overline{SQ}$.)

Symmetry of a graph about a point is defined in the expected manner. In particular, a graph $\mathscr{G}$ is said to be *symmetric with respect to the origin* if, whenever a point P lies on $\mathscr{G}$, the point Q symmetric to P with respect to the origin also lies on $\mathscr{G}$. The graph of an equation $f(x, y) = 0$ is symmetric with respect to the origin if and only if $f(x, y) = 0$ implies $f(-x, -y) = 0$.

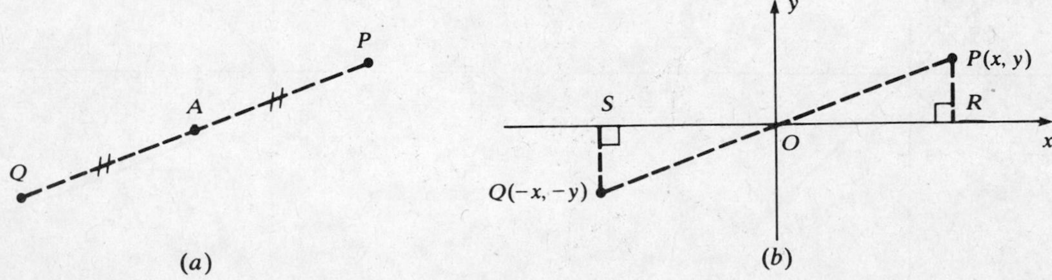

Fig. 6-5

EXAMPLES (*a*) The ellipse graphed in Fig. 6-4(*b*) is symmetric with respect to the origin, because

$$\frac{x^2}{4} + y^2 = 1 \quad \text{implies} \quad \frac{(-x)^2}{4} + (-y)^2 = 1$$

(*b*) The hyperbola $xy = 1$ (Fig. 6-6) is symmetric with respect to the origin, for, if $xy = 1$, then $(-x)(-y) = 1$. (*c*) If $y = ax$, then $-y = a(-x)$. Hence, any straight line through the origin is symmetric with respect to the origin.

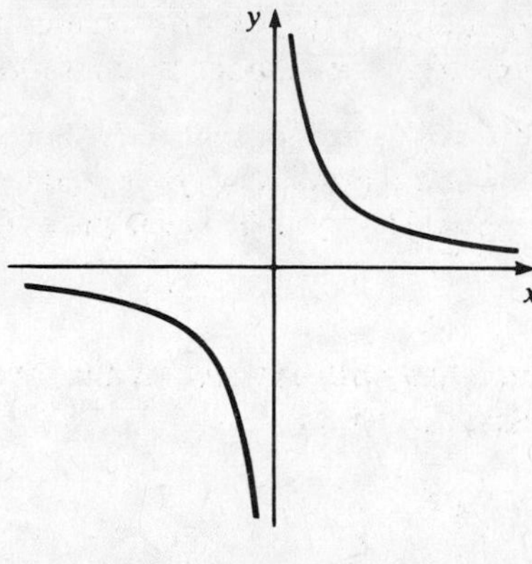

Fig. 6-6

Solved Problems

6.1 Determine whether the line $y = -x$ (Fig. 6-7) is symmetric with respect to (*a*) the *x*-axis, (*b*) the *y*-axis, (*c*) the origin.

(*a*) The line is not symmetric with respect to the *x*-axis, since $(-1, 1)$ is on the line, but $(-1, -1)$, the reflection of $(-1, 1)$ in the *x*-axis, is not on the line.

(*b*) The line is not symmetric with respect to the *y*-axis, since $(-1, 1)$ is on the line, but $(1, 1)$, the reflection of $(-1, 1)$ in the *y*-axis, is not on the line.

(*c*) The line is symmetric with respect to the origin, by Example (*c*) above.

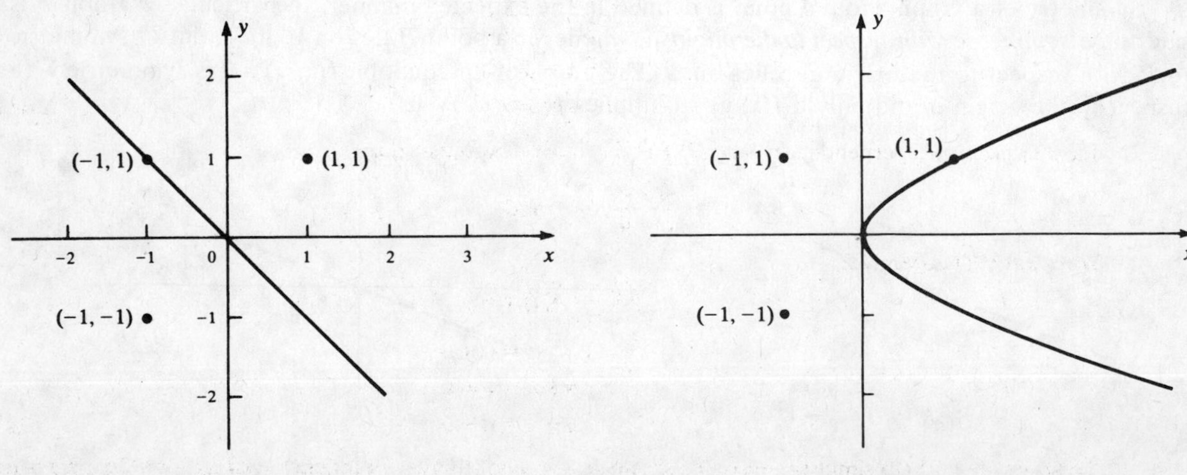

Fig. 6-7 Fig. 6-8

6.2 Determine whether the parabola $x = y^2$ (Fig. 6-8) is symmetric with respect to (*a*) the x-axis, (*b*) the y-axis, (*c*) the origin.

(*a*) The parabola is symmetric with respect to the x-axis, since $x = y^2$ implies $x = (-y)^2$.

(*b*) It is not symmetric with respect to the y-axis, since (1, 1) is on the parabola, but (−1, 1) is not.

(*c*) It is not symmetric with respect to the origin, since (1, 1) is on the parabola, but (−1, −1) is not.

6.3 Show that, if the graph of an equation $f(x, y) = 0$ is symmetric with respect to both the x-axis and the y-axis, then it is symmetric with respect to the origin. (However, the converse is false, as is shown by Problem 6.1.)

Assume that $f(x, y) = 0$; we must prove that $f(-x, -y) = 0$. Since $f(x, y) = 0$ and the graph is symmetric with respect to the x-axis, $f(x, -y) = 0$. Then, since $f(x, -y) = 0$ and the graph is symmetric with respect to the y-axis, $f(-x, -y) = 0$.

6.4 Let points P and Q be symmetric with respect to the line $\mathscr{L}$: $y = x$. If P has coordinates (a, b), show that Q has coordinates (b, a).

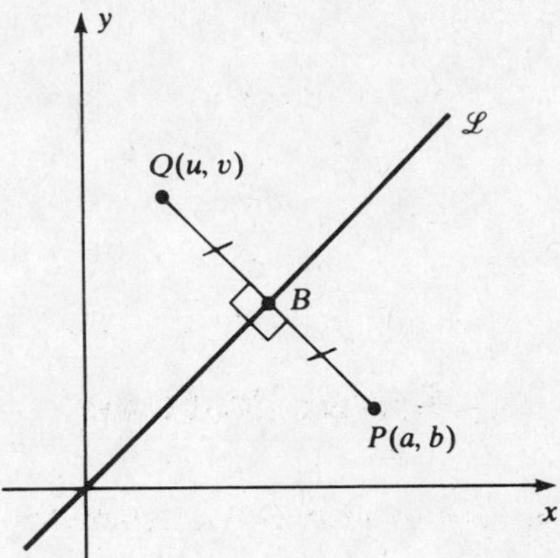

Fig. 6-9

Let Q have coordinates (u, v), and let B be the intersection point of $\mathscr{L}$ and the line PQ (see Fig. 6-9). B bisects PQ; hence, its coordinates are given by (*2.2*) as

$$\left(\frac{a+u}{2}, \frac{b+v}{2}\right)$$

from which, since B lies on $\mathscr{L}$,

$$\frac{b+v}{2} = \frac{a+u}{2}$$

$$b + v = a + u$$

$$v - u = a - b \tag{1}$$

Furthermore, the perpendicular lines PQ and $\mathscr{L}$ have respective slopes

$$\frac{b-v}{a-u} \quad \text{and} \quad 1$$

so that, by Theorem 4.2,

$$\left(\frac{b-v}{a-u}\right)(1) = -1$$

$$b - v = u - a$$

$$b + a = v + u$$

$$v + u = a + b \tag{2}$$

To solve (*1*) and (*2*) simultaneously for u and v, first add the two equations, yielding $2v = 2a$, or $v = a$; then subtract (*1*) from (*2*), yielding $2u = 2b$, or $u = b$. Thus, Q has coordinates (b, a).

Supplementary Problems

6.5 Determine whether the graphs of the following equations are symmetric with respect to the x-axis, the y-axis, the origin, or to none of these:

(*a*) $y = -x^2$ (*b*) $y = x^3$ (*c*) $\frac{x^2}{9} - \frac{y^2}{4} = 1$

(*d*) $x^2 + y^2 = 5$ (*e*) $(x-1)^2 + y^2 = 9$ (*f*) $y = (x-1)^2$

(*g*) $3x^2 - xy + y^2 = 4$ (*h*) $y = \frac{x+1}{x}$ (*i*) $y = (x^2+1)^2 - 4$

(*j*) $y = x^4 - 3x^2 + 5$ (*k*) $y = -x^5 + 7x$ (*l*) $y = (x-2)^3 + 1$

6.6 Find an equation of the new curve when (*a*) the graph of $x^2 - xy + y^2 = 1$ is reflected in the x-axis; (*b*) the graph of $y^3 - xy^2 + x^3 = 8$ is reflected in the y-axis; (*c*) the graph of $x^2 - 12x + 3y = 1$ is reflected in the origin (that is, each point is replaced by its symmetric point with respect to the origin).

Chapter 7

Functions and Their Graphs

7.1 THE NOTION OF A FUNCTION

To say that a quantity y is a *function* of some other quantity x means, in ordinary language, that the value of y depends on the value of x. For example, the volume V of a cube is a function of the length s of a side. In fact, the dependence of V on s can be made precise through the formula $V = s^3$. Such a specific association of a number s^3 with a given number s is what mathematicians usually mean by a function.

In Fig. 7-1, we picture a function f as some sort of process which, from a number x, produces a number $f(x)$; the number x is called an *argument* of f and the number $f(x)$ is called the *value* of f for the argument x.

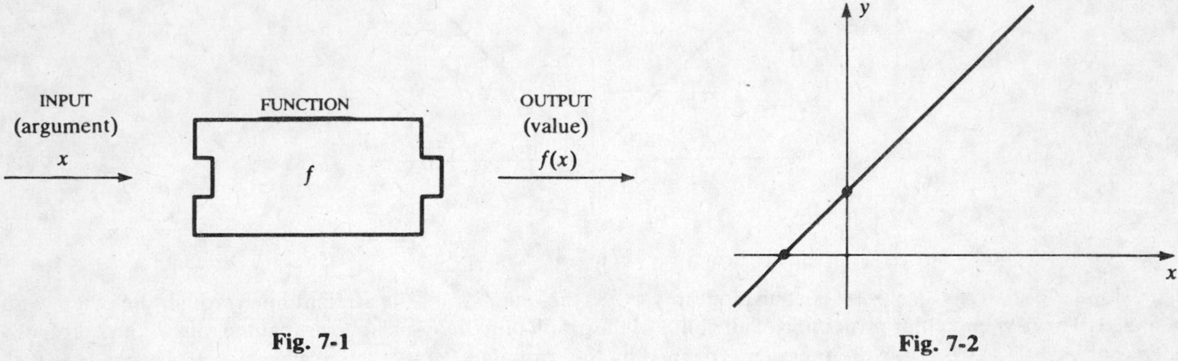

Fig. 7-1 **Fig. 7-2**

EXAMPLES (*a*) The square-root function associates with each nonnegative real number x the value $\sqrt{x}$; i.e., the unique nonnegative real number y such that $y^2 = x$. (*b*) The doubling function g associates with each real number x the value $2x$. Thus, $g(3) = 6$, $g(-1) = -2$, $g(\sqrt{2}) = 2\sqrt{2}$.

The *graph* of a function f consists of all points (x, y) such that $y = f(x)$.

EXAMPLES (*a*) Consider the function f such that $f(x) = x + 1$ for all x. The graph of f is the set of all points (x, y) such that $y = x + 1$. This is a straight line, with slope 1 and y-intercept 1 (Fig. 7-2). (*b*) The graph of the function f such that $f(x) = x^2$ for all x consists of all points (x, y) such that $y = x^2$. This is the parabola of Fig. 3-2. (*c*) Consider the function f such that $f(x) = x^3$ for all x. In Fig. 7-3(*a*) we have indicated a few points on the graph, which is sketched in Fig. 7-3(*b*).

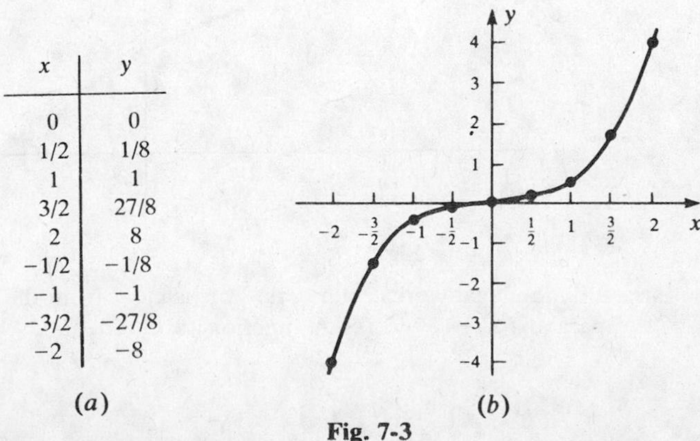

x	y
0	0
1/2	1/8
1	1
3/2	27/8
2	8
−1/2	−1/8
−1	−1
−3/2	−27/8
−2	−8

(*a*) (*b*)

Fig. 7-3

The numbers x for which a function f produces a value $f(x)$ form a collection of numbers called the *domain* of f. For example, the domain of the square-root function consists of all nonnegative real numbers; the function is not defined for negative arguments. On the other hand, the domain of the doubling function consists of all real numbers.

The numbers which are the values of a function form the *range* of the function.

The domain and range of a function f often can be determined easily by looking at the graph of f: the domain consists of all x-coordinates of points of the graph, and the range consists of all y-coordinates of points of the graph.

EXAMPLES (*a*) The range of the square-root function consists of all nonnegative real numbers. Indeed, for every nonnegative real number y, there is some number x such that $\sqrt{x} = y$; namely, the number y^2. (*b*) The range of the doubling function consists of all real numbers. Indeed, for every real number y, there exists a real number x such that $2x = y$; namely, the number $y/2$. (*c*) Consider the absolute-value function h, defined by $h(x) = |x|$. The domain consists of all real numbers, but the range is made up of all nonnegative real numbers. The graph is shown in Fig. 7-4. When $x \geq 0$, $y = |x|$ is equivalent to $y = x$, the equation of the straight line through the origin

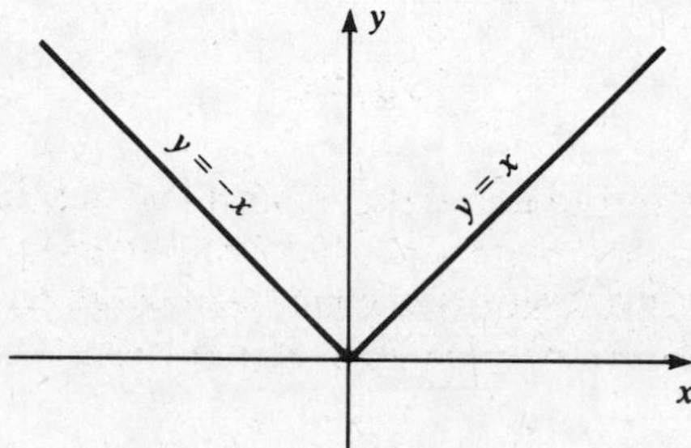

Fig. 7-4

with slope 1. When $x < 0$, $y = |x|$ is equivalent to $y = -x$, the equation of the straight line through the origin with slope -1. The perpendicular projection of all points of the graph onto the y-axis shows that the range consists of all y such that $y \geq 0$. (*d*) Consider the function g defined by the formula $g(x) = \sqrt{1-x}$, whenever this formula makes sense. Here, the value $\sqrt{1-x}$ is defined only when $1 - x \geq 0$; that is, only when $x \leq 1$. So, the domain of g consists of all real numbers x such that $x \leq 1$. It is a little harder to find the range of g. Consider a real number y; it will belong to the range of g if we can find some number x such that $y = \sqrt{1-x}$. Since $\sqrt{1-x}$ can't be negative (by definition of the square-root function), we must restrict our attention to nonnegative y. Now, if $y = \sqrt{1-x}$, then $y^2 = 1 - x$, and, therefore, $x = 1 - y^2$. Indeed, when $x = 1 - y^2$, then

$$\sqrt{1-x} = \sqrt{1-(1-y^2)} = \sqrt{y^2} = y$$

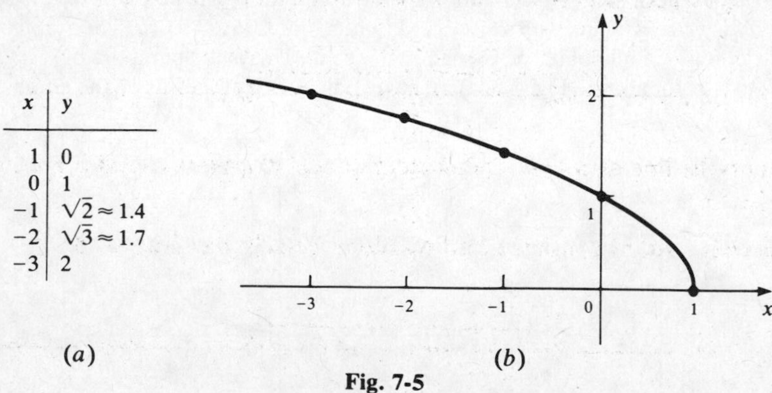

x	y
1	0
0	1
−1	$\sqrt{2} \approx 1.4$
−2	$\sqrt{3} \approx 1.7$
−3	2

(*a*)

(*b*)

Fig. 7-5

The range of g therefore consists of all nonnegative real numbers. This is clear from the graph of g, Fig. 7-5(*b*). The graph is the upper half of the parabola $x = 1 - y^2$. (*e*) A function can be defined "by cases"; for instance,

$$f(x) = \begin{cases} x^2 & \text{if } x < 0 \\ 1 + x & \text{if } 0 \leq x \leq 1 \end{cases}$$

Here, the value $f(x)$ is determined by two different formulas; the first formula applies when x is negative, and the second when $0 \leq x \leq 1$. The domain consists of all numbers x such that $x \leq 1$. The range turns out to be all positive real numbers. This can be seen from Fig. 7-6, in which projection of the graph onto the y-axis produces all y such that $y > 0$.

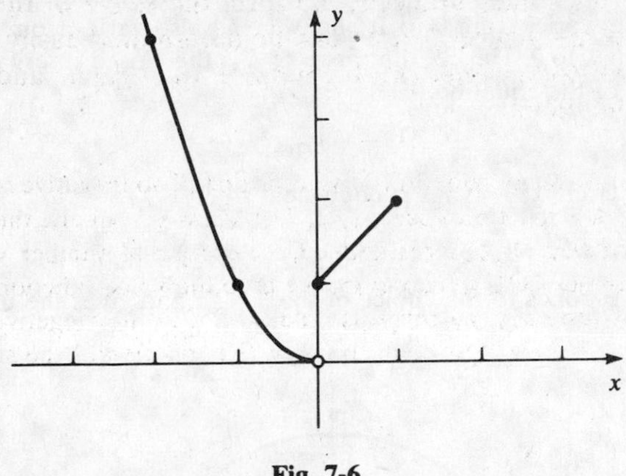

Fig. 7-6

NOTE. In many treatments of the foundations of mathematics, a function is identified with its graph. In other words, a function is defined to be a set f of ordered pairs such that f does not contain two pairs (a, b) and (a, c) with $b \neq c$. Then "$y = f(x)$" is defined to mean the same thing as "(x, y) belongs to f." However, this approach obscures the intuitive idea of a function.

7.2 INTERVALS

In dealing with the domains and ranges of functions, intervals of numbers occur so often that it is convenient to introduce special notation and terminology for them.

Closed interval. $[a, b]$ consists of all numbers x such that $a \leq x \leq b$.

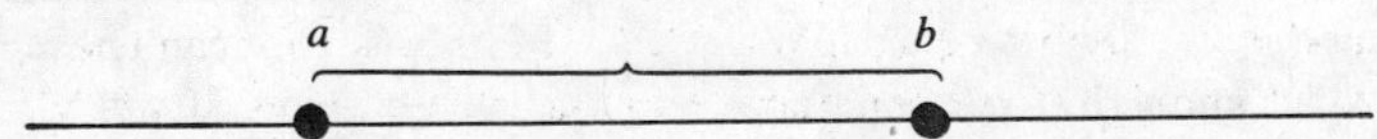

The solid dots on the line at a and b mean that a and b are included in the closed interval $[a, b]$.

Open interval. (a, b) consists of all numbers x such that $a < x < b$.

The open dots on the line at a and b indicate that the endpoints a and b are not included in the open interval (a, b).

Half-open intervals. $[a, b)$ consists of all numbers x such that $a \leq x < b$.

$(a, b]$ consists of all numbers x such that $a < x \leq b$.

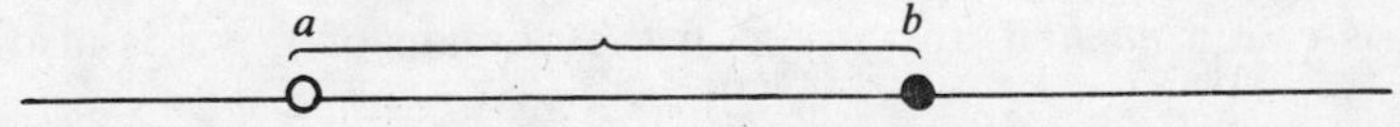

EXAMPLE Consider the function f such that $f(x) = \sqrt{1-x^2}$, whenever this formula makes sense. The domain of f consists of all numbers x such that

$$1 - x^2 \geq 0 \qquad \text{or} \qquad x^2 \leq 1 \qquad \text{or} \qquad -1 \leq x \leq 1$$

Thus, the domain of f is the closed interval $[-1, 1]$. To determine the range of f, notice that, as x varies from -1 to 0, x^2 varies from 1 to 0, $1 - x^2$ varies from 0 to 1, and $\sqrt{1-x^2}$ also varies from 0 to 1. Similarly, as x varies from 0 to 1, $\sqrt{1-x^2}$ varies from 1 to 0. Hence, the range of f is the closed interval $[0, 1]$. This is confirmed by looking at the graph in Fig. 7-7 of the equation

$$y = \sqrt{1-x^2}$$

This is a semicircle. In fact, the circle $x^2 + y^2 = 1$ is equivalent to

$$y^2 = 1 - x^2 \qquad \text{or} \qquad y = \pm\sqrt{1-x^2}$$

The value of the function f corresponds to the choice of the + sign and gives the upper half of the circle.

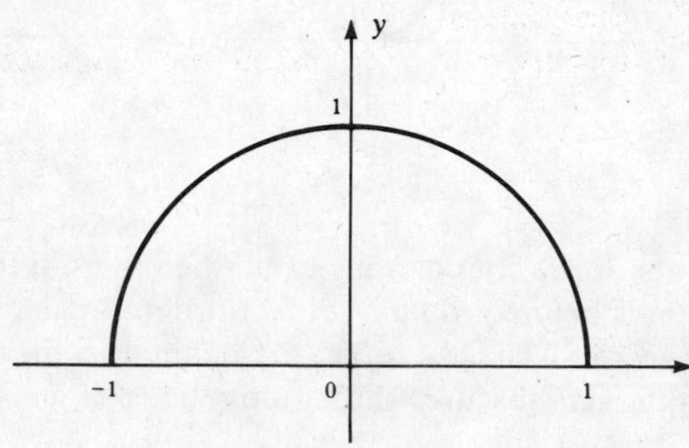

Fig. 7-7

Sometimes we deal with intervals that are unbounded on one side:

$[a, \infty)$ is made up of all x such that $a \leq x$;

(a, ∞) is made up of all x such that $a < x$;

$(-\infty, b]$ is made up of all x such that $x \leq b$;

$(-\infty, b)$ is made up of all x such that $x < b$.

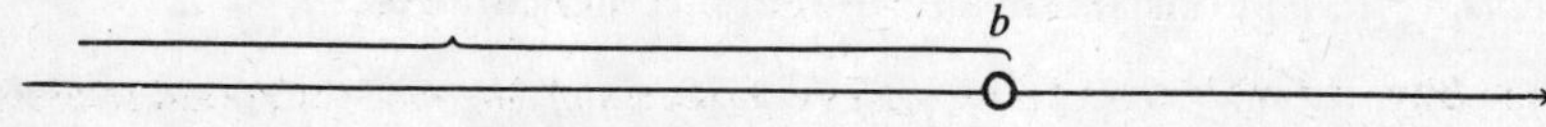

EXAMPLE The range of the function graphed in Fig. 7-6 is $(0, \infty)$.

7.3 EVEN AND ODD FUNCTIONS

A function f is called *even* if, for any x in the domain of f, $-x$ is also in the domain of f and $f(-x) = f(x)$.

EXAMPLES (*a*) Let $f(x) = x^2$ for all x. Then

$$f(-x) = (-x)^2 = x^2 = f(x)$$

and so f is even. (*b*) Let $f(x) = 3x^4 - 5x^2 + 2$ for all x. Then

$$f(-x) = 3(-x)^4 - 5(-x)^2 + 2 = 3x^4 - 5x^2 + 2 = f(x)$$

Thus, f is even. More generally, a function that is defined for all x and involves only even powers of x is an even function. (*c*) Let $f(x) = x^3 + 1$ for all x. Then

$$f(-x) = (-x)^3 + 1 = -x^3 + 1$$

Now, $-x^3 + 1$ is not equal to $x^3 + 1$ except when $x = 0$. Hence, f is not even.

A function f is even when and only when the graph of f is symmetric with respect to the y-axis. For the symmetry means that for any point $(x, f(x))$ on the graph, the image point $(-x, f(x))$ also lies on the graph; in other words, $f(-x) = f(x)$. See Fig. 7-8.

A function f is called *odd* if, for any x in the domain of f, $-x$ is also in the domain of f and $f(-x) = -f(x)$.

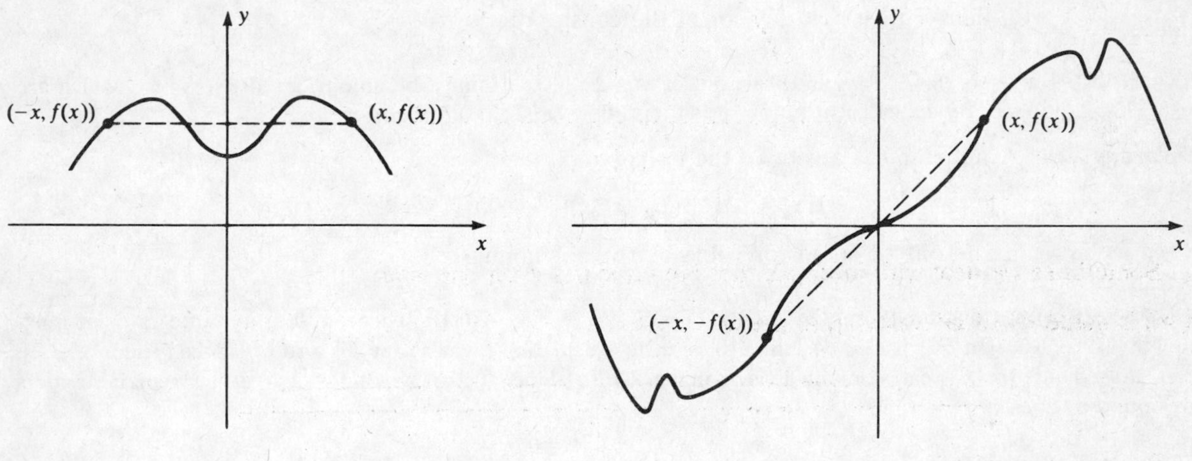

Fig. 7-8

Fig. 7-9

EXAMPLES (*a*) Let $f(x) = x^3$ for all x. Then

$$f(-x) = (-x)^3 = -x^3 = -f(x)$$

Thus, f is odd. (*b*) Let $f(x) = 4x$ for all x. Then

$$f(-x) = 4(-x) = -4x = -f(x)$$

Thus, f is odd. (*c*) Let $f(x) = 3x^5 - 2x^3 + x$ for all x. Then

$$\begin{aligned} f(-x) &= 3(-x)^5 - 2(-x)^3 + (-x) = 3(-x^5) - 2(-x^3) - x \\ &= -3x^5 + 2x^3 - x = -(3x^5 - 2x^3 + x) = -f(x) \end{aligned}$$

Thus, f is odd. More generally, if $f(x)$ is defined as a polynomial which involves only odd powers of x (and lacks a constant term), then $f(x)$ is an odd function. (*d*) The function $f(x) = x^3 + 1$, which was shown above to be not even, is not odd either. In fact,

$$f(1) = (1)^3 + 1 = 1 + 1 = 2 \quad \text{but} \quad -f(-1) = -[(-1)^3 + 1] = -[-1 + 1] = 0$$

A function f is odd if and only if the graph of f is symmetric with respect to the origin. For the symmetry means that for any point $(x, f(x))$ on the graph, the image point $(-x, -f(x))$ is also on the graph; i.e., $f(-x) = -f(x)$. See Fig. 7-9.

7.4 ALGEBRA REVIEW: ZEROS OF POLYNOMIALS

Functions defined by polynomials are so important in calculus that it is essential to review certain basic facts concerning polynomials.

Definition: For any function f, a *zero* or *root* of f is a number r such that $f(r) = 0$.

EXAMPLES (*a*) 3 is a root of the polynomial $2x^3 - 4x^2 - 8x + 6$, since

$$2(3)^3 - 4(3)^2 - 8(3) + 6 = 2(27) - 4(9) - 24 + 6 = 54 - 36 - 24 + 6 = 0$$

(*b*) 5 is a zero of

$$\frac{x^2 - 2x - 15}{x + 2}$$

because

$$\frac{(5)^2 - 2(5) - 15}{5 + 2} = \frac{25 - 10 - 15}{7} = \frac{0}{7} = 0$$

Theorem 7.1: If $f(x) = a_n x^n + a_{n-1}x^{n-1} + \cdots + a_1 x + a_0$ is a polynomial with integral coefficients (that is, the numbers $a_n, a_{n-1}, \ldots, a_1, a_0$ are integers), and if r is an integer that is a root of f, then r must be a divisor of the constant term a_0.

EXAMPLE By Theorem 7.1, any integral root of $x^3 - 2x^2 - 5x + 6$ must be among the divisors of 6, which are ±1, ±2, ±3, and ±6. By actual substitution, it is found that 1, -2, and 3 are roots.

Theorem 7.2: A number r is a root of the polynomial

$$f(x) = a_n x^n + a_{n-1}x^{n-1} + \cdots + a_1 x + a_0$$

if and only if $f(x)$ is divisible by the polynomial $x - r$.

EXAMPLES (*a*) Consider the polynomial $f(x) = x^3 - 4x^2 + 14x - 20$. By Theorem 7.1, any integral root must be a divisor of 20; that is, ±1, ±2, ±4, ±5, ±10, or ±20. Calculation reveals that 2 is a root. Hence (Theorem 7.2), $x - 2$ must divide $f(x)$; we carry out the division in Fig. 7-10(*a*). Since $f(x) = (x - 2)(x^2 - 2x + 10)$, its remaining roots are found by solving

$$x^2 - 2x + 10 = 0$$

Applying the quadratic formula (see Problem 5.2),

$$x = \frac{-(-2) \pm \sqrt{(-2)^2 - 4(10)}}{2} = \frac{2 \pm \sqrt{-36}}{2} = \frac{2 \pm \sqrt{36} \cdot \sqrt{-1}}{2}$$

$$= \frac{2 \pm 6\sqrt{-1}}{2} = 1 \pm 3\sqrt{-1}$$

Thus, the other two roots of $f(x)$ are the complex numbers $1 + 3\sqrt{-1}$ and $1 - 3\sqrt{-1}$. (*b*) Let us find the roots of $f(x) = x^3 - 5x^2 + 3x + 9$. The integral roots (if any) must be divisors of 9: ±1, ±3, ±9. 1 is not a root, but -1 is a root. Hence, $f(x)$ is divisible by $x + 1$; see Fig. 7-10(*b*). The other roots of $f(x)$ must be the roots of $x^2 - 6x + 9$.

$$\begin{array}{r|l}
 & x^2 - 2x + 10 \\
\hline
x - 2 & x^3 - 4x^2 + 14x - 20 \\
 & x^3 - 2x^2 \\
\hline
 & \quad - 2x^2 + 14x \\
 & \quad - 2x^2 + 4x \\
\hline
 & \qquad 10x - 20 \\
 & \qquad 10x - 20 \\
\hline
\end{array}$$

(*a*)

$$\begin{array}{r|l}
 & x^2 - 6x + 9 \\
\hline
x + 1 & x^3 - 5x^2 + 3x + 9 \\
 & x^3 + x^2 \\
\hline
 & \quad - 6x^2 + 3x \\
 & \quad - 6x^2 - 6x \\
\hline
 & \qquad 9x + 9 \\
 & \qquad 9x + 9 \\
\hline
\end{array}$$

(*b*)

Fig. 7-10

But $x^2 - 6x + 9 = (x - 3)^2$. Thus, -1 and 3 are the roots of $f(x)$; 3 is called a *repeated root*, because $(x - 3)^2$ is a factor of $x^3 - 5x^2 + 3x + 9$.

Theorem 7.3 (*Fundamental Theorem of Algebra*): If repeated roots are counted multiply, then every polynomial of degree n has exactly n roots.

As the complex roots of a polynomial (with real coefficients) occur in pairs, $a \pm b\sqrt{-1}$, the polynomial can have only an even number (possibly 0) of complex roots. Hence, a polynomial of odd degree must have at least one *real* root.

Solved Problems

7.1 Find the domain and range of the function f: $f(x) = -x^2$.

Since $-x^2$ is defined for every real number x, the domain of f consists of all real numbers. To find the range, notice that $x^2 \geq 0$ for all x, and, therefore, $-x^2 \leq 0$ for all x. Every nonpositive number y appears as a value $-x^2$ for a suitable argument x; namely, for the argument $x = \sqrt{-y}$ (and also for the argument $x = -\sqrt{-y}$). Thus, the range of f is $(-\infty, 0]$.

7.2 Find the domain and range of the function f defined by

$$f(x) = \begin{cases} x + 1 & \text{if } -1 < x < 0 \\ x & \text{if } \ 0 \leq x < 1 \end{cases}$$

The domain of f consists of all x such that either $-1 < x < 0$ or $0 \leq x < 1$. This makes up the open interval $(-1, 1)$. The range of f is easily found from the graph in Fig. 7-11, whose projection onto the y-axis is the half-open interval $[0, 1)$.

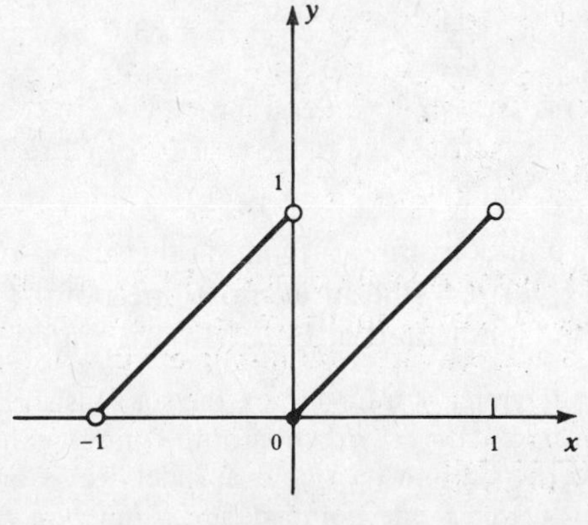

Fig. 7-11

7.3 Define $f(x)$ as the greatest integer less than or equal to x; this value is usually denoted $[x]$. Find the domain and range, and draw the graph of f.

Since $[x]$ is defined for all x, the domain is the set of all real numbers. The range of f consists of all integers. The graph is shown in Fig. 7-12. It consists of a sequence of horizontal, half-open, unit intervals. (A function whose graph consists of horizontal segments is called a *step function*.)

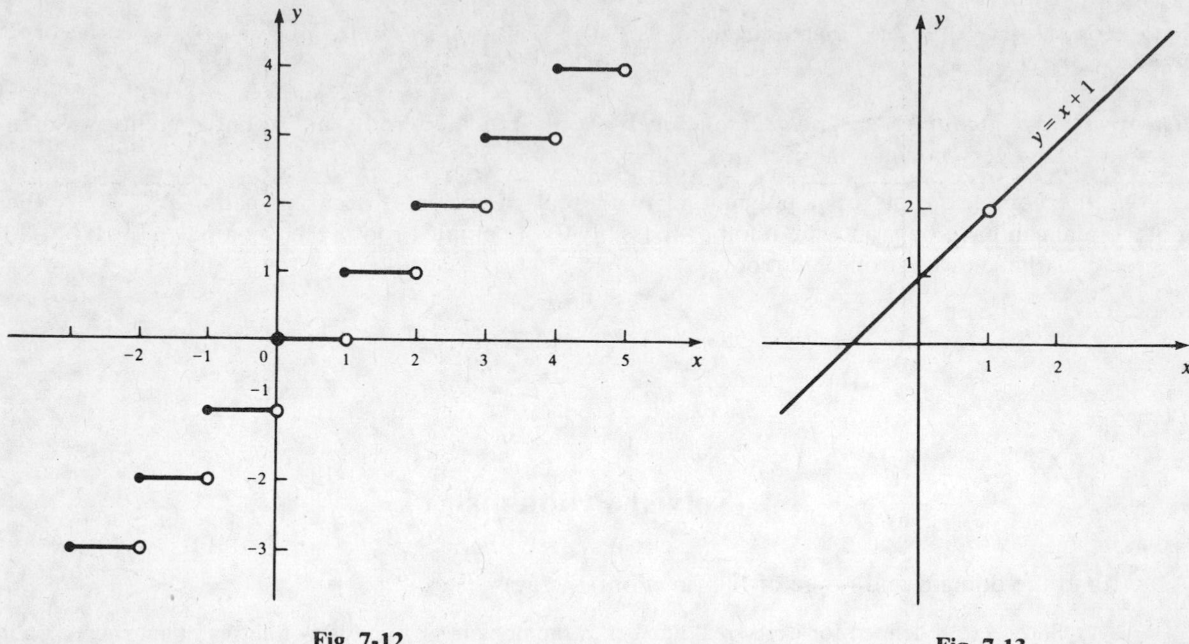

Fig. 7-12

Fig. 7-13

7.4 Consider the function f defined by the formula

$$f(x)=\frac{x^2-1}{x-1}$$

whenever this formula makes sense. Find the domain and range, and draw the graph of f.

The formula makes sense whenever the denominator $x-1$ is not 0. Hence, the domain of f is the set of all real numbers different from 1. Now,

$$x^2-1=(x-1)(x+1)$$

and so

$$\frac{x^2-1}{x-1}=x+1$$

Hence, the graph of $f(x)$ is the same as the graph of $y=x+1$, except that the point corresponding to $x=1$ is missing (see Fig. 7-13). Thus, the range consists of all real numbers except 2.

7.5 (*a*) Show that a set of points in the xy-plane is the graph of some function of x if and only if the set intersects every vertical line in at most one point (*Vertical Line Test*). (*b*) Determine whether the sets of points indicated in Fig. 7-14 are graphs of functions.

(*a*) If a set of points is the graph of a function f, the set consists of all points (x, y) such that $y=f(x)$. If (x_0, u) and (x_0, v) are points of intersection of the graph with the vertical line $x=x_0$, then $u=f(x_0)$ and $v=f(x_0)$. Hence, $v=u$, and the points are identical. Conversely, if a set $\mathscr{A}$ of points intersects each vertical line in at most one point, define a function f as follows: If $\mathscr{A}$ intersects the line $x=x_0$ at some point (x_0, w), let $f(x_0)=w$. Then $\mathscr{A}$ is the graph of f.

(*b*) (i) and (iv) are not graphs of functions, since they intersect certain vertical lines in more than one point.

7.6 Let $f(x)=x^2+2x-1$ for all x. Evaluate: (*a*) $f(2)$, (*b*) $f(-2)$, (*c*) $f(-x)$, (*d*) $f(x+1)$, (*e*) $f(x-1)$,

(*f*) $f(x+h)$ (*g*) $f(x+h)-f(x)$ (*h*) $\dfrac{f(x+h)-f(x)}{h}$

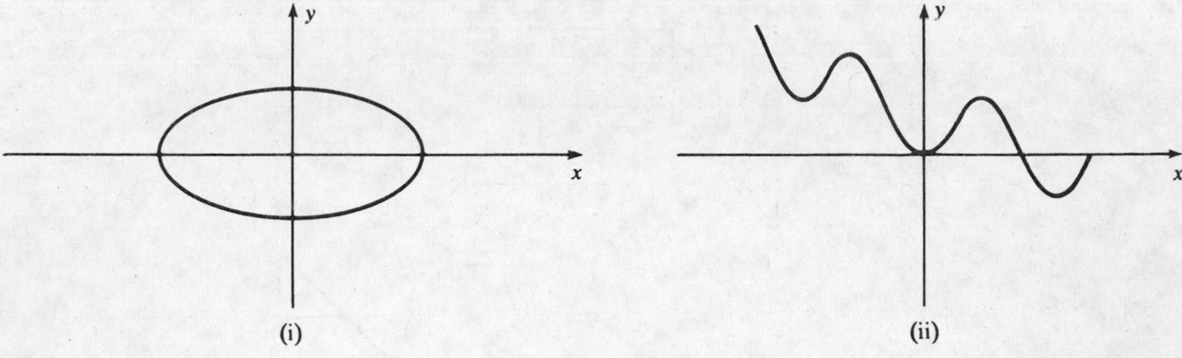

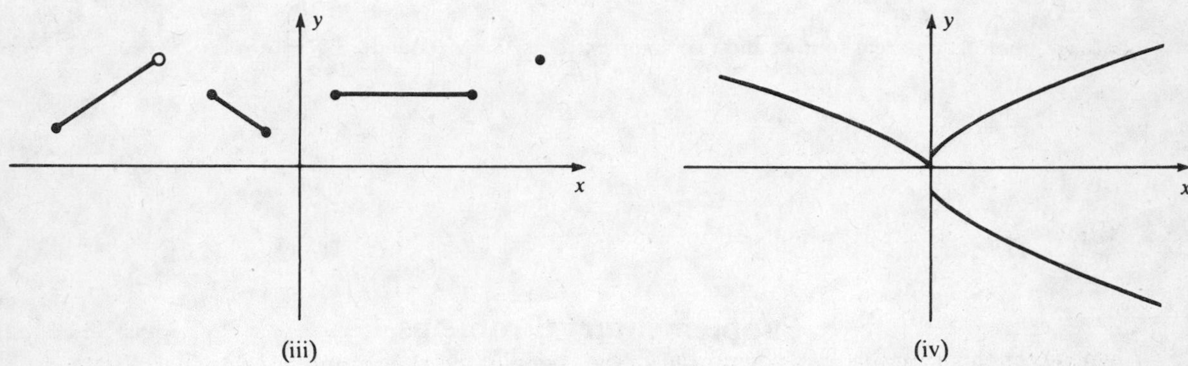

Fig. 7-14

In each case, we substitute the specified argument for all occurrences of "x" in the formula for $f(x)$.

(a) $$f(2) = (2)^2 + 2(2) - 1 = 4 + 4 - 1 = 7$$

(b) $$f(-2) = (-2)^2 + 2(-2) - 1 = 4 - 4 - 1 = -1$$

(c) $$f(-x) = (-x)^2 + 2(-x) - 1 = x^2 - 2x - 1$$

(d) $$f(x+1) = (x+1)^2 + 2(x+1) - 1 = (x^2 + 2x + 1) + 2x + 2 - 1 = x^2 + 4x + 2$$

(e) $$f(x-1) = (x-1)^2 + 2(x-1) - 1 = (x^2 - 2x + 1) + 2x - 2 - 1 = x^2 - 2$$

(f) $$\begin{aligned} f(x+h) &= (x+h)^2 + 2(x+h) - 1 = (x^2 + 2hx + h^2) + 2x + 2h - 1 \\ &= x^2 + 2hx + 2x + h^2 + 2h - 1 \end{aligned}$$

(g) Using the result of (f),

$$\begin{aligned} f(x+h) - f(x) &= (x^2 + 2hx + 2x + h^2 + 2h - 1) - (x^2 + 2x - 1) \\ &= x^2 + 2hx + 2x + h^2 + 2h - 1 - x^2 - 2x + 1 \\ &= 2hx + h^2 + 2h \end{aligned}$$

(h) Using the result of (g),

$$\frac{f(x+h) - f(x)}{h} = \frac{2hx + h^2 + 2h}{h} = \frac{h(2x + h + 2)}{h} = 2x + h + 2$$

7.7 Find all roots of the polynomial $f(x) = x^3 - 8x^2 + 21x - 20$.

The integral roots (if any) must be divisors of 20: ±1, ±2, ±4, ±5, ±10, ±20. Calculation shows that 4 is a root. Hence, $x - 4$ must be a factor of $f(x)$; see Fig. 7-15.

$$
\begin{array}{r|l}
 & x^2 - 4x + 5 \\
x - 4 & \overline{x^3 - 8x^2 + 21x - 20} \\
 & x^3 - 4x^2 \\
 & \overline{-4x^2 + 21x} \\
 & -4x^2 + 16x \\
 & \overline{5x - 20} \\
 & 5x - 20 \\
\end{array}
$$

Fig. 7-15

To find the roots of $x^2 - 4x + 5$, use the quadratic formula:

$$x = \frac{-(-4) \pm \sqrt{(-4)^2 - 4(5)}}{2} = \frac{4 \pm \sqrt{-4}}{2} = \frac{4 \pm \sqrt{4}\sqrt{-1}}{2} = \frac{4 \pm 2\sqrt{-1}}{2} = 2 \pm \sqrt{-1}$$

Hence, there is one real root, 4, and two complex roots, $2 + \sqrt{-1}$ and $2 - \sqrt{-1}$.

Supplementary Problems

7.8 Find the domain and range, and draw the graph of the functions determined by the following formulas (for all arguments x for which the formula makes sense):

(*a*) $h(x) = 4 - x^2$ (*b*) $G(x) = -2\sqrt{x}$ (*c*) $H(x) = \sqrt{4 - x^2}$ and $J(x) = -\sqrt{4 - x^2}$ (*d*) $U(x) = \sqrt{x^2 - 4}$

(*e*) $V(x) = |x - 1|$ (*f*) $f(x) = [2x]$ (*g*) $g(x) = \left[\frac{x}{3}\right]$ (*h*) $h(x) = \frac{1}{x}$

(*i*) $F(x) = \frac{1}{x - 1}$ (*j*) $H(x) = -\frac{1}{2}x^3$ (*k*) $J(x) = \begin{cases} x^2 & \text{if } x \le 0 \\ -x^2 & \text{if } x > 0 \end{cases}$

(*l*) $U(x) = \begin{cases} 3 - x & \text{for } x \le 1 \\ 5x - 3 & \text{for } x > 1 \end{cases}$ (*m*) $f(1) = -1$, $f(2) = 3$, $f(4) = -1$ (*n*) $G(x) = \frac{x^2 - 4}{x + 2}$

(*o*) $H(x) = \begin{cases} x & \text{if } x \le 2 \\ 4 & \text{if } x > 2 \end{cases}$ (*p*) $f(x) = \frac{|x|}{x}$ (*q*) $g(x) = \begin{cases} 1 - x & \text{if } x \le -1 \\ 2 & \text{if } -1 < x < 1 \\ x^2 + 1 & \text{if } x \ge 1 \end{cases}$

(*r*) $h(x) = \begin{cases} \dfrac{x^2 - 9}{x - 3} & \text{if } x \ne 3 \\ 6 & \text{if } x = 3 \end{cases}$ (*s*) $Z(x) = x - [x]$ (*t*) $f(x) = \sqrt[3]{x}$

7.9 In Fig. 7-16, determine which sets of points are graphs of functions.

7.10 Find a formula for the function f whose graph consists of all points (x, y) such that

(*a*) $x^3y - 2 = 0$ (*b*) $x = \frac{1 + y}{1 - y}$ (*c*) $x^2 - 2xy + y^2 = 0$

In each case, specify the domain of f.

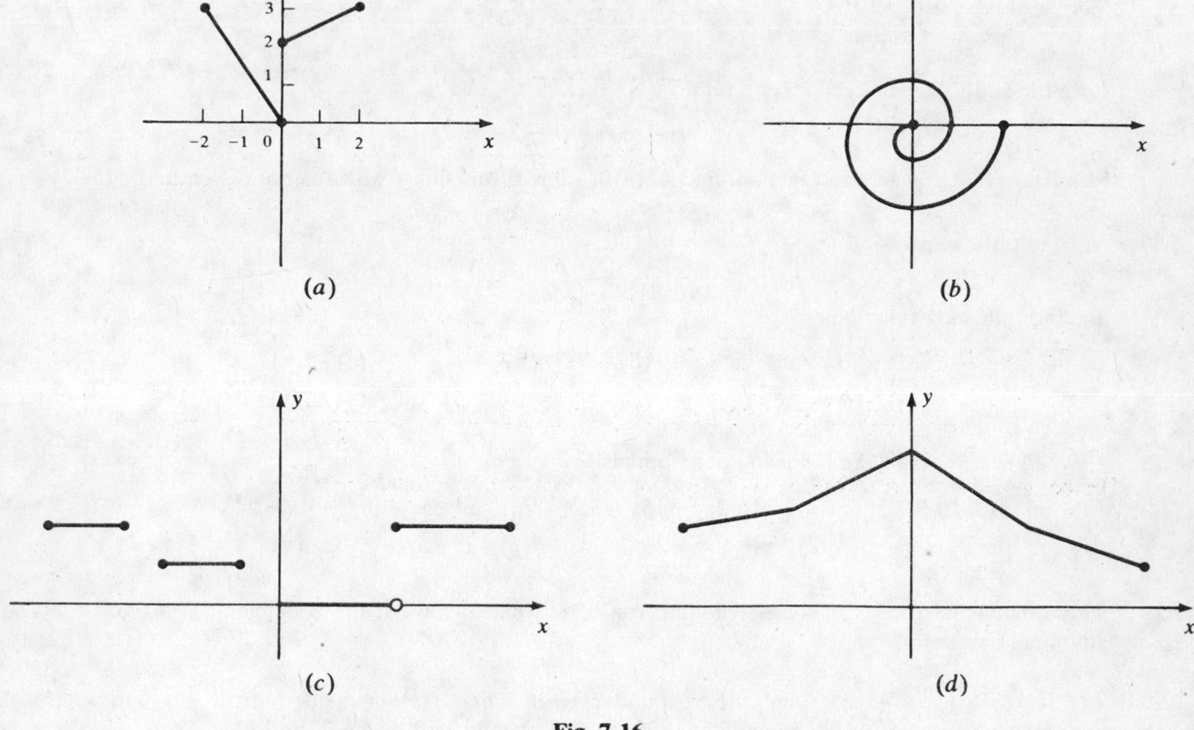

Fig. 7-16

7.11 For each of the following functions, specify the domain and range, using interval notation wherever possible.

(a) $f(x) = \dfrac{1}{(x-2)(x-3)}$ (b) $g(x) = \dfrac{1}{\sqrt{1-x^2}}$ (c) $h(x) = \begin{cases} x+1 & \text{if } -1 < x < 1 \\ 2 & \text{if } 1 \leq x \end{cases}$

(d) $F(x) = \begin{cases} x-1 & \text{if } 0 \leq x \leq 3 \\ x-2 & \text{if } 3 < x < 4 \end{cases}$ (e) $G(x) = |x| - x$

7.12 (a) Let $f(x) = x - 4$ and let

$$g(x) = \begin{cases} \dfrac{x^2-16}{x+4} & \text{if } x \neq -4 \\ k & \text{if } x = -4 \end{cases}$$

Determine k so that $f(x) = g(x)$ for all x. (b) Let

$$f(x) = \frac{x^2 - x}{x} \qquad g(x) = x - 1$$

Why is it wrong to assert that f and g are the same function?

7.13 In each of the following cases, define a function having the given set $\mathscr{D}$ as its domain and the given set $\mathscr{R}$ as its range: (a) $\mathscr{D} = (0, 1)$ and $\mathscr{R} = (0, 2)$, (b) $\mathscr{D} = [0, 1)$ and $\mathscr{R} = [-1, 4)$, (c) $\mathscr{D} = [0, \infty)$ and $\mathscr{R} = \{0, 1\}$, (d) $\mathscr{D} = (-\infty, 1) \cup (1, 2)$ [that is, $(-\infty, 1)$ together with $(1, 2)$] and $\mathscr{R} = (1, \infty)$.

7.14 For each of the functions in Problem 7.8, determine whether the graph of the function is symmetric with respect to the x-axis, the y-axis, the origin, or to none of these.

7.15 For each of the functions in Problem 7.8, determine whether the function is even, odd, or neither even nor odd.

7.16 (*a*) If f is an even function and $f(0)$ is defined, must $f(0)=0$? (*b*) If f is an odd function and $f(0)$ is defined, must $f(0)=0$? (*c*) If

$$f(x)=x^2+kx+1$$

for all x and if f is an even function, find k. (*d*) If

$$f(x)=x^3-(k-2)x^2+2x$$

for all x and if f is an odd function, find k. (*e*) Is there a function f which is both even and odd?

7.17 Evaluate the expression

$$\frac{f(x+h)-f(x)}{h}$$

for the following functions f:

(*a*) $f(x)=x^2-2x$ (*b*) $f(x)=x+4$ (*c*) $f(x)=x^3+1$
(*d*) $f(x)=\sqrt{x}$ (*e*) $f(x)=\sqrt{x}-5x$ (*f*) $f(x)=|x|$

7.18 Find all real roots of the following polynomials.

(*a*) x^4-10x^2+9 (*b*) $x^3+2x^2-16x-32$ (*c*) $x^4-x^3-10x^2+4x+24$
(*d*) x^3-2x^2+x-2 (*e*) $x^3+9x^2+26x+24$ (*f*) x^3-5x-2 (*g*) x^3-4x^2-2x+8

7.19 How many real roots can the polynomial ax^3+bx^2+cx+d have if the coefficients a, b, c, d are real numbers and $a\neq 0$?

7.20 (*a*) If $f(x)=(x+3)(x+k)$ and the remainder is 16 when $f(x)$ is divided by $x-1$, find k. (*b*) If $f(x)=(x+5)(x-k)$ and the remainder is 28 when $f(x)$ is divided by $x-2$, find k.

ALGEBRA The division of a polynomial $f(x)$ by another polynomial $g(x)$ yields an equation

$$f(x)=g(x)\,q(x)+r(x)$$

in which $q(x)$ (the quotient) and $r(x)$ (the remainder) are polynomials, with $r(x)$ either 0 or of lower degree than $g(x)$. In particular, for $g(x)=x-a$, we have

$$f(x)=(x-a)\,q(x)+r=(x-a)\,q(x)+f(a)$$

i.e., *the remainder when* $f(x)$ *is divided by* $x-a$ *is just* $f(a)$.

7.21 If the zeros of a function $f(x)$ are 3 and -4, what are the zeros of the function $g(x)=f(x/3)$?

7.22 If $f(x)=2x^3+Kx^2+Jx-5$, and if $f(2)=3$ and $f(-2)=-37$, which of the following is the value of $K+J$?

(i) 0 (ii) 1 (iii) -1 (iv) 2 (v) indeterminate

7.23 Express the set of solutions of each inequality below in terms of the notation for intervals.

(*a*) $2x+3<9$ (*b*) $5x+1\geq 6$ (*c*) $3x+4\leq 5$ (*d*) $7x-2>8$
(*e*) $3<4x-5<7$ (*f*) $-1\leq 2x+5<9$ (*g*) $|x+1|<2$ (*h*) $|3x-4|\leq 5$
(*i*) $\dfrac{2x-5}{x-2}<1$ (*j*) $x^2\leq 6$ (*k*) $(x-3)(x+1)<0$

7.24 For what values of x are the graphs of (*a*) $f(x)=(x-1)(x+2)$ and (*b*) $f(x)=x(x-1)(x+2)$ above the x-axis?

7.25 Prove Theorem 7.1. (*Hint*: Solve $f(r)=0$ for a_0.)

7.26 Prove Theorem 7.2. (*Hint*: Make use of the ALGEBRA following Problem 7.20.)

Chapter 8

Limits

8.1 INTRODUCTION

To a great extent, calculus is the study of the rates at which quantities change. It will be necessary to see how the value $f(x)$ of a function f behaves as the argument x approaches a given number. This leads to the idea of *limit.*

EXAMPLE Consider the function f such that

$$f(x) = \frac{x^2 - 9}{x - 3}$$

whenever this formula makes sense. Thus, f is defined for all x for which the denominator $x - 3$ is not 0; that is, for $x \neq 3$. What happens to the value $f(x)$ as x approaches 3? Well, x^2 approaches 9, and so $x^2 - 9$ approaches 0; moreover, $x - 3$ approaches 0. Since the numerator and the denominator both approach 0, it is not clear what happens to

$$\frac{x^2 - 9}{x - 3}$$

However, upon factoring the numerator, we observe that

$$\frac{x^2 - 9}{x - 3} = \frac{(x - 3)(x + 3)}{x - 3} = x + 3$$

Since $x + 3$ unquestionably approaches 6 as x approaches 3, we now know that our function approaches 6 as x approaches 3. The traditional mathematical way of expressing this fact is:

$$\lim_{x \to 3} \frac{x^2 - 9}{x - 3} = 6$$

This is read: "The limit of $\frac{x^2 - 9}{x - 3}$ as x approaches 3 is 6."

Notice that there is no problem when x approaches any number other than 3. For instance, when x approaches 4, $x^2 - 9$ approaches 7 and $x - 3$ approaches 1. Hence,

$$\lim_{x \to 4} \frac{x^2 - 9}{x - 3} = \frac{7}{1} = 7$$

8.2 PROPERTIES OF LIMITS

In the foregoing Example we assumed without explicit mention certain obvious properties of the notion of limit. Let us write them down explicitly.

PROPERTY I.

$$\lim_{x \to a} x = a$$

This follows directly from the meaning of the limit concept.

PROPERTY II. If c is a constant,

$$\lim_{x \to a} c = c$$

As x approaches a, the value of c remains c.

PROPERTY III. If c is a constant and f is a function,

$$\lim_{x\to a} c\cdot f(x) = c\cdot \lim_{x\to a} f(x)$$

EXAMPLES

$$\lim_{x\to 3} 5x = 5\cdot \lim_{x\to 3} x = 5\cdot 3 = 15$$

$$\lim_{x\to 3} -x = \lim_{x\to 3} (-1)x = (-1)\cdot \lim_{x\to 3} x = (-1)\cdot 3 = -3$$

PROPERTY IV. If f and g are functions,

$$\lim_{x\to a} (f(x)\cdot g(x)) = \lim_{x\to a} f(x)\cdot \lim_{x\to a} g(x)$$

The limit of a product is the product of the limits.

EXAMPLES

$$\lim_{x\to a} x^2 = \lim_{x\to a} x\cdot \lim_{x\to a} x = a\cdot a = a^2$$

More generally, for any positive integer n, $\lim_{x\to a} x^n = a^n$.

PROPERTY V. If f and g are functions,

$$\lim_{x\to a} (f(x)\pm g(x)) = \lim_{x\to a} f(x) \pm \lim_{x\to a} g(x)$$

The limit of a sum (difference) is the sum (difference) of the limits.

EXAMPLES (*a*)

$$\lim_{x\to 2} (3x^2+5x) = \lim_{x\to 2} 3x^2 + \lim_{x\to 2} 5x$$

$$= 3\cdot \lim_{x\to 2} x^2 + 5\cdot \lim_{x\to 2} x = 3(2)^2 + 5(2) = 22$$

(*b*) More generally, if $f(x) = a_n x^n + a_{n-1}x^{n-1} + \cdots + a_0$ is any polynomial function and k is any real number, then

$$\lim_{x\to k} f(x) = a_n k^n + a_{n-1}k^{n-1} + \cdots + a_0 = f(k)$$

PROPERTY VI. If f and g are functions and $\lim_{x\to a} g(x) \neq 0$, then

$$\lim_{x\to a} \frac{f(x)}{g(x)} = \frac{\lim_{x\to a} f(x)}{\lim_{x\to a} g(x)}$$

The limit of a quotient is the quotient of the limits.

EXAMPLE

$$\lim_{x\to 4} \frac{2x^3-5}{3x+2} = \frac{\lim_{x\to 4} (2x^3-5)}{\lim_{x\to 4} (3x+2)} = \frac{2(4)^3-5}{3(4)+2} = \frac{123}{14}$$

PROPERTY VII.

$$\lim_{x\to a} \sqrt{f(x)} = \sqrt{\lim_{x\to a} f(x)}$$

The limit of a square root is the square root of the limit.

EXAMPLE

$$\lim_{x\to 2} \sqrt{x^2+5} = \sqrt{\lim_{x\to 2} (x^2+5)} = \sqrt{9} = 3$$

Properties IV–VII have a common structure. Each tells us that, provided f and/or g has a limit as x approaches a (see Section 8.3), another, related function also has a limit as x approaches a, and this limit is as given by the indicated formula.

8.3 EXISTENCE OR NONEXISTENCE OF THE LIMIT

In certain cases, a function $f(x)$ will not approach a limit as x approaches a particular number.

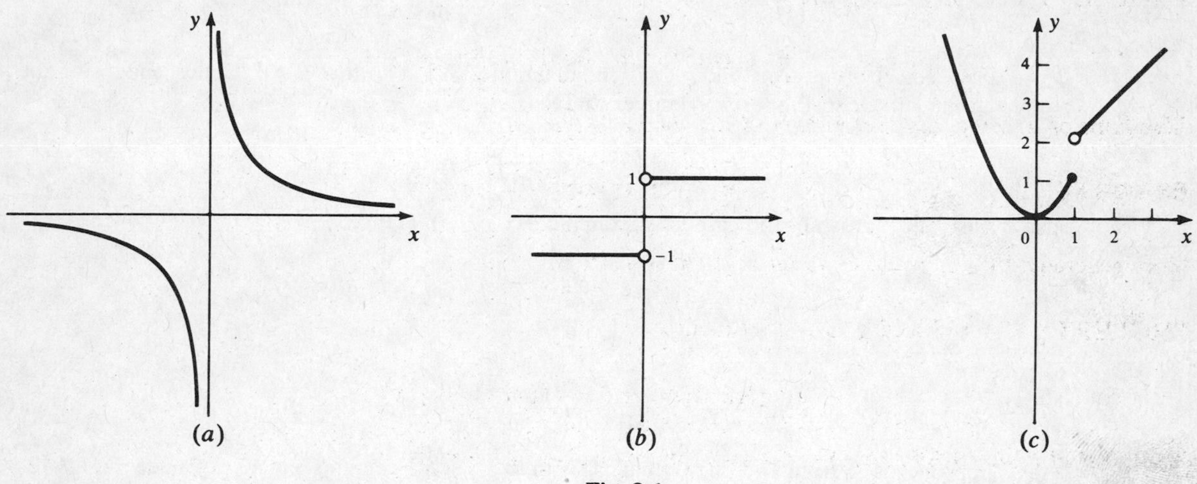

Fig. 8-1

EXAMPLES (*a*) Figure 8-1(*a*) indicates that

$$\lim_{x \to 0} \frac{1}{x}$$

does not exist. As x approaches 0, the magnitude of $1/x$ becomes larger and larger. (If $x > 0$, $1/x$ is positive and very large when x is close to 0. If $x < 0$, $1/x$ is negative and very "small" when x is close to 0.) (*b*) Figure 8-1(*b*) indicates that

$$\lim_{x \to 0} \frac{|x|}{x}$$

does not exist. When $x > 0$, $|x| = x$ and $|x|/x = 1$; when $x < 0$, $|x| = -x$ and $|x|/x = -1$. Thus, as x approaches 0, $|x|/x$ approaches two different values, 1 and -1, depending on whether x nears 0 through positive or through negative values. Since there is no ***unique*** limit as x approaches 0, we say that

$$\lim_{x \to 0} \frac{|x|}{x}$$

does not exist. (*c*) Let

$$f(x) = \begin{cases} x^2 & \text{if } x \le 1 \\ x+1 & \text{if } x > 1 \end{cases}$$

Then (see Fig. 8-1(*c*)), $\lim_{x \to 1} f(x)$ does not exist. As x approaches 1 from the left (that is, through values of $x < 1$), $f(x)$ approaches 1. But, as x approaches 1 from the right (that is, through values of $x > 1$), $f(x)$ approaches 2.

Notice that the existence or nonexistence of a limit for $f(x)$ as $x \to a$ does not depend upon the value $f(a)$, nor is it even required that f be defined at a. If $\lim_{x \to a} f(x) = L$, then L is a number to which $f(x)$ can be made arbitrarily close by letting x be sufficiently close to a. The value of L—or the very existence of L—is determined by the behavior of f *near a*, not by its value *at a* (if such a value even exists).

Solved Problems

8.1 Find the following limits (if they exist).

$$(a)\ \lim_{y\to 2}\left(y^2-\frac{1}{y}\right) \qquad (b)\ \lim_{x\to 0}\left(x^2-\frac{1}{x}\right) \qquad (c)\ \lim_{u\to 5}\frac{u^2-25}{u-5} \qquad (d)\ \lim_{x\to 2}[x]$$

(*a*) Both y^2 and $1/y$ have limits as $y\to 2$; so, by Property V,

$$\lim_{y\to 2}\left(y^2-\frac{1}{y}\right)=\lim_{y\to 2}y^2-\lim_{y\to 2}\frac{1}{y}=4-\frac{\lim_{y\to 2}1}{\lim_{y\to 2}y}=4-\frac{1}{2}=\frac{7}{2}$$

(*b*) Here it is necessary to proceed indirectly. The function x^2 has a limit as $x\to 0$. Hence, supposing the indicated limit to exist, Property V implies that

$$\lim_{x\to 0}\left(x^2-\left(x^2-\frac{1}{x}\right)\right)=\lim_{x\to 0}\frac{1}{x}$$

also exists. But that is not the case (see Example (*a*), p. 55). Hence,

$$\lim_{x\to 0}\left(x^2-\frac{1}{x}\right)$$

does not exist.

(*c*)
$$\lim_{u\to 5}\frac{u^2-25}{u-5}=\lim_{u\to 5}\frac{(u-5)(u+5)}{u-5}=\lim_{u\to 5}(u+5)=10$$

(*d*) As x approaches 2 from the right (that is, with $x>2$), $[x]$ remains equal to 2 (see Fig. 7-12). However, as x approaches 2 from the left (that is, with $x<2$), $[x]$ remains equal to 1. Hence, there is no unique number which is approached by $[x]$ as x approaches 2. Therefore, $\lim_{x\to 2}[x]$ does not exist.

8.2 Find

$$\lim_{h\to 0}\frac{f(x+h)-f(x)}{h}$$

(this limit will be important in the study of differential calculus) for each of the following functions:

$$(a)\ f(x)=3x-1 \qquad (b)\ f(x)=4x^2-x \qquad (c)\ f(x)=\frac{1}{x}$$

(*a*)
$$f(x+h)=3(x+h)-1=3x+3h-1$$
$$f(x)=3x-1$$
$$f(x+h)-f(x)=(3x+3h-1)-(3x-1)=3x+3h-1-3x+1=3h$$
$$\frac{f(x+h)-f(x)}{h}=\frac{3h}{h}=3$$

Hence
$$\lim_{h\to 0}\frac{f(x+h)-f(x)}{h}=\lim_{h\to 0}3=3$$

(*b*)
$$f(x+h)=4(x+h)^2-(x+h)=4(x^2+2hx+h^2)-x-h$$
$$=4x^2+8hx+4h^2-x-h$$
$$f(x)=4x^2-x$$
$$f(x+h)-f(x)=(4x^2+8hx+4h^2-x-h)-(4x^2-x)$$
$$=4x^2+8hx+4h^2-x-h-4x^2+x$$
$$=8hx+4h^2-h=h(8x+4h-1)$$
$$\frac{f(x+h)-f(x)}{h}=\frac{h(8x+4h-1)}{h}=8x+4h-1$$

Hence $$\lim_{h\to 0}\frac{f(x+h)-f(x)}{h} = \lim_{h\to 0}(8x+4h-1)$$

$$= \lim_{h\to 0}(8x-1) + \lim_{h\to 0} 4h = 8x-1+0 = 8x-1$$

(*c*)
$$f(x+h) = \frac{1}{x+h}$$

$$f(x) = \frac{1}{x}$$

$$f(x+h)-f(x) = \frac{1}{x+h} - \frac{1}{x}$$

ALGEBRA $$\frac{a}{b} - \frac{c}{d} = \frac{ad-bc}{bd}$$

$$= \frac{x-(x+h)}{(x+h)x} = \frac{x-x-h}{(x+h)x} = \frac{-h}{(x+h)x}$$

Hence $$\frac{f(x+h)-f(x)}{h} = \frac{-h}{(x+h)x} \div h = \frac{-h}{(x+h)x}\cdot\frac{1}{h} = \frac{-1}{(x+h)x}$$

and $$\lim_{h\to 0} = \frac{f(x+h)-f(x)}{h} = \lim_{h\to 0}\frac{-1}{(x+h)x} = -\lim_{h\to 0}\frac{1}{x+h}\cdot\lim_{h\to 0}\frac{1}{x}$$

$$= -\frac{1}{\lim\limits_{h\to 0}(x+h)}\cdot\frac{1}{x} = -\frac{1}{x}\cdot\frac{1}{x} = -\frac{1}{x^2}$$

8.3 Find $$\lim_{x\to 1}\frac{x^3-1}{x-1}$$

Both the numerator and the denominator approach 0 as x approaches 1. However, since 1 is a root of x^3-1, $x-1$ is a factor of x^3-1 (Theorem 7.2). Division of x^3-1 by $x-1$ produces the factorization

$$x^3-1 = (x-1)(x^2+x+1)$$

Hence $$\lim_{x\to 1}\frac{x^3-1}{x-1} = \lim_{x\to 1}\frac{(x-1)(x^2+x+1)}{x-1} = \lim_{x\to 1}(x^2+x+1) = 1^2+1+1 = 3$$

(cf. Example (*b*) following Property V in Section 8.2).

8.4 (*a*) Give a precise definition of the limit concept: $\lim\limits_{x\to a} f(x) = L$, and (*b*) using this definition, prove Property V of limits:

$$\lim_{x\to a} f(x) = L \quad\text{and}\quad \lim_{x\to a} g(x) = K \quad\text{imply}\quad \lim_{x\to a}(f(x)+g(x)) = L+K$$

(*a*) Intuitively, $\lim\limits_{x\to a} f(x) = L$ means that $f(x)$ can be made as close as we wish to L if x is taken sufficiently close to a. A mathematically precise version of this assertion is: For any real number $\epsilon > 0$, there exists a real number $\delta > 0$ such that

$$0 < |x-a| < \delta \quad\text{implies}\quad |f(x)-L| < \epsilon$$

for any x in the domain of f. Remember that $|x-a| < \delta$ means that the distance between x and a is smaller than δ, and that $|f(x)-L| < \epsilon$ means that the distance between $f(x)$ and L is smaller than ϵ. We assume that the domain of f is such that it contains at least one argument x within distance δ of a, for arbitrary $\delta > 0$. Observe also that the condition $0 < |x-a|$ excludes consideration of $x = a$, in line with the requirement that the value (if any) of $f(a)$ has nothing to do with $\lim\limits_{x\to a} f(x)$.

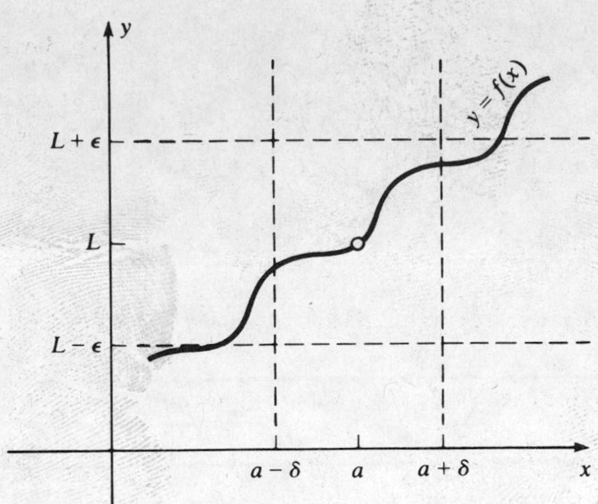

Fig. 8-2

A pictorial version of this definition is shown in Fig. 8-2. No matter how thin a horizontal strip (of width 2ϵ) that may be taken symmetrically above and below the line $y = L$, there is a thin vertical strip (of width 2δ) around the line $x = a$ such that, for any x-coordinate of a point in this strip except $x = a$, the corresponding point $(x, f(x))$ lies in the given horizontal strip. The number δ depends upon the given number ϵ; if ϵ is made smaller, then δ may also have to be chosen smaller.

The precise version just given for the limit concept is called an "epsilon-delta definition," because of the traditional use of the Greek letters ϵ and δ.

(*b*) Let $\epsilon > 0$ be given. Then $\epsilon/2 > 0$ and, since $\lim_{x \to a} f(x) = L$, there is a real number $\delta_1 > 0$ such that

$$0 < |x - a| < \delta_1 \quad \text{implies} \quad |f(x) - L| < \epsilon/2 \tag{1}$$

for all x in the domain of f. Likewise, since $\lim_{x \to a} g(x) = K$, there is a number $\delta_2 > 0$ such that:

$$0 < |x - a| < \delta_2 \quad \text{implies} \quad |g(x) - K| < \epsilon/2 \tag{2}$$

for all x in the domain of g. Let δ be the minimum of δ_1 and δ_2. Assume that

$$0 < |x - a| < \delta$$

and that x is in the domains of f and g. By the triangle inequality, (*1.10*),

$$|(f(x) + g(x)) - (L + K)| = |(f(x) - L) + (g(x) - K)| \leq |f(x) - L| + |g(x) - K| \tag{3}$$

For the number x under consideration, (*1*) and (*2*) show that the two terms on the right-hand side of (*3*) are respectively less than $\epsilon/2$. Hence,

$$|(f(x) + g(x)) - (L + K)| < \frac{\epsilon}{2} + \frac{\epsilon}{2} = \epsilon$$

and we have thereby established that $\lim_{x \to a} (f(x) + g(x)) = L + K$.

Supplementary Problems

8.5 Find the following limits (if they exist).

(*a*) $\lim_{x \to 2} 7$ (*b*) $\lim_{u \to 0} \frac{5u^2 - 4}{u + 1}$ (*c*) $\lim_{w \to -2} \frac{4 - w^2}{w + 2}$

(*d*) $\lim_{x\to 3/2} [x]$ (*e*) $\lim_{x\to 0} |x|$ (*f*) $\lim_{x\to 2} (7x^3 - 5x^2 + 2x - 4)$

(*g*) $\lim_{x\to 2} \dfrac{x^3-8}{x-2}$ (*h*) $\lim_{x\to 2} (x - [x])$ (*i*) $\lim_{x\to 4} \dfrac{x^2-x-12}{x-4}$

(*j*) $\lim_{x\to 3} \dfrac{x^3-x^2-x-15}{x-3}$ (*k*) $\lim_{x\to 2} \dfrac{2x^4-7x^2+x-6}{x-2}$ (*l*) $\lim_{x\to 1} \dfrac{x^4+3x^3-13x^2-27x+36}{x^2+3x-4}$

(*m*) $\lim_{x\to 0} \dfrac{\sqrt{x+3}-\sqrt{3}}{x}$ (*n*) $\lim_{x\to 1} \dfrac{\sqrt{x+3}-2}{x-1}$ (*o*) $\lim_{x\to 2} \left(\dfrac{1}{x-2} - \dfrac{4}{x^2-4}\right)$

8.6 Compute $\dfrac{f(x+h)-f(x)}{h}$ and then $\lim_{h\to 0} \dfrac{f(x+h)-f(x)}{h}$

(if the latter exists) for each of the following functions:

(*a*) $f(x) = 3x^2 + 5$ (*b*) $f(x) = \dfrac{1}{x+1}$ (*c*) $f(x) = 7x + 12$

(*d*) $f(x) = x^3$ (*e*) $f(x) = \sqrt{x}$ (*f*) $f(x) = 5x^2 - 2x + 4$

8.7 Give rigorous proofs of the following properties of the limit concept:

(*a*) $\lim_{x\to a} x = a$ (*b*) $\lim_{x\to a} c = c$ (*c*) $\lim_{x\to a} f(x) = L$ for at most one number L

8.8 Assuming that $\lim_{x\to a} f(x) = L$ and $\lim_{x\to a} g(x) = K$, prove rigorously:

(*a*) $\lim_{x\to a} c \cdot f(x) = c \cdot L$, where c is any real number.

(*b*) $\lim_{x\to a} (f(x) \cdot g(x)) = L \cdot K$.

(*c*) $\lim_{x\to a} \dfrac{f(x)}{g(x)} = \dfrac{L}{K}$, if $K \neq 0$.

(*d*) $\lim_{x\to a} \sqrt{f(x)} = \sqrt{L}$ if $L \geq 0$.

(*e*) If $\lim_{x\to a} (f(x) - L) = 0$, then $\lim_{x\to a} f(x) = L$.

(*f*) If $\lim_{x\to a} f(x) = L = \lim_{x\to a} h(x)$ and if $f(x) \leq g(x) \leq h(x)$ for all x near a, then

$$\lim_{x\to a} g(x) = L$$

(*Hints*: In (*d*), for $L > 0$,

$$|\sqrt{f(x)} - \sqrt{L}| = \left| (\sqrt{f(x)} - \sqrt{L}) \cdot \frac{\sqrt{f(x)}+\sqrt{L}}{\sqrt{f(x)}+\sqrt{L}} \right| = \left| \frac{f(x)-L}{\sqrt{f(x)}+\sqrt{L}} \right| \leq \frac{|f(x)-L|}{\sqrt{L}}$$

In (*f*), if $f(x)$ and $h(x)$ lie within the interval $(L-\epsilon, L+\epsilon)$, so must $g(x)$.

8.9 In an epsilon-delta proof of the fact that

$$\lim_{x\to 3} (2 + 5x) = 17$$

which of the following values of δ is the largest that can be used, given ϵ?

(*a*) ϵ (*b*) $\dfrac{\epsilon}{2}$ (*c*) $\dfrac{\epsilon}{4}$ (*d*) $\dfrac{\epsilon}{5}$ (*e*) $\dfrac{\epsilon}{8}$

Chapter 9

Special Limits

9.1 ONE-SIDED LIMITS

It is often useful to consider the limit of a function $f(x)$ as x approaches a given number either from the right or from the left.

EXAMPLE The function $f(x)$ of Example (*c*), Section 8.3, approaches 1 as x approaches 1 from the left, and approaches 2 as x approaches 1 from the right. We will notate these facts as follows:

$$\lim_{x\to 1^-} f(x) = 1 \qquad \text{and} \qquad \lim_{x\to 1^+} f(x) = 2$$

9.2 INFINITE LIMITS: VERTICAL ASYMPTOTES

If the value $f(x)$ of a function gets larger without bound as x approaches a, then $\lim_{x\to a} f(x)$ does not exist. However, we shall write

$$\lim_{x\to a} f(x) = +\infty$$

to indicate that $f(x)$ does get larger without bound.

EXAMPLE Let $f(x) = 1/x^2$ for all $x \neq 0$; the graph of f is shown in Fig. 9-1. As x approaches 0 from either side, $1/x^2$ grows without bound. Hence,

$$\lim_{x\to 0} \frac{1}{x^2} = +\infty$$

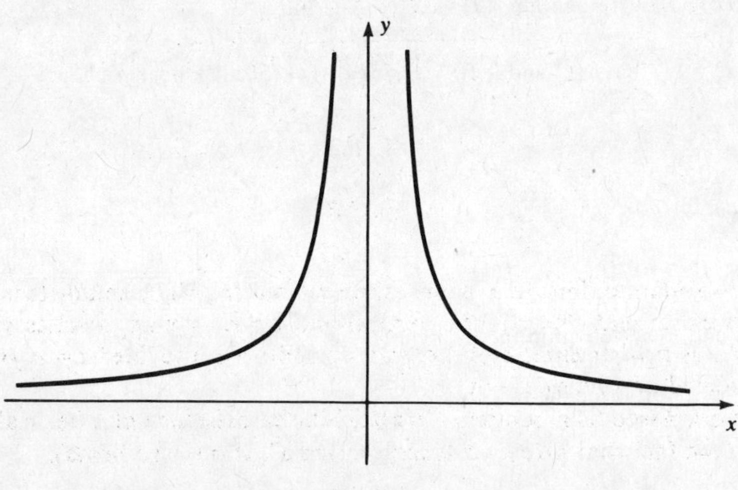

Fig. 9-1

The notation $\lim_{x\to a} f(x) = -\infty$ shall mean that $f(x)$ gets smaller without bound as x approaches a. That is,

$$\lim_{x\to a} f(x) = -\infty \quad \text{if and only if} \quad \lim_{x\to a} (-f(x)) = +\infty$$

EXAMPLE
$$\lim_{x\to 0}\left(-\frac{1}{x^2}\right) = -\infty$$

on the basis of the preceding Example.

Sometimes, the value $f(x)$ will get larger without bound or smaller without bound as x approaches a from one side ($x \to a^-$ or $x \to a^+$).

EXAMPLES (*a*) Let $f(x) = 1/x$ for all $x \neq 0$. Then we write

$$\lim_{x\to 0^+}\frac{1}{x} = +\infty$$

to indicate that $f(x)$ gets larger without bound as x approaches 0 from the right (see Fig. 8-1(*a*)). Similarly, we write

$$\lim_{x\to 0^-}\frac{1}{x} = -\infty$$

to express that fact that $f(x)$ decreases without bound as x approaches 0 from the left. (*b*) Let

$$f(x) = \begin{cases} \dfrac{1}{x} & \text{for } x > 0 \\ x & \text{for } x \leq 0 \end{cases}$$

Then,

$$\lim_{x\to 0^+} f(x) = +\infty \qquad \text{and} \qquad \lim_{x\to 0^-} f(x) = 0$$

(see Fig. 9-2).

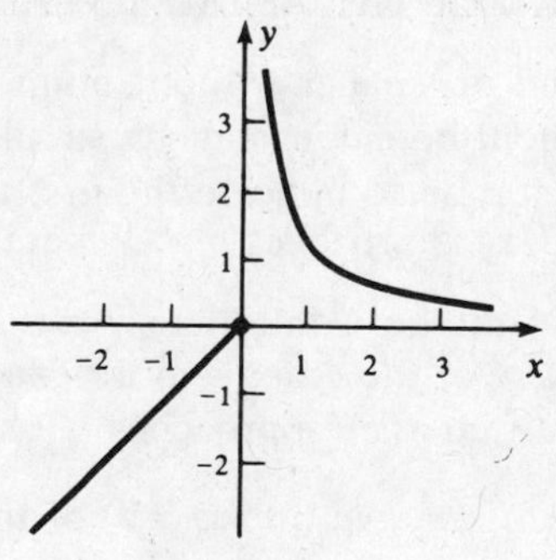

Fig. 9-2

When $f(x)$ has an infinite limit as x approaches a from the right and/or from the left, the graph of the function gets closer and closer to the vertical line $x = a$ as x approaches a. In such a case, the line $x = a$ is called a *vertical asymptote* of the graph. In Fig. 9-3, the lines $x = a$ and $x = b$ are vertical asymptotes (approached on one side only).

If a function is expressed as a quotient, $F(x)/G(x)$, the existence of a vertical asymptote $x = a$ is usually signaled by the fact that $G(a) = 0$ (except when $F(a) = 0$ also holds).

EXAMPLE Let

$$f(x) = \frac{x-2}{x-3}$$

for $x \neq 3$. Then $x = 3$ is a vertical asymptote of the graph of f, because

$$\lim_{x\to 3^+}\frac{x-2}{x-3} = +\infty \qquad \text{and} \qquad \lim_{x\to 3^-}\frac{x-2}{x-3} = -\infty$$

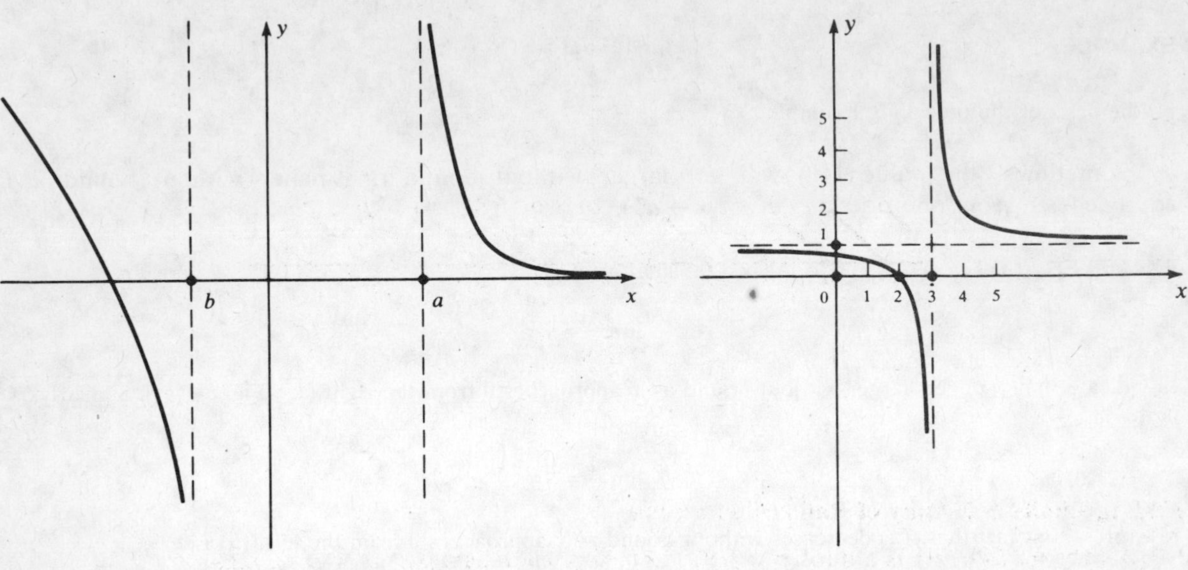

Fig. 9-3

Fig. 9-4

In this case, the asymptote $x = 3$ is approached from both the right and the left (see Fig. 9-4). (Notice that, by division,

$$\frac{x-2}{x-3} = 1 + \frac{1}{x-3}$$

Thus the graph of $f(x)$ is obtained by shifting the hyperbola $y = 1/x$ three units to the right and one unit up.)

9.3 LIMITS AT INFINITY: HORIZONTAL ASYMPTOTES

As x gets larger without bound or smaller without bound, the value $f(x)$ may approach a fixed real number, it may get larger without bound, it may get smaller without bound, or it may do none of these things. We now introduce notation to indicate the first three cases.

We shall write $\lim_{x\to+\infty} f(x) = c$ if $f(x)$ approaches c as x increases without bound. In such a case, the graph of f gets closer and closer to the horizontal line $y = c$ as x increases without bound. The line $y = c$ is called a *horizontal asymptote* of the graph—more exactly, a *horizontal asymptote to the right.* Similarly, $\lim_{x\to-\infty} f(x) = d$ shall mean that $f(x)$ approaches d as x decreases without bound. The line $y = d$ would be called a *horizontal asymptote* (*to the left*) of the graph of f.

EXAMPLES (*a*) For the function graphed in Fig. 9-2, the line $y = 0$ (the x-axis) is a horizontal asymptote to the right. (*b*) For the function

$$f(x) = 1 + \frac{1}{x-3}$$

(see Fig. 9-4), the line $y = 1$ is a horizontal asymptote to the right and to the left.

For a function that becomes definitely unbounded as x increases without bound, we shall replace c in the above notation by $+\infty$ or by $-\infty$, as appropriate. Similarly, we replace d by $\pm\infty$ if the function becomes definitely unbounded as x decreases without bound.

EXAMPLE For the function graphed in Fig. 9-2, $\lim_{x\to-\infty} f(x) = -\infty$.

EXAMPLE Consider the function f: $f(x) = x - [x]$ for all x. For each integer n, as x increases from n up to but not including $n + 1$, the value of $f(x)$ increases from 0 up to but not including 1. Thus, the graph consists of a sequence of line segments, as shown in Fig. 9-5. Hence, $\lim_{x\to+\infty} f(x)$ is undefined, since the value $f(x)$ neither approaches a fixed limit nor does it get larger or smaller without bound. Similarly, $\lim_{x\to-\infty} f(x)$ is undefined.

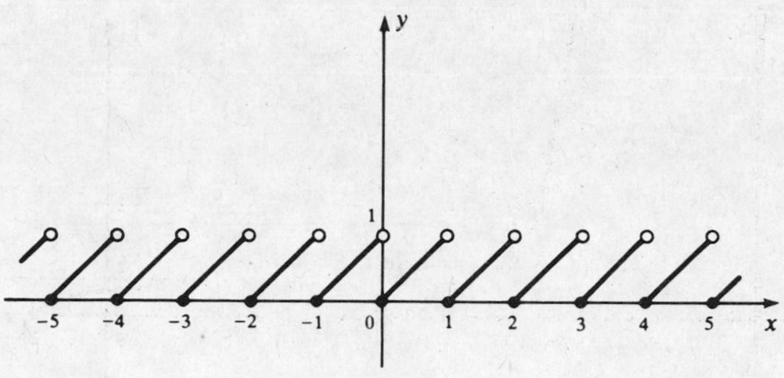

Fig. 9-5

Finding Limits at Infinity of Rational Functions

A *rational function* is a quotient, $f(x)/g(x)$, of polynomials $f(x)$ and $g(x)$; e.g.,

$$\frac{3x^2-5x+2}{x+7} \quad \text{and} \quad \frac{x^2-5}{4x^7+3x}$$

are rational functions.

RULE. To find

$$\lim_{x\to+\infty}\frac{f(x)}{g(x)} \quad \text{or} \quad \lim_{x\to-\infty}\frac{f(x)}{g(x)}$$

divide the numerator and denominator by the highest power of x in the denominator, and then use the fact that

$$\lim_{x\to\pm\infty}\frac{c}{x^r}=0 \tag{9.1}$$

for any positive real number r.

EXAMPLES

(*a*)

$$\lim_{x\to+\infty}\frac{2x+5}{x^2-7x+3}=\lim_{x\to+\infty}\frac{\frac{1}{x^2}(2x+5)}{\frac{1}{x^2}(x^2-7x+3)}=\lim_{x\to+\infty}\frac{\frac{2}{x}+\frac{5}{x^2}}{1-\frac{7}{x}+\frac{3}{x^2}}$$

$$=\frac{\lim\limits_{x\to+\infty}\frac{2}{x}+\lim\limits_{x\to+\infty}\frac{5}{x^2}}{\lim\limits_{x\to+\infty}1-\lim\limits_{x\to+\infty}\frac{7}{x}+\lim\limits_{x\to+\infty}\frac{3}{x^2}}=\frac{0+0}{1-0+0}=\frac{0}{1}=0$$

Thus, $y=0$ is a horizontal asymptote to the right for the graph of this rational function.

(*b*)

$$\lim_{x\to+\infty}\frac{3x^3-4x+2}{7x^3+5}=\lim_{x\to+\infty}\frac{\frac{1}{x^3}(3x^3-4x+2)}{\frac{1}{x^3}(7x^3+5)}=\lim_{x\to+\infty}\frac{3-\frac{4}{x^2}+\frac{2}{x^3}}{7+\frac{5}{x^3}}$$

$$=\frac{\lim\limits_{x\to+\infty}3-\lim\limits_{x\to+\infty}\frac{4}{x^2}+\lim\limits_{x\to+\infty}\frac{2}{x^3}}{\lim\limits_{x\to+\infty}7+\lim\limits_{x\to+\infty}\frac{5}{x^3}}$$

$$=\frac{3-0+0}{7+0}=\frac{3}{7}$$

(*c*)

$$\lim_{x\to+\infty}\frac{4x^5-1}{3x^3+7}=\lim_{x\to+\infty}\frac{\frac{1}{x^3}(4x^5-1)}{\frac{1}{x^3}(3x^3+7)}=\lim_{x\to+\infty}\frac{4x^2-\frac{1}{x^3}}{3+\frac{7}{x^3}}$$

$$=\frac{\lim\limits_{x\to+\infty}4x^2-\lim\limits_{x\to+\infty}\frac{1}{x^3}}{\lim\limits_{x\to+\infty}3+\lim\limits_{x\to+\infty}\frac{7}{x^3}}=\frac{\lim\limits_{x\to+\infty}4x^2-0}{3+0}$$

$$=\frac{1}{3}\lim_{x\to+\infty}4x^2=+\infty$$

(*d*)

$$\lim_{x\to-\infty}\frac{3x+7}{5x-3}=\lim_{x\to-\infty}\frac{\frac{1}{x}(3x+7)}{\frac{1}{x}(5x-3)}=\lim_{x\to-\infty}\frac{3+\frac{7}{x}}{5-\frac{3}{x}}$$

$$=\frac{\lim\limits_{x\to-\infty}3+\lim\limits_{x\to-\infty}\frac{7}{x}}{\lim\limits_{x\to-\infty}5-\lim\limits_{x\to-\infty}\frac{3}{x}}=\frac{3+0}{5-0}=\frac{3}{5}$$

Solved Problems

9.1 Evaluate the following limits:

(*a*) $\lim\limits_{x\to+\infty}(5x^3-20x^2+2x-14)$ (*b*) $\lim\limits_{x\to+\infty}(-2x^3+70x^2+50x+5)$

(*c*) $\lim\limits_{x\to-\infty}(7x^4-12x^3+4x-3)$ (*d*) $\lim\limits_{x\to-\infty}(7x^3+10x^2+3x+5)$

If $f(x)$ is a polynomial,

$$f(x)=a_nx^n+a_{n-1}x^{n-1}+a_{n-2}x^{n-2}+\cdots+a_1x+a_0$$

with $a_n\neq 0$, then

$$\frac{f(x)}{a_nx^n}=1+\frac{a_{n-1}/a_n}{x}+\frac{a_{n-2}/a_n}{x^2}+\cdots+\frac{a_1/a_n}{x^{n-1}}+\frac{a_0/a_n}{x^n}$$

It follows from (*9.1*) that, as $x\to\pm\infty$, $f(x)/a_nx^n$ becomes arbitrarily close to 1. Therefore, $f(x)$ must become unbounded exactly as does a_nx^n; i.e.,

$$\lim_{x\to\pm\infty}f(x)=\lim_{x\to\pm\infty}a_nx^n$$

Applying this rule to the given polynomials, we find:

(*a*) $\lim\limits_{x\to+\infty}5x^3=+\infty$ (*b*) $\lim\limits_{x\to+\infty}(-2x^3)=-\infty$

(*c*) $\lim\limits_{x\to-\infty}7x^4=+\infty$ (*d*) $\lim\limits_{x\to-\infty}7x^3=-\infty$

9.2 Evaluate

(*a*) $\lim\limits_{x\to1^+}\frac{x+3}{x-1}$ (*b*) $\lim\limits_{x\to1^-}\frac{x+3}{x-1}$

As x approaches 1, the denominator $x - 1$ approaches 0, while $x + 3$ approaches 4. Thus,

$$\left|\frac{x+3}{x-1}\right|$$

increases without bound.

(*a*) As x approaches 1 from the right, $x - 1$ is positive. Since $x + 3$ is positive when x is close to 1,

$$\frac{x+3}{x-1} > 0 \quad \text{and} \quad \lim_{x \to 1^+} \frac{x+3}{x-1} = +\infty$$

(*b*) As x approaches 1 from the left, $x - 1$ is negative, while $x + 3$ is positive. Therefore,

$$\frac{x+3}{x-1} < 0 \quad \text{and} \quad \lim_{x \to 1^-} \frac{x+3}{x-1} = -\infty$$

The line $x = 1$ is a vertical asymptote of the graph of the rational function; see Fig. 9-6.

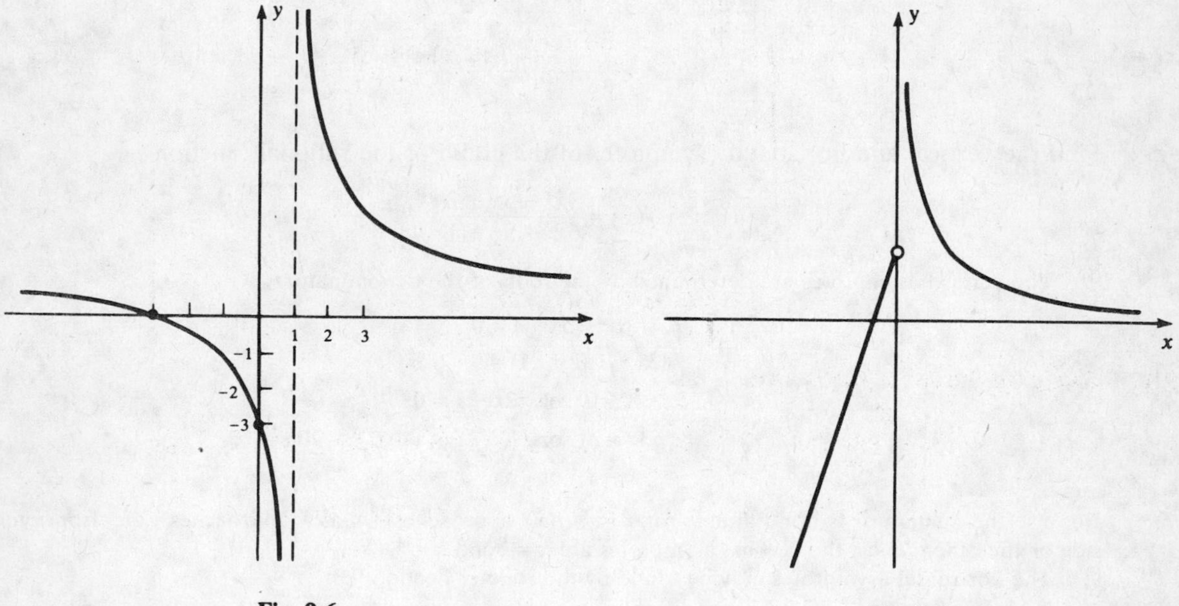

Fig. 9-6 **Fig. 9-7**

9.3 Given

$$f(x) = \begin{cases} 1/x & \text{if } x > 0 \\ 3x + 2 & \text{if } x < 0 \end{cases}$$

find (*a*) $\lim_{x \to 0^+} f(x)$, (*b*) $\lim_{x \to 0^-} f(x)$, (*c*) $\lim_{x \to +\infty} f(x)$, (*d*) $\lim_{x \to -\infty} f(x)$.

(*a*)
$$\lim_{x \to 0^+} f(x) = \lim_{x \to 0^+} \frac{1}{x} = +\infty$$

Hence, the line $x = 0$ (the y-axis) is a vertical asymptote of the graph of f (Fig. 9-7).

(*b*)
$$\lim_{x \to 0^-} f(x) = \lim_{x \to 0^-} (3x + 2) = 2$$

(*c*)
$$\lim_{x \to +\infty} f(x) = \lim_{x \to +\infty} \frac{1}{x} = 0$$

Hence, the line $y = 0$ (the x-axis) is a horizontal asymptote (to the right) of the graph of f.

(*d*)
$$\lim_{x \to -\infty} f(x) = \lim_{x \to -\infty} (3x + 2) = -\infty$$

9.4 Evaluate

$$(a)\ \lim_{x\to-\infty}\frac{3x^2-5}{4x^2+5} \qquad (b)\ \lim_{x\to+\infty}\frac{2x-7}{x^2-8} \qquad (c)\ \lim_{x\to+\infty}\frac{5x+2}{\sqrt{x^2-3x+1}}$$

(*a*) Apply the Rule of Section 9.3:

$$\lim_{x\to-\infty}\frac{3x^2-5}{4x^2+5}=\lim_{x\to-\infty}\frac{3-5/x^2}{4+5/x^2}=\frac{3-0}{4+0}=\frac{3}{4}$$

(*b*) Apply the Rule of Section 9.3:

$$\lim_{x\to+\infty}\frac{2x-7}{x^2-8}=\lim_{x\to+\infty}\frac{(2/x)-(7/x^2)}{1-(8/x^2)}=\frac{0-0}{1-0}=\frac{0}{1}=0$$

(*c*) By the rule developed in Problem 9.1, the denominator behaves like $\sqrt{x^2}=x$ as $x\to+\infty$. Thus,

$$\lim_{x\to+\infty}\frac{5x+2}{\sqrt{x^2-3x+1}}=\lim_{x\to+\infty}\frac{5x+2}{x}$$

$$=\lim_{x\to\infty}\left(5+\frac{2}{x}\right)=5+0=5$$

9.5 Find the vertical and horizontal asymptotes of the graph of the rational function

$$f(x)=\frac{3x^2-5x+2}{6x^2-5x+1}$$

The vertical asymptotes are determined by the roots of the denominator:

$$6x^2-5x+1=0$$
$$(3x-1)(2x-1)=0$$
$$3x-1=0 \quad\text{or}\quad 2x-1=0$$
$$3x=1 \quad\text{or}\quad 2x=1$$
$$x=\tfrac{1}{3} \quad\text{or}\quad x=\tfrac{1}{2}$$

Because the numerator is not 0 at $x=\frac{1}{3}$ or $x=\frac{1}{2}$, $|f(x)|$ approaches $+\infty$ as x approaches $\frac{1}{3}$ or $\frac{1}{2}$ from one side or the other. Thus, the vertical asymptotes are $x=\frac{1}{3}$ and $x=\frac{1}{2}$.

The horizontal asymptotes may be found by the Rule of Section 9.3:

$$\lim_{x\to+\infty}\frac{3x^2-5x+2}{6x^2-5x+1}=\lim_{x\to+\infty}\frac{3-\dfrac{5}{x}+\dfrac{2}{x^2}}{6-\dfrac{5}{x}+\dfrac{1}{x^2}}$$

$$=\frac{3-0+0}{6-0+0}=\frac{3}{6}=\frac{1}{2}$$

Thus, $y=\frac{1}{2}$ is a horizontal asymptote to the right. A similar procedure shows that

$$\lim_{x\to-\infty}f(x)=\frac{1}{2}$$

and so $y=\frac{1}{2}$ is also a horizontal asymptote to the left.

Supplementary Problems

9.6 Evaluate the following limits:

(a) $\lim_{x\to+\infty} (2x^{11} - 5x^6 + 3x^2 + 1)$ (b) $\lim_{x\to+\infty} (-4x^7 + 23x^3 + 5x^2 + 1)$

(c) $\lim_{x\to-\infty} (2x^4 - 12x^2 + x - 7)$ (d) $\lim_{x\to-\infty} (2x^3 - 12x^2 + x - 7)$

(e) $\lim_{x\to-\infty} (-x^8 + 2x^7 - 3x^3 + x)$ (f) $\lim_{x\to-\infty} (-2x^5 + 3x^4 - 2x^3 + x^2 - 4)$

9.7 Evaluate the following limits (if they exist):

(a) $\lim_{x\to2^+} f(x)$ and $\lim_{x\to2^-} f(x)$, if $f(x) = \begin{cases} 7x - 2 & \text{if } x \geq 2 \\ 3x + 5 & \text{if } x < 2 \end{cases}$ (b) $\lim_{x\to0^+} \frac{|x|}{x}$ and $\lim_{x\to0^-} \frac{|x|}{x}$.

(c) $\lim_{x\to0^+} \left(\frac{1}{x} - \frac{1}{x^2}\right)$ $\left(\textit{Hint}: \frac{1}{x} - \frac{1}{x^2} = \frac{x-1}{x^2}\right)$ (d) $\lim_{x\to3^+} \frac{x+3}{x^2-9}$ and $\lim_{x\to3^-} \frac{x+3}{x^2-9}$

(e) $\lim_{x\to3^+} \frac{x^2-5x+6}{x-3}$ and $\lim_{x\to3^-} \frac{x^2-5x+6}{x-3}$ (f) $\lim_{x\to3^+} \frac{1}{x^2-7x+12}$ and $\lim_{x\to3^-} \frac{1}{x^2-7x+12}$

(g) $\lim_{x\to+\infty} \frac{1}{3x-5}$ and $\lim_{x\to-\infty} \frac{1}{3x-5}$ (h) $\lim_{x\to+\infty} \frac{3x^3+x^2}{5x^3-1}$ and $\lim_{x\to-\infty} \frac{3x^3+x^2}{5x^3-1}$

(i) $\lim_{x\to+\infty} \frac{x^2+4x-5}{3x+1}$ and $\lim_{x\to-\infty} \frac{x^2+4x-5}{3x+1}$ (j) $\lim_{x\to+\infty} \frac{2x^3+x-5}{3x^4+4}$ and $\lim_{x\to-\infty} \frac{2x^3+x-5}{3x^4+4}$

(k) $\lim_{x\to+\infty} \frac{4x-1}{\sqrt{x^2+2}}$ and $\lim_{x\to-\infty} \frac{4x-1}{\sqrt{x^2+2}}$ (l) $\lim_{x\to+\infty} \frac{3x^3+2}{\sqrt{x^4-2}}$ and $\lim_{x\to-\infty} \frac{3x^3+2}{\sqrt{x^4-2}}$

(m) $\lim_{x\to+\infty} \frac{7x-4}{\sqrt{x^3+5}}$ and $\lim_{x\to-\infty} \frac{7x-4}{\sqrt{x^3+5}}$ (n) $\lim_{x\to+\infty} \frac{\sqrt{x^2+5}}{3x^2-2}$ and $\lim_{x\to-\infty} \frac{\sqrt{x^2+5}}{3x^2-2}$

(o) $\lim_{x\to+\infty} \frac{2x+5}{\sqrt[3]{x^3+4}}$ and $\lim_{x\to-\infty} \frac{2x+5}{\sqrt[3]{x^3+4}}$ (p) $\lim_{x\to+\infty} \frac{4x-3}{\sqrt[3]{x^2+1}}$ and $\lim_{x\to-\infty} \frac{4x-3}{\sqrt[3]{x^2+1}}$

(q) $\lim_{x\to+\infty} \frac{x^2-5}{3+5x-2x^2}$ (r) $\lim_{x\to3^+} \frac{\sqrt{x^2-9}}{x-3}$ (s) $\lim_{x\to-\infty} \frac{x^4-7x^3+4}{x^2-3}$

9.8 Find any vertical and horizontal asymptotes of the graphs of the following functions:

(a) $f(x) = x^3$ (b) $f(x) = \frac{2x-5}{3x+2}$ (c) $f(x) = \frac{4}{x^2+x-6}$

(d) $f(x) = \frac{2x-3}{2x^2+3x-5}$ (e) $f(x) = \frac{3x-1}{3x^2+5x-2}$ (f) $f(x) = \frac{x+3}{\sqrt{x^2+2x-8}}$

(g) $f(x) = \frac{2x+3}{\sqrt{x^2-2x+3}}$ (h) $f(x) = \sqrt{x+1} - \sqrt{x}$

9.9 Assume that f is a function defined for all x. Assume also that, for any real number c, there exists $\delta > 0$ such that $0 < |x| < \delta$ implies $f(x) > c$. Which of the following holds?

(a) $\lim_{x\to+\infty} f(x) = +\infty$ (b) $\lim_{x\to\delta} f(x) = c$ (c) $\lim_{x\to0} f(x) = +\infty$ (d) $\lim_{x\to0} f(x) = 0$

9.10 (a) Assume that $f(x) \geq 0$ for arguments to the right of and near a (that is, there exists some positive number δ such that $a < x < a + \delta$ implies $f(x) \geq 0$). Prove:

$$\lim_{x\to a^+} f(x) = L \quad \text{implies} \quad L \geq 0$$

(b) Assume that $f(x) \leq 0$ for arguments to the right of and near a. Prove:

$$\lim_{x\to a^+} f(x) = M \quad \text{implies} \quad M \leq 0$$

(c) Formulate and prove results similar to (a) and (b) for $\lim_{x\to a^-} f(x)$.

Chapter 10

Continuity

10.1 DEFINITION AND PROPERTIES

The notion of continuity refers to the absence of gaps or jumps in the graph of a function.

Definition: A function f is said to be *continuous at a* if the following three conditions hold:

(i) $\lim_{x \to a} f(x)$ exists;

(ii) $f(a)$ is defined;

(iii) $\lim_{x \to a} f(x) = f(a)$.

EXAMPLES (*a*) The function f such that $f(x) = [x]$ is *discontinuous* (that is, not continuous) at each integer, because condition (i) is not satisfied. See Problem 8.1(*d*). The discontinuities show up as jumps in the graph of the function (Fig. 7-12). (*b*) Let

$$f(x) = \begin{cases} 0 & \text{if } x = 0 \\ 1 & \text{if } x \neq 0 \end{cases}$$

This function is discontinuous at 0. Condition (i) is satisfied:

$$\lim_{x \to 0} f(x) = 1$$

Condition (ii) is satisfied: $f(0) = 0$. However, condition (iii) fails: $1 \neq 0$. There is a gap in the graph of f (Fig. 10-1) at the point (0, 1). (*c*) Let $f(x) = x^2$ for all x. This function is continuous at every a. As x gets closer and closer to a, x^2 gets closer and closer to a^2 (see Fig. 10-2). Notice that there are no gaps or jumps in the graph of f.

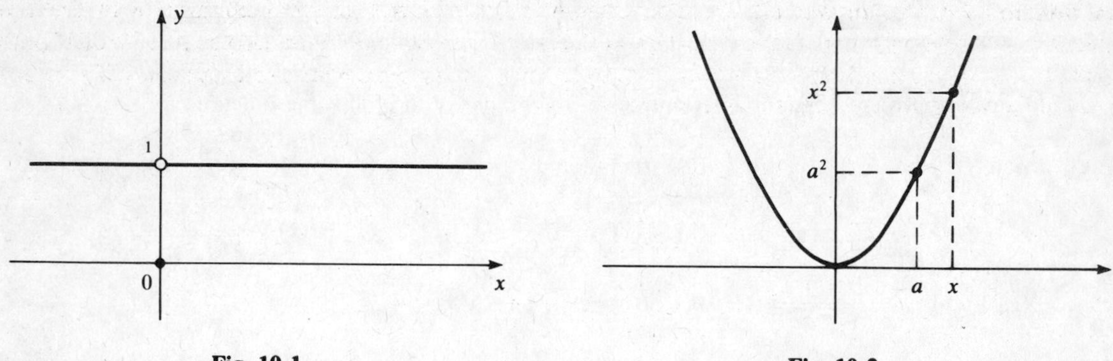

Fig. 10-1 **Fig. 10-2**

A function f is said to be *continuous on the set* $\mathscr{A}$ if f is continuous at each element of $\mathscr{A}$. If f is continuous at every number of its domain, then we simply say that f is *continuous*, or is a *continuous function.*

Every polynomial function is a continuous function (as was shown in Section 8.2, Example (*b*)).

Every rational function,

$$f(x) = \frac{g(x)}{h(x)}$$

where $g(x)$ and $h(x)$ are polynomials, is continuous at every real number a except the real roots (if any) of $h(x)$. For, if $h(a) \neq 0$, then

$$\lim_{x \to a} f(x) = \frac{\lim_{x \to a} g(x)}{\lim_{x \to a} h(x)} = \frac{g(a)}{h(a)} = f(a)$$

There are certain properties of continuous functions that follow directly from the standard properties of limits (Section 8.2). Assume f and g continuous at a. Then:

(1) The sum $f+g$ and the difference $f-g$ are continuous at a.

NOTATION $f \pm g$ are the functions such that $(f \pm g)(x) = f(x) \pm g(x)$ for every x in both the domain of f and the domain of g.

---

(2) If c is a constant, the function cf is continuous at a.

NOTATION cf is the function such that $(cf)(x) = c \cdot f(x)$ for every x in the domain of f.

---

(3) The product fg is continuous at a, and the quotient f/g is continuous at a provided $g(a) \neq 0$.

NOTATION fg is the function such that $(fg)(x) = f(x) \cdot g(x)$ for every x in both the domain of f and the domain of g; f/g is the function such that $(f/g)(x) = f(x)/g(x)$ for every x in both domains which is not a zero of $g(x)$.

---

(4) $\sqrt{f(x)}$ is continuous at a if $f(a) > 0$.

NOTE The restriction that $f(a) > 0$ guarantees that $f(x) > 0$—that the square root is defined—for x close to a.

10.2 ONE-SIDED CONTINUITY

A function f is *continuous on the right at a* if it satisfies the above Definition with the limit replaced by a right-hand limit (i.e., $x \to a^+$). Similarly, f is *continuous on the left at a* if it satisfies the Definition with a left-hand limit ($x \to a^-$). It is clear that f is continuous at a if and only if it is both continuous on the right at a and continuous on the left at a.

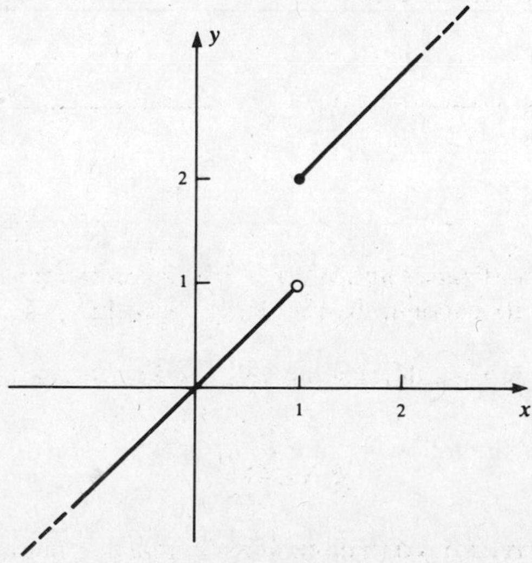

Fig. 10-3

EXAMPLE The function

$$f(x) = \begin{cases} x+1 & \text{if } x \geq 1 \\ x & \text{if } x < 1 \end{cases}$$

is continuous on the right at 1; for,

$$\lim_{x \to 1^+} f(x) = \lim_{x \to 1^+} (x+1) = 2 = f(1)$$

But, since

$$\lim_{x \to 1^-} f(x) = \lim_{x \to 1^-} x = 1 \neq f(1)$$

f is not continuous on the left at 1. Consequently, f is not continuous at 1, as is evidenced by the jump in its graph (Fig. 10-3).

10.3 CONTINUITY OVER A CLOSED INTERVAL

In the calculus, we shall often want to restrict our attention to a closed interval $[a, b]$ of the domain of a function, ignoring the function's behavior on the rest of its domain (if such exists). We thus make the following

Definition: A function f is continuous over $[a, b]$ if:

(i) f is continuous at each point of the open interval (a, b);
(ii) f is continuous on the right at a;
(iii) f is continuous on the left at b.

Figure 10-4 is the graph of a function that is continuous over $[a, b]$.

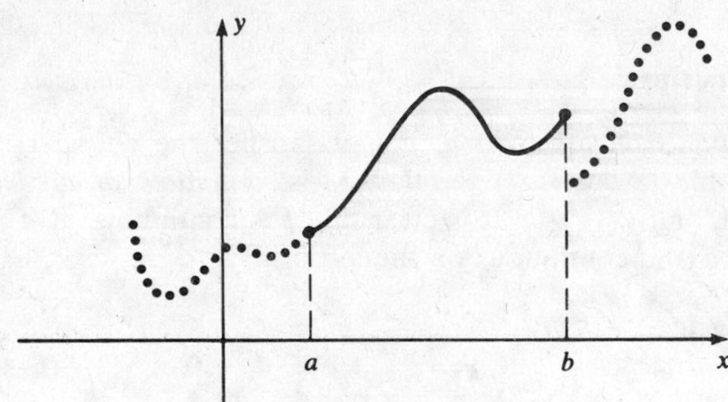

Fig. 10-4

Solved Problems

10.1 Find the points at which the following function f is continuous:

$$f(x) = \begin{cases} \dfrac{x^2 - 1}{x+1} & \text{if } x \neq -1 \\ -2 & \text{if } x = -1 \end{cases}$$

For $x \neq -1$, f is continuous, since f is the quotient of two continuous functions, with the denominator nonzero. In fact, for $x \neq -1$,

$$f(x) = \frac{x^2 - 1}{x + 1} = \frac{(x-1)(x+1)}{x+1} = x - 1$$

whence

$$\lim_{x \to -1} f(x) = \lim_{x \to -1} (x - 1) = -2 = f(-1)$$

Thus, f is also continuous at $x = -1$.

10.2 Consider the function f such that $f(x) = x - [x]$ (see Fig. 10-5). Find the points at which f is discontinuous. At those points, determine whether f is continuous on the right or continuous on the left (or neither).

For each integer n,

$$f(n) = n - [n] = n - n = 0$$

For $n < x < n + 1$, $f(x) = x - [x] = x - n$; hence,

$$\lim_{x \to n^+} f(x) = \lim_{x \to n^+} (x - n) = 0 = f(n)$$

Thus, f is continuous on the right at n. On the other hand,

$$\lim_{x \to n^-} f(x) = \lim_{x \to n^-} (x - (n-1)) = n - (n - 1)$$
$$= 1 \neq f(n)$$

so that f is not continuous on the left at n. It follows that f is discontinuous at each integer.

On each open interval $(n, n + 1)$, f coincides with the continuous function $x - n$. Therefore, there are no points of discontinuity other than the integers.

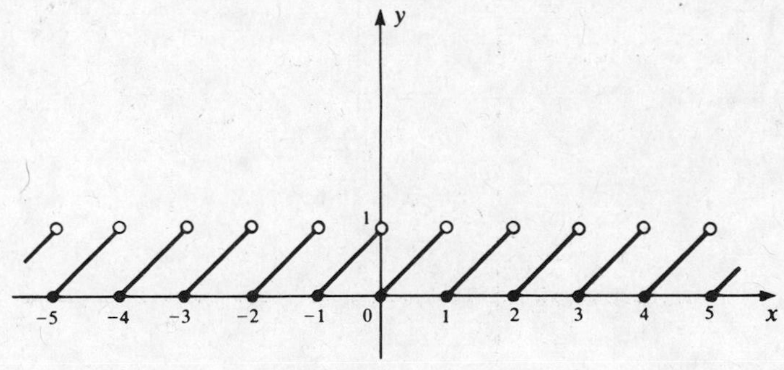

Fig. 10-5

10.3 For each function graphed in Fig. 10-6, find the points of discontinuity (if any). Also determine whether, at the points of discontinuity, the function is continuous on the right or continuous on the left (or neither).

(*a*) There are no points of discontinuity (no breaks in the graph).

(*b*) 0 is the only point of discontinuity. Continuity on the left holds at 0, since the value at 0 is the number approached by the values assumed to the left of 0.

(*c*) 1 is the only point of discontinuity. At 1, the function is continuous neither on the left nor on the right, since neither the limit on the left nor the limit on the right equals $f(1)$. (In fact, neither limit exists.)

(*d*) No points of discontinuity.

(*e*) 0 and 1 are points of discontinuity. Continuity on the left holds at 0, but neither continuity on the left nor on the right holds at 1.

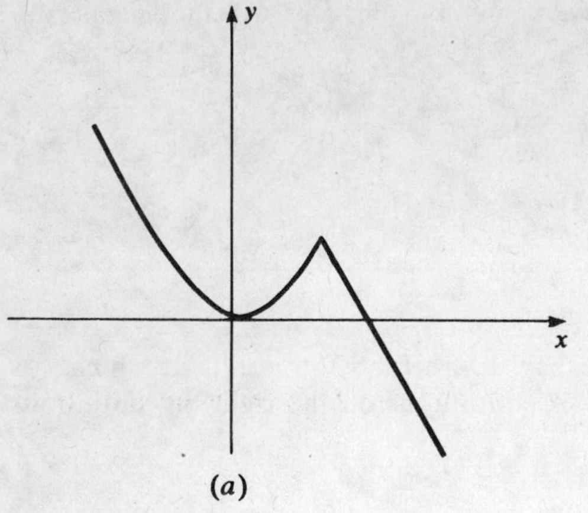

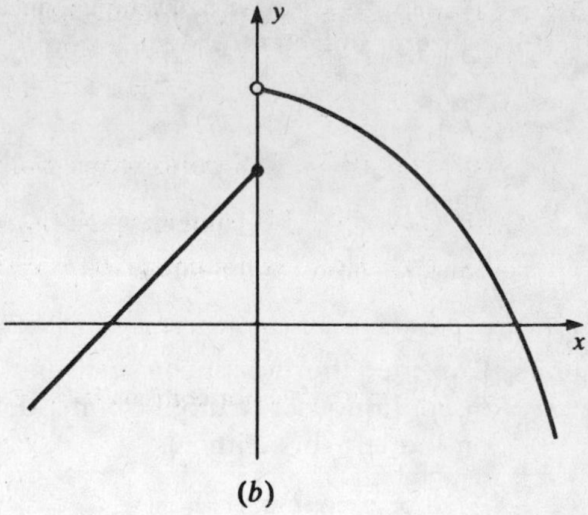

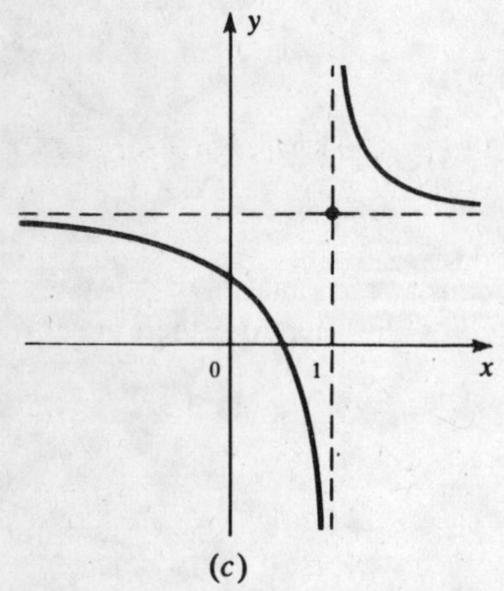

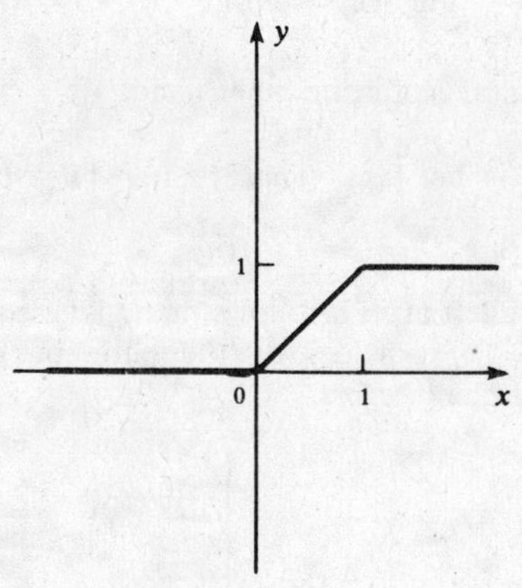

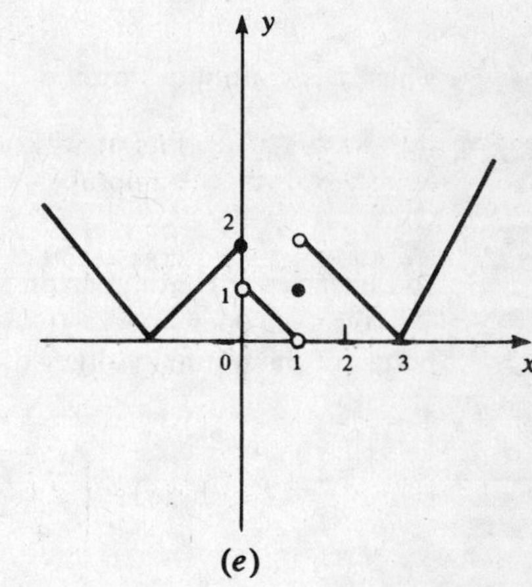

Fig. 10-6

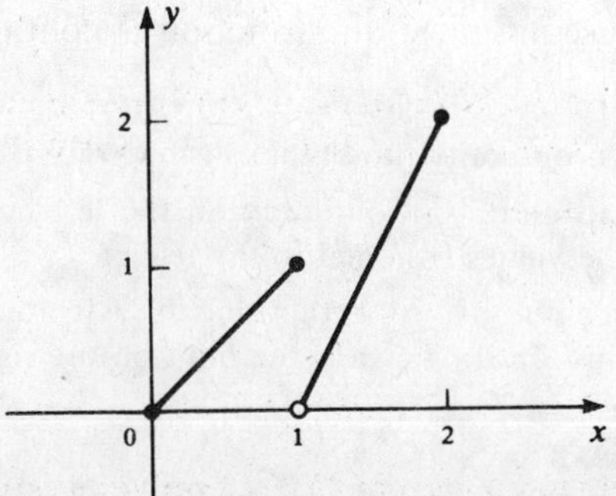

Fig. 10-7

10.4 Consider the function f such that

$$f(x)=\begin{cases} x & \text{for } 0\le x\le 1 \\ 2x-2 & \text{for } 1<x\le 2 \end{cases}$$

(see Fig. 10-7). Is f continuous over (*a*) $[0, 1]$? (*b*) $[1, 2]$? (*c*) $[0, 2]$?

(*a*) Yes, since f is continuous on the right at 0 and on the left at 1.

(*b*) No, since f is not continuous on the right at 1. In fact,

$$\lim_{x\to 1^+} f(x)=\lim_{x\to 1^+}(2x-2)=0\neq 1=f(1)$$

(*c*) No, since f is not continuous at $x=1$ and is therefore not continuous on the open interval $(0, 2)$.

Supplementary Problems

10.5 Determine the points at which each of the following functions is continuous. (Draw the graphs of the functions, if that is helpful.)

(*a*) $f(x)=\begin{cases} x^2 & \text{if } x\le 0 \\ x & \text{if } x>0 \end{cases}$ (*b*) $f(x)=\begin{cases} 1 & \text{if } x\ge 0 \\ -1 & \text{if } x<0 \end{cases}$

(*c*) $f(x)=|x|$ (*d*) $f(x)=\begin{cases} \dfrac{x^2-4}{x+2} & \text{if } x\neq -2 \\ 0 & \text{if } x=-2 \end{cases}$

(*e*) $f(x)=\begin{cases} \dfrac{x^2-4}{x+2} & \text{if } x\neq -2 \\ -4 & \text{if } x=-2 \end{cases}$ (*f*) $f(x)=\begin{cases} x+1 & \text{if } x\ge 2 \\ x & \text{if } 1<x<2 \\ x-1 & \text{if } x\le 1 \end{cases}$

10.6 Find the points of discontinuity (if any) of the functions whose graphs are shown in Fig. 10-8. At points of discontinuity, determine whether the function is continuous on the left or on the right (or neither).

10.7 Give simple examples of functions such that (*a*) f is defined on $[-2, 2]$, continuous over $[-1, 1]$, but not continuous over $[-2, 2]$; (*b*) g is defined on $[0, 1]$, continuous on the open interval $(0, 1)$, but not continuous over $[0, 1]$; (*c*) h is continuous at all points except $x=0$, where it is continuous on the right but not on the left.

10.8 Let f be the function defined by the formula

$$f(x)=\frac{3x+3}{x^2-3x-4}$$

(*a*) Find the arguments x at which f is discontinuous. (*b*) For each number a at which f is discontinuous, determine whether $\lim_{x\to a} f(x)$ exists; if it exists, find its value. (*c*) Write an equation for each vertical and horizontal asymptote of the graph of f.

10.9 Let

$$f(x)=x+\frac{1}{x}$$

for $x\neq 0$. (*a*) Find the points of discontinuity of f. (*b*) Determine all vertical and horizontal asymptotes of the graph of f.

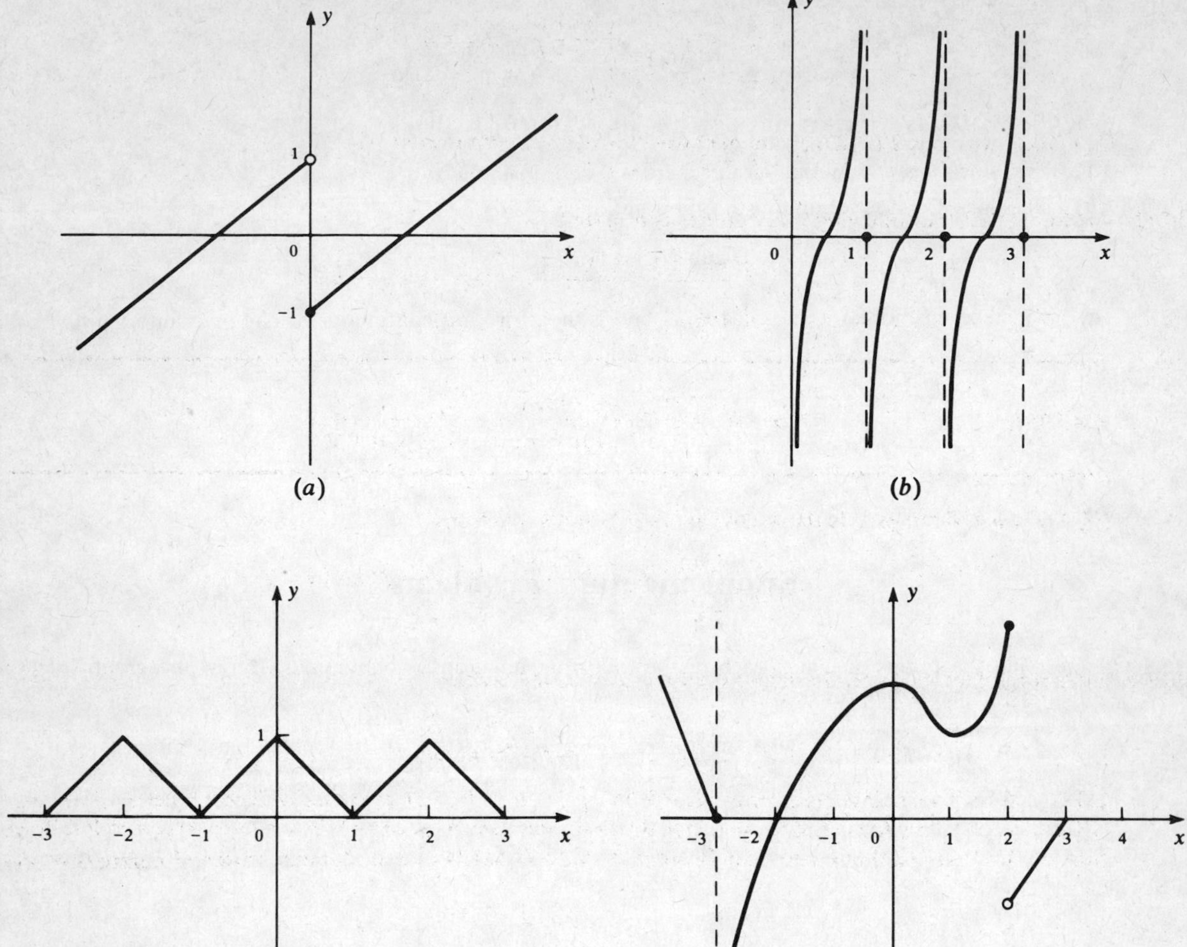

Fig. 10-8

10.10 For each of the following functions, determine whether it is continuous *over* the given interval:

(*a*) $f(x) = [x]$ over $[1, 2]$

(*b*) $f(x) = \begin{cases} \dfrac{1}{x} & \text{if } x > 0 \\ 0 & \text{if } x = 0 \end{cases}$ over $[0, 1]$

(*c*) $f(x) = \begin{cases} 2x & \text{if } 0 \le x \le 1 \\ x - 1 & \text{if } x > 1 \end{cases}$ over $[0, 1]$

(*d*) $f(x)$ as in (*c*) over $[1, 2]$

10.11 If the function

$$f(x) = \begin{cases} \dfrac{x^2 - 16}{x - 4} & \text{if } x \neq 4 \\ c & \text{if } x = 4 \end{cases}$$

is continuous, what is the value of c?

10.12 Let $b \neq 0$ and let g be the function such that

$$g(x) = \begin{cases} \dfrac{x^2 - b^2}{x - b} & \text{if } x \neq b \\ 0 & \text{if } x = b \end{cases}$$

(*a*) Does $g(b)$ exist? (*b*) Does $\lim_{x \to b} g(x)$ exist? (*c*) Is g continuous at b?

10.13 (*a*) Show that the following function f is continuous:

$$f(x) = \begin{cases} \dfrac{\sqrt{4x+1} - \sqrt{3x+7}}{x - 6} & \text{if } x \geq -1/4 \text{ and } x \neq 6 \\ 1/10 & \text{if } x = 6 \end{cases}$$

ALGEBRA $$\sqrt{u} - \sqrt{v} = \frac{(\sqrt{u} - \sqrt{v})(\sqrt{u} + \sqrt{v})}{\sqrt{u} + \sqrt{v}} = \frac{u - v}{\sqrt{u} + \sqrt{v}}$$

(*b*) For what value of k is the following a continuous function?

$$f(x) = \begin{cases} \dfrac{\sqrt{7x+2} - \sqrt{6x+4}}{x - 2} & \text{if } x \geq -2/7 \text{ and } x \neq 2 \\ k & \text{if } x = 2 \end{cases}$$

10.14 Determine the points of discontinuity of the following function f.

$$f(x) = \begin{cases} 1 & \text{if } x \text{ is rational} \\ 0 & \text{if } x \text{ is irrational} \end{cases}$$

(*Hint*: A rational number is an ordinary fraction, p/q, where p and q are integers. Recall Euclid's proof that $\sqrt{2}$ cannot be expressed in this form; it is an irrational number, as must be $\sqrt{2}/n$, for any integer n. It follows that any fixed rational number r can be approached arbitrarily closely through *irrational* numbers of the form

$$r + \frac{\sqrt{2}}{n}$$

Conversely, any fixed irrational number can be approached arbitrarily closely through *rational* numbers.)

10.15 Give an intuitive "proof" of the continuity of the function $f(x) = \sqrt[n]{x}$. (*Hint*: Figure 10-9(*a*) is the graph of the continuous function $g(x) = x^n$ (n odd). Interchanging the roles of the x- and y-axes—that is, viewing Fig. 10-9(*a*) sideways—produces the graph of $f(y)$, Fig. 10-9(b). Since there is no gap or jump in the first graph, there can be no jump or gap in the second. Similar "reasoning" holds when n is even.)

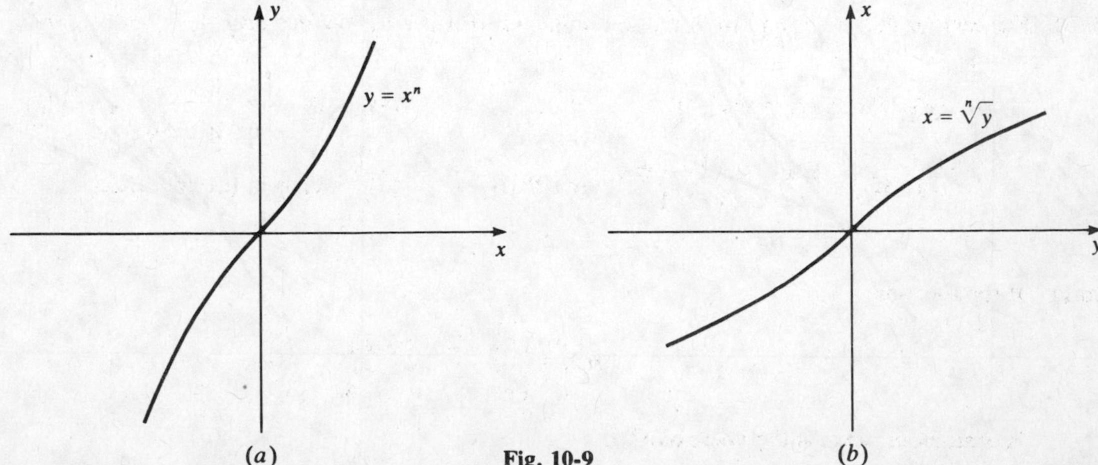

Fig. 10-9

Chapter 11

The Slope of a Tangent Line

The idea of a *tangent line* to a curve is familiar in the case of circles (see Fig. 11-1(*a*)). At each point P of a circle, there is a line $\mathscr{L}$ such that the circle touches the line at P and lies on one side of the line (*entirely* on one side, in the case of a circle). For the curve of Fig. 11-1(*b*), we would be inclined to call $\mathscr{L}_1$ the tangent line at P_1, $\mathscr{L}_2$ the tangent line at P_2, and $\mathscr{L}_3$ the tangent line at P_3. Let us develop a definition that corresponds to these intuitive ideas about tangent lines.

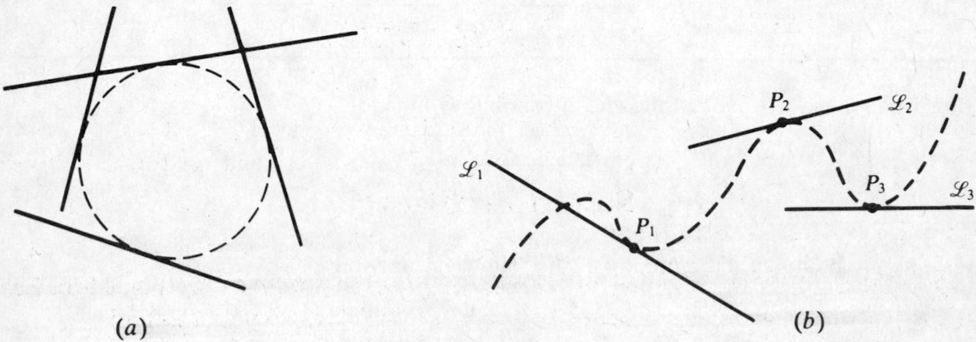

Fig. 11-1

Figure 11-2(*a*) is the graph of a continuous function f; remember that the graph consists of all points (x, y) such that $y = f(x)$. Let P be a point of the graph having abscissa x. Then the coordinates of P are $(x, f(x))$. Take a point Q on the graph having abscissa $x + h$; Q will be close to P if and only if h is close to 0 (because f is a continuous function). Since the x-coordinate of Q is $x + h$, the y-coordinate of Q must be $f(x + h)$. By the definition of slope, the line PQ will have slope

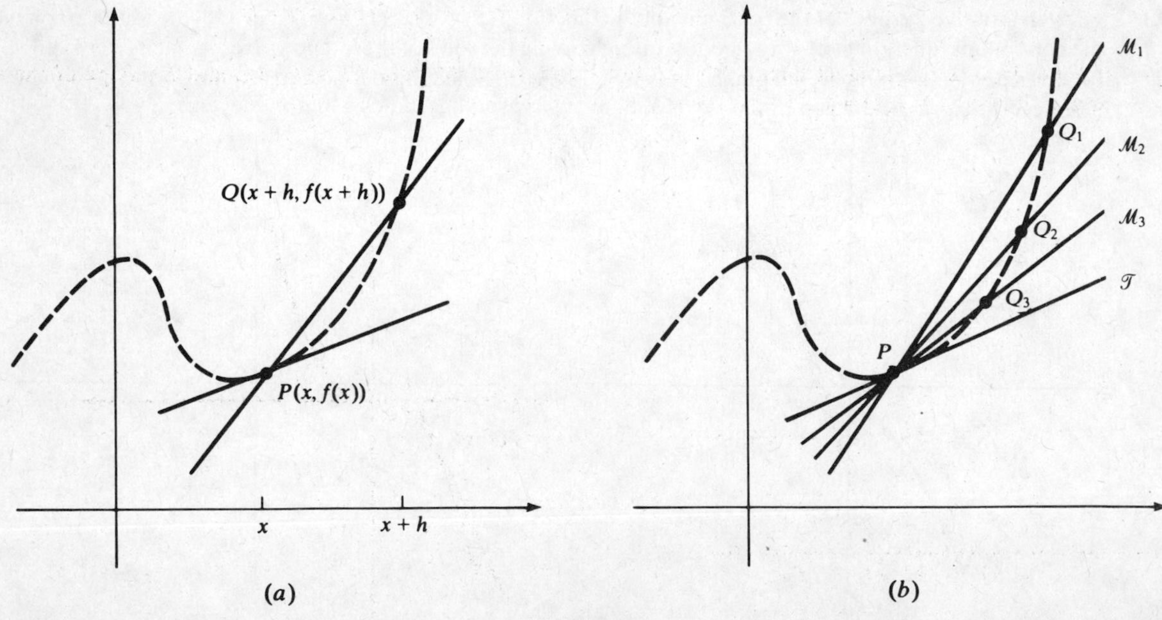

Fig. 11-2

$$\frac{f(x+h)-f(x)}{(x+h)-x}=\frac{f(x+h)-f(x)}{h}$$

Observe in Fig. 11-2(*b*) what happens to the line PQ as Q moves along the graph toward P. Some of the positions of Q have been designated as $Q_1, Q_2, Q_3, \ldots$, and the corresponding lines as $\mathcal{M}_1, \mathcal{M}_2, \mathcal{M}_3, \ldots$. These lines are getting closer and closer to what we think of as the tangent line $\mathcal{T}$ to the graph at P. Hence, the slope of the line PQ will approach the slope of the tangent line at P; that is to say, the slope of the tangent line at P will be given by

$$\lim_{h\to 0}\frac{f(x+h)-f(x)}{h}$$

What we have just said about tangent lines really amounts to a mathematical

Definition: Let a function f be continuous at x. By the *tangent line to the graph of f at the point* $P(x, f(x))$ is meant that line which passes through P and has slope

$$\lim_{h\to 0}\frac{f(x+h)-f(x)}{h}$$

Solved Problems

11.1 Consider the graph of the function f such that $f(x)=x^2$ (a parabola; see Fig. 11-3). For a point P on the parabola having abscissa x, perform the calculations needed to find

$$\lim_{h\to 0}\frac{f(x+h)-f(x)}{h}$$

We have:

$$f(x+h)=(x+h)^2=x^2+2xh+h^2$$
$$f(x)=x^2$$
$$f(x+h)-f(x)=(x^2+2xh+h^2)-x^2=2xh+h^2=h(2x+h)$$
$$\frac{f(x+h)-f(x)}{h}=\frac{h(2x+h)}{h}=2x+h$$

Thus

$$\lim_{h\to 0}\frac{f(x+h)-f(x)}{h}=\lim_{h\to 0}(2x+h)=\lim_{h\to 0}2x+\lim_{h\to 0}h=2x+0=2x$$

and the slope of the tangent line at P is $2x$. For example, at the point (2, 4), $x=2$, and the slope of the tangent line is $2x=2(2)=4$.

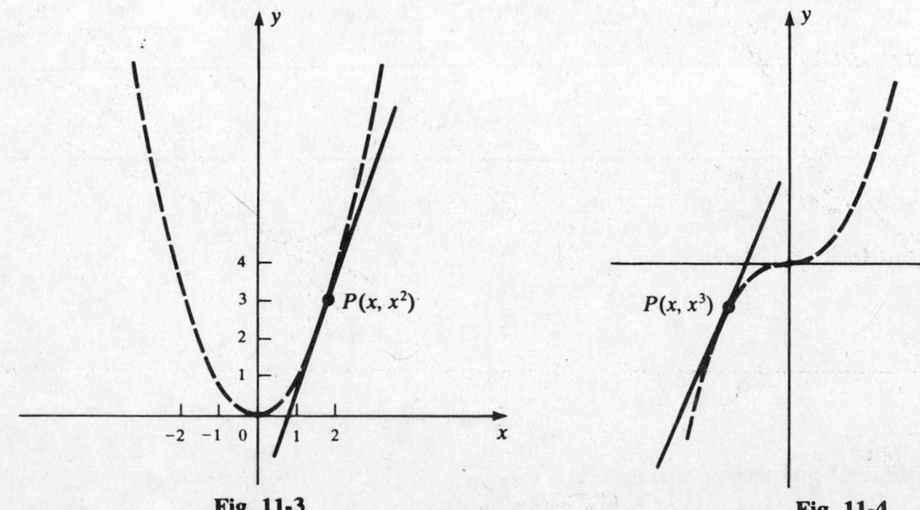

Fig. 11-3 **Fig. 11-4**

11.2 Consider the graph of the function f such that $f(x) = x^3$ (Fig. 11-4). For a point P on the graph having coordinates (x, x^3), compute the value of

$$\lim_{h\to 0}\frac{f(x+h)-f(x)}{h}$$

We have:

$$f(x+h) = (x+h)^3 = x^3 + 3x^2h + 3xh^2 + h^3 \qquad \text{and} \qquad f(x) = x^3$$

ALGEBRA For any u and v,

$$(u+v)^3 = [(u+v)(u+v)](u+v) = (u^2 + 2uv + v^2)(u+v)$$
$$= (u^3 + 2u^2v + uv^2) + (u^2v + 2uv^2 + v^3) = u^3 + 3u^2v + 3uv^2 + v^3$$

Hence,

$$f(x+h) - f(x) = (x^3 + 3x^2h + 3xh^2 + h^3) - x^3$$
$$= 3x^2h + 3xh^2 + h^3 = h(3x^2 + 3xh + h^2)$$
$$\frac{f(x+h)-f(x)}{h} = \frac{h(3x^2+3xh+h^2)}{h} = 3x^2 + 3xh + h^2$$

and

$$\lim_{h\to 0}\frac{f(x+h)-f(x)}{h} = \lim_{h\to 0}(3x^2 + 3xh + h^2)$$
$$= 3x^2 + 3x(0) + 0^2 = 3x^2$$

(Remember that x is fixed, so that we are finding the limit of a polynomial function of h—see Section 8.2, Example (*b*).)

This shows that the slope of the tangent line at P is $3x^2$. For example, the slope of the tangent line at (2, 8) is

$$3x^2 = 3(2)^2 = 3(4) = 12$$

11.3 (*a*) Find a formula for the slope of the tangent line at any point of the graph of the function f such that $f(x) = 1/x$ (a hyperbola; Fig. 11-5). (*b*) Find the slope-intercept equation of the tangent line to the graph of f at the point $(2, \frac{1}{2})$.

(*a*)

$$f(x+h) = \frac{1}{x+h} \qquad \text{and} \qquad f(x) = \frac{1}{x}$$

$$f(x+h) - f(x) = \frac{1}{x+h} - \frac{1}{x} = \frac{x-(x+h)}{(x+h)x} = \frac{x-x-h}{(x+h)x} = -\frac{h}{(x+h)x}$$

ALGEBRA

$$\frac{a}{b} - \frac{c}{d} = \frac{ad}{bd} - \frac{bc}{bd} = \frac{ad-bc}{bd}$$

Hence,

$$\frac{f(x+h)-f(x)}{h} = -\frac{h}{(x+h)x} \div h = -\frac{h}{(x+h)x}\cdot\frac{1}{h} = -\frac{1}{(x+h)x}$$

and

$$\lim_{h\to 0}\frac{f(x+h)-f(x)}{h} = \lim_{h\to 0} -\frac{1}{(x+h)x} = -\frac{\lim_{h\to 0} 1}{\lim_{h\to 0}(x+h)x}$$
$$= -\frac{1}{\lim_{h\to 0}(x+h)\cdot\lim_{h\to 0} x} = -\frac{1}{x\cdot x} = -\frac{1}{x^2}$$

Thus, the slope of the tangent line at $(x, 1/x)$ is $-1/x^2$.

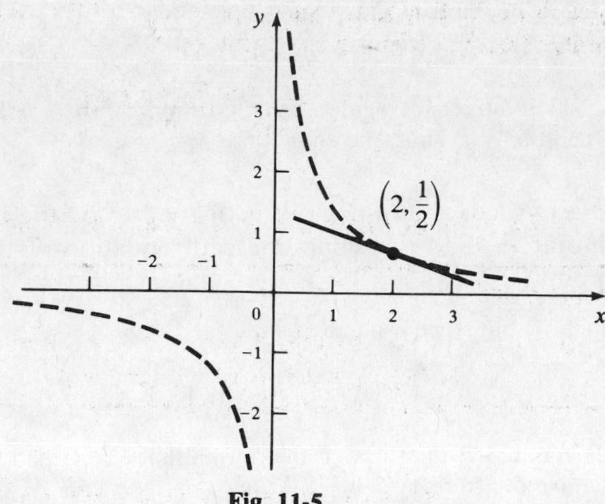

Fig. 11-5

(*b*) From (*a*), the slope of the tangent line at $(2, \frac{1}{2})$ is

$$-\frac{1}{x^2} = -\frac{1}{2^2} = -\frac{1}{4}$$

Hence, the slope-intercept equation of the tangent line $\mathscr{L}$ has the form

$$y = -\frac{1}{4}x + b$$

Since $\mathscr{L}$ passes through $(2, \frac{1}{2})$, substitution of 2 for x and $\frac{1}{2}$ for y yields

$$\frac{1}{2} = -\frac{1}{4}(2) + b \quad \text{or} \quad \frac{1}{2} = -\frac{1}{2} + b \quad \text{or} \quad b = 1$$

Hence, the equation of $\mathscr{L}$ is

$$y = -\frac{1}{4}x + 1$$

11.4 (*a*) Find a formula for the slope of the tangent line at any point of the graph of the function f such that

$$f(x) = 3x^2 - 6x + 4$$

(*b*) Find the slope-intercept equation of the tangent line at the point $(0, 4)$ of the graph.

(*a*) Without bothering to draw the graph of f, we may compute $f(x + h)$ by replacing all occurrences of "x" in the formula for $f(x)$ by "$x + h$":

$$\begin{aligned} f(x+h) &= 3(x+h)^2 - 6(x+h) + 4 \\ &= 3(x^2 + 2xh + h^2) - 6x - 6h + 4 \\ &= 3x^2 + 6xh + 3h^2 - 6x - 6h + 4 \end{aligned}$$

$$f(x) = 3x^2 - 6x + 4$$

$$\begin{aligned} f(x+h) - f(x) &= (3x^2 + 6xh + 3h^2 - 6x - 6h + 4) - (3x^2 - 6x + 4) \\ &= 3x^2 + 6xh + 3h^2 - 6x - 6h + 4 - 3x^2 + 6x - 4 \\ &= 6xh + 3h^2 - 6h = h(6x + 3h - 6) \end{aligned}$$

$$\frac{f(x+h) - f(x)}{h} = \frac{h(6x + 3h - 6)}{h} = 6x + 3h - 6$$

$$\begin{aligned} \lim_{h \to 0} \frac{f(x+h) - f(x)}{h} &= \lim_{h \to 0} (6x + 3h - 6) \\ &= 6x + 0 - 6 = 6x - 6 \end{aligned}$$

Thus, the slope of the tangent line at $(x, f(x))$ is $6x - 6$.

(*b*) From (*a*), the slope of the tangent line at (0, 4) is

$$6x - 6 = 6(0) - 6 = 0 - 6 = -6$$

Hence, the slope-intercept equation has the form $y = -6x + b$. Since the line passes through (0, 4), the y-intercept, b, is 4. Thus, the equation is $y = -6x + 4$.

11.5 The *normal line* to a curve at a point P is defined to be the line through P perpendicular to the tangent line at P. Find the slope-intercept equation of the normal line to the parabola $y = x^2$ at the point $(\frac{1}{2}, \frac{1}{4})$.

By Problem 11.1, the tangent line has slope

$$2\left(\frac{1}{2}\right) = 1$$

Therefore, by Theorem 4.2, the slope of the normal line is -1, and the slope-intercept equation of the normal line will have the form $y = -x + b$. When $x = \frac{1}{2}$,

$$y = x^2 = \left(\frac{1}{2}\right)^2 = \frac{1}{4}$$

whence $$\frac{1}{4} = -\left(\frac{1}{2}\right) + b \qquad \text{or} \qquad b = \frac{3}{4}$$

Thus, the equation is

$$y = -x + \frac{3}{4}$$

Supplementary Problems

11.6 For each function f and argument $x = a$ below, (i) find a formula for the slope of the tangent line at an arbitrary point $P(x, f(x))$ of the graph of f; (ii) find the slope-intercept equation of the tangent line corresponding to the given argument a; (iii) draw the graph of f and show the tangent line found in (ii).

(*a*) $f(x) = 2x^2 + x;\ a = \frac{1}{4}$ (*b*) $f(x) = \frac{1}{3}x^3 + 1;\ a = 2$

(*c*) $f(x) = x^2 - 2x;\ a = 1$ (*d*) $f(x) = 4x^2 + 3;\ a = \frac{1}{2}$

11.7 Find the point(s) on the graph of $y = x^2$ at which the tangent line is parallel to the line $y = 6x - 1$.

11.8 Find the point(s) on the graph of $y = x^3$ at which the tangent line is perpendicular to the line $3x + 9y = 4$.

11.9 Find the slope-intercept equation of the normal line to the graph of $y = x^3$ at the point at which $x = \frac{1}{3}$.

11.10 At what point(s) does the normal line to the curve $y = x^2 - 3x + 5$ at the point (3, 5) intersect the curve?

11.11 At any point (x, y) of the straight line having the slope-intercept equation

$$y = mx + b$$

show that the tangent line is the straight line itself.

11.12 Find the point(s) on the graph of $y = x^2$ at which the tangent line is a line passing through the point (2, −12). (*Hint*: Write an equation of the tangent line at the point (x_0, x_0^2).

11.13 Find the slope-intercept equation of the tangent line to the graph of $y = \sqrt{x}$ at the point (4, 2). (*Hint*: See ALGEBRA for Problem 10.13.)

Chapter 12

The Derivative

The expression for the slope of the tangent line,

$$\lim_{h\to 0}\frac{f(x+h)-f(x)}{h}$$

determines a number which depends upon x. Thus, the expression defines a function, called the *derivative* of f.

Definition: The derivative f' of f is the function defined by the formula

$$f'(x)=\lim_{h\to 0}\frac{f(x+h)-f(x)}{h}$$

NOTATION There are other notations traditionally used for the derivative:

$$D_x f(x) \qquad \text{and} \qquad \frac{dy}{dx}$$

When a variable y represents $f(x)$, the derivative is denoted y', $D_x y$, or $\frac{dy}{dx}$. We shall use whichever notation is most convenient or customary in a given case.

The derivative is so important in all parts of pure and applied mathematics that we must devote a great deal of effort to finding formulas for the derivatives of various kinds of functions. If the limit in the above Definition exists, the function f is said to be *differentiable* at x, and the process of calculating f' is called *differentiation* of f.

EXAMPLES (*a*) Let $f(x)=3x+5$ for all x. Then:

$$f(x+h)=3(x+h)+5=3x+3h+5$$
$$f(x+h)-f(x)=(3x+3h+5)-(3x+5)=3x+3h+5-3x-5=3h$$
$$\frac{f(x+h)-f(x)}{h}=\frac{3h}{h}=3$$

Hence
$$f'(x)=\lim_{h\to 0}\frac{f(x+h)-f(x)}{h}=\lim_{h\to 0}3=3$$

or, in another notation, $D_x(3x+5)=3$. In this case, the derivative is independent of x. (*b*) Generalizing (*a*) to the case of the function $f(x)=Ax+B$, where A and B are constants, we find:

$$\frac{f(x+h)-f(x)}{h}=A \qquad \text{and so} \qquad f'(x)=\lim_{h\to 0}\frac{f(x+h)-f(x)}{h}=A$$

or $D_x(Ax+B)=A$. In particular (choose $A=0$; then choose $A=1$, $B=0$),

$$D_x(B)=0 \qquad \textit{(the derivative of a constant function is zero)}$$
$$D_x(x)=1$$

(*c*)
$$D_x(x^2)=2x \qquad D_x(x^3)=3x^2$$

as follows from Problems 11.1 and 11.2, respectively.

(*d*)
$$D_x\left(\frac{1}{x}\right)=-\frac{1}{x^2}$$

as follows from Problem 11.3(*a*).

We shall need to know how to differentiate functions built up by arithmetic operations on simpler functions. For this purpose, several rules of differentiation will be proved.

RULE 1.
$$D_x(f(x) \pm g(x)) = D_x f(x) \pm D_x g(x)$$

The derivative of a sum (difference) is the sum (difference) of the derivatives. (For a proof, see Problem 12.1(*a*).)

EXAMPLES

(*a*)
$$D_x(x^3 + x^2) = D_x(x^3) + D_x(x^2) = 3x^2 + 2x$$

(*b*)
$$D_x\left(x^2 - \frac{1}{x}\right) = D_x(x^2) - D_x\left(\frac{1}{x}\right) = 2x - \left(-\frac{1}{x^2}\right) = 2x + \frac{1}{x^2}$$

RULE 2.
$$D_x(c \cdot f(x)) = c \cdot D_x f(x)$$

where c is a constant. (For a proof, see Problem 12.1(*b*).)

EXAMPLES

(*a*)
$$D_x(7x^2) = 7 \cdot D_x(x^2) = 7 \cdot 2x = 14x$$

(*b*)
$$D_x(12x^3) = 12 \cdot D_x(x^3) = 12(3x^2) = 36x^2$$

(*c*)
$$D_x\left(-\frac{4}{x}\right) = D_x\left((-4)\frac{1}{x}\right) = -4 \cdot D_x\left(\frac{1}{x}\right) = -4\left(-\frac{1}{x^2}\right) = \frac{4}{x^2}$$

(*d*)
$$\begin{aligned} D_x(3x^3 + 5x^2 + 2x + 4) &= D_x(3x^3) + D_x(5x^2) + D_x(2x) + D_x(4) \\ &= 3 \cdot D_x(x^3) + 5 \cdot D_x(x^2) + 2 \cdot D_x(x) + 0 \\ &= 3(3x^2) + 5(2x) + 2(1) = 9x^2 + 10x + 2 \end{aligned}$$

RULE 3 (Product Rule).
$$D_x(f(x) \cdot g(x)) = f(x) \cdot D_x g(x) + g(x) \cdot D_x f(x)$$

(A proof will be given in Problem 13.1.)

EXAMPLES

(*a*)
$$D_x(x^4) = D_x(x^3 \cdot x)$$

ALGEBRA $\quad u^a \cdot u^b = u^{a+b} \quad$ and $\quad \dfrac{u^a}{u^b} = u^{a-b}$

$$\begin{aligned} &= x^3 \cdot D_x(x) + x \cdot D_x(x^3) \quad \text{[by the Product Rule]} \\ &= x^3(1) + x(3x^2) = x^3 + 3x^3 = 4x^3 \end{aligned}$$

(*b*)
$$\begin{aligned} D_x(x^5) &= D_x(x^4 \cdot x) \\ &= x^4 \cdot D_x(x) + x \cdot D_x(x^4) \quad \text{[by the Product Rule]} \\ &= x^4(1) + x(4x^3) \quad \text{[by Example (a)]} \\ &= x^4 + 4x^4 = 5x^4 \end{aligned}$$

(*c*)
$$\begin{aligned} &D_x((x^3 + x)(x^2 - x + 2)) \\ &\quad = (x^3 + x) \cdot D_x(x^2 - x + 2) + (x^2 - x + 2) \cdot D_x(x^3 + x) \\ &\quad = (x^3 + x)(2x - 1) + (x^2 - x + 2)(3x^2 + 1) \end{aligned}$$

The reader may have noticed a pattern in the derivatives of the powers of x:

$$D_x(x) = 1 = 1 \cdot x^0 \qquad D_x(x^2) = 2x \qquad D_x(x^3) = 3x^2 \qquad D_x(x^4) = 4x^3 \qquad D_x(x^5) = 5x^4$$

This pattern does in fact hold for all powers of x.

RULE 4. $$D_x(x^n) = nx^{n-1}$$

where n is any positive integer. (For a proof, see Problem 12.2.)

EXAMPLES

$$(a)\quad D_x(x^9) = 9x^8 \qquad (b)\quad D_x(5x^{11}) = 5\cdot D_x(x^{11}) = 5(11x^{10}) = 55x^{10}$$

Using Rules 1, 2, and 4, we have an easy method for differentiating *any polynomial.*

EXAMPLE

$$\begin{aligned} D_x\left(\frac{3}{5}x^3 - 4x^2 + 2x - \frac{1}{2}\right) &= D_x\left(\frac{3}{5}x^3\right) - D_x(4x^2) + D_x(2x) - D_x\left(\frac{1}{2}\right) \quad \text{[by Rule 1]} \\ &= \frac{3}{5}\cdot D_x(x^3) - 4\cdot D_x(x^2) + 2\cdot D_x(x) - 0 \quad \text{[by Rule 2]} \\ &= \frac{3}{5}(3x^2) - 4(2x) + 2(1) = \frac{9}{5}x^2 - 8x + 2 \quad \text{[by Rule 4]} \end{aligned}$$

More concisely, we have:

RULE 5. To differentiate a polynomial, change each nonconstant term a_kx^k to ka_kx^{k-1}; the constant term (if any) is simply dropped.

EXAMPLES

(a) $$D_x(8x^5 - 2x^4 + 3x^2 + 5x + 7) = 40x^4 - 8x^3 + 6x + 5$$

(b) $$D_x\left(3x^7 + \sqrt{2}x^5 - \frac{4}{3}x^2 + 7x - \pi\right) = 21x^6 + 5\sqrt{2}x^4 - \frac{8}{3}x + 7$$

Solved Problems

12.1 Prove (*a*) Rule 1 and (*b*) Rule 2 of differentiation. Assume that $D_xf(x)$ and $D_xg(x)$ are defined.

(*a*)
$$\begin{aligned} D_x(f(x) \pm g(x)) &= \lim_{h\to 0}\frac{(f(x+h) \pm g(x+h)) - (f(x) \pm g(x))}{h} \\ &= \lim_{h\to 0}\frac{(f(x+h) - f(x)) \pm (g(x+h) - g(x))}{h} \\ &= \lim_{h\to 0}\left(\frac{f(x+h) - f(x)}{h} \pm \frac{g(x+h) - g(x)}{h}\right) \\ &= \lim_{h\to 0}\frac{f(x+h) - f(x)}{h} \pm \lim_{h\to 0}\frac{g(x+h) - g(x)}{h} \quad \text{[by Section 8.2, Property V]} \\ &= D_xf(x) \pm D_xg(x) \end{aligned}$$

(*b*)
$$\begin{aligned} D_x(c\cdot f(x)) &= \lim_{h\to 0}\frac{c\cdot f(x+h) - c\cdot f(x)}{h} = \lim_{h\to 0}\frac{c(f(x+h) - f(x))}{h} \\ &= c\cdot\lim_{h\to 0}\frac{f(x+h) - f(x)}{h} \quad \text{[by Section 8.2, Property III]} \\ &= c\cdot D_xf(x) \end{aligned}$$

12.2 Prove Rule 4, $D_x(x^n) = nx^{n-1}$, for any positive integer n.

We already know that Rule 4 holds when $n = 1$:

$$D_x(x^1) = D_x(x) = 1 = 1 \cdot x^0$$

(remember that $x^0 = 1$). We can prove the rule by mathematical induction. This involves showing that, *if the rule holds for any particular positive integer k, then the rule also must hold for the next integer k + 1*. Since we know that the rule holds for $n = 1$, it would then follow that it holds for all positive integers.

Assume, then, that $D_x(x^k) = kx^{k-1}$. We have:

$$\begin{aligned} D_x(x^{k+1}) &= D_x(x^k \cdot x) \quad [\text{since } x^{k+1} = x^k \cdot x^1 = x^k \cdot x] \\ &= x^k \cdot D_x(x) + x \cdot D_x(x^k) \quad [\text{by the Product Rule}] \\ &= x^k \cdot 1 + x(kx^{k-1}) \quad [\text{by the assumption that } D_x(x^k) = kx^{k-1}] \\ &= x^k + kx^k \quad [\text{since } x \cdot x^{k-1} = x^1 \cdot x^{k-1} = x^k] \\ &= (1+k)x^k = (k+1)x^{(k+1)-1} \end{aligned}$$

and the proof by induction is complete.

12.3 Find the derivative of the polynomial

$$5x^9 - 12x^6 + 4x^5 - 3x^2 + x - 2$$

By Rule 5,

$$D_x(5x^9 - 12x^6 + 4x^5 - 3x^2 + x - 2) = 45x^8 - 72x^5 + 20x^4 - 6x + 1$$

12.4 Find the slope-intercept equations of the tangent lines to the graphs of the following functions at the given points:

(*a*) $f(x) = 3x^2 - 5x + 1$, at $x = 2$ (*b*) $f(x) = x^7 - 12x^4 + 2x$, at $x = 1$

(*a*) For $f(x) = 3x^2 - 5x + 1$, Rule 5 gives $f'(x) = 6x - 5$. Then,

$$f'(2) = 6(2) - 5 = 12 - 5 = 7$$

and

$$f(2) = 3(2)^2 - 5(2) + 1 = 3(4) - 10 + 1 = 12 - 9 = 3$$

Thus, the slope of the tangent line to the graph at $(2, f(2)) = (2, 3)$ is $f'(2) = 7$, and we have as a point-slope equation of the tangent line

$$\frac{y-3}{x-2} = 7$$

Solving for y yields the slope-intercept equation:

$$\begin{aligned} y - 3 &= 7(x-2) \\ y - 3 &= 7x - 14 \\ y &= 7x - 11 \end{aligned}$$

(*b*)

$$f(x) = x^7 - 12x^4 + 2x \quad \text{and} \quad f'(x) = 7x^6 - 48x^3 + 2$$

which give

$$f(1) = (1)^7 - 12(1)^4 + 2(1) = 1 - 12 + 2 = -9$$

and

$$f'(1) = 7(1)^6 - 48(1)^3 + 2 = 7 - 48 + 2 = -39$$

Hence, for the tangent line,

$$\frac{y-(-9)}{x-1} = -39 \quad \text{or} \quad \frac{y+9}{x-1} = -39$$

or $y + 9 = -39(x - 1)$ or $y + 9 = -39x + 39$ or $y = -39x + 30$.

12.5 At what point(s) of the graph of $y = x^5 + 4x - 3$ does the tangent line to the graph also pass through the point (0, 1)?

The slope of the tangent line at a point $(x_0, y_0) = (x_0, x_0^5 + 4x_0 - 3)$ of the graph is the value of the derivative dy/dx at $x = x_0$. By Rule 5,

$$\frac{dy}{dx} = 5x^4 + 4 \qquad \text{and so} \qquad \left.\frac{dy}{dx}\right|_{x=x_0} = 5x_0^4 + 4$$

Now, a straight line through the points $(x_0, x_0^5 + 4x_0 - 3)$ and (0, 1) will have slope $5x_0^4 + 4$ if and only if

$$\frac{(x_0^5 + 4x_0 - 3) - 1}{x_0 - 0} = 5x_0^4 + 4$$

$$x_0^5 + 4x_0 - 4 = 5x_0^5 + 4x_0$$

$$-4 = 4x_0^5$$

$$x_0^5 = -1$$

$$x_0 = -1$$

Thus, the required point of the graph is

$$(-1, (-1)^5 + 4(-1) - 3) = (-1, -1 - 4 - 3) = (-1, -8)$$

Supplementary Problems

12.6 Use the basic definition of $f'(x)$ as a limit to calculate the derivatives of the following functions:

(a) $f(x) = 2x - 5$ (b) $f(x) = \frac{1}{3}x^2 - 7x + 4$

(c) $f(x) = 2x^3 + 3x - 1$ (d) $f(x) = x^4$

12.7 Use Rule 5 to find the derivatives of the following polynomials:

(a) $3x^3 - 4x^2 + 5x - 2$ (b) $-8x^5 + \sqrt{3}x^3 + 2\pi x^2 - 12$

(c) $3x^{13} - 5x^{10} + 10x^2$ (d) $2x^{51} + 3x^{12} - 14x^2 + \sqrt[3]{7}x + \sqrt{5}$

12.8 (*Notational*)

(a) Find $D_x\left(3x^7 - \frac{1}{5}x^5\right)$. (b) Find $\frac{d(3x^2 - 5x + 1)}{dx}$.

(c) If $y = \frac{1}{2}x^4 + 5x$, find $\frac{dy}{dx}$. (d) Find $\frac{d(3t^7 - 12t^2)}{dt}$.

(e) If $u = \sqrt{2}x^5 - x^3$, find $D_x u$.

12.9 Find slope-intercept equations of the tangent lines to the graphs of the following functions f at the specified points: (a) $f(x) = x^2 - 5x + 2$, at $x = -1$; (b) $f(x) = 4x^3 - 7x^2$, at $x = 3$; (c) $f(x) = -x^4 + 2x^2 + 3$, at $x = 0$.

12.10 Specify all straight lines (a) through the point (0, 2) and tangent to the curve

$$y = x^4 - 12x + 50$$

(b) through the point (1, 5) and tangent to the curve

$$y = 3x^3 + x + 4$$

12.11 Find the slope-intercept equation of the normal line to the graph of $y = x^3 - x^2$ at the point where $x = 1$.

12.12 Find the point(s) on the graph of $y = \frac{1}{2}x^2$ at which the normal line passes through the point (4, 1).

12.13 Recalling the definition of the derivative, evaluate

$$(a)\ \lim_{h\to 0} \frac{(3+h)^5 - 3^5}{h} \qquad (b)\ \lim_{h\to 0} \frac{5\left(\frac{1}{3}+h\right)^4 - 5\left(\frac{1}{3}\right)^4}{h}$$

12.14 A function f, defined for all real numbers, is such that: (i) $f(1) = 2$, (ii) $f(2) = 8$, and

$$\text{(iii)}\quad f(u+v) - f(u) = kuv - 2v^2$$

for all u and v, where k is some constant. Find $f'(x)$ for arbitrary x.

12.15 Let $f(x) = 2x^2 + \sqrt{3}x$ for all x. (*a*) Find the nonnegative value(s) of x for which the tangent line to the graph of f at $(x, f(x))$ is perpendicular to the tangent line to the graph at $(-x, f(-x))$. (*b*) Find the point of intersection of each pair of perpendicular lines found in (*a*).

12.16 If the line $4x - 9y = 0$ is tangent in the first quadrant to the graph of

$$y = \frac{1}{3}x^3 + c$$

what is the value of c?

12.17 For what nonnegative value of b is the line

$$y = -\frac{1}{12}x + b$$

normal to the graph of $y = x^3 + \frac{1}{3}$?

12.18 Let f be differentiable (that is, f' exists). Define a function $f^{\#}$ by the equation

$$f^{\#}(x) = \lim_{h\to 0} \frac{f(x+h) - f(x-h)}{h}$$

(*a*) Find $f^{\#}(x)$ if $f(x) = x^2 - x$. (*b*) Find the relationship between $f^{\#}$ and the derivative f'. (*Hint*:

$$\frac{f(x+h) - f(x-h)}{h} = \frac{f(x+h) - f(x)}{h} + \frac{f(x+k) - f(x)}{k}$$

where $k = -h$.)

12.19 Let $f(x) = x^3 + x^2 - 9x - 9$. (*a*) Find the zeros of f. (*b*) Find the slope-intercept equation of the tangent to the graph of f at the point where $x = 1$. (*c*) A certain point (x_0, y_0) is on the graph of f, and the line tangent to the graph at (x_0, y_0) passes through the point $(4, -1)$. Find x_0 and y_0.

12.20 Let $f(x) = 3x^3 - 11x^2 - 15x + 63$. (*a*) Find the zeros of f. (*b*) Write an equation of the line normal to the graph of f at $x = 0$. (*c*) Find all points on the graph of f where the tangent line to the graph is horizontal.

Chapter 13

More on the Derivative

13.1 DIFFERENTIABILITY AND CONTINUITY

It would be difficult to comprehend what was meant by "the tangent line" to a curve at a point where the curve has a break. Thus, in Chapter 11, when we defined the tangent line to the graph of f at the point corresponding to $x = a$ to have slope

$$\lim_{x \to a} \frac{f(x) - f(a)}{x - a} \equiv f'(a)$$

(in the Definition, replace x by a, and h by $x - a$), we required that f be continuous at a. Indeed, if f were discontinuous at a, then, as $x \to a$, the denominator $x - a$ would approach zero, but the numerator $f(x) - f(a)$ would not. Consequently, the limit could not exist, and $f'(a)$ would not be defined. We summarize this result—that f is differentiable at a *only if* it is continuous at a—in the following.

Theorem 13.1: If f is differentiable at a, then f is continuous at a.

EXAMPLE Differentiability is a *stronger condition* than continuity; in other words, the converse of Theorem 13.1 is not true. To see this, consider the absolute-value function, $f(x) = |x|$ (Fig. 13-1). f is obviously continuous at $x = 0$; but it is not differentiable at $x = 0$. In fact,

$$\lim_{h \to 0^+} \frac{f(0+h) - f(0)}{h} = \lim_{h \to 0^+} \frac{h - 0}{h} = \lim_{h \to 0^+} 1 = 1$$

$$\lim_{h \to 0^-} \frac{f(0+h) - f(0)}{h} = \lim_{h \to 0^-} \frac{-h - 0}{h} = \lim_{h \to 0^-} -1 = -1$$

and so the two-sided limit needed to define $f'(0)$ does not exist. (The sharp corner in the graph is a tip-off: where there is no unique tangent line, there can be no derivative.)

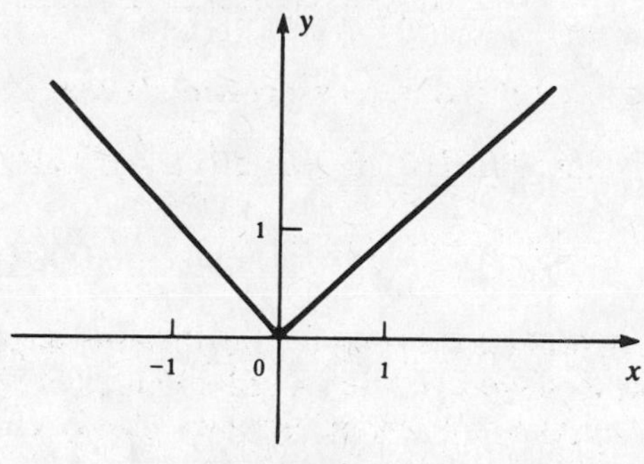

Fig. 13-1

13.2 FURTHER RULES FOR DERIVATIVES

Theorem 13.1 enables us to justify Rule 3, the Product Rule, of Chapter 12, and to establish two additional rules.

RULE 6 (Quotient Rule). If f and g are differentiable at x and if $g(x) \neq 0$, then

$$D_x\left(\frac{f(x)}{g(x)}\right) = \frac{g(x) \cdot D_x f(x) - f(x) \cdot D_x g(x)}{(g(x))^2}$$

(For a proof, see Problem 13.2.)

EXAMPLES

(*a*)
$$D_x\left(\frac{x+1}{x^2-2}\right)=\frac{(x^2-2)\cdot D_x(x+1)-(x+1)\cdot D_x(x^2-2)}{(x^2-2)^2}$$
$$=\frac{(x^2-2)(1)-(x+1)(2x)}{(x^2-2)^2}=\frac{x^2-2-2x^2-2x}{(x^2-2)^2}$$
$$=\frac{-x^2-2x-2}{(x^2-2)^2}=-\frac{x^2+2x+2}{(x^2-2)^2}$$

(*b*)
$$D_x\left(\frac{1}{x^2}\right)=\frac{x^2\cdot D_x(1)-1\cdot D_x(x^2)}{(x^2)^2}=\frac{x^2(0)-1(2x)}{x^4}=-\frac{2x}{x^4}=-\frac{2}{x^3}$$

The Quotient Rule allows us to extend Rule 4 of Chapter 12:

RULE 7. $D_x(x^k)=kx^{k-1}$ for any integer k (positive, zero, or negative). (For a proof, see Problem 13.3.)

EXAMPLES

(*a*)
$$D_x\left(\frac{1}{x}\right)=D_x(x^{-1})=(-1)x^{-2}=(-1)\frac{1}{x^2}=-\frac{1}{x^2}$$

(*b*)
$$D_x\left(\frac{1}{x^2}\right)=D_x(x^{-2})=-2x^{-3}=-\frac{2}{x^3}$$

Solved Problems

13.1 Prove the Product Rule: If f and g are differentiable at x, then

$$D_x(f(x)\cdot g(x))=f(x)\cdot D_xg(x)+g(x)\cdot D_xf(x)$$

Since, algebraically,

$$f(x+h)g(x+h)-f(x)g(x)=f(x+h)(g(x+h)-g(x))+g(x)(f(x+h)-f(x))$$

we have:

$$D_x(f(x)\cdot g(x))=\lim_{h\to 0}\frac{f(x+h)g(x+h)-f(x)g(x)}{h}$$
$$=\lim_{h\to 0}\frac{f(x+h)(g(x+h)-g(x))+g(x)(f(x+h)-f(x))}{h}$$
$$=\lim_{h\to 0}\frac{f(x+h)(g(x+h)-g(x))}{h}+\lim_{h\to 0}\frac{g(x)(f(x+h)-f(x))}{h}$$
$$=\lim_{h\to 0}f(x+h)\cdot\lim_{h\to 0}\frac{g(x+h)-g(x)}{h}+\lim_{h\to 0}g(x)\cdot\lim_{h\to 0}\frac{f(x+h)-f(x)}{h}$$
$$=f(x)\cdot D_xg(x)+g(x)\cdot D_xf(x)$$

In the last step, $\lim_{h\to 0}f(x+h)=f(x)$ follows from the fact that f is continuous at x, by Theorem 13.1.

13.2 Prove Rule 6, the Quotient Rule: If f and g are differentiable at x and if $g(x)\neq 0$, then

$$D_x\left(\frac{f(x)}{g(x)}\right)=\frac{g(x)D_xf(x)-f(x)D_xg(x)}{(g(x))^2}$$

If $g(x)\neq 0$, then $1/g(x)$ is defined. Moreover, since g is continuous at x (by Theorem 13.1), $g(x+h)\neq 0$ for all sufficiently small values of h; hence, $1/g(x+h)$ is defined for those same values of h.

We may then calculate:

$$\lim_{h\to 0}\frac{\frac{1}{g(x+h)}-\frac{1}{g(x)}}{h}=\lim_{h\to 0}\frac{g(x)-g(x+h)}{hg(x)g(x+h)}\quad\text{[by algebra: multiply top and bottom by } g(x)g(x+h)\text{]}$$

$$=\lim_{h\to 0}\left(\frac{-1/g(x)}{g(x+h)}\right)\left(\frac{g(x+h)-g(x)}{h}\right)\quad\text{[by algebra]}$$

$$=\lim_{h\to 0}\frac{-1/g(x)}{g(x+h)}\cdot\lim_{h\to 0}\frac{g(x+h)-g(x)}{h}\quad\text{[by Property IV of limits]}$$

$$=\frac{\lim_{h\to 0}(-1/g(x))}{\lim_{h\to 0}g(x+h)}\cdot D_xg(x)\quad\text{[by Property VI of limits and differentiability of } g\text{]}$$

$$=\frac{-1/g(x)}{g(x)}\cdot D_xg(x)\quad\text{[by Property II of limits and continuity of } g\text{]}$$

$$=\frac{-1}{(g(x))^2}D_xg(x)$$

Having thus proved that

$$D_x\left(\frac{1}{g(x)}\right)=\frac{-1}{(g(x))^2}D_xg(x)\tag{1}$$

we may substitute in the Product Rule (proved in Problem 13.1) to obtain:

$$D_x\left(\frac{f(x)}{g(x)}\right)=D_x\left(f(x)\cdot\frac{1}{g(x)}\right)=f(x)\frac{-1}{(g(x))^2}D_xg(x)+\frac{1}{g(x)}D_xf(x)$$

$$=\frac{-f(x)D_xg(x)}{(g(x))^2}+\frac{g(x)D_xf(x)}{(g(x))^2}$$

$$=\frac{g(x)D_xf(x)-f(x)D_xg(x)}{(g(x))^2}$$

which is the desired Quotient Rule.

13.3 Prove Rule 7: $D_x(x^k)=kx^{k-1}$ for any integer k.

When k is positive, this is just Rule 4 (Chapter 12). When $k=0$,

$$D_x(x^k)=D_x(x^0)=D_x(1)=0=0\cdot x^{-1}=kx^{k-1}$$

Now assume k is negative: $k=-n$, where n is positive.

ALGEBRA By definition,

$$x^k=x^{-n}=\frac{1}{x^n}$$

By (*1*) of Problem 13.2,

$$D_x(x^k)=D_x\left(\frac{1}{x^n}\right)=\frac{-1}{(x^n)^2}D_x(x^n)$$

But $(x^n)^2=x^{2n}$ and, by Rule 4, $D_x(x^n)=nx^{n-1}$. Therefore:

$$D_x(x^k)=\frac{-1}{x^{2n}}nx^{n-1}=-nx^{(n-1)-2n}=-nx^{-n-1}=kx^{k-1}$$

ALGEBRA We have used the law of exponents

$$\frac{x^a}{x^b}=x^{a-b}$$

13.4 Find the derivative of the function f such that

$$f(x)=\frac{x^2+x-2}{x^3+4}$$

Use the Quotient Rule, and then Rule 5 (Chapter 12):

$$\begin{aligned}
D_x\left(\frac{x^2+x-2}{x^3+4}\right) &= \frac{(x^3+4)D_x(x^2+x-2)-(x^2+x-2)D_x(x^3+4)}{(x^3+4)^2}\\
&= \frac{(x^3+4)(2x+1)-(x^2+x-2)(3x^2)}{(x^3+4)^2}\\
&= \frac{(2x^4+x^3+8x+4)-(3x^4+3x^3-6x^2)}{(x^3+4)^2}\\
&= \frac{-x^4-2x^3+6x^2+8x+4}{(x^3+4)^2}
\end{aligned}$$

13.5 Find the slope-intercept equation of the tangent line to the graph of

$$y=\frac{1}{x^3}$$

when $x=\frac{1}{2}$.

The slope of the tangent line is the derivative.

$$\frac{dy}{dx}=D_x\left(\frac{1}{x^3}\right)=D_x(x^{-3})=-3x^{-4}=-\frac{3}{x^4}$$

When $x=\frac{1}{2}$,

$$\left.\frac{dy}{dx}\right|_{x=1/2}=-\frac{3}{(\frac{1}{2})^4}=-\frac{3}{\frac{1}{16}}=-3(16)=-48$$

So, the tangent line has slope-intercept equation $y=-48x+b$. When $x=\frac{1}{2}$, the y-coordinate of the point on the graph is

$$\frac{1}{(\frac{1}{2})^3}=\frac{1}{\frac{1}{8}}=8$$

Substituting 8 for y and $\frac{1}{2}$ for x in $y=-48x+b$, we have:

$$8=-48(\tfrac{1}{2})+b \qquad \text{or} \qquad 8=-24+b \qquad \text{or} \qquad b=32$$

Thus, the equation is $y=-48x+32$.

Supplementary Problems

13.6 Find the derivatives of the functions defined by the following formulas:

(*a*) $(x^{100}+2x^{50}-3)(7x^8+20x+5)$ (*b*) $\frac{x^2-3}{x+4}$ (*c*) $\frac{x^5-x+2}{x^3+7}$

(*d*) $\frac{3}{x^5}$ (*e*) $8x^3-x^2+5-\frac{2}{x}+\frac{4}{x^3}$ (*f*) $\frac{3x^7+x^5-2x^4+x-3}{x^4}$

13.7 Find the slope-intercept equation of the tangent line to the graph of the function at the indicated point:

(*a*) $f(x)=\frac{1}{x^2}$, at $x=2$ (*b*) $f(x)=\frac{x+2}{x^3-1}$, at $x=-1$

13.8 Let

$$f(x) = \frac{x+2}{x-2}$$

for all $x \neq 2$. Find $f'(-2)$.

13.9 Determine the points at which the function $f(x) = |x - 3|$ is differentiable.

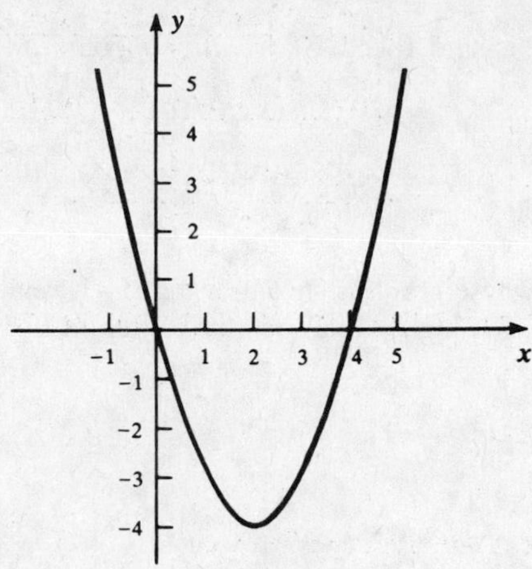

Fig. 13-2

13.10 The parabola in Fig. 13-2 is the graph of the function $f(x) = x^2 - 4x$. (a) Draw the graph of

$$y = |f(x)|$$

(b) Where does the derivative of $|f(x)|$ fail to exist?

Chapter 14

Maximum and Minimum Problems

14.1 RELATIVE EXTREMA

A (continuous) function f is said to have a *relative maximum* at $x = c$ if

$$f(x) \leq f(c)$$

for all x near c. (More precisely, f achieves a relative maximum at c if there exists $\delta > 0$ such that $|x - c| < \delta$ implies $f(x) \leq f(c)$.)

EXAMPLE For the function f whose graph is shown in Fig. 14-1, relative maxima occur at $x = c_1$ and $x = c_2$. This is obvious, since the point A is higher than nearby points on the graph, and the point B is higher than nearby points on the graph.

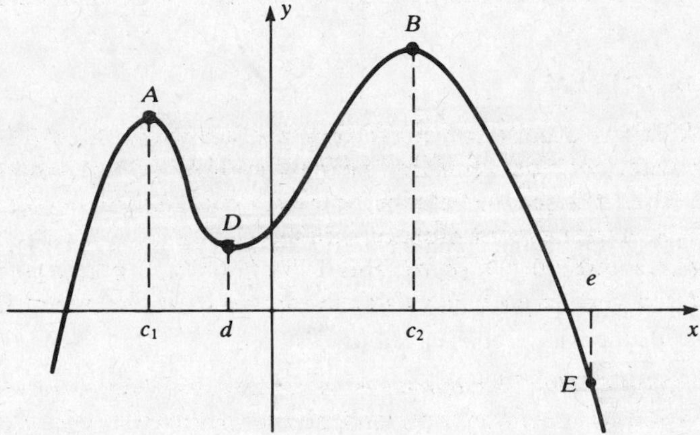

Fig. 14-1

The word "relative" is used to modify "maximum" because the value of a function at a relative maximum is not necessarily the greatest value of the function. Thus, in Fig. 14-1, the value $f(c_1)$ at c_1 is smaller than many other values of $f(x)$; in particular, $f(c_1) < f(c_2)$. In this example, the value $f(c_2)$ *is* the greatest value of the function.

A function f is said to have a *relative minimum* at $x = c$ if

$$f(x) \geq f(c)$$

for x near c. In Fig. 14-1, f achieves a relative minimum at $x = d$, since point D is lower than nearby points on the graph. The value at a relative minimum need not be the smallest value of the function; e.g., in Fig. 14-1, the value $f(e)$ is smaller than $f(d)$.

By a *relative extremum* is meant either a relative maximum or a relative minimum. Points at which a relative extremum exists possess the following characteristic property.

Theorem 14.1: If f has a relative extremum at $x = c$, and if $f'(c)$ exists, then $f'(c) = 0$.

The theorem is intuitively obvious. If $f'(c)$ exists, then there is a well-defined tangent line at the point on the graph of f where $x = c$. But, at a relative maximum or relative minimum, the tangent line is horizontal (see Fig. 14-2), and so its slope, $f'(c)$, is zero. (For a rigorous proof, see Problem 14.28.)

The converse of Theorem 14.1 does not hold: if $f'(c) = 0$, then f need not have a relative extremum at $x = c$. In Chapter 23, a method will be given that often will enable us to determine whether a relative extremum actually exists when $f'(c) = 0$.

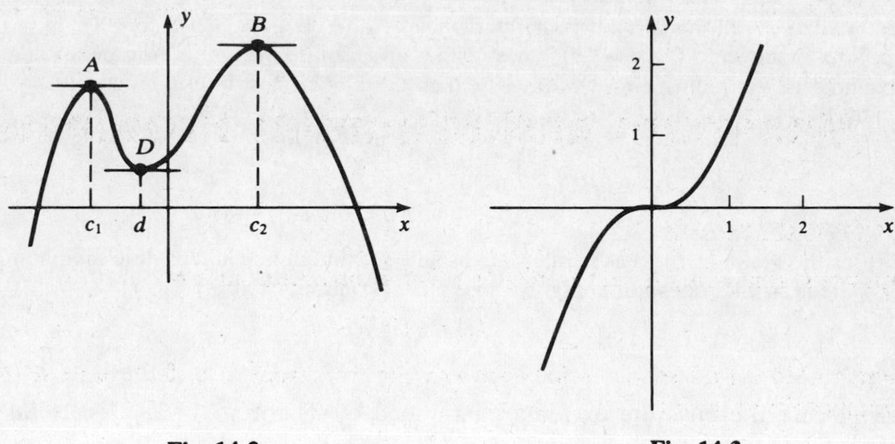

Fig. 14-2

Fig. 14-3

EXAMPLE Consider the function $f(x) = x^3$. Because $f'(x) = 3x^2$, $f'(x) = 0$ when and only when $x = 0$. But, from the graph of f in Fig. 14-3, it is clear that f has neither a relative maximum nor a relative minimum at $x = 0$.

14.2 ABSOLUTE EXTREMA

Practical applications usually call for finding the *absolute* maximum or *absolute* minimum of a function on a given set. Let f be a function defined on a set $\mathscr{E}$ (and possibly at other points, too), and let c belong to $\mathscr{E}$. Then f is said to achieve an *absolute maximum on* $\mathscr{E}$ at c if $f(x) \leq f(c)$ for all x in $\mathscr{E}$. Similarly, f is said to achieve an *absolute minimum on* $\mathscr{E}$ at d if $f(x) \geq f(d)$ for all x in $\mathscr{E}$.

If the set $\mathscr{E}$ is a closed interval $[a, b]$, and if the function f is continuous over $[a, b]$—see Section 10.3—then we have a very important existence theorem (which cannot be proved in an elementary way).

Theorem 14.2 *(Extreme-Value Theorem)*: Any continuous function f over a closed interval $[a, b]$ has an absolute maximum and an absolute minimum on $[a, b]$.

EXAMPLES (*a*) Let $f(x) = x + 1$ for all x in the closed interval $[0, 2]$. The graph of f is shown in Fig. 14-14(*a*). Then f achieves an absolute maximum on $[0, 2]$ at $x = 2$; this absolute maximum value is 3. In addition, f achieves

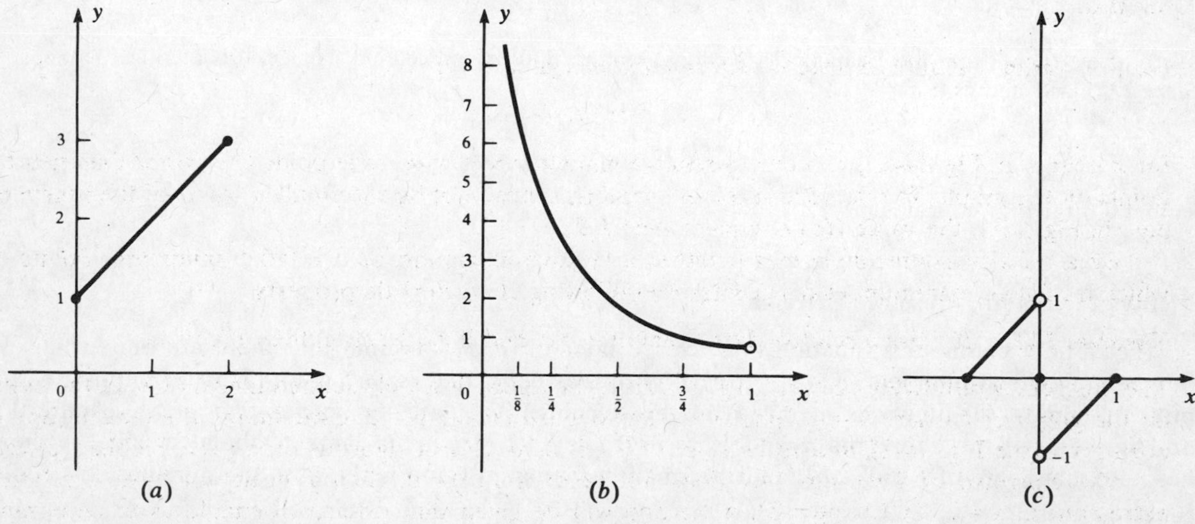

Fig. 14-4

an absolute minimum at $x = 0$; this absolute minimum value is 1. (*b*) Let $f(x) = 1/x$ for all x in the *open* interval (0, 1). The graph of f is shown in Fig. 14-4(*b*). f has neither an absolute maximum nor an absolute minimum on (0, 1). (If we extend f to the half-open interval (0, 1), then there is an absolute minimum at $x = 1$, but still no absolute maximum.) (*c*) Let

$$f(x) = \begin{cases} x+1 & \text{if } -1 \le x < 0 \\ 0 & \text{if } \; x = 0 \\ x-1 & \text{if } \; 0 < x \le 1 \end{cases}$$

(See Fig. 14-4(*c*) for the graph of f.) f has neither an absolute maximum nor an absolute minimum on the closed interval $[-1, 1]$. (Theorem 14.2 does not apply, because f is discontinuous at 0.)

Critical Numbers

To actually locate the absolute extrema guaranteed by Theorem 14.2, it is useful to have the following notion.

Definition: A *critical number* of a function f is a number c in the domain of f for which either $f'(c) = 0$ or $f'(c)$ is not defined.

EXAMPLES (*a*) Let $f(x) = 3x^2 - 2x + 4$. Then $f'(x) = 6x - 2$. Since $6x - 2$ is defined for all x, the only critical numbers are given by

$$6x - 2 = 0 \qquad \text{or} \qquad 6x = 2 \qquad \text{or} \qquad x = \frac{2}{6} = \frac{1}{3}$$

The sole critical number is 1/3. (*b*) Let $f(x) = x^3 - x^2 - 5x + 3$. Then

$$f'(x) = 3x^2 - 2x - 5$$

and, since $3x^2 - 2x - 5$ is defined for all x, the only critical numbers are the solutions of:

$$3x^2 - 2x - 5 = 0$$
$$(3x - 5)(x + 1) = 0$$
$$3x - 5 = 0 \quad \text{or} \quad x + 1 = 0$$
$$3x = 5 \quad \text{or} \quad x = -1$$
$$x = \frac{5}{3} \quad \text{or} \quad x = -1$$

Hence, there are two critical numbers, -1 and 5/3. (*c*) Let $f(x) = |x|$, or

$$f(x) = \begin{cases} x & \text{if } 0 \le x \\ -x & \text{if } x < 0 \end{cases}$$

We already know (from the Example in Section 13.1) that $f'(0)$ is not defined. Hence, 0 is a critical number. Since $D_x(x) = 1$ and $D_x(-x) = -1$,

$$f'(x) = \begin{cases} 1 & \text{if } 0 < x \\ -1 & \text{if } x < 0 \end{cases}$$

so that 0 is the only critical number.

Method for Finding Absolute Extrema

Let f be a continuous function on a closed interval $[a, b]$. Assume that there are only a finite number of critical numbers, $c_1, c_2, \ldots, c_k$, of f *inside* $[a, b]$; that is, in (a, b). (This assumption will hold for almost all functions met with in the calculus.) Tabulate the values of f at these critical numbers and at the endpoints a and b, as in Table 14-1. Then the largest tabulated value is the absolute maximum of f on $[a, b]$, and the smallest tabulated value is the absolute minimum of f on $[a, b]$.

Table 14-1

x	$f(x)$
c_1	$f(c_1)$
c_2	$f(c_2)$
...	...
c_k	$f(c_k)$
a	$f(a)$
b	$f(b)$

Table 14-2

x	$f(x)$
1/3	40/27
0	1
1	0

EXAMPLE Find the absolute maximum and minimum values of

$$f(x) = x^3 - 5x^2 + 3x + 1$$

on [0, 1], and find the arguments at which these values are achieved.

The function is continuous everywhere; in particular, on [0, 1]. As

$$f'(x) = 3x^2 - 10x + 3$$

is defined for all x, the only critical numbers are the solutions of:

$$3x^2 - 10x + 3 = 0$$
$$(3x - 1)(x - 3) = 0$$
$$3x - 1 = 0 \quad \text{or} \quad x - 3 = 0$$
$$3x = 1 \quad \text{or} \quad x = 3$$
$$x = \frac{1}{3} \quad \text{or} \quad x = 3$$

Hence, the only critical number in the open interval (0, 1) is $\frac{1}{3}$.

Now construct Table 14-2:

$$f(\tfrac{1}{3}) = (\tfrac{1}{3})^3 - 5(\tfrac{1}{3})^2 + 3(\tfrac{1}{3}) + 1$$
$$= \frac{1}{27} - \frac{5}{9} + 1 + 1$$
$$= \frac{1}{27} - \frac{15}{27} + 2 = 2 - \frac{14}{27} = \frac{40}{27}$$
$$f(0) = 0^3 - 5(0)^2 + 3(0) + 1 = 1$$
$$f(1) = 1^3 - 5(1)^2 + 3(1) + 1 = 1 - 5 + 3 + 1 = 0$$

The absolute maximum is the largest value in the second column, 40/27, and it is achieved at $x = 1/3$. The absolute minimum is the smallest value, 0, which is achieved at $x = 1$.

Solved Problems

14.1 Justify the tabular method for locating the absolute maximum and minimum of a function on a closed interval.

By the Extreme-Value Theorem (Theorem 14.2), a function f continuous on $[a, b]$ must have an absolute maximum and an absolute minimum on $[a, b]$. Let p be an argument at which the absolute maximum is achieved.

Case 1: p is one of the endpoints, a or b. Then $f(p)$ will be one of the values in our table; in fact, it will be the largest value in the table (since $f(p)$ is the absolute maximum of f on $[a, b]$).

Case 2: p is not an endpoint and $f'(p)$ is not defined. Then p is a critical number and will be one of the numbers $c_1, \ldots, c_k$ in our list. Hence, $f(p)$ will appear as a tabulated value, and it will be the largest of the tabulated values.

Case 3: p is not an endpoint and $f'(p)$ is defined. Since $f(p)$ is the absolute maximum of f on $[a, b]$, $f(p) \geq f(x)$ for all x near p. Thus, f has a relative maximum at p, and Theorem 14.1 gives $f'(p) = 0$. But then p is a critical number, and the conclusion follows as in Case 2.

A completely analogous argument shows that the method yields the absolute minimum.

14.2 Find the absolute maximum and minimum of each function on the given interval:

$$(a)\quad f(x) = 2x^3 - 5x^2 + 4x - 1, \text{ on } [-1, 2] \qquad (b)\quad f(x) = \frac{x^2+3}{x+1}, \text{ on } [0, 3]$$

(*a*) Since $f'(x) = 6x^2 - 10x + 4$, the critical numbers are the solutions of:

$$6x^2 - 10x + 4 = 0$$
$$3x^2 - 5x + 2 = 0$$
$$(3x-2)(x-1) = 0$$
$$3x - 2 = 0 \quad \text{or} \quad x - 1 = 0$$
$$3x = 2 \quad \text{or} \quad x = 1$$
$$x = \frac{2}{3} \quad \text{or} \quad x = 1$$

Thus, the critical numbers are $\frac{2}{3}$ and 1, both of which are in $(-1, 2)$.

Now construct Table 14-3:

$$f(\tfrac{2}{3}) = 2(\tfrac{2}{3})^3 - 5(\tfrac{2}{3})^2 + 4(\tfrac{2}{3}) - 1 = \frac{16}{27} - \frac{20}{9} + \frac{8}{3} - 1 = \frac{16}{27} - \frac{60}{27} + \frac{72}{27} - \frac{27}{27} = \frac{1}{27}$$
$$f(1) = 2(1)^3 - 5(1)^2 + 4(1) - 1 = 2 - 5 + 4 - 1 = 0$$
$$f(-1) = 2(-1)^3 - 5(-1)^2 + 4(-1) - 1 = -2 - 5 - 4 - 1 = -12$$
$$f(2) = 2(2)^3 - 5(2)^2 + 4(2) - 1 = 16 - 20 + 8 - 1 = 3$$

Thus, the absolute maximum is 3, achieved at $x = 2$, and the absolute minimum is -12, achieved at $x = -1$.

Table 14-3

x	$f(x)$
2/3	1/27
1	0
−1	−12 *min*
2	3 *max*

Table 14-4

x	$f(x)$
1	2 *min*
0	3 *max*
3	3 *max*

(*b*)

$$f'(x) = \frac{(x+1)D_x(x^2+3) - (x^2+3)D_x(x+1)}{(x+1)^2} = \frac{(x+1)(2x) - (x^2+3)(1)}{(x+1)^2}$$
$$= \frac{2x^2 + 2x - x^2 - 3}{(x+1)^2} = \frac{x^2 + 2x - 3}{(x+1)^2}$$

$f'(x)$ is not defined when $(x+1)^2 = 0$; that is, when $x = -1$. But, since -1 is not in $(0, 3)$, the only critical numbers that need be considered are the zeros of $x^2 + 2x - 3$ in $(0, 3)$:

$$x^2 + 2x - 3 = 0$$
$$(x+3)(x-1) = 0$$
$$x + 3 = 0 \quad \text{or} \quad x - 1 = 0$$
$$x = -3 \quad \text{or} \quad x = 1$$

Thus, 1 is the only critical number in $(0, 3)$.

Now construct Table 14-4:

$$f(1) = \frac{(1)^2+3}{1+1} = \frac{1+3}{2} = \frac{4}{2} = 2$$

$$f(0) = \frac{(0)^2+3}{0+1} = \frac{3}{1} = 3$$

$$f(3) = \frac{(3)^2+3}{3+1} = \frac{9+3}{4} = \frac{12}{4} = 3$$

Thus, the absolute maximum, achieved at 0 and 3, is 3; and the absolute minimum, achieved at 1, is 2.

14.3 Among all pairs of positive real numbers u and v whose sum is 10, which gives the greatest product uv?

Let $P = uv$. Since $u + v = 10$, $v = 10 - u$ and so

$$P = u(10-u) = 10u - u^2$$

Here, $0 < u < 10$, but we are free to include the cases $u = 0$ and $u = 10$ (we know in advance that neither yields the absolute maximum). Thus, we must find the absolute maximum of $P = 10u - u^2$ on the closed interval $[0, 10]$. The derivative

$$\frac{dP}{du} = 10 - 2u$$

vanishes only at $u = 5$; and this critical point must yield the maximum. Thus, the absolute maximum is

$$P(5) = 5(10-5) = 5(5) = 25$$

attained for $u = v = 5$.

ALGEBRA Calculus was not really needed in this problem, for

$$P = \frac{(u+v)^2 - (u-v)^2}{4} = \frac{10^2 - (u-v)^2}{4}$$

14.4 An open box is to be made from a rectangular piece of cardboard that is 8 feet by 3 feet, by cutting out four equal squares from the corners and then folding up the flaps (see Fig. 14-5). What length of the side of a square will yield the box with the largest volume?

Let x be the side of the square that is removed from each corner. The volume is $V = \ell wh$, where ℓ, w, and h are the length, width, and height of the box. Now,

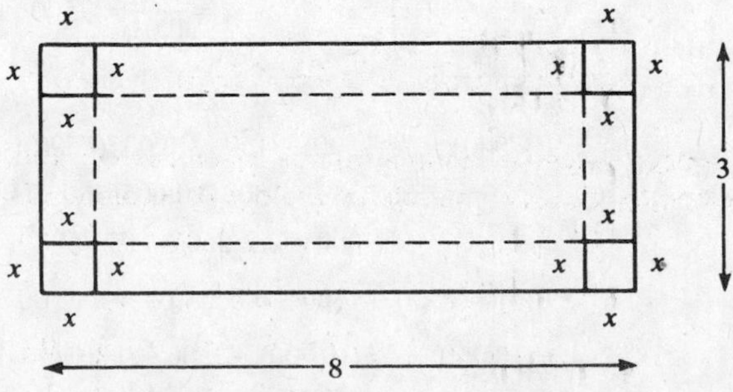

Fig. 14-5

$$\ell = 8 - 2x \qquad w = 3 - 2x \qquad h = x$$

giving
$$V(x) = (8-2x)(3-2x)x = (4x^2 - 22x + 24)x = 4x^3 - 22x^2 + 24x$$

Now, the width must be positive. Hence,

$$3 - 2x > 0 \qquad \text{or} \qquad 3 > 2x \qquad \text{or} \qquad \tfrac{3}{2} > x$$

Furthermore, $x > 0$. But, we can also admit the values $x = 0$ and $x = \frac{3}{2}$, which make $V = 0$ and which therefore cannot yield the maximum volume. Thus, we have to maximize $V(x)$ on the interval $[0, \frac{3}{2}]$. Since

$$\frac{dV}{dx} = 12x^2 - 44x + 24$$

the critical numbers are the solutions of:

$$12x^2 - 44x + 24 = 0$$
$$3x^2 - 11x + 6 = 0$$
$$(3x-2)(x-3) = 0$$
$$3x - 2 = 0 \quad \text{or} \quad x - 3 = 0$$
$$3x = 2 \quad \text{or} \quad x = 3$$
$$x = \frac{2}{3} \quad \text{or} \quad x = 3$$

The only critical number in $(0, \frac{3}{2})$ is $\frac{2}{3}$. Hence, the volume is greatest when $x = \frac{2}{3}$ foot.

14.5 A manufacturer sells each of his TV sets for \$85. The cost C (in dollars) of manufacturing and selling x TV sets per week is

$$C = 1500 + 10x + 0.0005x^2$$

If at most 10 000 sets can be produced per week, how many sets should be made and sold to maximize the weekly profit?

For x sets per week, the total income is $85x$. The profit is the income minus the cost:

$$P = 85x - (1500 + 10x + 0.005x^2) = 75x - 1500 - 0.005x^2$$

We wish to maximize P on the interval $[0, 10\,000]$, since the output is at most 10 000.

$$\frac{dP}{dx} = 75 - 0.01x$$

and the critical number is the solution of:

$$75 - 0.01x = 0$$
$$0.01x = 75$$
$$x = \frac{75}{0.01} = 7500$$

We now construct Table 14-5:

$$P(7500) = 75(7500) - 1500 - 0.005(7500)^2$$
$$= 562\,500 - 1500 - 0.005(56\,250\,000)$$
$$= 561\,000 - 281\,250 = 279\,750$$

$$P(0) = 75(0) - 1500 - 0.0005(0)^2 = -1500$$

$$P(10\,000) = 75(10\,000) - 1500 - 0.005(10\,000)^2$$
$$= 750\,000 - 1500 - 0.005(100\,000\,000)$$
$$= 748\,500 - 500\,000 = 248\,500$$

Table 14-5

x	$P(x)$
7500	279 750
0	−1500
10 000	248 500

Thus, the maximum profit is achieved when 7500 TV sets are produced and sold per week.

14.6 An orchard has an average yield of 25 bushels per tree when there are at most 40 trees per acre. When there are more than 40 trees per acre, the average yield per tree decreases by $\frac{1}{2}$ bushel per tree for every tree over 40. Find the number of trees per acre that will give the greatest yield per acre.

Let x be the number of trees per acre, and let $f(x)$ be the total yield, in bushels per acre. When $0 \le x \le 40$, $f(x) = 25x$. If $x > 40$, the number of bushels produced by each tree becomes $25 - \frac{1}{2}(x - 40)$ (here, $x - 40$ is the number of trees over 40, and $\frac{1}{2}(x - 40)$ is the corresponding decrease in bushels per tree). Hence, for $x > 40$, $f(x)$ is given by

$$\left(25 - \frac{1}{2}(x - 40)\right)x = \left(25 - \frac{1}{2}x + 20\right)x = \left(45 - \frac{1}{2}x\right)x = \frac{x}{2}(90 - x)$$

Thus
$$f(x) = \begin{cases} 25x & \text{if } 0 \le x \le 40 \\ \frac{x}{2}(90 - x) & \text{if } x > 40 \end{cases}$$

$f(x)$ is continuous everywhere, since $25x = (x/2)(90 - x)$ when $x = 40$. Clearly, $f(x) < 0$ when $x > 90$; hence, we can restrict attention to the interval $[0, 90]$.

For $0 < x < 40$, $f(x) = 25x$, and $f'(x) = 25$. Thus, there are no critical numbers in the open interval $(0, 40)$. For $40 < x < 90$,

$$f(x) = \frac{x}{2}(90 - x) = 45x - \frac{x^2}{2} \qquad \text{and} \qquad f'(x) = 45 - x$$

Thus, $x = 45$ is a critical number. In addition, 40 is also a critical number, since $f'(40)$ happens not to exist. We do not have to verify this fact, since there is no harm in adding 40 (or any other number in $(0, 90)$) to the list for which we compute $f(x)$.

We now construct Table 14-6:

$$f(45) = \frac{45}{2}(90 - 45) = \frac{45}{2}(45) = \frac{2025}{2} = 1012.5$$
$$f(40) = 25(40) = 1000$$
$$f(0) = 25(0) = 0$$
$$f(90) = \frac{90}{2}(90 - 90) = \frac{90}{2}(0) = 0$$

Table 14-6

x	$f(x)$
45	1012.5
40	1000
0	0
90	0

The maximum yield per acre is realized when there are 45 trees per acre.

Supplementary Problems

14.7 Find the absolute maxima and minima of the following functions on the indicated intervals:

(*a*) $f(x) = -4x + 5$, on $[-2, 3]$

(*b*) $f(x) = 2x^2 - 7x - 10$, on $[-1, 3]$

(*c*) $f(x) = x^3 + 2x^2 + x - 1$, on $[-1, 1]$

(*d*) $f(x) = 4x^3 - 8x^2 + 1$, on $[-1, 1]$

(*e*) $f(x) = x^4 - 2x^3 - x^2 - 4x + 3$, on $[0, 4]$

(*f*) $f(x) = \dfrac{2x+5}{x^2-4}$, on $[-5, -3]$

(*g*) $f(x) = \dfrac{x^2}{16} + \dfrac{1}{x}$, on $[1, 4]$

(*h*) $f(x) = \dfrac{x^3}{x+2}$, on $[-1, 1]$

(*i*) $f(x) = \begin{cases} x^3 - \frac{1}{3}x & \text{for } 0 \le x \le 1 \\ x^2 + x - \frac{4}{3} & \text{for } 1 < x \le 2 \end{cases}$, on $[0, 2]$

14.8 A farmer wishes to fence in a rectangular field. If north-south fencing costs $3 per yard, and east-west fencing costs $2 per yard, what are the dimensions of the field of maximum area that can be fenced in for $600?

14.9 A farmer has to fence in a rectangular field alongside a straight-running stream. If the farmer has 120 yards of fencing, and the side of the field along the stream does not have to be fenced, what dimensions of the field will yield the largest area?

14.10 The distance by bus from New York to Boston is 225 miles. The bus driver gets paid $12.50 per hour, while the other costs of running the bus at a steady speed of x miles per hour amount to $90 + 0.5x$ cents per mile. The minimum and maximum legal speeds on the bus route are 40 and 55 miles per hour. At what steady speed should the bus be driven to minimize the total cost?

14.11 A charter airline is planning a flight for which it is considering a price between $150 and $300 per person. The airline estimates that the number of passengers taking the flight will be $200 - 0.5x$, depending on the price of x dollars that will be set. What price will maximize the income?

14.12 Suppose that a company can sell x radios per week if it charges $100 - 0.1x$ dollars per radio. Its production cost is $30x + 5000$ dollars when x radios are produced per week. How many radios should be produced to maximize the profit, and what will be the selling price per radio?

14.13 A box with square base and vertical sides is to be made from 150 square feet of cardboard. What dimensions will provide the greatest volume (*a*) if the box has a top surface? (*b*) if the box has an open top?

14.14 A farmer wishes to fence in a rectangular field, and also to divide the field in half by another fence (AB in Fig. 14-6). The outside fence costs $2 per foot, and the fence in the middle costs $3 per foot. If the farmer has $840 to spend, what dimensions will maximize the total area?

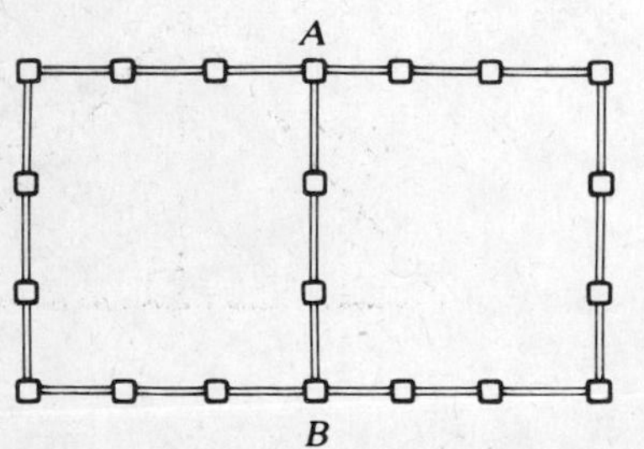

Fig. 14-6

Fig. 14-7

14.15 On a charter flight, the price per passenger is \$250 for any number of passengers up to 100. The flight will be canceled if there are fewer than 50 passengers. However, for every passenger over 100, the price per passenger will be decreased by \$1; the maximum number of passengers that can be flown is 225. What number of passengers will yield the maximum income?

14.16 Among all pairs x, y of nonnegative numbers whose sum, $x + y$, is 100, find those pairs (*a*) the sum of whose squares, $x^2 + y^2$, is a minimum; (*b*) the sum of whose squares, $x^2 + y^2$, is a maximum; (*c*) the sum of whose cubes, $x^3 + y^3$, is a minimum.

14.17 A sports complex is to be built in the form of a rectangular field with two equal semicircular areas at each end (Fig. 14-7). If the border of the entire complex is to be a running track 1256 meters long, what should be the dimensions of the complex so that the area of the rectangular field is a maximum?

14.18 A wire of length L is cut into two pieces. The first piece is bent into a circle and the second piece into a square. Where should the wire be cut so that the total area of the circle plus the square is (*a*) a maximum? (*b*) a minimum?

GEOMETRY The area of a circle of radius r is πr^2, and the circumference is $2\pi r$.

14.19 A wire of length L is cut into two pieces. The first piece is bent into a square and the second into an equilateral triangle. At what point should the cut be made so that the total area of the square and triangle is greatest?

GEOMETRY The area of an equilateral triangle of side s is $\sqrt{3}s^2/4$.

14.20 A company earns a profit of \$40 on every TV set it makes when it produces at most 1000 sets. If the profit per item decreases by 5 cents for every TV set over 1000, what production level maximizes the total profit?

14.21 Find the radius and height of the right circular cylinder of greatest volume that can be inscribed in a right circular cone having a radius of 3 feet and a height of 5 feet (Fig. 14-8).

GEOMETRY The volume of a circular cylinder is $\pi r^2 h$; and, at a point on the lateral surface of the cone,

$$\frac{5-h}{r} = \frac{5}{3}$$

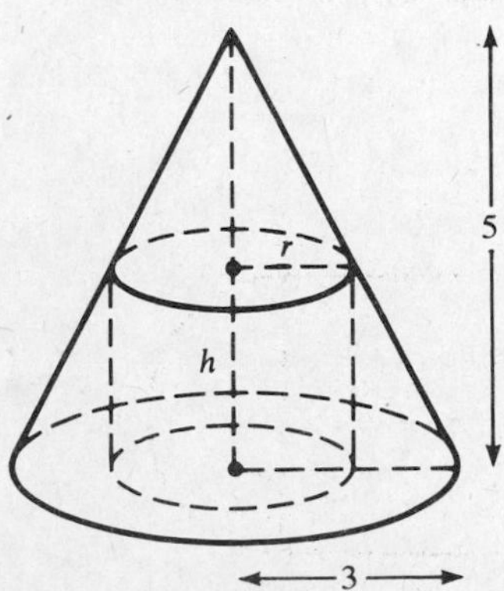

Fig. 14-8

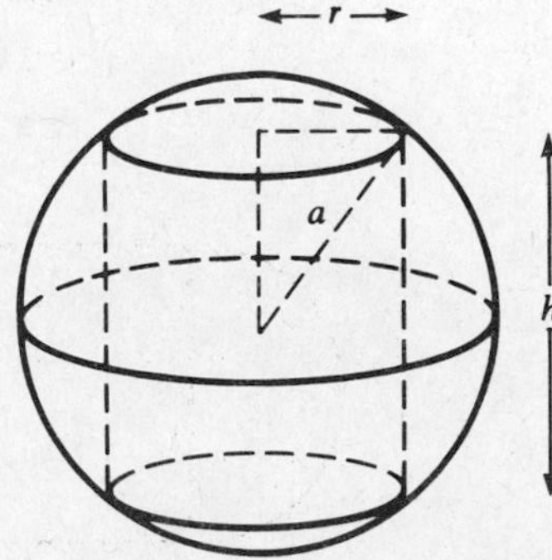

Fig. 14-9

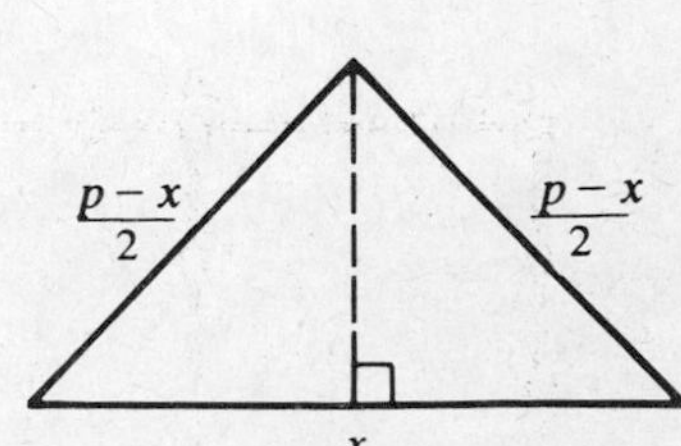

Fig. 14-10

14.22 Find the height h and radius r of the right circular cylinder of greatest volume that can be inscribed in a sphere of radius a. (*Hint*: See Fig. 14-9. The Pythagorean Theorem relates $h/2$ and r, and provides bounds on each.)

14.23 Among all isosceles triangles with a fixed perimeter p, which has the largest area? (*Hint*: See Fig. 14-10, and solve the equivalent problem of maximizing the square of the area.)

14.24 Find the point(s) on the ellipse

$$\frac{x^2}{25}+\frac{y^2}{9}=1$$

that is (are) (a) closest to the point $(1, 0)$, (b) farthest from the point $(1, 0)$. (*Hint*: It is easier to find the extrema of the square of the distance from $(1, 0)$ to (x, y). Notice that $-5 \le x \le 5$ for points (x, y) on the ellipse.)

14.25 A rectangular swimming pool is to be built with 6-foot borders at the north and south ends, and 10-foot borders at the east and west ends. If the total area available is 6000 square feet, what are the dimensions of the largest possible water area?

14.26 A farmer has to enclose two fields. One is to be a rectangle with the length twice the width, and the other is to be a square. The rectangle is required to contain at least 882 square meters, and the square has to contain at least 400 square meters. There are 680 meters of fencing available. (a) If x is the width of the rectangular field, what are the maximum and minimum possible values of x? (b) What is the maximum possible total area?

14.27 It costs a company $0.1x^2 + 4x + 3$ dollars to produce x tons of gold. If more than 10 tons is produced, the need for additional labor raises the cost by $2(x - 10)$ dollars. If the price per ton is \$9, regardless of the production level, and if the maximum production capacity is 20 tons, what output maximizes the profit?

14.28 Prove Theorem 14.1. (*Hint*: Take the case of a relative minimum at $x = c$. Then

$$\frac{f(c+h)-f(c)}{h} \ge 0$$

for h sufficiently small and positive, and

$$\frac{f(c+h)-f(c)}{h} \le 0$$

for h sufficiently small and negative.)

Chapter 15

The Chain Rule

15.1 COMPOSITE FUNCTIONS

There are still many functions whose derivatives we do not know how to calculate; for example,

$$\text{(i)}\quad \sqrt{x^3 - x + 2} \qquad \text{(ii)}\quad \sqrt[3]{x+4} \qquad \text{(iii)}\quad (x^2 + 3x - 1)^{23}$$

In case (iii), we could, of course, multiply $x^2 + 3x - 1$ by itself twenty-two times and then differentiate the resulting polynomial; but, without a computer, this would be extremely arduous.

The above three functions have the common feature that they are combinations of simpler functions:

(i) $\sqrt{x^3 - x + 2}$ is the result of starting with the function

$$f(x) = x^3 - x + 2$$

and then applying the function $g(x) = \sqrt{x}$ to the result. Thus,

$$\sqrt{x^3 - x + 2} = g(f(x))$$

(ii) $\sqrt[3]{x+4}$ is the result of starting with the function $F(x) = x + 4$ and then applying the function $G(x) = \sqrt[3]{x}$. Thus,

$$\sqrt[3]{x+4} = G(F(x))$$

(iii) $(x^2 + 3x - 1)^{23}$ is the result of beginning with the function

$$H(x) = x^2 + 3x - 1$$

and then applying the function $K(x) = x^{23}$. Thus,

$$(x^2 + 3x - 1)^{23} = K(H(x))$$

Functions that are put together this way out of simpler functions are called *composite* functions.

Definition: If f and g are any functions, then the *composition* $g \circ f$ of f and g is the function such that

$$(g \circ f)(x) = g(f(x))$$

The "process" of composition is diagramed in Fig. 15-1.

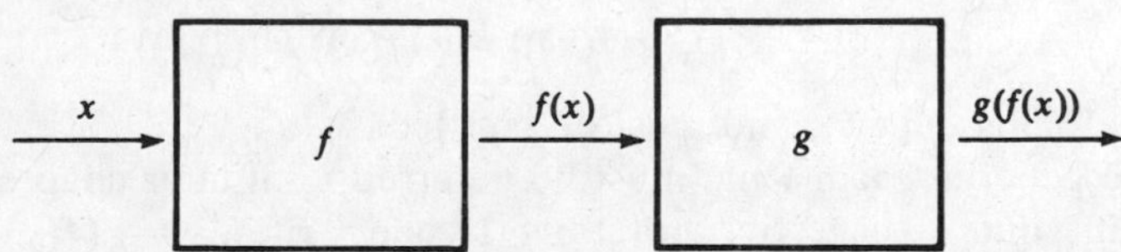

Fig. 15-1

EXAMPLES (*a*) Let $f(x) = x - 1$ and $g(x) = x^2$. Then,

$$(g \circ f)(x) = g(f(x)) = g(x-1) = (x-1)^2$$

On the other hand,

$$(f \circ g)(x) = f(g(x)) = f(x^2) = x^2 - 1$$

Thus, $f \circ g$ and $g \circ f$ are not necessarily the same function (and usually are not the same). (*b*) Let $f(x) = x^2 + 2x$ and $g(x) = \sqrt{x}$. Then,

$$(g \circ f)(x) = g(f(x)) = g(x^2 + 2x) = \sqrt{x^2 + 2x}$$
$$(f \circ g)(x) = f(g(x)) = f(\sqrt{x}) = (\sqrt{x})^2 + 2(\sqrt{x}) = x + 2\sqrt{x}$$

Again, $g \circ f$ and $f \circ g$ are different.

A composite function $g \circ f$ is defined only for those x for which $f(x)$ is defined and $g(f(x))$ is defined. In other words, the domain of $g \circ f$ consists of those x in the domain of f for which $f(x)$ is in the domain of g.

Theorem 15.1: The composition of continuous functions is a continuous function: If f is continuous at a, and g is continuous at $f(a)$, then $g \circ f$ is continuous at a.

For a proof, see Problem 15.25.

15.2 DIFFERENTIATION OF COMPOSITE FUNCTIONS

First, let us treat an important special case. The function $(f(x))^n$ is the composition $g \circ f$ of f and the function $g(x) = x^n$. We have:

Theorem 15.2 (*Power Chain Rule*): Let f be differentiable and let n be any integer; then,

$$D_x((f(x))^n) = n(f(x))^{n-1}\, D_x(f(x)) \tag{15.1}$$

EXAMPLES

(*a*)
$$D_x((x^2-5)^3) = 3(x^2-5)^2\, D_x(x^2-5) = 3(x^2-5)^2(2x) = 6x(x^2-5)^2$$

(*b*)
$$\begin{aligned} D_x((x^3-2x^2+3x-1)^7) &= 7(x^3-2x^2+3x-1)^6\, D_x(x^3-2x^2+3x-1) \\ &= 7(x^3-2x^2+3x-1)^6(3x^2-4x+3) \end{aligned}$$

(*c*)
$$\begin{aligned} D_x\left(\frac{1}{(3x-5)^4}\right) &= D_x((3x-5)^{-4}) = -4(3x-5)^{-5}\, D_x(3x-5) \\ &= -\frac{4}{(3x-5)^5}(3) = -\frac{12}{(3x-5)^5} \end{aligned}$$

Theorem 15.3 (*Chain Rule*): Assume that f is differentiable at x and that g is differentiable at $f(x)$. Then the composition $g \circ f$ is differentiable at x, and its derivative, $(g \circ f)'$, is given by

$$(g \circ f)'(x) = g'(f(x))\, f'(x)$$

that is,

$$D_x(g(f(x))) = g'(f(x))\, D_x(f(x)) \tag{15.2}$$

The proof of Theorem 15.3 is difficult, and must be omitted.

Applications of the general Chain Rule must be deferred until later chapters. Before leaving it, however, we shall point out a suggestive notation. If one writes $y = g(f(x))$ and $u = f(x)$, then $y = g(u)$, and (*15.2*) may be expressed in the form

$$\frac{dy}{dx} = \frac{dy}{du}\frac{du}{dx} \tag{15.3}$$

—just as though derivatives were fractions (which they are not) and as though the Chain Rule were an identity, obtained by cancellation of the du's on the right-hand side. While this "identity" makes for an easy way to remember the Chain Rule, it must be borne in mind that y on the left-hand side of (*15.3*) stands for a certain function of x (i.e., $(g \circ f)(x)$), while on the right-hand side it stands for a *different function* of u (i.e., $g(u)$).

LET U= $(2x-5)^{-4}$

EXAMPLE Using (*15.3*) to rework the preceding Example (*c*), we write

$$y = (3x-5)^{-4} = u^{-4}$$

where $u = 3x - 5$. Then,

$$\frac{dy}{dx} = \frac{dy}{du}\frac{du}{dx} = (-4u^{-5})(3) = \frac{-12}{u^5} = \frac{-12}{(3x-5)^5}$$

Differentiation of Rational Powers

We want to be able to differentiate the function $f(x) = x^r$, where r is a rational number. The special case of r an integer is already covered by Rule 7 of Chapter 13.

ALGEBRA A rational number r is one that can be represented in the form $r = n/k$, where n and k are integers, with k positive. By definition,

$$a^{n/k} = (\sqrt[k]{a})^n$$

except when a is negative and k is even (in which case the kth root of a is undefined). For instance,

$$8^{2/3} = (\sqrt[3]{8})^2 = (2)^2 = 4$$

$$32^{-2/5} = (\sqrt[5]{32})^{-2} = (2)^{-2} = \frac{1}{2^2} = \frac{1}{4}$$

$$(-27)^{4/3} = (\sqrt[3]{-27})^4 = (-3)^4 = 81$$

$$(-4)^{7/8} \text{ is not defined}$$

Observe that

$$(\sqrt[k]{a})^n = \sqrt[k]{a^n}$$

whenever both sides are defined. In fact,

$$((\sqrt[k]{a})^n)^k = (\sqrt[k]{a})^{nk} = (\sqrt[k]{a})^{kn} = ((\sqrt[k]{a})^k)^n = a^n$$

which shows that $(\sqrt[k]{a})^n$ is the kth root of a^n. In calculations, we are free to choose whichever expression for $a^{n/k}$ is the more convenient. Thus: (i) $64^{2/3}$ is easier to compute as

$$(\sqrt[3]{64})^2 = (4)^2 = 16$$

than as

$$\sqrt[3]{(64)^2} = \sqrt[3]{4096}$$

but (ii) $(\sqrt{8})^{2/3}$ is easier to compute as

$$\sqrt[3]{(\sqrt{8})^2} = \sqrt[3]{8} = 2$$

than as

$$(\sqrt[3]{\sqrt{8}})^2$$

The usual laws of exponents hold for rational exponents:

(1) $a^r \cdot a^s = a^{r+s}$

(2) $\dfrac{a^r}{a^s} = a^{r-s}$

(3) $(a^r)^s = a^{rs}$

(4) $(ab)^r = a^r b^r$

where r and s are any rational numbers.

Theorem 15.4: For any rational number r, $D_x(x^r) = rx^{r-1}$.

For a proof, see Problem 15.6.

EXAMPLES

$$(a)\quad D_x(\sqrt{x}) = D_x(x^{1/2}) = \frac{1}{2}x^{-1/2} = \frac{1}{2}\frac{1}{x^{1/2}} = \frac{1}{2\sqrt{x}}$$

$$(b)\quad D_x(x^{3/2}) = \frac{3}{2}x^{1/2} = \frac{3}{2}\sqrt{x} \qquad (c)\quad D_x(x^{3/4}) = \frac{3}{4}x^{-1/4} = \frac{3}{4}\frac{1}{x^{1/4}} = \frac{3}{4\sqrt[4]{x}}$$

Theorem 15.4, together with the Chain Rule (Theorem 15.3), allows us to extend the Power Chain Rule (Theorem 15.2) to rational exponents.

Corollary 15.5: If f is differentiable, (*15.1*) holds for rational n.

EXAMPLES

(*a*)
$$D_x(\sqrt{x^2-3x+1}) = D_x((x^2-3x+1)^{1/2}) = \frac{1}{2}(x^2-3x+1)^{-1/2}D_x(x^2-3x+1)$$
$$= \frac{1}{2}\frac{1}{(x^2-3x+1)^{1/2}}(2x-3) = \frac{2x-3}{2\sqrt{x^2-3x+1}}$$

(*b*)
$$D_x((3x^2-1)^{5/4}) = \frac{5}{4}(3x^2-1)^{1/4}D_x(3x^2-1)$$
$$= \frac{5}{4}\sqrt[4]{3x^2-1}\,(6x) = \frac{15x}{2}\sqrt[4]{3x^2-1}$$

(*c*)
$$D_x\left(\frac{1}{\sqrt[3]{7x+2}}\right) = D_x\left(\frac{1}{(7x+2)^{1/3}}\right) = D_x((7x+2)^{-1/3}) = -\frac{1}{3}(7x+2)^{-4/3}D_x(7x+2)$$
$$= -\frac{1}{3}\frac{1}{(7x+2)^{4/3}}(7) = -\frac{7}{3(\sqrt[3]{7x+2})^4}$$

Solved Problems

15.1 For each pair of functions f and g, find formulas for $g \circ f$ and $f \circ g$, and determine the domains of $g \circ f$ and $f \circ g$.

$$(a)\quad g(x) = \sqrt{x} \text{ and } f(x) = x + 1 \qquad (b)\quad g(x) = x^2 \text{ and } f(x) = x - 1$$

(*a*)
$$(g \circ f)(x) = g(f(x)) = g(x+1) = \sqrt{x+1}$$

Because $\sqrt{x+1}$ is defined when and only when $x \ge -1$, the domain of $g \circ f$ is $[-1, \infty)$.

$$(f \circ g)(x) = f(g(x)) = f(\sqrt{x}) = \sqrt{x} + 1$$

Because $\sqrt{x} + 1$ is defined when and only when $x \ge 0$, the domain of $f \circ g$ is $[0, \infty)$.

(*b*)
$$(g \circ f)(x) = g(f(x)) = g(x-1) = (x-1)^2$$
$$(f \circ g)(x) = f(g(x)) = f(x^2) = x^2 - 1$$

Both composite functions are polynomials, and so the domain of each is the set of all real numbers.

15.2 Calculate the derivatives of

$$(a)\quad (x^4 - 3x^2 + 5x - 2)^3 \qquad (b)\quad \sqrt{7x^3 - 2x^2 + 5} \qquad (c)\quad \frac{1}{(5x^2+4)^3}$$

The Power Chain Rule is used in each case.

(*a*)
$$D_x((x^4-3x^2+5x-2)^3) = 3(x^4-3x^2+5x-2)^2D_x(x^4-3x^2+5x-2)$$
$$= 3(x^4-3x^2+5x-2)^2(4x^3-6x+5)$$

(*b*) $$D_x(\sqrt{7x^3-2x^2+5}) = D_x((7x^3-2x^2+5)^{1/2}) = \frac{1}{2}(7x^3-2x^2+5)^{-1/2}D_x(7x^3-2x^2+5)$$
$$= \frac{1}{2}\frac{1}{(7x^3-2x^2+5)^{1/2}}(21x^2-4x) = \frac{x(21x-4)}{2\sqrt{7x^3-2x^2+5}}$$

(*c*) $$D_x\left(\frac{1}{(5x^2+4)^3}\right) = D_x((5x^2+4)^{-3}) = -3(5x^2+4)^{-4}D_x(5x^2+4)$$
$$= \frac{-3}{(5x^2+4)^4}(10x) = -\frac{30x}{(5x^2+4)^4}$$

15.3 Find the derivative of the function

$$f(x) = \sqrt{1+\sqrt{x+1}} = (1+(x+1)^{1/2})^{1/2}$$

By the Power Chain Rule, used twice,

$$f'(x) = \frac{1}{2}(1+(x+1)^{1/2})^{-1/2}D_x(1+(x+1)^{1/2})$$
$$= \frac{1}{2}(1+(x+1)^{1/2})^{-1/2}\left(\frac{1}{2}(x+1)^{-1/2}D_x(x+1)\right)$$
$$= \frac{1}{4}((x+1)+(x+1)^{1/2}(x+1))^{-1/2}(1) \quad [\text{by } (ab)^r = a^r b^r]$$
$$= \frac{1}{4}(x+1+(x+1)^{3/2})^{-1/2} \quad [\text{by } a^r a^s = a^{r+s}]$$

15.4 Find the absolute extrema of $f(x) = x\sqrt{1-x^2}$ on $[0, 1]$.

$$D_x(x\sqrt{1-x^2}) = xD_x(\sqrt{1-x^2}) + \sqrt{1-x^2}\,D_x(x) \quad [\text{Product Rule}]$$
$$= xD_x((1-x^2)^{1/2}) + \sqrt{1-x^2}$$
$$= x\left(\frac{1}{2}(1-x^2)^{-1/2}D_x(1-x^2)\right) + \sqrt{1-x^2} \quad [\text{Power Chain Rule}]$$
$$= \frac{x}{2}\frac{1}{(1-x^2)^{1/2}}(-2x) + \sqrt{1-x^2} = \frac{-x^2}{\sqrt{1-x^2}} + \sqrt{1-x^2}$$
$$= \frac{-x^2+(1-x^2)}{\sqrt{1-x^2}} = \frac{1-2x^2}{\sqrt{1-x^2}} \quad \left[\text{by } \frac{a}{c}+b = \frac{a+bc}{c}\right]$$

The right-hand side is not defined when the denominator is 0; that is, when $x^2 = 1$. Hence, 1 and -1 are critical numbers. The right-hand side is 0 when the numerator is 0; that is, when

$$2x^2 = 1 \qquad \text{or} \qquad x^2 = \tfrac{1}{2} \qquad \text{or} \qquad x = \pm\sqrt{\tfrac{1}{2}}$$

Thus, $\sqrt{\frac{1}{2}}$ and $-\sqrt{\frac{1}{2}}$ are also critical numbers. The only critical number in $(0, 1)$ is $\sqrt{\frac{1}{2}}$,

---

ALGEBRA $$\sqrt{\frac{1}{2}} = \sqrt{\frac{2}{4}} = \frac{\sqrt{2}}{2} \approx 0.707$$

---

and $$f(\sqrt{\tfrac{1}{2}}) = \sqrt{\tfrac{1}{2}}\sqrt{1-\tfrac{1}{2}} = \sqrt{\tfrac{1}{2}}\sqrt{\tfrac{1}{2}} = \tfrac{1}{2}$$

At the endpoints, $f(0) = f(1) = 0$. Hence, $\frac{1}{2}$ is the absolute maximum (achieved at $x = \sqrt{\frac{1}{2}}$) and 0 is the absolute minimum (achieved at $x = 0$ and $x = 1$).

15.5 A spy on a submarine S, 6 kilometers off a straight shore, has to reach a point B which is 9 kilometers down the shore from the point A opposite S (Fig. 15-2). The spy must row a boat to some point C on the shore and then walk the rest of the way to B. If he rows at 4 kilometers per hour and walks at 5 kilometers per hour, at what point C should he land in order to reach B as soon as possible?

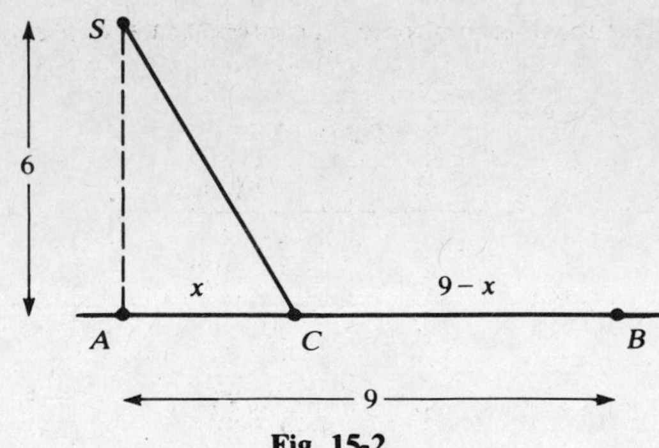

Fig. 15-2

Let $x = \overline{AC}$; then $\overline{BC} = 9 - x$. By the Pythagorean Theorem,

$$\overline{SC}^2 = (6)^2 + x^2 \qquad \text{or} \qquad \overline{SC} = \sqrt{36 + x^2}$$

The time spent rowing is, therefore,

$$T_1 = \frac{\sqrt{36 + x^2}}{4} \quad \text{(hours)}$$

and the time spent walking will be $T_2 = (9 - x)/5$ (hours). The total time $T(x)$ is given by the formula

$$T(x) = T_1 + T_2 = \frac{\sqrt{36 + x^2}}{4} + \frac{9 - x}{5}$$

We have to minimize $T(x)$ on the interval $[0, 9]$, since x can vary from 0 (at A) to 9 (at B).

$$\begin{aligned} T'(x) &= D_x\left(\frac{(36 + x^2)^{1/2}}{4}\right) + D_x\left(\frac{9}{5} - \frac{x}{5}\right) \\ &= \frac{1}{4}\cdot\frac{1}{2}(36 + x^2)^{-1/2}\cdot D_x(36 + x^2) - \frac{1}{5} \quad \text{[Power Chain Rule]} \\ &= \frac{1}{8}\frac{1}{(36 + x^2)^{1/2}}(2x) - \frac{1}{5} = \frac{x}{4\sqrt{36 + x^2}} - \frac{1}{5} \end{aligned}$$

The only critical numbers are the solutions of:

$$\begin{aligned} \frac{x}{4\sqrt{36 + x^2}} - \frac{1}{5} &= 0 \\ \frac{x}{4\sqrt{36 + x^2}} &= \frac{1}{5} \\ 5x &= 4\sqrt{36 + x^2} \quad \text{[cross multiply]} \\ 25x^2 &= 16(36 + x^2) \quad \text{[square both sides]} \\ 25x^2 &= 576 + 16x^2 \\ 9x^2 &= 576 \\ x^2 &= 64 \\ x &= \pm 8 \end{aligned}$$

The only critical number in $(0, 9)$ is 8. Computing the values for the tabular method,

$$T(8) = \frac{\sqrt{36 + (8)^2}}{4} + \frac{9 - 8}{5} = \frac{\sqrt{100}}{4} + \frac{1}{5} = \frac{10}{4} + \frac{1}{5} = \frac{5}{2} + \frac{1}{5} = \frac{27}{10}$$

$$T(0) = \frac{\sqrt{36 + (0)^2}}{4} + \frac{9 - 0}{5} = \frac{\sqrt{36}}{4} + \frac{9}{5} = \frac{6}{4} + \frac{9}{5} = \frac{3}{2} + \frac{9}{5} = \frac{33}{10}$$

$$T(9) = \frac{\sqrt{36 + (9)^2}}{4} + \frac{9 - 9}{5} = \frac{\sqrt{36 + 81}}{4} + 0 = \frac{\sqrt{117}}{4} = \frac{\sqrt{9}\sqrt{13}}{4} = \frac{3}{4}\sqrt{13}$$

we generate Table 15-1. The absolute minimum is achieved at $x = 8$; the spy should land 8 kilometers down the shore from A.

Table 15-1

x	$T(x)$
8	$\frac{27}{10}$ min
0	$\frac{33}{10}$
9	$\frac{3}{4}\sqrt{13}$

ALGEBRA $T(9) > T(8)$; for, assuming the contrary,

$$\frac{3}{4}\sqrt{13} \le \frac{27}{10}$$

$$\sqrt{13} \le \frac{36}{10} \quad \text{[multiply by 4/3]}$$

$$13 \le \frac{1296}{100} = 12.96 \quad \text{[by squaring]}$$

which is false.

15.6 Prove Theorem 15.4: $D_x(x^r) = rx^{r-1}$ for any rational number r.

Let $r = n/k$, where n is an integer and k is a positive integer. That $x^{n/k}$ is differentiable is not easy to prove; see Problem 15.26. Assuming this, let us now derive the formula for the derivative. Let

$$f(x) = x^{n/k} = \sqrt[k]{x^n}$$

Then, since $(f(x))^k = x^n$,

$$D_x((f(x))^k) = D_x(x^n) = nx^{n-1}$$

But, by Theorem 15.2, $D_x((f(x))^k) = k(f(x))^{k-1}f'(x)$. Hence,

$$k(f(x))^{k-1}f'(x) = nx^{n-1}$$

and solving for $f'(x)$, we obtain

$$f'(x) = \frac{nx^{n-1}}{k(f(x))^{k-1}} = \frac{n}{k}\frac{x^{rk-1}}{(x^r)^{k-1}}$$

$$= r\frac{x^{rk-1}}{x^{rk-r}} = rx^{r-1}$$

Supplementary Problems

15.7 For each pair of functions $f(x)$ and $g(x)$, find formulas for $(f \circ g)(x)$ and $(g \circ f)(x)$.

(a) $f(x) = \frac{2}{x+1}$, $g(x) = 3x$ (b) $f(x) = x^2 + 2x - 5$, $g(x) = x^3$

(c) $f(x) = 2x^3 - x^2 + 4$, $g(x) = 3$ (d) $f(x) = x^3$, $g(x) = x^2$

(e) $f(x) = \frac{1}{x}$, $g(x) = \frac{1}{x}$ (f) $f(x) = x$, $g(x) = x^2 - 4$

15.8 For each pair of functions f and g, find the set of solutions of the equation $(f \circ g)(x) = (g \circ f)(x)$.

(a) $f(x) = x^3,\ g(x) = x^2$ (b) $f(x) = \dfrac{2}{x+1},\ g(x) = 3x$

(c) $f(x) = 2x,\ g(x) = \dfrac{1}{x-1}$ (d) $f(x) = x^2,\ g(x) = \dfrac{1}{x+1}$

(e) $f(x) = x^2,\ g(x) = \dfrac{1}{x^2-3}$

15.9 Express each of the following functions as the composition $(g \circ f)(x)$ of two simpler functions. (The functions $f(x)$ and $g(x)$ obviously will not be unique.)

(a) $(x^3 - x^2 + 2)^7$ (b) $(8-x)^4$ (c) $\sqrt{1+x^2}$ (d) $\dfrac{1}{x^2-4}$

15.10 Find the derivatives of the following functions:

(a) $(x^3 - 2x^2 + 7x - 3)^4$ (b) $(7+3x)^5$ (c) $(2x-3)^{-2}$

(d) $(3x^2+5)^{-3}$ (e) $(4x^2-3)^2(x+5)^3$ (f) $\left(\dfrac{x+2}{x-3}\right)^3$

(g) $\left(\dfrac{x^2-2}{2x^2+1}\right)^2$ (h) $\dfrac{4}{3x^2-x+5}$

15.11 Find the derivatives of the following functions:

(a) $2x^{3/4}$ (b) $x^2(1-3x^3)^{1/3}$ (c) $\dfrac{x}{\sqrt{x^2+1}}$

(d) $(7x^3-4x^2+2)^{1/4}$ (e) $\dfrac{\sqrt{x+2}}{\sqrt{x-1}}$ (f) $8x^{3/4} + 4x^{1/4} - x^{-1/3}$

(g) $\sqrt[3]{(4x^2+3)^2}$ (h) $\sqrt{\dfrac{4}{x}} - \sqrt{3x}$ (i) $\sqrt{4-\sqrt{4+x}}$

15.12 Find the slope-intercept equation of the tangent line to the graph of

$$y = \frac{\sqrt{x-1}}{x^2+1}$$

at the point $(2, \frac{1}{5})$.

15.13 Find the slope-intercept equation of the normal line to the curve

$$y = \sqrt{x^2+16}$$

at the point (3, 5).

15.14 Let $$g(x) = x^2 - 4 \quad \text{and} \quad f(x) = \frac{x+2}{x-2}$$

(a) Find a formula for $(g \circ f)(x)$, and then compute $(g \circ f)'(x)$. (b) Show that the Chain Rule gives the same answer for $(g \circ f)'(x)$ as was found in (a).

15.15 Find the absolute extrema of

(a) $f(x) = \dfrac{x}{\sqrt{1+x^2}}$ on $[-1, 1]$ (b) $f(x) = (x-2)^2(x+3)^3$ on $[-4, 3]$

(c) $f(x) = \sqrt{5-4x}$ on $[-1, 1]$

15.16 Two towns, P and Q, are located 2 miles and 3 miles, respectively, from a railroad line, as shown in Fig. 15-3. What point R on the line should be chosen for a new station in order to minimize the sum of the distances from P and Q to the station, if the distance between A and B is 4 miles?

15.17 Assume that F and G are differentiable functions such that $F'(x) = -G(x)$ and $G'(x) = -F(x)$. If $H(x) = (F(x))^2 - (G(x))^2$, find a formula for $H'(x)$.

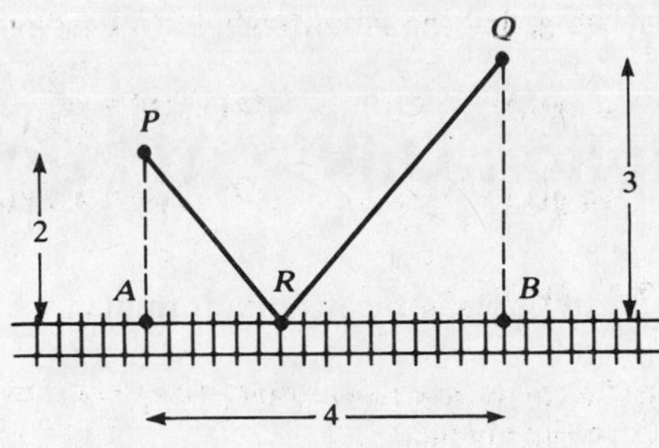

Fig. 15-3

15.18 If $y = x^3 - 2$ and $z = 3x + 5$, then y can be considered a function of z. Express $\dfrac{dy}{dz}$ in terms of x.

15.19 Let F be a differentiable function, and let $G(x) = F'(x)$. Express $D_x(F(x^3))$ in terms of G and x.

15.20 If $g(x) = x^{1/5}(x-1)^{3/5}$, find the domain of $g'(x)$.

15.21 Let f be a differentiable odd function (Section 7.3). Find the relationship between $f'(-x)$ and $f'(x)$.

15.22 Let F and G be differentiable functions such that

$$\begin{array}{lll} F(3) = 5 & F'(3) = 13 & F'(7) = 2 \\ G(3) = 7 & G'(3) = 6 & G'(7) = 0 \end{array}$$

If $H(x) = F(G(x))$, find $H'(3)$.

15.23 Let $F(x) = \sqrt{1+3x}$. (*a*) Find the domain and range of F. (*b*) Find the slope-intercept equation of the tangent line to the graph of F at $x = 5$. (*c*) Find the coordinates of the point(s) on the graph of F such that the normal line there is parallel to the line $4x + 3y = 1$.

15.24 Find the dimensions of the rectangle of largest area that can be inscribed in a semicircle of radius 1, if a side of the rectangle is on the diameter.

15.25 If f is continuous at a and g is continuous at $f(a)$, prove that $g \circ f$ is continuous at a. (*Hint*: For arbitrary $\epsilon > 0$, let $\delta_1 > 0$ be such that

$$|g(u) - g(f(a))| < \epsilon$$

whenever $|u - f(a)| < \delta_1$. Then choose $\delta > 0$ such that

$$|f(x) - f(a)| < \delta_1$$

whenever $|x - a| < \delta$.)

15.26 Prove that $x^{n/k}$ is differentiable. (*Hint*: It is enough to show that $f(x) = x^{1/k}$ $(k > 1)$ is differentiable.) Proceed as follows:

(1) By direct multiplication, establish that

$$a^k - b^k = (a-b)(a^{k-1} + a^{k-2}b + \cdots + ab^{k-2} + b^{k-1})$$

or

$$\frac{a-b}{a^k - b^k} = \frac{1}{a^{k-1} + a^{k-2}b + \cdots + ab^{k-2} + b^{k-1}}$$

(2) Substitute $a = (x+h)^{1/k}$ and $b = x^{1/k}$ in (1):

$$\frac{(x+h)^{1/k} - x^{1/k}}{h} = \frac{1}{(x+h)^{(k-1)/k} + (x+h)^{(k-2)/k}x^{1/k} + \cdots + (x+h)^{1/k}x^{(k-2)/k} + x^{(k-1)/k}}$$

(3) Let $h \to 0$ in (2).

Chapter 16

Implicit Differentiation

A function is usually defined *explicitly*, by means of a formula.

EXAMPLES

$$(a)\ \ f(x) = x^2 - x + 2 \qquad (b)\ \ f(x) = \sqrt{x} \qquad (c)\ \ f(x) = \begin{cases} x^2 - 1 & \text{if } x \geq 1 \\ 1 - x & \text{if } x < 1 \end{cases}$$

However, sometimes the value $y = f(x)$ is not given by such a direct formula.

EXAMPLES (*a*) The equation $y^3 - x = 0$ *implicitly* determines y as a function of x. In this case, we can solve for y explicitly:

$$y^3 = x \qquad \text{or} \qquad y = \sqrt[3]{x}$$

(*b*) The equation

$$y^3 + 12y^2 + 48y - 8x + 64 = 0$$

is satisfied when $y = 2\sqrt[3]{x} - 4$, but it is not easy to find this solution. In more complicated cases, it will be impossible to find a formula for y in terms of x. (*c*) The equation $x^2 + y^2 = 1$ implicitly determines *two* functions of x,

$$y = \sqrt{1 - x^2} \qquad \text{and} \qquad y = -\sqrt{1 - x^2}$$

The question of how many functions an equation determines and of the properties of these functions is too complex to be considered here. We shall content ourselves with learning a method for finding the derivatives of functions determined implicitly by equations.

EXAMPLE Let us find the derivative of a function y determined by the equation $x^2 + y^2 = 4$. Since y is assumed to be some function of x, the two sides of the equation represent the same function of x, and so must have the same derivative:

$$D_x(x^2 + y^2) = D_x(4)$$

$$2x + 2y\,D_xy = 0 \quad \text{[by the Power Chain Rule]}$$

$$2y\,D_xy = -2x$$

$$D_xy = -\frac{2x}{2y} = -\frac{x}{y}$$

Thus, D_xy has been found in terms of x and y. Sometimes, this is all the information we may need. For example, if we want to know the slope of the tangent line to the graph of $x^2 + y^2 = 4$ at the point $(\sqrt{3}, 1)$, then this slope is the derivative

$$D_xy = -\frac{x}{y} = -\frac{\sqrt{3}}{1} = -\sqrt{3}$$

The process by which D_xy has been found, without first solving explicitly for y, is called *implicit differentiation.*

Note that the given equation could, in this case, have been solved explicitly for y:

$$y = \pm\sqrt{4 - x^2}$$

and from this, using the Power Chain Rule,

$$D_xy = D_x(\pm(4 - x^2)^{1/2}) = \pm\tfrac{1}{2}(4 - x^2)^{-1/2}\,D_x(4 - x^2)$$

$$= \pm\frac{1}{2\sqrt{4 - x^2}}(-2x) = \pm\frac{-x}{\sqrt{4 - x^2}} = \frac{x}{\mp\sqrt{4 - x^2}}$$

Solved Problems

16.1 Consider the curve $3x^2 - xy + 4y^2 = 141$. (*a*) Find a formula in x and y for the slope of the tangent line at any point (x, y) of the curve. (*b*) Write the slope-intercept equation of the line tangent to the curve at the point (1, 6). (*c*) Find the coordinates of all other points on the curve where the slope of the tangent line is the same as the slope of the tangent line at (1, 6).

(*a*) We may assume that y is some function of x such that

$$3x^2 - xy + 4y^2 = 141$$

Hence,

$$D_x(3x^2 - xy + 4y^2) = D_x(141)$$
$$6x - D_x(xy) + D_x(4y^2) = 0$$
$$6x - \left(x\frac{dy}{dx} + y \cdot 1\right) + 8y\frac{dy}{dx} = 0$$
$$-x\frac{dy}{dx} + 8y\frac{dy}{dx} = y - 6x$$
$$(-x + 8y)\frac{dy}{dx} = y - 6x$$
$$\frac{dy}{dx} = \frac{y - 6x}{8y - x}$$

which is the slope of the tangent line at (x, y).

(*b*) The slope of the tangent line at (1, 6) is obtained by substituting 1 for x and 6 for y in the result of (*a*). Thus, the slope is

$$\frac{6 - 6(1)}{8(6) - 1} = \frac{6 - 6}{48 - 1} = \frac{0}{47} = 0$$

and the slope-intercept equation is $y = b = 6$.

(*c*) If (x, y) is a point on the curve where the tangent line has slope 0, then,

$$\frac{y - 6x}{8y - x} = 0 \qquad \text{or} \qquad y - 6x = 0 \qquad \text{or} \qquad y = 6x$$

Substitute $6x$ for y in the equation of the curve:

$$3x^2 - x(6x) + 4(6x)^2 = 141$$
$$3x^2 - 6x^2 + 144x^2 = 141$$
$$141x^2 = 141$$
$$x^2 = 1$$
$$x = \pm 1$$

Hence, $(-1, -6)$ is another point for which the slope of the tangent line is zero.

16.2 If $y = f(x)$ is a function satisfying the equation

$$x^3y^2 - 2x + y^3 = 36$$

find a formula for the derivative $\dfrac{dy}{dx}$.

$$D_x(x^3y^2 - 2x + y^3) = D_x(36)$$
$$D_x(x^3y^2) - 2D_x(x) + D_x(y^3) = 0$$
$$x^3D_x(y^2) + y^2D_x(x^3) - 2(1) + 3y^2\frac{dy}{dx} = 0$$
$$x^3\left(2y\frac{dy}{dx}\right) + y^2(3x^2) - 2 + 3y^2\frac{dy}{dx} = 0$$

$$(2x^3y + 3y^2)\frac{dy}{dx} = 2 - 3x^2y^2$$

$$\frac{dy}{dx} = \frac{2 - 3x^2y^2}{2x^3y + 3y^2}$$

16.3 If $y = f(x)$ is a differentiable function satisfying the equation $x^2y^3 - 5xy^2 - 4y = 4$ and if $f(3) = 2$, find the slope of the tangent line to the graph of f at the point (3, 2).

$$D_x(x^2y^3 - 5xy^2 - 4y) = D_x(4)$$

$$x^2(3y^2y') + y^3(2x) - 5(x(2yy') + y^2) - 4y' = 0$$

$$3x^2y^2y' + 2xy^3 - 10xyy' - 5y^2 - 4y' = 0$$

Substitute 3 for x and 2 for y:

$$108y' + 48 - 60y' - 20 - 4y' = 0$$

$$44y' + 28 = 0$$

$$y' = -\frac{28}{44} = -\frac{7}{11}$$

Hence, the slope of the tangent line at (3, 2) is $-7/11$.

Supplementary Problems

16.4 (*a*) Find a formula for the slope of the tangent line to the curve $x^2 - xy + y^2 = 12$ at any point (x, y). Also, find the coordinates of all points on the curve where the tangent line is (*b*) horizontal, (*c*) vertical.

16.5 Consider the hyperbola $5x^2 - 2y^2 = 130$ (*a*) Find a formula for the slope of the tangent line to this hyperbola at (x, y). (*b*) For what value(s) of k will the line $x - 3y + k = 0$ be normal to the hyperbola at a point of intersection?

16.6 Find y' by implicit differentiation.

(*a*) $x^2 + y^2 = 25$ (*b*) $x^3 = \frac{2x + y}{2x - y}$ (*c*) $\frac{1}{x} + \frac{1}{y} = 1$

(*d*) $\sqrt{x} + \sqrt{y} = 1$ (*e*) $x^3 - y^3 = 2xy$ (*f*) $(7x - 1)^3 = 2y^4$

(*g*) $\frac{x^2}{9} + \frac{y^2}{4} = 1$

16.7 Use implicit differentiation to find the slope-intercept equation of the tangent line at the indicated point.

(*a*) $y^3 - xy = 2$, at (3, 2) (*b*) $\frac{x^2}{16} + y^2 = 1$, at $(2, \sqrt{3}/2)$

(*c*) $(y - x)^2 + y^3 = xy + 7$, at (1, 2) (*d*) $x^3 - y^3 = 7xy$, at (4, 2)

(*e*) $4xy^2 + 98 = 2x^4 - y^4$, at (3, 2)

16.8 Use implicit differentiation to find the slope-intercept equation of the normal line at the indicated point.

(*a*) $y^3x + 2y = x^2$, at (2, 1) (*b*) $2x^3y + 2y^4 - x^4 = 2$, at (2, 1)

(*c*) $y\sqrt{x} - x\sqrt{y} = 12$, at (9, 16) (*d*) $x^2 + y^2 = 25$, at (3, 4)

16.9 Use implicit differentiation to find the slope of the tangent line to the graph of

$$y = \sqrt{1 - \sqrt{1 - x}}$$

at $x = 7/16$. (*Hint*: First eliminate the radicals by squaring twice.)

Chapter 17

The Mean-Value Theorem. The Sign of the Derivative

17.1 ROLLE'S THEOREM AND THE MEAN-VALUE THEOREM

Let us consider a function f that is continuous over a closed interval $[a, b]$ and differentiable at every point of the open interval (a, b). We also suppose that $f(a) = f(b) = 0$. Graphs of some examples of such a function are shown in Fig. 17-1. It seems clear that there must always be some point between $x = a$ and $x = b$ at which the tangent line is horizontal, and, therefore, at which the derivative of f is 0.

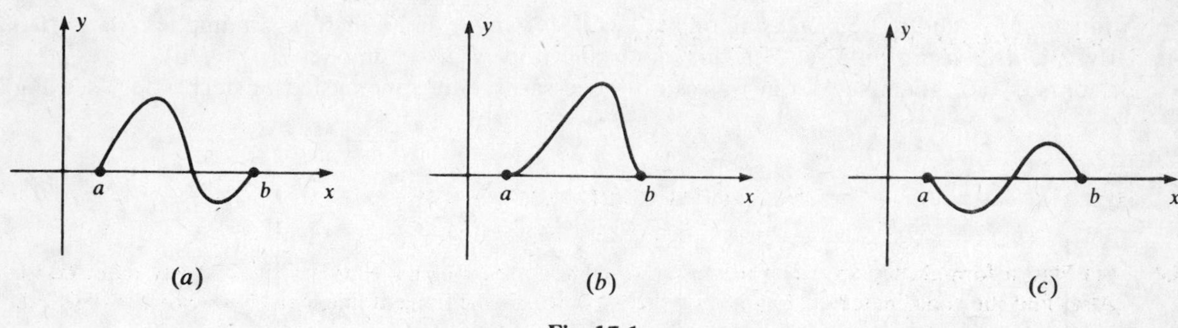

Fig. 17-1

Theorem 17.1 (*Rolle's Theorem*): If f is continuous over a closed interval $[a, b]$, differentiable on the open interval (a, b), and if $f(a) = f(b) = 0$, then there is at least one number c in (a, b) such that $f'(c) = 0$.

See Problem 17.6 for the proof.

Rolle's Theorem enables us to prove the following basic theorem (which is also referred to as the *Law of the Mean for Derivatives*).

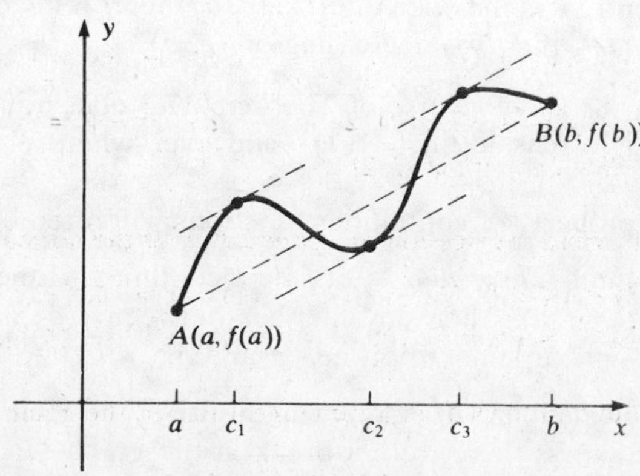

Fig. 17-2

Theorem 17.2 (*Mean-Value Theorem*): Let f be continuous over the closed interval $[a, b]$ and differentiable on the open interval (a, b). Then there is a number c in the open interval (a, b) such that

$$f'(c) = \frac{f(b) - f(a)}{b - a}$$

For a proof, see Problem 17.7.

EXAMPLE In graphical terms, the Mean-Value Theorem states that at some point along an arc of a curve, the tangent is parallel to the chord. This can be seen in Fig. 17-2, where there are three numbers (c_1, c_2, and c_3) between a and b for which the slope of the tangent line to the graph, $f'(c)$, is equal to the slope of the chord AB,

$$\frac{f(b) - f(a)}{b - a}$$

17.2 THE SIGN OF THE DERIVATIVE

A function f is said to be *increasing* on a set $\mathscr{A}$ if, for any u and v in $\mathscr{A}$, $u < v$ implies $f(u) < f(v)$. Similarly, f is *decreasing* on a set $\mathscr{A}$ if, for any u and v in $\mathscr{A}$, $u < v$ implies $f(u) > f(v)$.

Of course, on a given set, a function is not necessarily either increasing or decreasing; see Fig. 17-3.

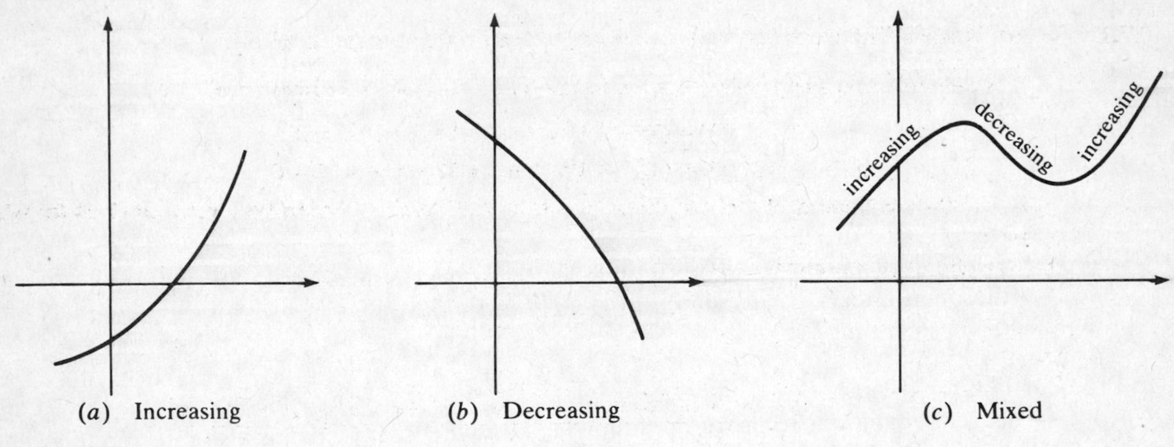

(*a*) Increasing (*b*) Decreasing (*c*) Mixed

Fig. 17-3

Theorem 17.3: If $f'(x) > 0$ for all x in the open interval (a, b) then f is increasing on (a, b). If $f'(x) < 0$ for all x in (a, b), then f is decreasing on (a, b).

For the proof, see Problem 17.8. The converse of Theorem 17.3 does not hold. In fact, the function $f(x) = x^3$ is differentiable and increasing on $(-1, 1)$—and everywhere else—but $f'(x) = 3x^2$ is zero for $x = 0$ (see the graph, Fig. 7-3(*b*)).

The following important property of continuous functions will often be useful.

Theorem 17.4 (*Intermediate-Value Theorem*): Let f be a continuous function over a closed interval $[a, b]$, with $f(a) \neq f(b)$. Then any number between $f(a)$ and $f(b)$ is assumed as the value of f for some argument between a and b.

While Theorem 17.4 is not elementary, its content is intuitively obvious: The function could not "skip" an intermediate value unless there were a break in the graph—i.e., unless the function were discontinuous. As illustrated in Fig. 17-4, a function f satisfying Theorem 17.4 may also take on values that are not between $f(a)$ and $f(b)$. In Problem 17.9, we prove

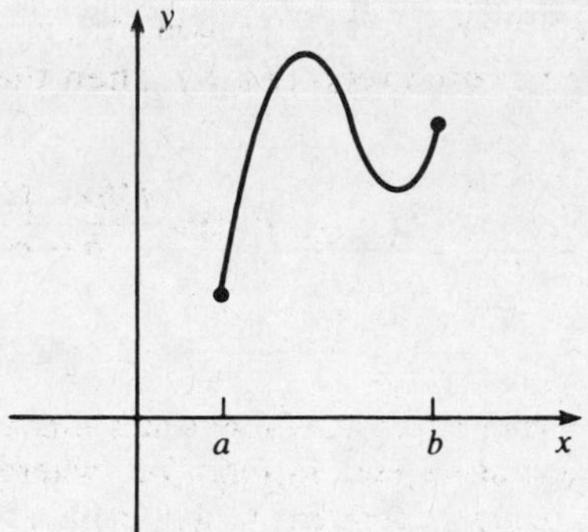

Fig. 17-4

Corollary 17.5: If f is a continuous function with domain $[a, b]$, then the range of f is either a closed interval or a point.

Solved Problems

17.1 Verify Rolle's Theorem for $f(x) = x^3 - 3x^2 - x + 3$ on the interval $[1, 3]$.

f is differentiable everywhere, and, therefore, also continuous. Furthermore,

$$f(1) = (1)^3 - 3(1)^2 - 1 + 3 = 1 - 3 - 1 + 3 = 0$$
$$f(3) = (3)^3 - 3(3)^2 - 3 + 3 = 27 - 27 - 3 + 3 = 0$$

so that all the hypotheses of Rolle's Theorem are valid. There must then be some c in $(1, 3)$ for which $f'(c) = 0$.

Now, by the quadratic formula, the roots of

$$f'(x) = 3x^2 - 6x - 1 = 0$$

are

$$x = \frac{6 \pm \sqrt{6^2 - 4(3)(-1)}}{2(3)} = \frac{6 \pm \sqrt{36 + 12}}{6} = \frac{6 \pm \sqrt{48}}{6} = \frac{6 \pm \sqrt{16}\sqrt{3}}{6}$$
$$= \frac{6 \pm 4\sqrt{3}}{6} = 1 \pm \frac{2}{3}\sqrt{3}$$

Consider the root $c = 1 + \frac{2}{3}\sqrt{3}$. Since $\sqrt{3} < 3$,

$$1 < 1 + \frac{2}{3}\sqrt{3} < 1 + \frac{2}{3}(3) = 1 + 2 = 3$$

Thus, c is in $(1, 3)$ and $f'(c) = 0$.

17.2 Verify the Mean-Value Theorem for $f(x) = x^3 - 6x^2 - 4x + 30$ on the interval $[4, 6]$.

f is differentiable, and therefore continuous, for all x.

$$f(6) = (6)^3 - 6(6)^2 - 4(6) + 30 = 216 - 216 - 24 + 30 = 6$$
$$f(4) = (4)^3 - 6(4)^2 - 4(4) + 30 = 64 - 96 - 16 + 30 = -18$$

whence
$$\frac{f(6) - f(4)}{6 - 4} = \frac{6 - (-18)}{6 - 4} = \frac{24}{2} = 12$$

We must therefore find some c in $(4, 6)$ such that $f'(c) = 12$.

Now, $f'(x) = 3x^2 - 12x - 4$, so that c will be a solution of

$$3x^2 - 12x - 4 = 12 \quad \text{or} \quad 3x^2 - 12x - 16 = 0$$

By the quadratic formula,

$$x = \frac{12 \pm \sqrt{144 - 4(3)(-16)}}{2(3)} = \frac{12 \pm \sqrt{144 + 192}}{6} = \frac{12 \pm \sqrt{336}}{6}$$
$$= \frac{12 \pm \sqrt{16}\sqrt{21}}{6} = \frac{12 \pm 4\sqrt{21}}{6} = 2 \pm \tfrac{2}{3}\sqrt{21}$$

Choose $c = 2 + \frac{2}{3}\sqrt{21}$. Since $4 < \sqrt{21} < 5$,

$$4 < 2 + \frac{8}{3} = 2 + \frac{2}{3}(4) < 2 + \frac{2}{3}\sqrt{21} < 2 + \frac{2}{3}(5) < 2 + 4 = 6$$

Thus, c is in $(4, 6)$ and $f'(c) = 12$.

17.3 Determine when the function $f(x) = x^3 - 6x^2 + 9x + 2$ is increasing and when it is decreasing.

We have:

$$f'(x) = 3x^2 - 12x + 9 = 3(x^2 - 4x + 3) = 3(x-1)(x-3)$$

When $x < 1$, $x - 1 < 0$ and $x - 3 < 0$. Hence, $f'(x) > 0$, since the product of two negative numbers is positive.

When $1 < x < 3$, $x - 1 > 0$ and $x - 3 < 0$. Hence, $f'(x) < 0$, since the product of a negative and a positive number is negative.

When $x > 3$, $x - 1 > 0$ and $x - 3 > 0$. Hence, $f'(x) > 0$.

Thus, by Theorem 17.3, f is increasing on $(-\infty, 1)$, decreasing on $(1, 3)$, and increasing on $(3, \infty)$. See Fig. 17-5.

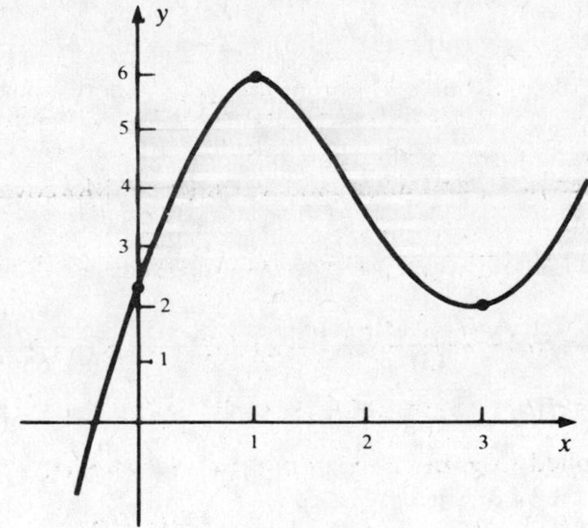

Fig. 17-5

17.4 Verify Rolle's Theorem for

$$f(x) = 2x^6 - 8x^5 + 6x^4 - x^3 + 6x^2 - 11x + 6$$

on $[1, 3]$.

f is differentiable everywhere, and

$$f(1) = 2 - 8 + 6 - 1 + 6 - 11 + 6 = 0$$
$$f(3) = 1458 - 1944 + 486 - 27 + 54 - 33 + 6 = 0$$

It is difficult to compute a value of x in $(1, 3)$ for which

$$f'(x) = 12x^5 - 40x^4 + 24x^3 - 3x^2 + 12x - 11 = 0$$

However, $f'(x)$ is itself a continuous function such that

$$f'(1) = 12 - 40 + 24 - 3 + 12 - 11 = -6 < 0$$
$$f'(3) = 2916 - 3240 + 648 - 27 + 36 - 11 = 322 > 0$$

Hence, the Intermediate-Value Theorem assures us that there must be some number c between 1 and 3 for which $f'(c) = 0$.

17.5 Show that $f(x) = 2x^3 + x - 4 = 0$ has exactly one real solution.

Since $f(0) = -4$ and $f(2) = 16 + 2 - 4 = 14$, the Intermediate-Value Theorem guarantees that f has a zero between 0 and 2; call it x_0.

Because $f'(x) = 6x^2 + 1 > 0$, $f(x)$ is increasing everywhere (Theorem 17.3). Therefore, when $x > x_0$, $f(x) > 0$; and when $x < x_0$, $f(x) < 0$. In other words, there is no zero other than x_0.

17.6 Prove Rolle's Theorem (Theorem 17.1).

Case 1: $f(x) = 0$ for all x in $[a, b]$. Then $f'(x) = 0$ for all x in (a, b), since the derivative of a constant function is 0.

Case 2: $f(x) > 0$ for some x in (a, b). Then, by the Extreme-Value Theorem (Theorem 14.2), an absolute maximum of f on $[a, b]$ exists, and must be positive (since $f(x) > 0$ for some x in (a, b)). Because $f(a) = f(b) = 0$, the maximum is achieved at some point c in the open interval (a, b). Thus, the absolute maximum is also a relative maximum, and, by Theorem 14.1, $f'(c) = 0$.

Case 3: $f(x) < 0$ for some x in (a, b). Let $g(x) = -f(x)$. Then, by Case 2, $g'(c) = 0$ for some c in (a, b); consequently, $f'(c) = -g'(c) = 0$.

17.7 Prove the Mean-Value Theorem (Theorem 17.2).

Let

$$g(x) = f(x) - \frac{f(b) - f(a)}{b - a}(x - a) - f(a)$$

Then g is continuous over $[a, b]$ and differentiable on (a, b). Moreover,

$$g(a) = f(a) - \frac{f(b) - f(a)}{b - a}(a - a) - f(a) = f(a) - 0 - f(a) = 0$$

$$\begin{aligned} g(b) &= f(b) - \frac{f(b) - f(a)}{b - a}(b - a) - f(a) = f(b) - (f(b) - f(a)) - f(a) \\ &= f(b) - f(b) + f(a) - f(a) = 0 \end{aligned}$$

By Rolle's Theorem, applied to g, there exists c in (a, b) for which $g'(c) = 0$. But,

$$g'(x) = f'(x) - \frac{f(b) - f(a)}{b - a}$$

whence

$$0 = g'(c) = f'(c) - \frac{f(b) - f(a)}{b - a} \qquad \text{or} \qquad f'(c) = \frac{f(b) - f(a)}{b - a}$$

17.8 Prove Theorem 17.3.

Assume that $f'(x) > 0$ for all x in (a, b) and that $a < u < v < b$; we must show that $f(u) < f(v)$. By the Mean-Value Theorem, applied to f on the closed interval $[u, v]$, there is some number c in (u, v) such that

$$f'(c) = \frac{f(v) - f(u)}{v - u} \qquad \text{or} \qquad f(v) - f(u) = f'(c)(v - u)$$

But, $f'(c) > 0$ and $v - u > 0$; hence, $f(v) - f(u) > 0$, $f(u) < f(v)$.

The case $f'(x) < 0$ is handled similarly.

17.9 Prove Corollary 17.5.

By the Extreme-Value Theorem, f has an absolute maximum value, $f(d)$, at some argument d in $[a, b]$, and an absolute minimum value, $f(c)$, at some argument c in $[a, b]$. If $f(c) = f(d) = k$, then f is constant on $[a, b]$, and its range is the single point k. If $f(c) \neq f(d)$, then the Intermediate-Value Theorem, applied to the closed subinterval bounded by d and c, ensures that f assumes every value between $f(c)$ and $f(d)$. The range of f is then the closed interval $[f(c), f(d)]$ (which includes the values assumed on that part of $[a, b]$ which lies outside the subinterval).

Supplementary Problems

17.10 Determine whether the hypotheses of Rolle's Theorem hold for each function f, and, if they do, verify the conclusion of the theorem.

(*a*) $f(x) = x^2 - 2x - 3$, on $[-1, 3]$ (*b*) $f(x) = x^3 - x$, on $[0, 1]$
(*c*) $f(x) = 9x^3 - 4x$, on $[-\frac{2}{3}, \frac{2}{3}]$ (*d*) $f(x) = x^3 - 3x^2 + x + 1$, on $[1, 1 + \sqrt{2}]$
(*e*) $f(x) = \dfrac{x^2 - x - 6}{x - 1}$, on $[-2, 3]$ (*f*) $f(x) = \begin{cases} \dfrac{x^3 - 2x^2 - 5x + 6}{x - 1} & \text{if } x \neq 1 \\ -6 & \text{if } x = 1 \end{cases}$, on $[-2, 3]$
(*g*) $f(x) = x^{2/3} - 2x^{1/3}$, on $[0, 8]$ (*h*) $f(x) = \begin{cases} x^2 & \text{if } 0 \le x \le 1 \\ 2 - x & \text{if } 1 < x \le 2 \end{cases}$, on $[0, 2]$

17.11 Verify that the hypotheses of the Mean-Value Theorem hold for each function f on the given interval, and find a value c satisfying the conclusion of the theorem.

(*a*) $f(x) = 2x + 3$, on $[1, 4]$ (*b*) $f(x) = 3x^2 - 5x + 1$, on $[2, 5]$
(*c*) $f(x) = x^{3/4}$, on $[0, 16]$ (*d*) $f(x) = \dfrac{x + 3}{x - 4}$, on $[1, 3]$
(*e*) $f(x) = \sqrt{25 - x^2}$, on $[-3, 4]$ (*f*) $f(x) = \dfrac{1}{x - 4}$, on $[0, 2]$

17.12 Determine where the function f is increasing and where it is decreasing. Then sketch the graph of f.

(*a*) $f(x) = 3x + 1$ (*b*) $f(x) = -2x + 2$ (*c*) $f(x) = x^2 - 4x + 7$
(*d*) $f(x) = 1 - 4x - x^2$ (*e*) $f(x) = \sqrt{1 - x^2}$ (*f*) $f(x) = \frac{1}{3}\sqrt{9 - x^2}$
(*g*) $f(x) = x^3 - 9x^2 + 15x - 3$ (*h*) $f(x) = x + \dfrac{1}{x}$ (*i*) $f(x) = x^3 - 12x + 20$

17.13 Consider the polynomial $f(x) = 5x^3 - 2x^2 + 3x - 4$. (*a*) Show that $f(x)$ has a zero between 0 and 1. (*b*) show that $f(x)$ has only one real zero. (*Hint*: First solve Problem 17.18.)

17.14 Assume f continuous over $[0, 1]$ and assume that $f(0) = f(1)$. Which of the following assertions must be true? (*a*) If f has an absolute maximum at c in $(0, 1)$, then $f'(c) = 0$. (*b*) f' exists on $(0, 1)$. (*c*) $f'(c) = 0$ for some c in $(0, 1)$. (*d*) $\lim_{x \to c} f(x) = f(c)$ for all c in $(0, 1)$. (*e*) f has an absolute maximum at some point c in $(0, 1)$.

17.15 Let f and g be differentiable functions. (*a*) If $f(a) = g(a)$ and $f(b) = g(b)$, where $a < b$, show that $f'(c) = g'(c)$ for some c in (a, b). (*b*) If $f(a) \ge g(a)$ and $f'(x) > g'(x)$ for all x, show that $f(x) > g(x)$ for all $x > a$. (*c*) If $f'(x) > g'(x)$ for all x, show that the graphs of f and g intersect at most once. (*Hint*: In each part, apply the appropriate theorem to the function $h(x) = f(x) - g(x)$.)

17.16 Let f be a differentiable function on an open interval (a, b). (*a*) If f is increasing on (a, b), prove that $f'(x) \geq 0$ for every x in (a, b). (*Hint*:

$$f'(x) = \lim_{h \to 0^+} \frac{f(x+h) - f(x)}{h}$$

and Problem 9.10(*a*) applies.) (*b*) If f is decreasing on (a, b), prove that $f'(x) \leq 0$ for every x in (a, b).

17.17 The Mean-Value Theorem ensures the existence of a certain point on the graph of $y = \sqrt[3]{x}$ between (27, 3) and (125, 5). Find the x-coordinate of the point.

17.18 Let f be a differentiable function such that $f'(x) \neq 0$ for all x in the open interval (a, b). Prove that there is at most one zero of $f(x)$ in (a, b).

17.19 Let $f(x) = x^3 - 4x^2 + 4x$ and $g(x) = 1$, for all x. (*a*) Find the intersection of the graphs of f and g. (*b*) Find the zeros of f. (*c*) If the domain of f is restricted to the closed interval [0, 3], what is the range of f?

17.20 Prove that $8x^3 - 6x^2 - 2x + 1$ has a zero between 0 and 1. (*Hint*: Apply Rolle's Theorem to the function $2x^4 - 2x^3 - x^2 + x$.)

17.21 Show that $x^3 + 2x - 5 = 0$ has exactly one real root.

Chapter 18

Rectilinear Motion. Instantaneous Velocity

Rectilinear motion is motion along a straight line. Consider, for instance, an automobile moving along a straight road. We can imagine a coordinate system imposed on the line containing the road (see Fig. 18-1). (On many highways there actually is such a coordinate system, with markers along the side of the road indicating the distance from one end of the highway.) If s designates the coordinate of the automobile and t denotes the time, then the motion of the automobile is specified by expressing s, its position, as a function of t: $s = f(t)$.

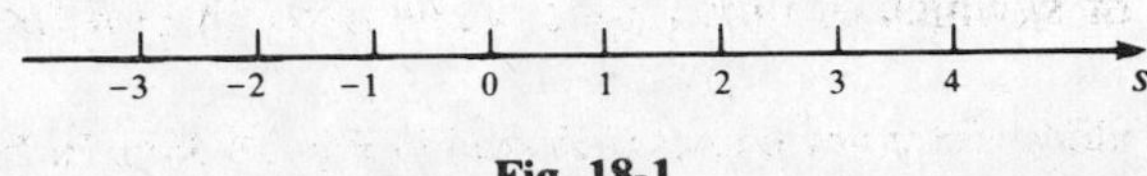

Fig. 18-1

The speedometer indicates how fast the automobile is moving. Since the speedometer reading often varies continuously, it is obvious that the speedometer indicates how fast the car is moving *at the moment* when it is read. Let us analyze this notion in order to find the mathematical concept that lies behind it.

If the automobile moves according to the equation $s = f(t)$, its position at time t is $f(t)$; and at time $t + h$ very close to time t, its position is $f(t + h)$. The *distance* (or, more precisely, the *displacement*) between its position at time t and its position at time $t + h$ is $f(t + h) - f(t)$ (which is possibly negative). The time elapsed between t and $t + h$ is h. Hence, the *average speed* (or *average velocity*) during this time interval is

$$\frac{f(t+h) - f(t)}{h}$$

(Remember that average speed = distance ÷ time.) Now, as the elapsed time h gets closer to 0, the average speed approaches what we intuitively think of as the *instantaneous velocity* v at time t. Thus,

$$v = \lim_{h \to 0} \frac{f(t+h) - f(t)}{h}$$

In other words, the instantaneous velocity v is the derivative $f'(t)$.

EXAMPLES (*a*) The height s of a water column is observed to follow the law $s = f(t) = 3t + 2$. Thus, the instantaneous velocity v of the top surface is $f'(t) = 3$. (*b*) The position s of an automobile along a highway is given by $s = f(t) = t^2 - 2t$. Hence, its instantaneous velocity is $v = f'(t) = 2t - 2$. At time $t = 3$, its velocity v is $2(3) - 2 = 4$.

The sign of the instantaneous velocity v indicates the direction in which the object is moving. If $v = ds/dt > 0$ over a time interval, Theorem 17.3 tells us that s is increasing in that interval. Thus, if the s-axis is horizontal and directed to the right, as in Fig. 18-2(*a*), then the object is moving to the right; but, if the s-axis is vertical and directed upward, as in Fig. 18-2(*b*), then the object is moving upward. On the other hand, if $v = ds/dt < 0$ over a time interval, then s must be decreasing in that interval. In Fig. 18-2(*a*), the object would be moving to the left (in the direction of decreasing s); in Fig. 18-2(*b*), the object would be moving downward.

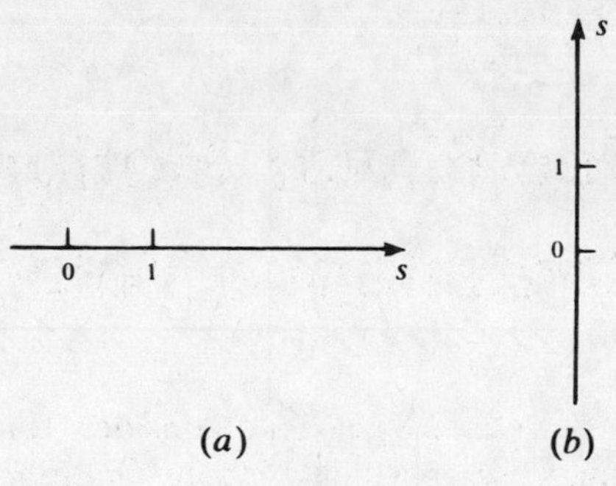

Fig. 18-2

A consequence of these facts is that *at an instant t when a continuously moving object reverses direction, its instantaneous velocity v must be* 0. For if v were, say, positive at t, it would be positive in a small interval of time surrounding t; the object would therefore be moving in the same direction just before and just after t. Or, to say the same thing in a slightly different way, a reversal in direction means a relative extremum of s, which in turn implies (Theorem 14.1) $ds/dt = 0$.

EXAMPLE An object moves along a straight line as indicated in Fig. 18-3(a). In functional form,

$$s = f(t) = (t+2)^2 \quad (s \text{ in meters, } t \text{ in seconds})$$

as graphed in Fig. 18-3(b). The object's instantaneous velocity is

$$v = f'(t) = 2(t+2) \quad \text{(meters per second)}$$

For $t+2<0$, or $t<-2$, v is negative and the object is moving to the left; for $t+2>0$, or $t>-2$, v is positive and the object is moving to the right. The object reverses direction at $t=-2$, and at that instant $v=0$. (Note that $f(t)$ has a relative minimum at $t=-2$.)

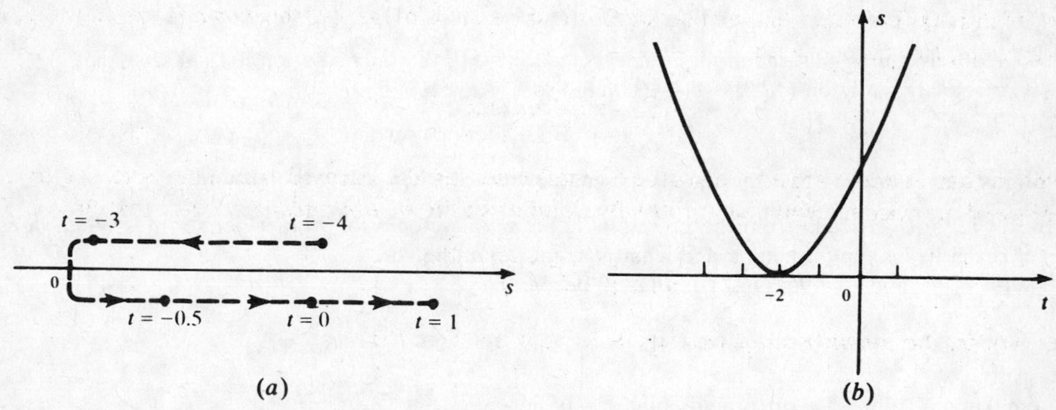

Fig. 18-3

Free Fall

Consider an object that has been thrown straight up or down, or has been dropped from rest, and which is acted upon solely by the gravitational pull of the earth. The ensuing rectilinear motion is called *free fall.*

Let us put a coordinate system on the vertical line along which the object moves, such that the s-axis is directed upward, away from the earth, with $s=0$ located at the surface of the earth (Fig. 18-4). Then the equation of free fall is

$$s = s_0 + v_0t - 16t^2 \qquad (18.1)$$

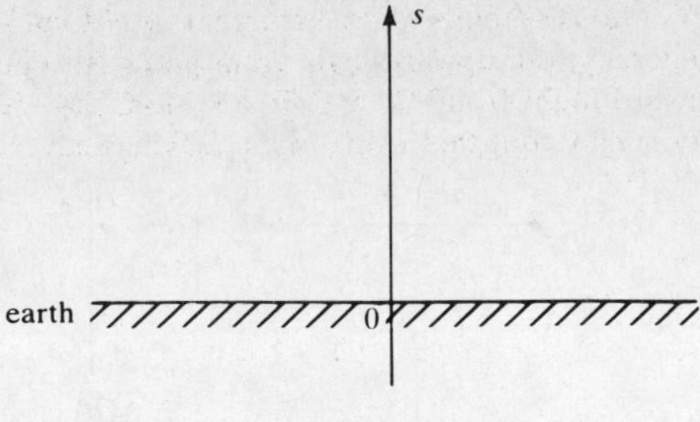

Fig. 18-4

where s is measured in feet and t in seconds. Here, s_0 and v_0 are the position (height) and velocity of the object at time $t = 0$. The instantaneous velocity v is obtained by differentiating (*18.1*):

$$v = \frac{ds}{dt} = v_0 - 32t \tag{18.2}$$

EXAMPLES

(*a*) At $t = 0$, a rock is dropped from rest from the top of a building 256 feet high. When, and with what velocity, does it strike the ground?

With $s_0 = 256$ feet and $v_0 = 0$, (*18.1*) becomes

$$s = 256 - 16t^2$$

and the time of striking the ground is given by

$$\begin{aligned} 0 &= 256 - 16t^2 \\ 16t^2 &= 256 \\ t^2 &= 16 \\ t &= \pm 4 \text{ seconds} \end{aligned}$$

Since we are assuming that the motion takes place for $t \geq 0$, the only solution is $t = 4$ seconds.

The velocity equation (*18.2*) is $v = -32t$, and so, for $t = 4$,

$$v = -32(4) = -128 \text{ feet per second}$$

the minus sign indicating that the rock is moving downward when it hits the ground.

(*b*) A rocket is shot vertically from the ground with an initial velocity of 96 feet per second. When does the rocket reach its maximum height, and what is its maximum height?

With $s_0 = 0$ and $v_0 = 96$, (*18.1*) and (*18.2*) become

$$s = 96t - 16t^2 \qquad \text{and} \qquad v = \frac{ds}{dt} = 96 - 32t$$

At a maximum value of s, or turning point, $v = 0$; hence

$$0 = 96 - 32t \qquad \text{or} \qquad 32t = 96 \qquad \text{or} \qquad t = 3$$

Thus, it takes 3 seconds for the rocket to reach its maximum height, which is

$$s = 96(3) - 16(3)^2 = 288 - 16(9) = 288 - 144 = 144 \text{ feet}$$

(*c*) When does the rocket of (*b*) hit the ground? It suffices to set $s = 0$ in the free-fall equation:

$$\begin{aligned} 0 &= 96t - 16t^2 \\ 0 &= 6t - t^2 \quad \text{[divide by 16]} \\ 0 &= t(6 - t) \end{aligned}$$

from which $t = 0$ or $t = 6$. Hence, after 6 seconds the rocket hits the ground again.

Notice that the rocket rose for 3 seconds to its maximum height, and then took 3 more seconds to fall back to the ground. More generally, the upward flight from point P to point Q will take exactly the same amount of time as the downward flight from Q to P. In addition, the rocket will return to a given height with the same speed (velocity magnitude) that it had upon leaving that height.

Solved Problems

18.1 A stone is thrown straight down from the top of an 80-foot tower. If the initial speed is 64 feet per second, how long does it take to hit the ground, and with what speed does it hit the ground?

Here, $s_0 = 80$ feet and $v_0 = -64$ feet per second (the minus sign for v_0 indicates that the object is moving downward). Hence,

$$s = 80 - 64t - 16t^2 \qquad \text{and} \qquad v = \frac{ds}{dt} = -64 - 32t$$

The stone hits the ground when $s = 0$:

$$\begin{aligned} 0 &= 80 - 64t - 16t^2 \\ 0 &= t^2 + 4t - 5 \quad [\text{divide by } -16] \\ 0 &= (t+5)(t-1) \\ t + 5 = 0 \quad &\text{or} \quad t - 1 = 0 \\ t = -5 \quad &\text{or} \quad t = 1 \end{aligned}$$

Since the time of fall must be positive, $t = 1$ second. The velocity v when the stone hits the ground is

$$v(1) = -64 - 32(1) = -64 - 32 = -96 \text{ feet per second.}$$

18.2 A rocket, shot straight up from the ground, reaches a height of 256 feet after 2 seconds. What was its initial velocity, what will be its maximum height, and when does it reach its maximum height?

With $s_0 = 0$,

$$s = v_0 t - 16t^2 \qquad \text{and} \qquad v = \frac{ds}{dt} = v_0 - 32t$$

When $t = 2$, $s = 256$:

$$256 = v_0(2) - 16(2)^2 \quad \text{or} \quad 256 = 2v_0 - 64 \quad \text{or} \quad 320 = 2v_0 \quad \text{or} \quad 160 = v_0$$

The initial velocity was 160 feet per second, so that

$$s = 160t - 16t^2 \qquad \text{and} \qquad v = 160 - 32t$$

To find the time when the maximum height is reached, set $v = 0$:

$$0 = 160 - 32t \quad \text{or} \quad 32t = 160 \quad \text{or} \quad t = 5 \text{ seconds}$$

To find the maximum height, substitute $t = 5$ in the formula for s:

$$s = 160(5) - 16(5)^2 = 800 - 16(25) = 800 - 400 = 400 \text{ feet}$$

ALGEBRA 88 feet per second = 60 miles per hour; and so the initial velocity of the rocket was

$$v_0 = (160 \text{ ft/sec})\left(\frac{60 \text{ mph}}{88 \text{ ft/sec}}\right)$$
$$= 20\left(\frac{60 \text{ mph}}{11}\right) \quad [\text{cancel 8 ft/sec}]$$
$$\approx 109 \text{ mph}$$

18.3 A car is moving along a straight road according to the equation

$$s = f(t) = 2t^3 - 3t^2 - 12t$$

Describe its motion by indicating when and where the car is moving to the right, and when and where it is moving to the left.

We have

$$v = f'(t) = 6t^2 - 6t - 12 = 6(t^2 - t - 2) = 6(t-2)(t+1)$$

The car is moving to the right when and only when $v > 0$; that is, when and only when

$$\begin{cases} t-2>0 \\ t+1>0 \end{cases} \quad \text{or} \quad \begin{cases} t-2<0 \\ t+1<0 \end{cases}$$
$$\begin{cases} t>2 \\ t>-1 \end{cases} \quad \text{or} \quad \begin{cases} t<2 \\ t<-1 \end{cases}$$
$$t>2 \quad \text{or} \quad t<-1$$

NOTE ($t>2$ and $t>-1$) is equivalent to $t>2$ alone; ($t<2$ and $t<-1$) is equivalent to $t<-1$ alone.

When $t = -1$,

$$s = 2(-1)^3 - 3(-1)^2 - 12(-1) = -2 - 3 + 12 = 7$$

When $t = 2$,

$$s = 2(2)^3 - 3(2)^2 - 12(2) = 16 - 12 - 24 = -20$$

Thus, the car moves to the right until, at $t = -1$, it reaches $s = 7$, where it reverses direction and moves left until, at $t = 2$, it reaches $s = -20$, where it reverses direction again and keeps moving to the right thereafter. See Fig. 18-5.

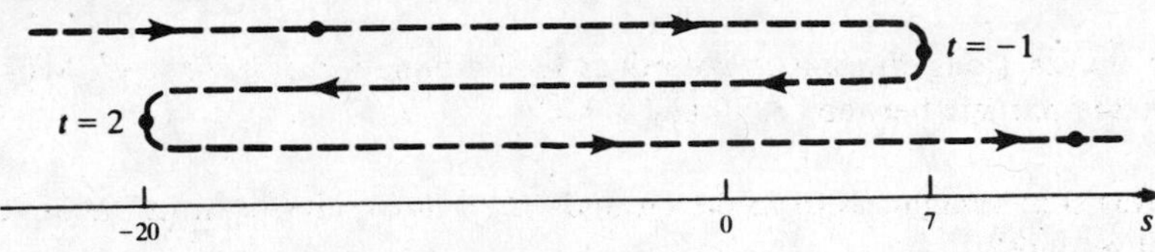

Fig. 18-5

Supplementary Problems

18.4 (*a*) If an object is released from rest at any given height, show that, after t seconds, it has dropped $16t^2$ feet (assuming that it has not yet struck the ground). (*b*) How many seconds does it take the object in (*a*) to fall (i) 1 foot? (ii) 16 feet? (iii) 64 feet? (iv) 100 feet?

18.5 A rock is dropped down a well that is 256 feet deep. When will it hit the bottom of the well?

18.6 Assuming that one story of a building is 10 feet, with what speed, in miles per hour, does an object dropped from the top of a 40-story building hit the ground?

18.7 A rocket is shot straight up into the air with an initial velocity of 128 feet per second. (*a*) How far has it traveled in 1 second? in 2 seconds? (*b*) When does it reach its maximum height? (*c*) What is its maximum height? (*d*) When does it hit the ground again? (*e*) What is its speed when it hits the ground?

18.8 A rock is thrown straight down from a height of 480 feet with an initial velocity of 16 feet per second. (*a*) How long does it take to hit the ground? (*b*) With what speed does it hit the ground? (*c*) How long does it take before the rock is moving at a speed of 112 feet per second? (*d*) When has the rock traveled a distance of 60 feet?

18.9 An automobile moves along a straight highway, with its position given by

$$s = 12t^3 - 18t^2 + 9t - \frac{3}{2} \quad (s \text{ in miles, } t \text{ in hours})$$

(*a*) Describe the motion of the car: when it is moving to the right, when to the left, where and when it changes direction. (*b*) What distance has it traveled in the one hour from $t = 0$ to $t = 1$?

18.10 The position of a moving object on a line is given by the formula

$$s = (t-1)^3(t-5)$$

(*a*) When is the object moving to the right? (*b*) When is it moving to the left? (*c*) When is it changing direction? (*d*) When is it at rest? (*e*) What is the farthest to the left of the origin that it moves?

18.11 A particle moves on a straight line so that its position s (miles) at time t (hours) is given by

$$s = (4t-1)(t-1)^2$$

(*a*) When is the particle moving to the right? (*b*) When is the particle moving to the left? (*c*) When does it change direction? (*d*) When the particle is moving to the left, what is the maximum *speed* that it achieves? (The speed is the absolute value of the velocity.)

18.12 A particle moves along the x-axis according to the equation $x = 10t - 2t^2$. What is the *total* distance covered by the particle between $t = 0$ and $t = 3$?

18.13 A rocket was shot straight up from the ground. What must have been its initial velocity if it returned to earth in 20 seconds?

18.14 Two particles move along the x-axis. Their positions $f(t)$ and $g(t)$ are given by

$$f(t) = 6t - t^2 \qquad \text{and} \qquad g(t) = t^2 - 4t$$

(*a*) When do they have the same position? (*b*) When do they have the same velocity? (*c*) When they have the same position, are they moving in the same direction?

Chapter 19

Instantaneous Rate of Change

One quantity, y, may be related to another quantity, x, by a function f: $y = f(x)$. A change in the value of x usually induces a corresponding change in the value of y.

EXAMPLE Let x be the length of the side of a cube, and let y be the volume of the cube. Then $y = x^3$. In the case where the side has length $x = 2$ units, consider a small change Δx in the length.

NOTATION Δx (read "delta-ex") is the traditional symbol in calculus for a small change in x. Δx is considered a single symbol, *not* a product of Δ and x. In earlier chapters, the role of Δx often was taken by the symbol "h".

The new volume will be $(2 + \Delta x)^3$, and so the *change* in the value of the volume y is $(2 + \Delta x)^3 - 2^3$. This change in y is denoted traditionally by Δy:

$$\Delta y = (2 + \Delta x)^3 - 2^3$$

Now, the natural way to compare the change Δy in y to the change Δx in x is to calculate the ratio $\Delta y/\Delta x$. This ratio depends of course upon Δx, but, if we let Δx approach 0, then the limit of $\Delta y/\Delta x$ will define the *instantaneous rate of change* of y compared to x, when $x = 2$. We have (ALGEBRA, Problem 11.2),

$$\begin{aligned}\Delta y = (2 + \Delta x)^3 - 2^3 &= ((2)^3 + 3(2)^2(\Delta x)^1 + 3(2)^1(\Delta x)^2 + (\Delta x)^3) - 2^3 \\ &= 12\Delta x + 6(\Delta x)^2 + (\Delta x)^3 = (\Delta x)(12 + 6\Delta x + (\Delta x)^2)\end{aligned}$$

Hence
$$\frac{\Delta y}{\Delta x} = 12 + 6\Delta x + (\Delta x)^2$$

and
$$\lim_{\Delta x \to 0} \frac{\Delta y}{\Delta x} = \lim_{\Delta x \to 0} (12 + 6\Delta x + (\Delta x)^2) = 12$$

Therefore, when the side is 2, the rate of change of the volume with respect to the side is 12. This means that, for sides close to 2, the change Δy in the volume is approximately 12 times the change Δx in the side (since $\Delta y/\Delta x$ is close to 12). Let us look at a few numerical cases:

If $\Delta x = 0.1$, then the new side $x + \Delta x$ is 2.1, and the new volume is $(2.1)^3 = 9.261$. So, $\Delta y = 9.261 - 8 = 1.261$, and

$$\frac{\Delta y}{\Delta x} = \frac{1.261}{0.1} = 12.61$$

If $\Delta x = 0.01$, then the new side $x + \Delta x$ is 2.01, and the new volume is $(2.01)^3 = 8.120601$. So, $\Delta y = 8.120601 - 8 = 0.120601$, and

$$\frac{\Delta y}{\Delta x} = \frac{0.120601}{0.01} = 12.0601$$

If $\Delta x = 0.001$, a similar computation yields

$$\frac{\Delta y}{\Delta x} = 12.006001$$

Let us extend the result of the above Example from $y = x^3$ to an arbitrary differentiable function $y = f(x)$. Consider a small change Δx in the value of the argument x. The new value of the argument is then $x + \Delta x$, and the new value of y will be $f(x + \Delta x)$. Hence, the change Δy in the value of the function is

$$\Delta y = f(x + \Delta x) - f(x)$$

The ratio of the change in the function value to the change in the argument is

$$\frac{\Delta y}{\Delta x} = \frac{f(x + \Delta x) - f(x)}{\Delta x}$$

The *instantaneous rate of change of y with respect to x* is defined to be:

$$\lim_{\Delta x \to 0} \frac{\Delta y}{\Delta x} = \lim_{\Delta x \to 0} \frac{f(x + \Delta x) - f(x)}{\Delta x} = f'(x)$$

The instantaneous rate of change is evaluated by the derivative. It follows that, for Δx close to 0, $\Delta y/\Delta x$ will be close to $f'(x)$, so that

$$\Delta y \approx f'(x)\,\Delta x \tag{19.1}$$

Solved Problems

19.1 The weekly profit P, in dollars, of a corporation is determined by the number x of radios produced per week, according to the formula

$$P = 75x - 0.03x^2 - 15\,000$$

(*a*) Find the rate at which the profit is changing when the production level x is 1000 radios per week. (*b*) Find the change in weekly profit when the production level x is increased to 1001 radios per week.

(*a*) The rate of change of the profit P with respect to the production level x is

$$\frac{dP}{dx} = 75 - 0.06x$$

When $x = 1000$,

$$\frac{dP}{dx} = 75 - 0.06(1000) = 75 - 60 = 15 \text{ dollars per radio}$$

(*b*) In economics, the rate of change of profit with respect to production level is called the *marginal profit.* According to (*19.1*), the marginal profit is an approximate measure of how much the profit will change when the production level is increased by one unit. In the present case, we have:

$$\begin{aligned} P(1000) &= 75(1000) - 0.03(1000)^2 - 15\,000 \\ &= 75\,000 - 30\,000 - 15\,000 = 30\,000. \\ P(1001) &= 75(1001) - 0.03(1001)^2 - 15\,000 \\ &= 75\,075 - 30\,060.03 - 15\,000 = 30\,014.97 \\ \Delta P &= P(1001) - P(1000) = 14.97 \text{ dollars per week} \end{aligned}$$

which is very closely approximated by the marginal profit, 15.00, as computed in (*a*).

19.2 The volume V of a sphere of radius r is given by the formula $V = 4\pi r^3/3$. (*a*) How fast is the volume changing relative to the radius when the radius is 10 millimeters? (*b*) What is the change in volume when the radius changes from 10 to 10.1 millimeters?

(*a*)
$$\frac{dV}{dr} = D_r\left(\frac{4}{3}\pi r^3\right) = \frac{4}{3}\pi(3r^2) = 4\pi r^2$$

When $r = 10$,

$$\frac{dV}{dr} = 4\pi(10)^2 = 400\pi \approx 400(3.14) = 1256$$

(*b*)

$$V(10) = \frac{4}{3}\pi(10)^3 = \frac{4000\pi}{3}$$

$$V(10.1) = \frac{4}{3}\pi(10.1)^3 = \frac{4}{3}\pi(1030.301) = \frac{4121.204\,\pi}{3}$$

$$\Delta V = V(10.1) - V(10) = \frac{4121.204\,\pi}{3} - \frac{4000\pi}{3}$$

$$= \frac{\pi}{3}(4121.204 - 4000) = \frac{\pi}{3}(121.204) \approx \frac{3.14}{3}(121.204)$$

$$= 126.86 \text{ mm}^3$$

The change predicted from (*19.1*) and (*a*) is

$$\Delta V \approx \frac{dV}{dr}\Delta r = (1256)(0.1) = 125.6 \text{ mm}^3$$

19.3 An oil tank is being filled; the oil volume V, in gallons, after t minutes is given by

$$V = 1.5t^2 + 2t$$

How fast is the volume increasing when there is 10 gallons of oil in the tank? [To answer the question "How fast?", you must always find the derivative with respect to *time*.]

When there are 10 gallons in the tank,

$$1.5t^2 + 2t = 10 \qquad \text{or} \qquad 1.5t^2 + 2t - 10 = 0$$

Solving by the quadratic formula,

$$t = \frac{-2 \pm \sqrt{4 - 4(1.5)(-10)}}{2(1.5)} = \frac{-2 \pm \sqrt{4 + 60}}{3} = \frac{-2 \pm \sqrt{64}}{3} = \frac{-2 \pm 8}{3} = 2 \text{ or } -\frac{10}{3}$$

Since t must be positive, $t = 2$ min. The rate at which the oil volume is growing is

$$\frac{dV}{dt} = D_t(1.5t^2 + 2t) = 3t + 2$$

Hence, at the instant $t = 2$ min when $V = 10$ gal,

$$\frac{dV}{dt} = 3(2) + 2 = 6 + 2 = 8 \text{ gal/min}$$

Supplementary Problems

19.4 The cost C, in dollars per day, of producing x TV sets per day is given by the formula

$$C = 7000 + 50x - 0.05x^2$$

Find the rate of change of C with respect to x (called the *marginal cost*) when 200 sets are being produced each day.

19.5 The profit P, in dollars per day, resulting from making x units per day of an antibiotic is

$$P = 5x + 0.02x^2 - 120$$

Find the marginal profit when the production level x is 50 units per day.

19.6 Find the rate at which the surface area of a cube of side x is changing with respect to x, when x is 2 feet.

19.7 The number of kilometers from earth of a rocket ship is given by the formula

$$E = 30t + 0.005t^2$$

where t is measured in seconds. How fast is the distance changing when the rocket ship is 35 000 kilometers from earth?

19.8 As a gasoline tank is being emptied, the number G of gallons left after t seconds is given by $G = 3(15 - t)^2$. (*a*) How fast is gasoline being emptied after 12 seconds? (*b*) What was the average rate at which the gasoline was being drained from the tank over the first 12 seconds? (*Hint*: The average rate is the total amount emptied divided by the time during which it was emptied.)

19.9 If $y = 3x^2 - 2$, find (*a*) the average rate at which y changes with respect to x over the interval $[1, 2]$; (*b*) the instantaneous rate of change of y with respect to x when $x = 1$.

19.10 If $y = f(x)$ is a function such that $f'(x) \neq 0$ for any x, find those values of y for which the rate of increase of y^4 with respect to x is 32 times that of y with respect to x.

Chapter 20

Related Rates

Most quantities encountered in science or in everyday life vary with time. If two such quantities are related by an equation, and if we know the rate at which one of them changes, then, by differentiating the equation with respect to time, we can find the rate at which the other quantity changes.

EXAMPLES

(*a*) A 6-foot man is running away from the base of a streetlight that is 15 feet high (Fig. 20-1). If he moves at the rate of 18 feet per second, how fast is the length of his shadow changing?

Let x be the distance of the man from the base A of the streetlight, and let y be the length of the man's shadow.

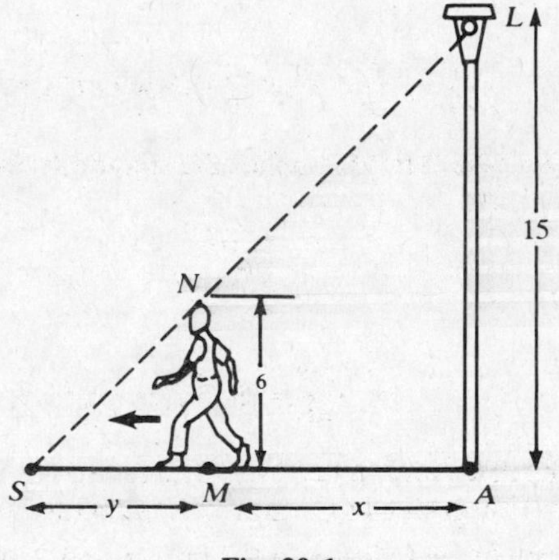

Fig. 20-1

GEOMETRY Two triangles,

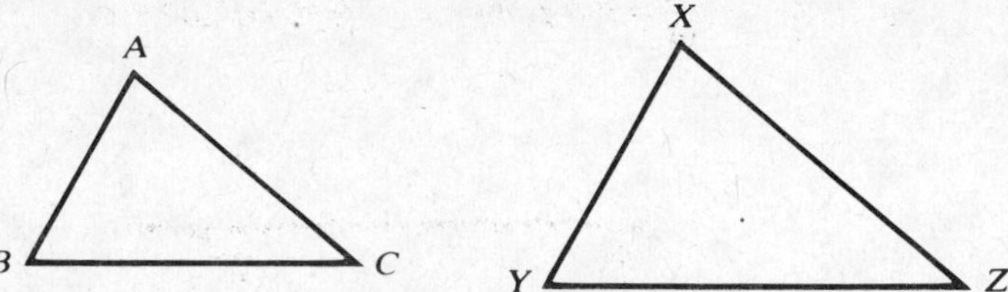

are *similar* if their angles are equal in pairs: $\angle A = \angle X$, $\angle B = \angle Y$, $\angle C = \angle Z$. (For this condition to hold, it suffices that *two* angles of the one triangle be equal to two angles of the other.) Similar triangles have corresponding sides in fixed ratio:

$$\frac{\overline{AB}}{\overline{XY}} = \frac{\overline{AC}}{\overline{XZ}} = \frac{\overline{BC}}{\overline{YZ}}$$

In Fig. 20-1, $\triangle SMN$ and $\triangle SAL$ are similar, whence

$$\frac{\overline{SM}}{\overline{SA}} = \frac{\overline{NM}}{\overline{LA}} \qquad \text{or} \qquad \frac{y}{y+x} = \frac{6}{15} \tag{1}$$

which is the desired relation between x and y. In this case, it is convenient to solve (*1*) for y in terms of x:

$$\frac{y}{y+x} = \frac{2}{5}$$
$$5y = 2y + 2x$$
$$3y = 2x$$
$$y = \frac{2}{3}x \qquad (2)$$

Differentiation of (*2*) with respect to t gives

$$\frac{dy}{dt} = \frac{2}{3}\frac{dx}{dt} \qquad (3)$$

Now, because the man is running away from A at the rate of 18 ft/sec, x is increasing at that rate. Hence,

$$\frac{dx}{dt} = 18 \text{ ft/sec} \qquad \text{and} \qquad \frac{dy}{dt} = \frac{2}{3}(18 \text{ ft/sec}) = 12 \text{ ft/sec}$$

i.e., the shadow is lengthening at the rate of 12 feet per second.

(*b*) A cube of ice is melting; the side s of the cube is decreasing at the constant rate of 2 inches per minute. How fast is the volume V decreasing?

Since $V = s^3$,

$$\frac{dV}{dt} = \frac{d(s^3)}{dt} = 3s^2\frac{ds}{dt} \quad \text{[Power Chain Rule]}$$

The fact that s is ***decreasing*** at the rate of 2 inches per minute translates into the mathematical statement

$$\frac{ds}{dt} = -2$$

Hence,

$$\frac{dV}{dt} = 3s^2(-2) = -6s^2$$

Thus, although s is decreasing at a constant rate, V is decreasing at a rate proportional to the square of s. For instance, when $s = 3$ inches, V is decreasing at a rate of 54 cubic inches per minute.

(*c*) Two small airplanes start from a common point A at the same time. One flies east at the rate of 300 kilometers per hour and the other flies south at the rate of 400 kilometers per hour. After 2 hours, how fast is the distance between them changing?

Refer to Fig. 20-2. We are given that

$$\frac{dx}{dt} = 300 \text{ km/h} \qquad \frac{dy}{dt} = 400 \text{ km/h}$$

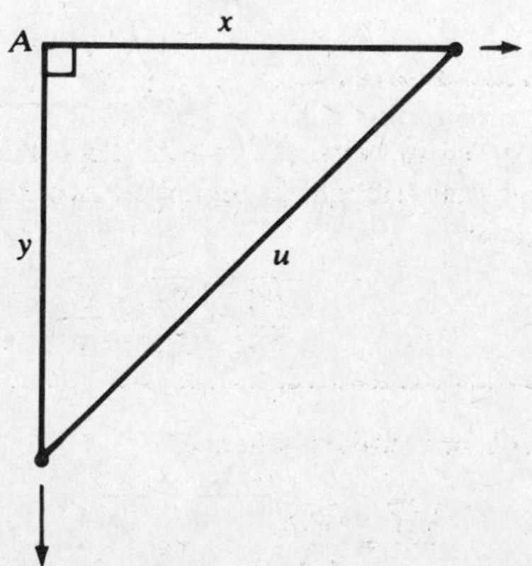

Fig. 20-2

and wish to find the value of du/dt at $t = 2$ h. The necessary relation between u, x, and y is furnished by the Pythagorean Theorem:

$$u^2 = x^2 + y^2$$

Therefore,

$$\frac{d(u^2)}{dt} = \frac{d(x^2 + y^2)}{dt}$$

$$2u\frac{du}{dt} = \frac{d(x^2)}{dt} + \frac{d(y^2)}{dt} \quad \text{[Power Chain Rule]}$$

$$2u\frac{du}{dt} = 2x\frac{dx}{dt} + 2y\frac{dy}{dt} \quad \text{[Power Chain Rule]}$$

$$u\frac{du}{dt} = x\frac{dx}{dt} + y\frac{dy}{dt} = 300x + 400y \tag{4}$$

Now we must find x, y, and u after two hours. Since x is increasing at the constant rate of 300 km/h and t is measured from the beginning of the flight, $x = 300t$ (distance = speed × time, when speed is constant). Similarly, $y = 400t$. Hence, at $t = 2$,

$$x = 300(2) = 600 \qquad y = 400(2) = 800$$

and $$u^2 = (600)^2 + (800)^2 = 360\,000 + 640\,000 = 1\,000\,000 \qquad \text{or} \qquad u = 1000$$

Substituting in (4),

$$1000\frac{du}{dt} = 300(600) + 400(800) = 180\,000 + 320\,000 = 500\,000$$

$$\frac{du}{dt} = \frac{500\,000}{1000} = 500 \text{ km/h}$$

Solved Problems

20.1 Air is leaking out of a spherical balloon at the rate of 3 cubic inches per minute. When the radius is 5 inches, how fast is the radius decreasing?

Since air is leaking out at the rate of 3 in³/min, the volume V of the balloon is decreasing at that rate:

$$\frac{dV}{dt} = -3 \text{ in}^3/\text{min}$$

But the volume of a sphere of radius r is $V = \frac{4}{3}\pi r^3$; hence,

$$-3 = \frac{dV}{dt} = \frac{d}{dt}\left(\frac{4}{3}\pi r^3\right) = \frac{4}{3}\pi\frac{d(r^3)}{dt} = \frac{4}{3}\pi\left(3r^2\frac{dr}{dt}\right) = 4\pi r^2\frac{dr}{dt}$$

or $$\frac{dr}{dt} = -\frac{3}{4\pi r^2}$$

Substituting $r = 5$,

$$\frac{dr}{dt} = -\frac{3}{100\pi} \approx -\frac{3}{314} = -0.00955$$

Thus, when the radius is 5 inches, the radius is decreasing at about 0.01 inch per minute.

20.2 A 13-foot ladder leans against a vertical wall (Fig. 20-3). If the bottom of the ladder is slipping away from the base of the wall at the rate of 2 feet per second, how fast is the top of the ladder moving down the wall when the bottom of the ladder is 5 feet from the base?

Let x be the distance of the bottom of the ladder from the base of the wall, and let y be the distance of the top of the ladder from the base of the wall. Since the bottom of the ladder is moving away from the base of the wall at 2 ft/sec, $dx/dt = 2$. We wish to compute dy/dt when $x = 5$ ft. Now, by the Pythagorean Theorem,

$$(13)^2 = x^2 + y^2 \tag{1}$$

Differentiation of this, as in Example (*c*), gives:

$$0 = x\frac{dx}{dt} + y\frac{dy}{dt} = 2x + y\frac{dy}{dt} \tag{2}$$

But, when $x = 5$, (*1*) gives

$$y = \sqrt{(13)^2 - (5)^2} = \sqrt{169 - 25} = \sqrt{144} = 12$$

so that (*2*) becomes

$$0 = 2(5) + 12\frac{dy}{dt}$$

$$\frac{dy}{dt} = -\frac{2(5)}{12} = -\frac{5}{6}$$

Hence, the top of the ladder is moving *down* the wall ($dy/dt < 0$) at $\frac{5}{6}$ ft/sec when the bottom of the ladder is 5 feet from the wall.

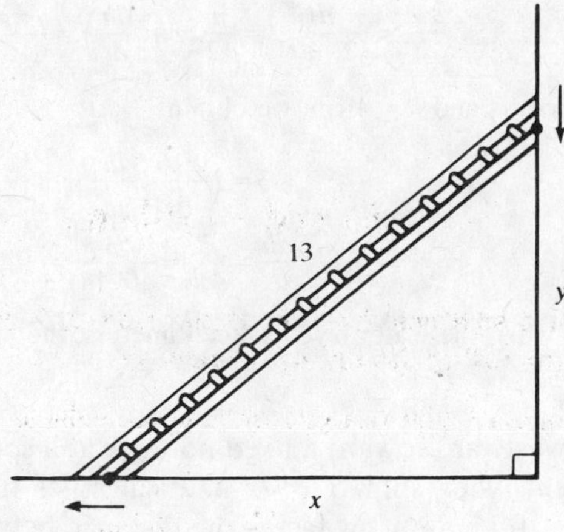

Fig. 20-3

20.3 A cone-shaped paper cup (Fig. 20-4) is being filled with water at the rate of 3 cubic centimeters per second. The height of the cup is 10 centimeters and the radius of the base is 5 centimeters. How fast is the water level rising when the level is 4 centimeters?

At time t (seconds), when the water depth is h, the volume of water in the cup is given by the cone formula

$$V = \frac{1}{3}\pi r^2 h$$

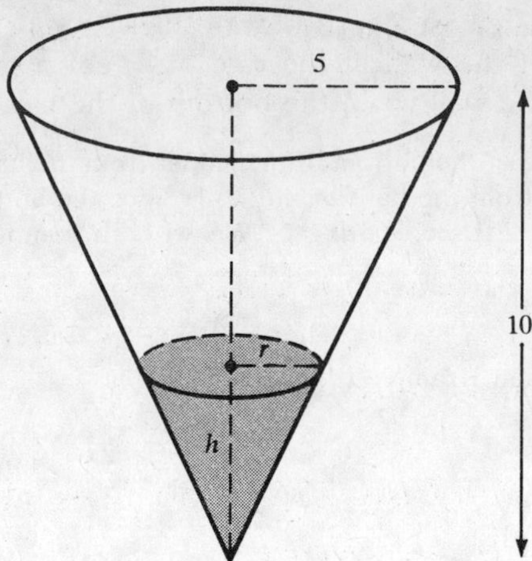

Fig. 20-4

where r is the radius of the top surface. But, by similar triangles in Fig. 20-4,

$$\frac{r}{5}=\frac{h}{10} \qquad \text{or} \qquad r=\frac{5h}{10}=\frac{h}{2}$$

(Only h is of interest, so we are eliminating r.) Thus,

$$V=\frac{1}{3}\pi\left(\frac{h}{2}\right)^2 h=\frac{1}{3}\pi\left(\frac{h^2}{4}\right)h=\frac{\pi}{12}h^3$$

and, by the Power Chain Rule,

$$\frac{dV}{dt}=\frac{\pi}{12}\frac{d(h^3)}{dt}=\frac{\pi}{12}\left(3h^2\frac{dh}{dt}\right)=\left(\frac{\pi h^2}{4}\right)\frac{dh}{dt}$$

Substituting $dV/dt = 3\ \text{cm}^3/\text{s}$ and $h = 4$ cm, we obtain

$$3=\left(\frac{\pi 16}{4}\right)\frac{dh}{dt}$$

$$\frac{dh}{dt}=\frac{3}{4\pi}\approx\frac{3}{4(3.14)}\approx 0.24\ \text{cm/s}$$

Hence, at the moment when the water level is 4 centimeters, the level is rising at about 0.24 centimeters per second.

20.4 A ship B is moving westward toward a fixed point A at a speed of 12 knots (nautical miles per hour). At the moment when ship B is 72 nautical miles from A, ship C passes through A, heading due south at 10 knots. How fast is the distance between the ships changing two hours after ship C has passed through A?

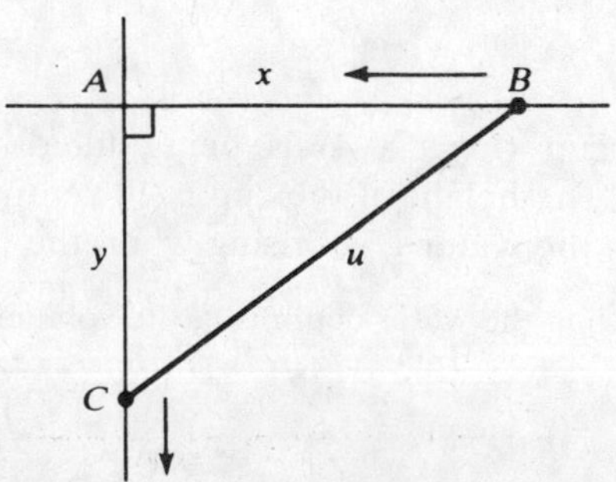

Fig. 20-5

Figure 20-5 shows the situation at time $t > 0$; at $t = 0$, ship C was at A. Since

$$u^2 = x^2 + y^2 \tag{1}$$

we have, as in Example (c),

$$u\frac{du}{dt} = x\frac{dx}{dt} + y\frac{dy}{dt} = -12x + 10y \tag{2}$$

(x decreases at 12 knots, y increases at 10 knots.) At $t = 2$ hr, we have (distance = speed × time):

$$y = (10 \text{ knots})(2 \text{ hr}) = 20 \text{ mi}$$

$$(72 \text{ mi}) - x = (12 \text{ knots})(2 \text{ hr}) \qquad \text{or} \qquad x = 72 - 24 = 48 \text{ mi}$$

and, by (1),

$$u = \sqrt{(48)^2 + (20)^2} = \sqrt{2304 + 400} = \sqrt{2704} = 52 \text{ mi}$$

Substitution of these three values in (2) gives:

$$52\frac{du}{dt} = -12(48) + 10(20) = -576 + 200 = -376$$

$$\frac{du}{dt} = -\frac{376}{52} \approx -7.23 \text{ knots}$$

which shows that, two hours after the crossing, the distance between ships B and C is *decreasing* at the rate of 7.23 knots.

Supplementary Problems

20.5 The top of a 25-foot ladder, leaning against a vertical wall, is slipping down the wall at the rate of 1 foot per minute. How fast is the bottom of the ladder slipping along the ground when the bottom of the ladder is 7 feet away from the base of the wall?

20.6 A cylindrical tank of radius 10 feet is being filled with wheat at the rate of 314 cubic feet per minute. How fast is the depth of the wheat increasing? (The volume of a cylinder is $\pi r^2 h$, where r is its radius and h its height.)

20.7 A 5-foot girl is walking toward a 20-foot lamppost at the rate of 6 feet per second. How fast is the tip of her shadow (cast by the lamp) moving?

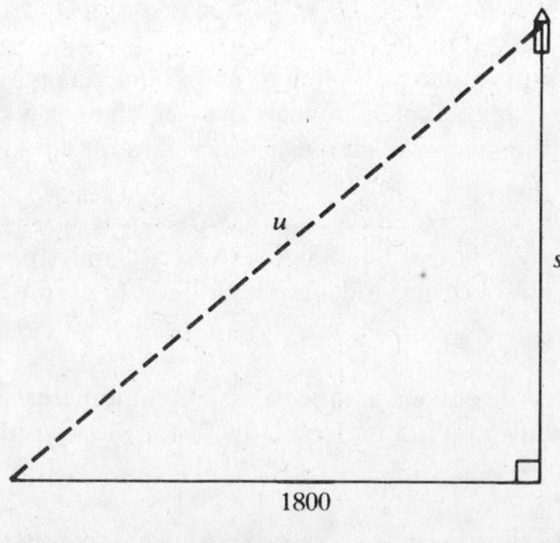

Fig. 20-6

20.8 A rocket is shot vertically upward with an initial velocity of 400 feet per second. Its height s after t seconds is

$$s = 400t - 16t^2$$

How fast is the distance changing from the rocket to an observer on the ground 1800 feet away from the launching site, when the rocket is still rising and is 2400 feet above the ground? (See Fig. 20-6.)

20.9 A small funnel in the shape of a cone is being emptied of fluid at the rate of 12 cubic millimeters per second. The height of the funnel is 20 millimeters and the radius of the base is 4 millimeters. How fast is the fluid level dropping when the level stands 5 millimeters above the vertex of the cone? (Remember that the volume of a cone is $\frac{1}{3}\pi r^2 h$.)

20.10 A balloon is being inflated by pumping air in at the rate of 2 cubic inches per second. How fast is the diameter of the balloon increasing when the radius is one-half inch?

20.11 Oil from an uncapped oil well in the ocean is radiating outward in the form of a circular film on the surface of the water. If the radius of the circle is increasing at the rate of 2 meters per minute, how fast is the area of the oil film growing when the radius is 100 meters?

20.12 The length of a rectangle of constant area 800 square millimeters is increasing at the rate of 4 millimeters per second. (*a*) What is the width of the rectangle at the moment when the width is decreasing at the rate of 0.5 millimeters per second? (*b*) How fast is the diagonal of the rectangle changing when the width is 20 millimeters?

20.13 A particle moves on the hyperbola $x^2 - 18y^2 = 9$ in such a way that its y-coordinate increases at a constant rate of 9 units per second. How fast is its x-coordinate changing when $x = 9$?

20.14 An object moves along the graph of $y = f(x)$. At a certain point, the slope of the curve (that is, the slope of the tangent line to the curve) is 1/2 and the abscissa (x-coordinate) of the object is decreasing at the rate of 3 units per second. At that point, how fast is the ordinate (y-coordinate) of the object changing? (*Hint*: $y = f(x)$, and x is a function of t. So, y is a composite function of t which may be differentiated by the Chain Rule.)

20.15 If the radius of a sphere is increasing at the constant rate of 3 millimeters per second, how fast is the volume changing when the surface area ($4\pi r^2$) is 10 square millimeters?

20.16 What is the radius of an expanding circle at a moment when the rate of change of its area is numerically twice as large as the rate of change of its radius?

20.17 A particle moves along the curve $y = 2x^3 - 3x^2 + 4$. At a certain moment, when $x = 2$, the particle's x-coordinate is increasing at the rate of 0.5 units per second. How fast is its y-coordinate changing at that moment?

20.18 A plane flying parallel to the ground at a height of 4 kilometers passes over a radar station R (Fig. 20-7). A short time later, the radar equipment reveals that the plane is 5 kilometers away and that the distance between the plane and the station is increasing at a rate of 300 kilometers per hour. At that moment, how fast is the plane moving horizontally?

20.19 A boat passes a fixed buoy at 9 A.M., heading due west at 3 miles per hour. Another boat passes the same buoy at 10 A.M., heading due north at 5 miles per hour. How fast is the distance between the boats changing at 11:30 A.M.?

20.20 Water is pouring into an inverted cone at the rate of 3.14 cubic meters per minute. The height of the cone is 10 meters and the radius of its base is 5 meters. How fast is the water level rising when the water stands 7.5 meters in the cone?

20.21 A particle moves along the curve $y = \frac{1}{2}x^2 + 2x$. At what point(s) on the curve are the abscissa and ordinate of the particle changing at the same rate?

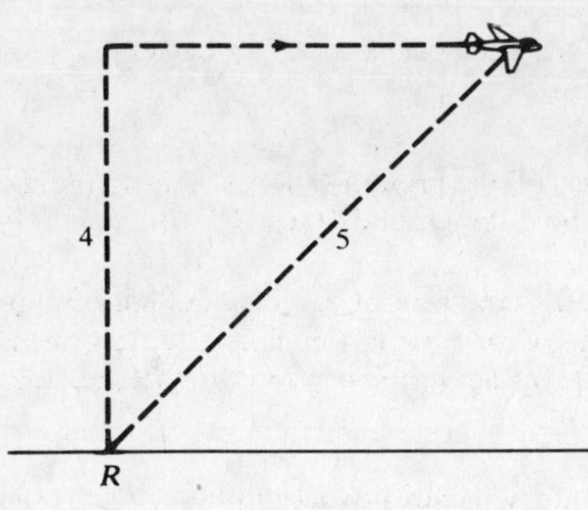

Fig. 20-7

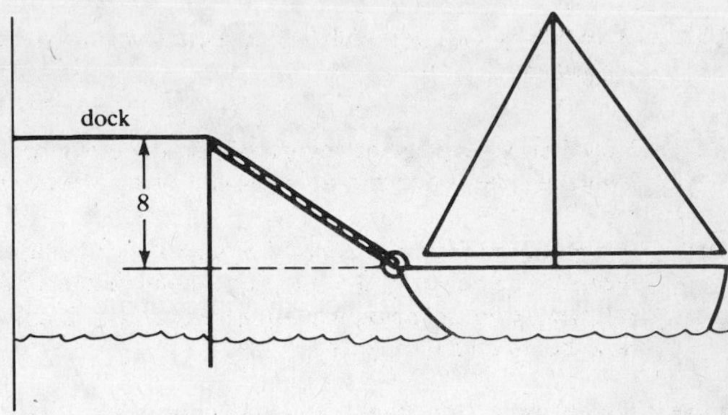

Fig. 20-8

20.22 A boat is being pulled into a dock by a rope that passes through a ring on the bow of the boat (Fig. 20-8). The dock is 8 feet higher than the bow ring. How fast is the boat approaching the dock when the length of rope between the dock and the boat is 10 feet, if the rope is being pulled in at the rate of 3 feet per second?

20.23 A girl is flying a kite, which is at a height of 120 feet. The wind is carrying the kite horizontally away from the girl at a speed of 10 feet per second. How fast must the string be let out when the kite is 150 feet away from the girl?

Chapter 21

Approximation by Differentials

21.1 INTRODUCTION

In Chapter 19, we obtained an approximate relation between the change

$$\Delta y \equiv f(x + \Delta x) - f(x)$$

in a differentiable function f and the change Δx in the argument of f. For convenience, we repeat (*19.1*) here as the *Approximation Principle*

$$\Delta y \approx f'(x)\,\Delta x \tag{21.1}$$

The present chapter will give further applications of the Approximation Principle, and will introduce a new concept associated with it.

21.2 ESTIMATING THE VALUE OF A FUNCTION

Many practical problems involve finding the value $f(c)$ of some function $f(x)$ at a specified argument $x = c$. A direct calculation of $f(c)$ may be difficult or, quite often, impossible. However, if an argument x *close to c*—the closer the better—can be found, such that $f(x)$ and $f'(x)$ can be computed exactly, then the Approximation Principle, with $\Delta x = c - x$, can be applied to give

$$\begin{aligned} f(x + \Delta x) - f(x) &\approx f'(x)\,\Delta x \\ f(c) - f(x) &\approx f'(x)\,\Delta x \\ f(c) &\approx f(x) + f'(x)\,\Delta x \end{aligned} \tag{21.2}$$

EXAMPLE Let us estimate $\sqrt{9.2}$; that is, estimate $f(c)$, where $f(x) = \sqrt{x}$ and $c = 9.2$. For the nearby argument $x = 9$, we have

$$\begin{aligned} f(x) &= \sqrt{9} = 3 \\ f'(x) &= D_x x^{1/2} = \frac{1}{2}x^{-1/2} = \frac{1}{2\sqrt{x}} = \frac{1}{2\sqrt{9}} \\ &= \frac{1}{2(3)} = \frac{1}{6} \end{aligned}$$

and (*21.2*) gives

$$f(9.2) \approx 3 + \frac{1}{6}(9.2 - 9) = 3 + \frac{0.2}{6} = 3.0333\cdots$$

which is correct to three decimal places ($\sqrt{9.2} = 3.0331\cdots$).

21.3 THE DIFFERENTIAL

The product on the right side of (*21.1*) is called the *differential* of f, denoted df.

Definition: Let f be a differentiable function. Then, for a given argument x and increment Δx, the *differential* of f is

$$df \equiv f'(x)\,\Delta x$$

Notice that df depends on two quantities, x and Δx. Although Δx is usually taken to be small, this is not explicitly required in the Definition; thus, there is no reason to suppose that df is a small quantity. However, if Δx *is* small, then the content of the Approximation Principle is that

$$f(x+\Delta x)-f(x) \approx df \tag{21.3}$$

EXAMPLE A graphical picture of this last form of the Approximation Principle is given in Fig. 21-1. Line $\mathscr{L}$ is tangent to the graph of f at P; its slope is therefore $f'(x)$. But then

$$f'(x)=\frac{\overline{RT}}{\overline{PR}}=\frac{\overline{RT}}{\Delta x}$$

or

$$\overline{RT}=f'(x)\,\Delta x \equiv df$$

Now, it is clear that, for Δx very small, $\overline{RT} \approx \overline{RQ}$; i.e.,

$$df \approx f(x+\Delta x)-f(x)$$

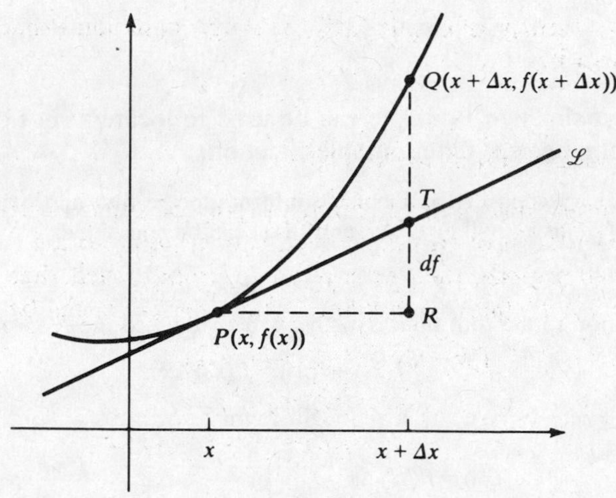

Fig. 21-1

If the value of a function f is given by a formula—say, $f(x)=x^2+2x^{-3}$—let us agree that the differential df may also be written $d(x^2+2x^{-3})$:

$$d(x^2+2x^{-3})=df=f'(x)\,\Delta x=(2x-6x^{-4})\,\Delta x$$

In particular, if $f(x)=x$, we shall write

$$dx=df=f'(x)\,\Delta x=1\cdot\Delta x=\Delta x$$

that is, *we have the new notation* dx *for* Δx. In this new notation, the definition of the differential reads:

$$df \equiv f'(x)\,dx$$

Assuming that $dx=\Delta x \neq 0$, we may divide both sides of the identity by dx, and obtain the following result: *The derivative* $f'(x)$ *is expressible as a fraction whose numerator is the differential of* f *and whose denominator is the differential of* x.

EXAMPLE Ever since Chapter 12 we have been using the *purely formal* fraction

$$\frac{df}{dx}$$

—wherein neither "df" nor "dx" had a separate existence—as one way of representing the derivative. Now we can see a reason for this odd choice of notation: when differentials are introduced, the formal fraction becomes a real fraction.

Solved Problems

21.1 Estimate the value of $\sqrt{62}$.

Letting $f(x) = \sqrt{x}$ and $c = 62$, choose $x = 64$ (the perfect square closest to 62) and

$$\Delta x = c - x = 62 - 64 = -2$$

Then,

$$f(x) = \sqrt{64} = 8 \qquad f'(x) = \frac{1}{2\sqrt{x}} = \frac{1}{2\sqrt{64}} = \frac{1}{2(8)} = \frac{1}{16}$$

and the Approximation Principle, (*21.2*), gives

$$\sqrt{62} \approx 8 + \frac{1}{16}(-2) = 8 - \frac{1}{8} = 7\tfrac{7}{8} = 7.875$$

Actually, $\sqrt{62} = 7.8740\cdots$.

21.2 Show how the Approximation Principle can be used to locate a root of the equation $f(x) = 0$, where f is some (complicated) differentiable function.

Assume that, by inspection or from outside information, a first approximation, x_0, to the desired root is known. Thus, $f(x_0)$ is a small quantity, and $f'(x_0)$ can be computed. We wish to select Δx such that

$$x_1 \equiv x_0 + \Delta x$$

is a better approximation to the root than is x_0. Now, according to the Approximation Principle,

$$f(x_1) - f(x_0) \approx f'(x_0)\,\Delta x$$

Hence, if $f(x_1)$ is to be zero (very nearly), we ought to make Δx satisfy

$$-f(x_0) = f'(x_0)\,\Delta x \qquad \text{or} \qquad \Delta x = -\frac{f(x_0)}{f'(x_0)}$$

which we are free to do, provided $f'(x_0) \neq 0$. Thus, our improved approximation to the root is

$$x_1 = x_0 - \frac{f(x_0)}{f'(x_0)}$$

The repetition of the process, generating

$$x_2 = x_1 - \frac{f(x_1)}{f'(x_1)}, \quad x_3 = x_2 - \frac{f(x_2)}{f'(x_2)}, \quad \cdots$$

constitutes *Newton's method* of solving an equation. The method will give better and better approximations, so long as the initial guess x_0 was good enough and so long as $f'(x)$ does not hit zero at some stage.

21.3 A measurement of the side of a square room yields the result 18.5 ft; hence, the area is

$$A = (18.5)^2 = 342.25 \text{ ft}^2$$

If the measuring device has a possible error of at most 0.05 ft, estimate the maximum possible error in the area?

The formula for the area is $A = x^2$, where x is the side of the room. Hence,

$$\frac{dA}{dx} = 2x$$

Let $x = 18.5$, and let $18.5 + \Delta x$ be the true length of the side of the room. By assumption, $|\Delta x| \leq 0.05$.

The Approximation Principle yields:

$$A(x+\Delta x) - A(x) \approx \frac{dA}{dx}\Delta x$$

$$A(18.5+\Delta x) - 342.25 \approx 2(18.5)(\Delta x)$$

or

$$|A(18.5+\Delta x) - 342.25| \approx |37\,\Delta x| \leq 37(0.05) = 1.85$$

Hence, the error in the area should be at most 1.85 ft^2, putting the actual area in the range $(342.25 \pm 1.85)\ \text{ft}^2$. See Problem 21.11.

Supplementary Problems

21.4 Use the Approximation Principle to estimate the following quantities:

(*a*) $\sqrt{51}$ (*b*) $\sqrt{78}$ (*c*) $\sqrt[3]{123}$

(*d*) $(8.35)^{2/3}$ (*e*) $(33)^{-1/5}$ (*f*) $\sqrt[4]{\frac{17}{81}}$

(*g*) $\sqrt[3]{0.065}$

21.5 Measurement of the side of a cubical container yields the result 8.14 cm, with a possible error of at most 0.005 cm. Give an estimate of the maximum possible error in the value

$$V = (8.14)^3 = 539.35314\ \text{cm}^3$$

for the volume of the container.

21.6 It is desired to give a spherical tank of diameter 20 feet (240 inches) a coat of paint 0.1 inch thick. Use the Approximation Principle to estimate how many gallons of paint will be required. ($V = \frac{4}{3}\pi r^3$, and one gallon is about 231 cubic inches.)

21.7 A solid steel cylinder has a radius of 2.5 cm and a height of 10 cm. A tight-fitting sleeve is to be made that will extend the radius to 2.6 cm. Find the amount of steel needed for the sleeve, (*a*) by the Approximation Principle, (*b*) by an exact calculation.

21.8 If the side of a cube is measured with a percentage error of at most 3%, estimate the maximum percentage error in the volume of the cube. (If ΔQ is the error in measurement of a quantity Q, then

$$\left|\frac{\Delta Q}{Q}\right| \times 100\%$$

is the *percentage error*.)

21.9 Assume, contrary to fact, that the Earth is a perfect sphere, with a radius of 4000 miles. The volume of ice at the North and South Poles is estimated to be about 8 000 000 cubic miles. If this ice were melted and if the resulting water were distributed uniformly over the globe, approximately what would be the depth of the added water at any point of the Earth?

21.10 (*a*) Let $y = x^{3/2}$. When $x = 4$ and $dx = 2$, find the value of dy. (*b*) Let

$$y = 2x\sqrt{1+x^2}$$

When $x = 0$ and $dx = 3$, find the value of dy.

21.11 For Problem 21.3, calculate exactly the largest possible error in the area.

21.12 Establish the very useful approximation formula

$$(1+u)^r \approx 1 + ru$$

where r is any rational exponent and $|u|$ is small compared to 1. (*Hint*: Apply the Approximation Principle to $f(x) = x^r$, letting $x = 1$ and $\Delta x = u$.)

Chapter 22

Higher-Order Derivatives

The derivative f' of a function f is itself a function—in fact, it is (almost always) a differentiable function. Hence, f' has a derivative, which is denoted f''; f'' has a derivative, which is denoted f'''; etc.

Definition:

$$\begin{aligned} f''(x) &= D_x(f'(x)) \\ f'''(x) &= D_x(f''(x)) \\ f^{(4)}(x) &= D_x(f'''(x)) \\ &\cdots\cdots\cdots \end{aligned}$$

We call f' the *first derivative* of f, f'' the *second derivative* of f, f''' the *third derivative* of f. If the *order* n exceeds three, we write $f^{(n)}$ for the n*th derivative* of f.

EXAMPLES (*a*) If $f(x) = 3x^4 - 7x^3 + 5x^2 + 2x - 1$, then

$$\begin{aligned} f'(x) &= 12x^3 - 21x^2 + 10x + 2 \\ f''(x) &= 36x^2 - 42x + 10 \\ f'''(x) &= 72x - 42 \\ f^{(4)}(x) &= 72 \\ f^{(n)} &= 0 \quad \text{for all } n \ge 5 \end{aligned}$$

(*b*) If $f(x) = \frac{1}{2}x^3 - 5x^2 + x + 4$, then

$$\begin{aligned} f'(x) &= \frac{3}{2}x^2 - 10x + 1 \\ f''(x) &= 3x - 10 \\ f'''(x) &= 3 \\ f^{(n)}(x) &= 0 \quad \text{for all } n \ge 4 \end{aligned}$$

It is clear that if f is a polynomial of degree k, then the nth derivative $f^{(n)}$ will be 0 for all $n > k$. (*c*) If $f(x) = 1/x = x^{-1}$, then

$$\begin{aligned} f'(x) &= -x^{-2} = \frac{-1}{x^2} \\ f''(x) &= 2x^{-3} = \frac{2}{x^3} \\ f'''(x) &= -6x^{-4} = \frac{-6}{x^4} \\ f^{(4)}(x) &= 24x^{-5} = \frac{24}{x^5} \\ &\cdots\cdots\cdots \end{aligned}$$

In this case, the nth derivative will never be the constant function 0.

Alternative Notation

first derivative $\qquad f'(x) \equiv D_x f(x) \equiv \dfrac{df}{dx} \equiv \dfrac{dy}{dx} \equiv D_x y \equiv y'$

second derivative $\qquad f''(x) \equiv D_x^2 f(x) \equiv \dfrac{d^2 f}{dx^2} \equiv \dfrac{d^2 y}{dx^2} \equiv D_x^2 y \equiv y''$

third derivative $\qquad f'''(x) \equiv D_x^3 f(x) \equiv \dfrac{d^3 f}{dx^3} \equiv \dfrac{d^3 y}{dx^3} \equiv D_x^3 y \equiv y'''$

***n*th derivative** $\qquad f^{(n)}(x) \equiv D_x^n f(x) \equiv \dfrac{d^n f}{dx^n} \equiv \dfrac{d^n y}{dx^n} \equiv D_x^n y \equiv y^{(n)}$

Higher-Order Implicit Differentiation

EXAMPLE Let $y = f(x)$ be a differentiable function satisfying the equation

$$x^2 + y^2 = 9 \tag{0}$$

(We know that $y = \sqrt{9-x^2}$ or $y = -\sqrt{9-x^2}$; their graphs are shown in Fig. 22-1.)

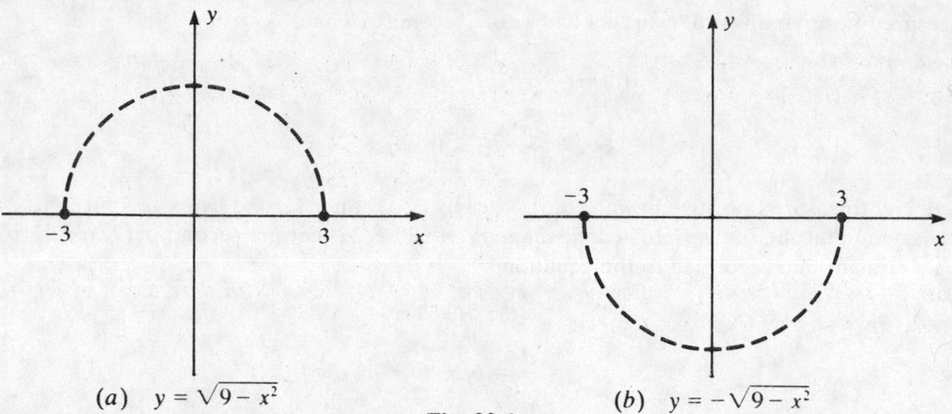

Fig. 22-1

Let us find a formula for the second derivative y'' (where y stands for either of the two functions).

$$D_x(x^2 + y^2) = D_x(9)$$

$$2x + 2yy' = 0 \quad [D_xy^2 = 2yy' \text{ by the Power Chain Rule}]$$

$$x + yy' = 0 \tag{1}$$

Next, differentiate (*1*) with respect to x:

$$D_x(x + yy') = D_x(0)$$

$$1 + yy'' + (y')^2 = 0 \tag{2}$$

Equations (*1*) and (*2*) compose a "triangular system": (*1*) can be solved for y' in terms of x and y, the result can be substituted in (*2*), and then (*2*) can be solved for y'' in terms of x and y. Thus:

$$y' = -\frac{x}{y} \quad [\text{solving } (1)]$$

$$1 + yy'' + \frac{x^2}{y^2} = 0 \quad [\text{substituting in } (2)]$$

$$\left.\begin{aligned} y^2 + y^3y'' + x^2 &= 0 \\ y'' &= -\frac{x^2+y^2}{y^3} \end{aligned}\right\} \quad [\text{solving for } y'']$$

It is clear that nth order implicit differentiation could be handled the same way, by solving a triangular system of n equations successively for $y', y'', \ldots, y^{(n-1)}, y^{(n)}$. In the particular problem just treated, the original equation (*0*) may be used to eliminate x from the final result:

$$y'' = -\frac{9}{y^3}$$

Such an occurrence should not be expected in general.

Acceleration

Let an object move along a coordinate axis according to an equation $s = f(t)$, where s is the coordinate of the object and t is the time. From Chapter 18, the object's velocity is given by

$$v = \frac{ds}{dt} = f'(t)$$

The rate at which the velocity changes is called the *acceleration, a*.

Definition:
$$a = \frac{dv}{dt} = \frac{d^2s}{dt^2} = f''(t)$$

EXAMPLES (*a*) For an object in free fall,

$$s = s_0 + v_0t - 16t^2 \tag{18.1}$$

where s, measured positive upward, is in feet and t is in seconds. Hence,

$$v = \frac{ds}{dt} = v_0 - 32t$$

$$a = \frac{dv}{dt} = -32 \text{ ft/sec}^2$$

Thus, the velocity (measured positive upward) decreases by 32 feet per second every second. This is sometimes expressed by saying that the (downward) *acceleration of gravity* is 32 feet per second per second. (*b*) An object moves along a straight line according to the equation

$$s = 2t^3 - 3t^2 + t - 1$$

Then

$$v = \frac{ds}{dt} = 6t^2 - 6t + 1$$

$$a = \frac{dv}{dt} = 12t - 6$$

In this case, the acceleration is not constant. Notice that $a > 0$ when $12t - 6 > 0$, or $t > \frac{1}{2}$. This implies (Theorem 17.3) that the velocity is increasing for $t > \frac{1}{2}$.

Solved Problems

22.1 Describe all the derivatives (first, second, etc.) of

$$(a)\quad f(x) = \frac{1}{4}x^4 - 5x - \pi \qquad (b)\quad f(x) = \frac{x}{x+1}$$

(*a*)
$$f'(x) = x^3 + 5 \qquad f''(x) = 3x^2 \qquad f'''(x) = 6x \qquad f^{(4)}(x) = 6$$

and $f^{(n)}(x) = 0$ for $n \geq 5$.

(*b*)
$$f'(x) = \frac{(x+1)D_x x - xD_x(x+1)}{(x+1)^2} \quad \text{[Quotient Rule]}$$

$$= \frac{(x+1)(1) - x(1)}{(x+1)^2} = \frac{x+1-x}{(x+1)^2} = \frac{1}{(x+1)^2} = (x+1)^{-2}$$

$$f''(x) = (-2)(x+1)^{-3}D_x(x+1) \quad \text{[Power Chain Rule]}$$

$$= (-2)(x+1)^{-3}(1) = \frac{-2}{(x+1)^3}$$

$$f'''(x) = (-2)(-3)(x+1)^{-4} = \frac{+6}{(x+1)^4}$$

$$f^{(4)}(x) = (-2)(-3)(-4)(x+1)^{-5} = \frac{-24}{(x+1)^5}$$

....

$$f^{(n)}(x) = (-2)(-3)(-4)(-5)\cdots(-n)(x+1)^{-(n+1)} = \frac{(-1)^{n-1}n!}{(x+1)^{n+1}}$$

ALGEBRA $(-1)^{n-1}$ will be $+1$ when n is odd and -1 when n is even. $n!$ stands for the product

$$1 \times 2 \times 3 \times \cdots \times n$$

of the first n positive integers.

22.2 Find y'' if

$$y^3 - xy = 1 \tag{0}$$

Differentiation of (*0*), using the Power Chain Rule for $D_x y^3$ and the Product Rule for $D_x(xy)$, gives

$$3y^2y' - (xy' + y) = 0$$
$$3y^2y' - xy' - y = 0$$
$$(3y^2 - x)y' - y = 0 \tag{1}$$

Next, differentiate (*1*):

$$(3y^2 - x)D_xy' + y'D_x(3y^2 - x) - y' = 0 \quad \text{[Product Rule]}$$
$$(3y^2 - x)y'' + y'(6yy' - 1) - y' = 0 \quad \text{[Power Chain Rule]}$$
$$(3y^2 - x)y'' + y'((6yy' - 1) - 1) = 0 \quad \text{[factor } y'\text{]}$$
$$(3y^2 - x)y'' + y'(6yy' - 2) = 0 \tag{2}$$

Now solve (*1*) for y',

$$y' = \frac{y}{3y^2 - x}$$

substitute in (*2*), and solve for y'':

$$(3y^2 - x)y'' + \frac{y}{3y^2 - x}\cdot\left(\frac{6y^2}{3y^2 - x} - 2\right) = 0$$
$$(3y^2 - x)^3y'' + (3y^2 - x)\frac{y}{3y^2 - x}\cdot(3y^2 - x)\left(\frac{6y^2}{3y^2 - x} - 2\right) = 0 \quad \text{[multiply by } (3y^2 - x)^2\text{]}$$
$$(3y^2 - x)^3y'' + y\cdot(6y^2 - 2(3y^2 - x)) = 0 \quad \left[a\left(\frac{b}{a} - c\right) = b - ca\right]$$
$$(3y^2 - x)^3y'' + y(6y^2 - 6y^2 + 2x) = 0$$
$$(3y^2 - x)^3y'' + 2xy = 0$$
$$y'' = \frac{-2xy}{(3y^2 - x)^3}$$

22.3 If y is a function of x such that

$$x^3 - 2xy + y^3 = 8 \tag{0}$$

and such that $y = 2$ when $x = 2$ (note that these values satisfy (*0*)), find the values of y' and y'' when $x = 2$.

Proceed as in Problem 22.2.

$$x^3 - 2xy + y^3 = 8$$
$$D_x(x^3 - 2xy + y^3) = D_x(8)$$
$$3x^2 - 2(xy' + y) + 3y^2y' = 0$$
$$3x^2 - 2xy' - 2y + 3y^2y' = 0 \tag{1}$$
$$D_x(3x^2 - 2xy' - 2y + 3y^2y') = D_x(0)$$
$$6x - 2(xy'' + y') - 2y' + 3(y^2y'' + y'(2yy')) = 0$$
$$6x - 2xy'' - 2y' - 2y' + 3y^2y'' + 6y(y')^2 = 0 \tag{2}$$

Substitute 2 for x and 2 for y in (*1*):

$$12 - 4y' - 4 + 12y' = 0 \quad \text{or} \quad 8y' + 8 = 0 \quad \text{or} \quad y' = -1$$

Substitute 2 for x, 2 for y, and -1 for y' in (*2*):

$$12 - 4y'' + 2 + 2 + 12y'' + 12 = 0 \quad \text{or} \quad 8y'' + 28 = 0 \quad \text{or} \quad y'' = -\frac{28}{8} = -\frac{7}{2}$$

22.4 Let $s = t^3 - 9t^2 + 24t$ describe the position s at time t of an object moving on a straight line. (*a*) Find the velocity and acceleration. (*b*) Determine when the velocity is positive and when it is negative. (*c*) Determine when the acceleration is positive and when it is negative. (*d*) Describe the motion of the object.

(*a*)
$$v = \frac{ds}{dt} = 3t^2 - 18t + 24 = 3(t^2 - 6t + 8)$$
$$= 3(t-2)(t-4)$$
$$a = \frac{dv}{dt} = 6t - 18 = 6(t-3)$$

(*b*) v is positive when:

$$(t-2>0 \text{ and } t-4>0) \quad \text{or} \quad (t-2<0 \text{ and } t-4<0)$$

that is, when:

$$(t>2 \text{ and } t>4) \quad \text{or} \quad (t<2 \text{ and } t<4)$$

which is equivalent to: $t>4$ or $t<2$. $v = 0$ when and only when $t = 2$ or $t = 4$; hence, $v<0$ when $2<t<4$.

(*c*) $a>0$ when $t>3$, and $a<0$ when $t<3$.

(*d*) Assuming that s increases to the right, positive velocity indicates movement to the right, and negative velocity movement to the left. The object moves right until, at $t = 2$, it is at $s = 20$, where it reverses direction. It then moves left until, at $t = 4$, it is at $s = 16$, where it reverses direction again. Thereafter, it continues to move right. See Fig. 22-2.

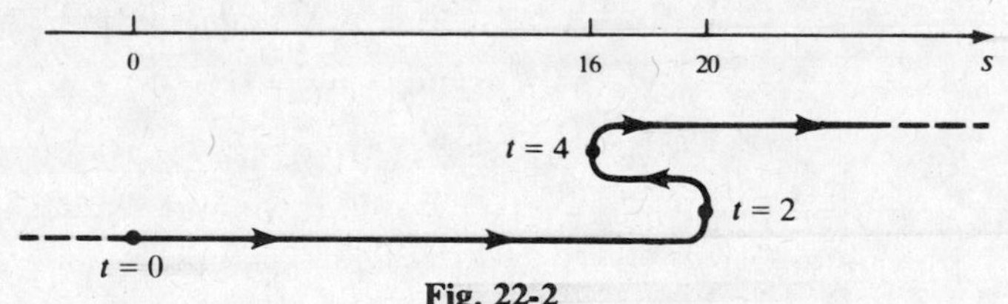

Fig. 22-2

Supplementary Problems

22.5 Find the second derivative $D_x^2 y$ of the following functions y:

(*a*) $y = x - \frac{1}{x}$ (*b*) $y = \pi x^3 - 7x$ (*c*) $y = \sqrt{x+5}$ (*d*) $y = \sqrt[3]{x-1}$

(*e*) $y = \sqrt[4]{x^2+1}$ (*f*) $y = (x+1)^2(x-3)^3$ (*g*) $y = \frac{x}{(1-x)^2}$

22.6 Use implicit differentiation to find the second derivative y'' in the following cases:

(*a*) $x^2 + y^2 = 1$ (*b*) $x^2 - y^2 = 1$ (*c*) $x^3 - y^3 = 1$ (*d*) $xy + y^2 = 1$

22.7 If $\sqrt{x} + \sqrt{y} = 1$, calculate y'' (*a*) by explicitly solving for y and then differentiating twice; (*b*) by implicit differentiation. (*c*) Which of methods (*a*) and (*b*) is the simpler?

22.8 Find all derivatives (first, second, etc.) of y:

(*a*) $y = 4x^4 - 2x^2 + 1$ (*b*) $y = 2x^2 + x - 1 + \dfrac{1}{x}$ (*c*) $y = \sqrt{x}$

(*d*) $y = \dfrac{x+1}{x-1}$ (*e*) $y = \dfrac{1}{3+x}$

22.9 Find the velocity the first time that the acceleration is 0, if the equation of motion is:

(*a*) $s = t^2 - 5t + 7$ (*b*) $s = t^3 - 3t + 2$ (*c*) $s = t^4 - 4t^3 + 6t^2 - 4t + 3$

22.10 At the point (1, 2) of the curve $x^2 - xy + y^2 = 3$, find the rate of change with respect to x of the slope of the tangent line to the curve.

22.11 If $x^2 + 2xy + 3y^2 = 2$, find the values of dy/dx and d^2y/dx^2 when $y = 1$.

22.12 Let

$$f(x) = \begin{cases} 1 + 3K(x-2) + (x-2)^2 & \text{if } x \le 2 \\ Lx + K & \text{if } x > 2 \end{cases}$$

where L and K are constants. (*a*) If $f(x)$ is differentiable at $x = 2$, find L and K. (*b*) With L and K as found in (*a*), is $f''(x)$ continuous for all x?

Chapter 23

Applications of the Second Derivative

23.1 CONCAVITY

If a curve has the shape of a cup or part of a cup (as the curves in Fig. 23-1), then we say that it is *concave upward*. (To remember the sense of "concave upward," notice that the letters "c" and "up" form the word "cup.") A mathematical description of this notion can be given: a curve is concave upward if the entire curve lies above the tangent line at any given point of the curve. Thus, in Fig. 23-1(a), the curve lies above all three tangent lines.

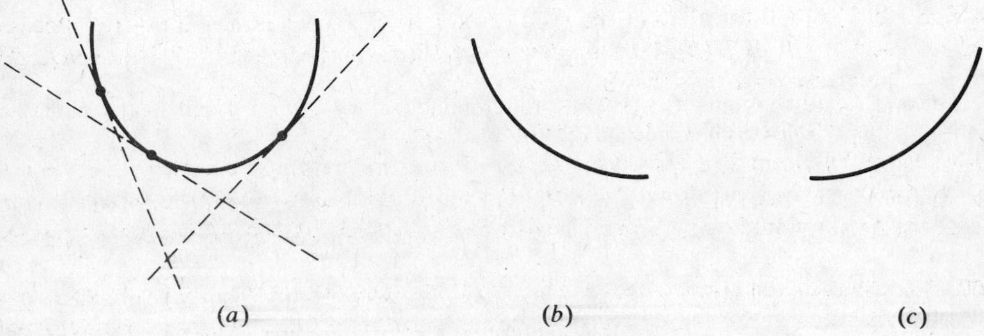

Fig. 23-1. Concavity Upward

A curve is said to be *concave downward* if it has the shape of a cap or part of a cap (Fig. 23-2). In mathematical terms, a curve is concave downward if it lies wholly below the tangent line at an arbitrary point of the curve. (See Fig. 23-2(a).)

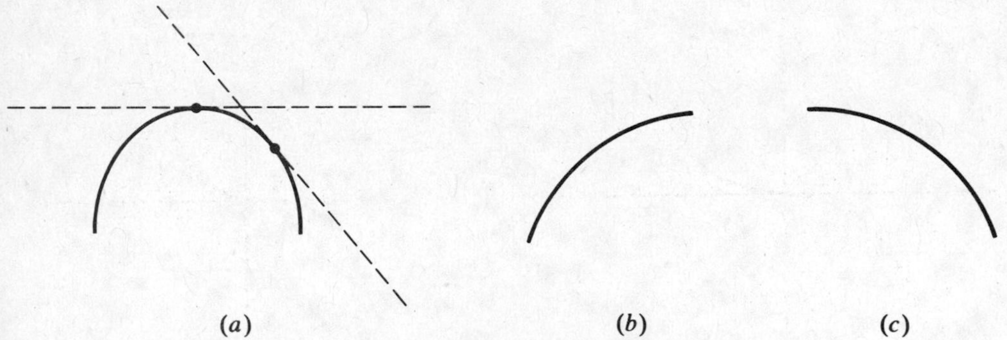

Fig. 23-2. Concavity Downward

NOMENCLATURE You may find in some books the word "convexity" used to denote concavity upward, and "concavity" to denote concavity downward.

A curve may, of course, be composed of parts of different concavity. The curve in Fig. 23-3 is concave downward from A to B, concave upward from B to C, concave downward from C to D, and concave upward from D to E. A point on a curve at which the concavity changes is called an *inflection point*; B, C, and D are inflection points in Fig. 23-3.

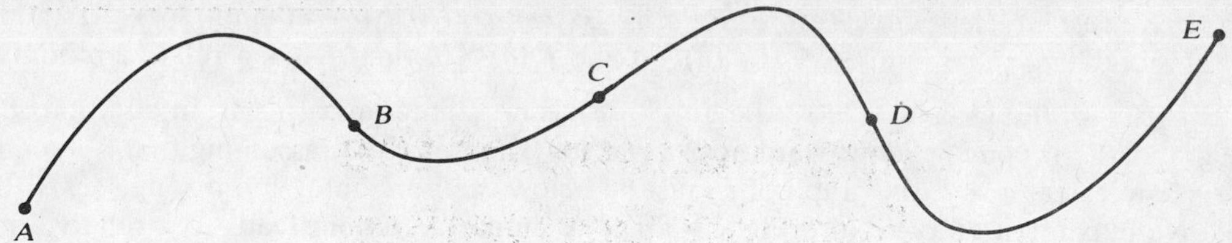

Fig. 23-3

It is obvious from Fig. 23-1 that if a cuplike curve is traversed in a generally rightward direction, the slope of the tangent line will steadily increase (become less negative or more positive)—or, at least, it will never decrease. Conversely, if the tangent line has this property, the curve must be cuplike. Now, for the curve $y = f(x)$, the tangent line will certainly have this property if $f''(x) > 0$ (for then, by Theorem 17.3, the slope $f'(x)$ of the tangent line will be an increasing function). By a similar consideration, we see that if $f''(x) < 0$, the curve $y = f(x)$ must be caplike. To summarize:

Theorem 23.1: If $f''(x) > 0$ for all x in (a, b), then the graph of f is concave upward between $x = a$ and $x = b$. If $f''(x) < 0$ for all x in (a, b), then the graph of f is concave downward between $x = a$ and $x = b$.

For a rigorous proof of Theorem 23.1, see Problem 23.17.

It follows from Theorem 23.1 that *if the graph of f has an inflection point at $x = c$, then $f''(c) = 0$.* Indeed, if $f''(c)$ were, say, negative, then $f''(x)$ would be negative over a small open interval containing c, the graph would be concave downward there, and c could not be an inflection point.

EXAMPLES (*a*) Consider the graph of $y = x^3$ (Fig. 23-4(*a*)). Here, $y' = 3x^2$ and $y'' = 6x$. Since $y'' > 0$ when $x > 0$, and $y'' < 0$ when $x < 0$, the curve is concave upward when $x > 0$, and concave downward when $x < 0$. There is an inflection point at the origin, where the concavity changes. This is the only possible inflection point, for, if $y'' = 6x = 0$, then x must be 0. (*b*) If $f''(c) = 0$, the graph of f need not have an inflection point at $x = c$. For instance, the graph of $f(x) = x^4$ (Fig. 23-4(*b*)) has a relative minimum, not an inflection point, at $x = 0$, where

$$f''(x) = 12x^2 = 0$$

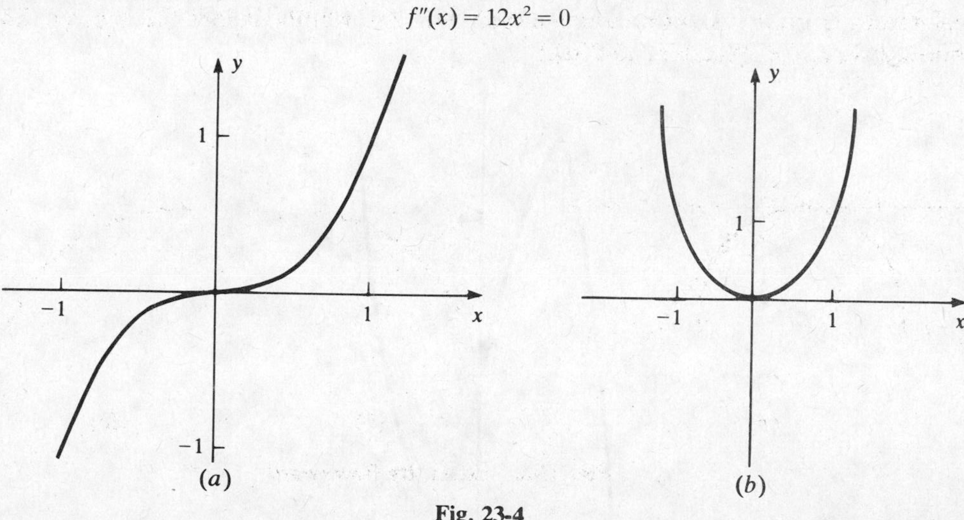

Fig. 23-4

23.2 TESTS FOR RELATIVE EXTREMA

We already know, from Chapter 14, that the condition $f'(c) = 0$ is necessary, but not sufficient, for a differentiable function $f(x)$ to have a relative maximum or minimum at $x = c$. We need some additional information which will tell us whether a function actually has a relative extremum at a point where its derivative is zero.

Theorem 23.2 *(Second-Derivative Test for Relative Extrema)*: If $f'(c) = 0$ and $f''(c) < 0$, then f has a relative maximum at c. If $f'(c) = 0$ and $f''(c) > 0$, then f has a relative minimum at c.

Proof. If $f'(c) = 0$, the tangent line to the graph of f is horizontal at $x = c$. If, in addition, $f''(c) < 0$, then the graph of f is concave downward near $x = c$ (by Theorem 23.1, assuming that f'' is continuous at $x = c$). Hence, near $x = c$, the graph of f must lie below the horizontal line through $(c, f(c))$; f thus has a relative maximum at $x = c$ (see Fig. 23-5(a)). A similar argument leads to a relative minimum when $f''(c) > 0$ (see Fig. 23-5(b)).

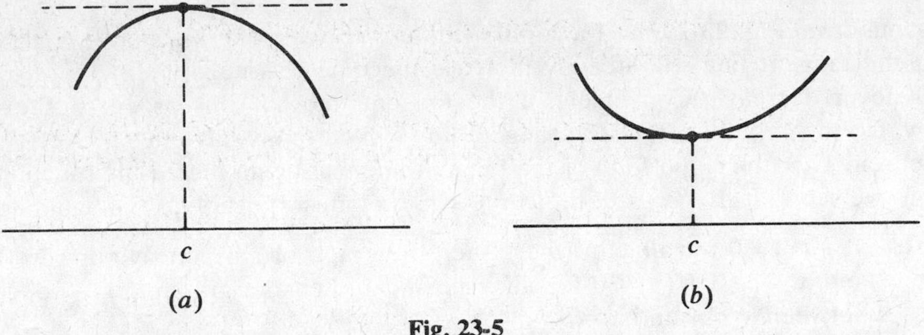

Fig. 23-5

EXAMPLE Consider the function $f(x) = 2x^3 + x^2 - 4x + 2$. Then

$$f'(x) = 6x^2 + 2x - 4 = 2(3x^2 + x - 2) = 2(3x - 2)(x + 1)$$

Hence, if $f'(x) = 0$, then $3x - 2 = 0$ or $x + 1 = 0$; that is, $x = \frac{2}{3}$ or $x = -1$. Now, $f''(x) = 12x + 2$; hence,

$$f''(-1) = 12(-1) + 2 = -12 + 2 = -10 < 0$$
$$f''(\tfrac{2}{3}) = 12(\tfrac{2}{3}) + 2 = 8 + 2 = 10 > 0$$

Since $f''(-1) < 0$, f has a relative maximum at $x = -1$, with

$$f(-1) = 2(-1)^3 + (-1)^2 - 4(-1) + 2 = -2 + 1 + 4 + 2 = 5$$

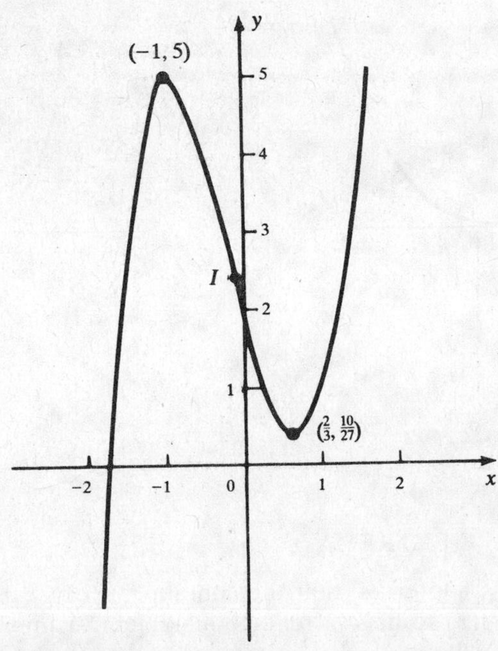

Fig. 23-6

Since $f''(\frac{2}{3}) > 0$, f has a relative minimum at $x = \frac{2}{3}$, with

$$f(\tfrac{2}{3}) = 2(\tfrac{2}{3})^3 + (\tfrac{2}{3})^2 - 4(\tfrac{2}{3}) + 2 = \frac{16}{27} + \frac{4}{9} - \frac{8}{3} + 2$$
$$= \frac{16}{27} + \frac{12}{27} - \frac{72}{27} + \frac{54}{27} = \frac{10}{27}$$

The graph of f is shown in Fig. 23-6.

Now, because

$$f''(x) = 12x + 2 = 12\left(x + \frac{1}{6}\right) = 12\left(x - \left(-\frac{1}{6}\right)\right)$$

$f''(x) > 0$ when $x > -\frac{1}{6}$, and $f''(x) < 0$ when $x < -\frac{1}{6}$. Hence, the curve is concave upward for $x > -\frac{1}{6}$ and concave downward for $x < -\frac{1}{6}$. So, there must be an inflection point I, where $x = -\frac{1}{6}$.

From Problem 9.1, we know that

$$\lim_{x\to+\infty} f(x) = \lim_{x\to+\infty} 2x^3 = +\infty \qquad \lim_{x\to-\infty} f(x) = \lim_{x\to-\infty} 2x^3 = -\infty$$

Thus, the curve moves upward without bound toward the right, and downward without bound toward the left.

The Second-Derivative Test tells us *nothing* when $f'(c) = 0$ and $f''(c) = 0$. This is shown by the examples in Fig. 23-7, where each function is such that $f = f' = f'' = 0$ at $x = 0$.

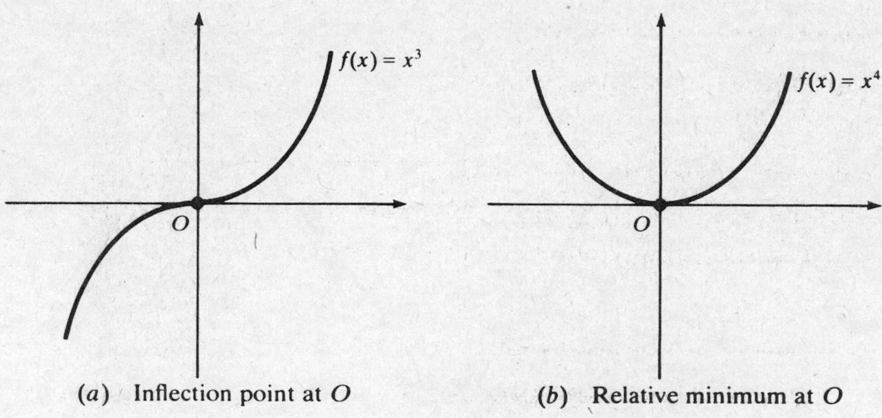

(*a*) Inflection point at O (*b*) Relative minimum at O

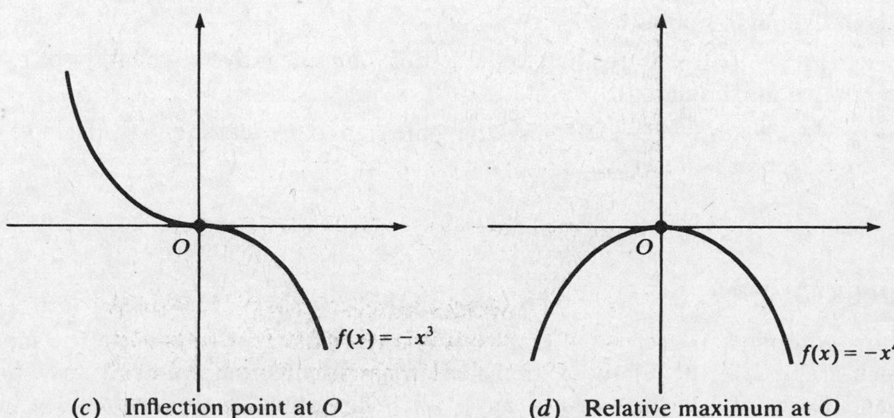

(*c*) Inflection point at O (*d*) Relative maximum at O

Fig. 23-7

To distinguish among the four cases shown in Fig. 23-7, consider the sign of the derivative just to the left and just to the right of the critical point (the origin, in the present case). Recalling that the sign of the derivative is the sign of the slope of the tangent line, we have the four combinations shown in Fig. 23-8; these lead directly to Theorem 23.3.

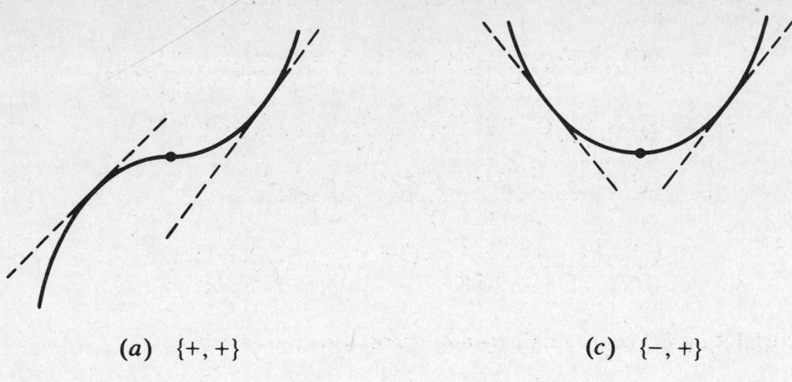

(*a*) {+, +} (*c*) {−, +}

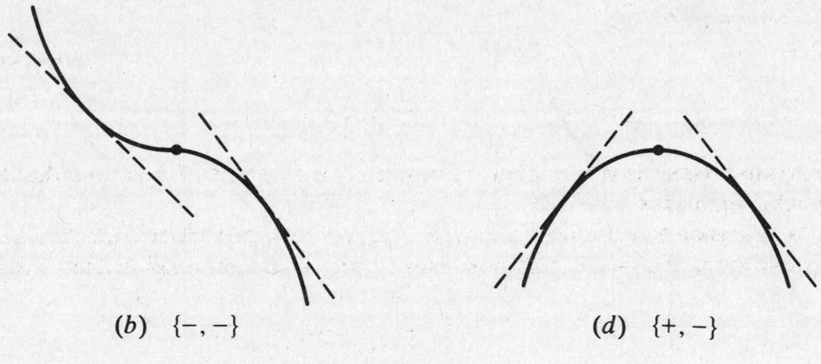

(*b*) {−, −} (*d*) {+, −}

Fig. 23-8

Theorem 23.3 (*First-Derivative Test for Relative Extrema*): Assume $f'(c) = 0$.

{−, +} If f' is negative to the left of c and positive to the right of c, then f has a **relative minimum** at c.

{+, −} If f' is positive to the left of c and negative to the right of c, then f has a **relative maximum** at c.

{+, +} or {−, −} If f' has the same sign to the left and to the right of c, then f has an **inflection point** at c.

23.3 GRAPH SKETCHING

We are now equipped to sketch the graphs of a great variety of functions. The most important features of such graphs are: (i) relative extrema (if any), (ii) inflection points (if any), (iii) concavity, (iv) vertical and horizontal asymptotes (if any), (v) behavior as x approaches $+\infty$ and $-\infty$. The procedure was illustrated for the function $f(x) = 2x^3 + x^2 - 4x + 2$ in Section 23.2; an additional example follows.

EXAMPLE Sketch the graph of the rational function

$$f(x) = \frac{x}{x^2+1}$$

First of all, the function is seen to be odd (Section 7.3), so that it need be graphed only for positive x; the graph is then completed by reflection in the origin. Compute the first two derivatives of f:

$$f'(x) = \frac{(x^2+1)D_x(x) - xD_x(x^2+1)}{(x^2+1)^2} = \frac{x^2+1-x(2x)}{(x^2+1)^2} = \frac{1-x^2}{(x^2+1)^2}$$

$$\begin{aligned} f''(x) = D_x f'(x) &= \frac{(x^2+1)^2 D_x(1-x^2) - (1-x^2)D_x((x^2+1)^2)}{(x^2+1)^4} \\ &= \frac{(x^2+1)^2(-2x) - (1-x^2)(2(x^2+1)(2x))}{(x^2+1)^4} \\ &= \frac{(-2x)(x^2+1)(x^2+1+2(1-x^2))}{(x^2+1)^4} = \frac{(-2x)(3-x^2)}{(x^2+1)^3} \end{aligned}$$

Since $(1-x^2) = (1-x)(1+x)$, $f'(x)$ has a single positive root, $x = 1$, at which

$$f''(1) = \frac{(-2)(2)}{(2)^3} = -\frac{1}{2}$$

Then, by Theorem 23.2, f has a relative maximum at $x = 1$; the maximum value is $f(1) = 1/2$.

Application of Theorem 23.1 to the expression for $f''(x)$ shows that the graph of f is concave downward from $x = 0$ to $x = \sqrt{3}$ and concave upward for $x > \sqrt{3}$. At $x = \sqrt{3}$, and also at $x = 0$, where the concavity changes, there is an inflection point I.

Finally, calculate

$$\lim_{x\to+\infty} \frac{x}{x^2+1} = \lim_{x\to+\infty} \frac{1/x}{1+(1/x^2)} = \frac{0}{1+0} = 0$$

which shows that the positive x-axis is a horizontal asymptote to the right. It shows, too, that the value 1/2 is an absolute, as well as a relative, maximum value.

The graph, with its extension to negative x (dashed), is sketched in Fig. 23-9. Note the minimum at $x = -1$, corresponding to the maximum at $x = +1$, and note how concavity of one kind reflects into concavity of the other kind.

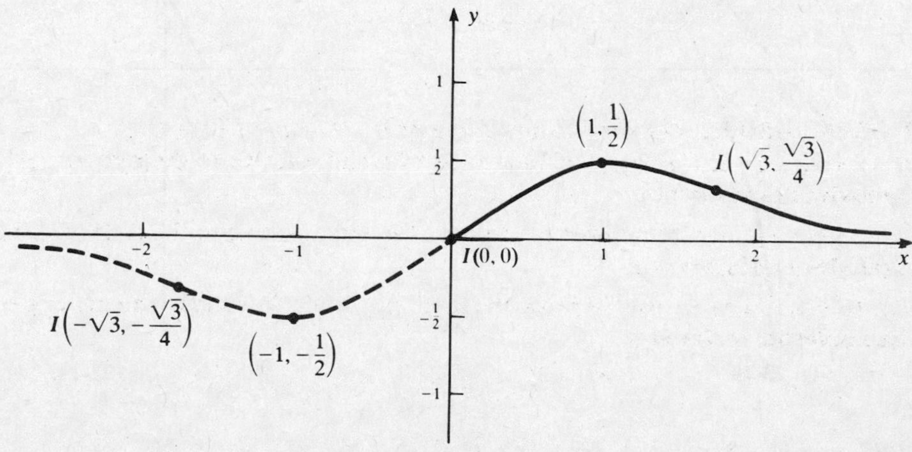

Fig. 23-9

Solved Problems

23.1 Sketch the graph of $f(x) = x - \frac{1}{x}$ $(x \neq 0)$.

The function is odd, with

$$f'(x) = D_x(x - x^{-1}) = 1 - (-1)x^{-2} = 1 + \frac{1}{x^2}$$

$$f''(x) = D_x(1 + x^{-2}) = -2x^{-3} = -\frac{2}{x^3}$$

Because x is an increasing function, $1/x$ is a decreasing function and $-1/x$ an increasing function. Thus, as the sum of increasing functions, $f(x)$ is an increasing function (confirmed by the positivity of $f'(x)$). Moreover, for $x > 0$, the graph of f is concave downward—because $f'' < 0$—and has the line $y = x$ as an asymptote—because

$$\lim_{x \to +\infty} (x - f(x)) = \lim_{x \to +\infty} \frac{1}{x} = 0$$

Finally, since

$$\lim_{x \to 0^+} f(x) = \lim_{x \to 0^+} \left(x - \frac{1}{x}\right) = \lim_{x \to 0^+} \left(-\frac{1}{x}\right) = -\infty$$

the graph of f has the negative y-axis as a vertical asymptote. Notice that $x = 0$, at which f' is undefined, is the sole critical number of f.

The graph is sketched, for all x, in Fig. 23-10. Although the concavity changes at $x = 0$, there is no inflection point there, since $f(0)$ is not defined.

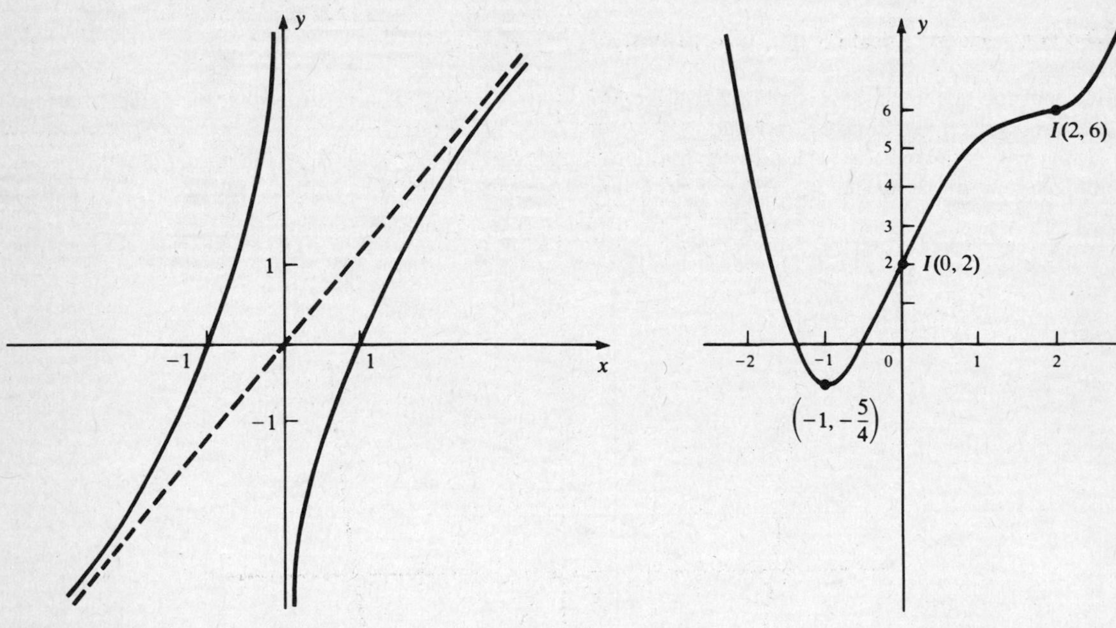

Fig. 23-10 **Fig. 23-11**

23.2 Sketch the graph of $f(x) = \frac{1}{4}x^4 - x^3 + 4x + 2$.

$$f'(x) = x^3 - 3x^2 + 4$$

By inspection, -1 is a root of $x^3 - 3x^2 + 4$.

ALGEBRA When trying to find roots of a polynomial, first test the integral factors of the constant term. In this case, the factors are ± 1, ± 2, ± 4.

So (Theorem 7.2), $f'(x)$ is divisible by $x+1$; the division yields:

$$x^3 - 3x^2 + 4 = (x+1)(x^2 - 4x + 4) = (x+1)(x-2)^2$$

Hence, the critical numbers are $x = -1$ and $x = 2$. Now,

$$f''(x) = 3x^2 - 6x = 3x(x-2)$$

So, $f''(-1) = 3(-1)(-1-2) = 9$. Hence, by the Second-Derivative Test, f has a relative minimum at $x = -1$.

Since $f''(2) = 3(2)(2-2) = 0$, we make the First-Derivative Test at $x = 2$.

$$f'(x) = (x+1)(x-2)^2$$

On both sides of $x = 2$, $f'(x) > 0$, since $x + 1 > 0$ and $(x-2)^2 > 0$. This is the case $\{+, +\}$: there is an inflection point at (2, 6). Furthermore, $f''(x)$ changes sign at $x = 0$, so that there is also an inflection point at (0, 2). Because

$$\lim_{x\to\pm\infty} f(x) = \lim_{x\to\pm\infty} \tfrac{1}{4}x^4 = +\infty$$

the graph moves upward without bound on the left and the right. The graph is shown in Fig. 23-11.

23.3 Sketch the graph of $f(x) = x^4 - 8x^2$.

As the function is even, we restrict attention to $x \ge 0$.

$$f'(x) = 4x^3 - 16x = 4x(x^2 - 4) = 4x(x-2)(x+2)$$

$$f''(x) = 12x^2 - 16 = 4(3x^2 - 4) = 12\left(x^2 - \frac{4}{3}\right) = 12\left(x + \frac{2}{\sqrt{3}}\right)\left(x - \frac{2}{\sqrt{3}}\right)$$

The nonnegative critical numbers are $x = 0$ and $x = 2$. Testing,

x	$f(x)$	$f''(x)$	
0	0	-16	*rel. max.*
2	-16	32	*rel. min.*
$+\frac{2}{\sqrt{3}}$	$-\frac{80}{9}$	0	*infl. pt.*

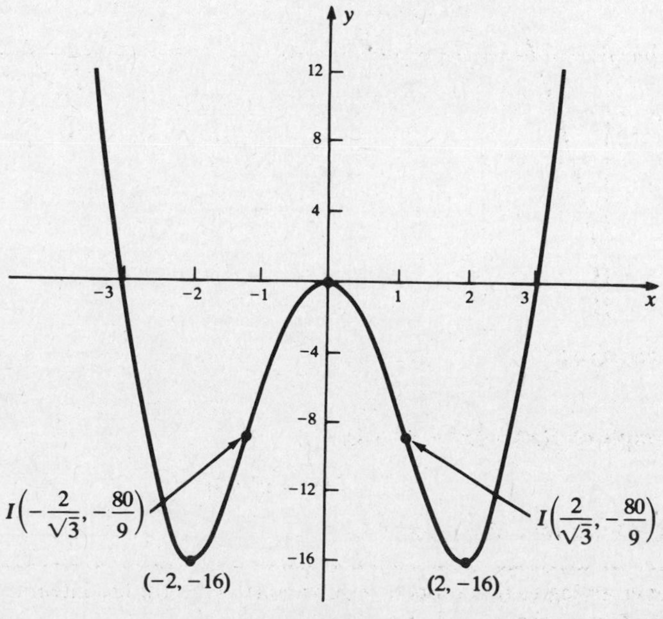

Fig. 23-12

Checking the sign of $f''(x)$, we see that the graph will be concave downward for $0 < x < 2/\sqrt{3}$ and concave upward for $x > 2/\sqrt{3}$. Because $\lim_{x\to+\infty} f(x) = +\infty$, the graph moves upward without bound on the right.

The graph is sketched in Fig. 23-12. Observe that, on the set of all real numbers, f has an absolute minimum value, -16, assumed at $x = \pm 2$, but no absolute maximum value.

Supplementary Problems

23.4 Determine the intervals where the graphs of the following functions are concave upward, and the intervals where they are concave downward. Find all inflection points.

(a) $f(x) = x^2 - x + 12$ (b) $f(x) = x^3 + 15x^2 + 6x + 1$

(c) $f(x) = x^4 + 18x^3 + 120x^2 + x + 1$ (d) $f(x) = \dfrac{x}{2x-1}$ (e) $f(x) = 5x^4 - x^5$

23.5 Find the critical numbers of the following functions, and determine whether they yield relative maxima, relative minima, inflection points, or none of these.

(a) $f(x) = 8 - 3x + x^2$ (b) $f(x) = x^4 - 18x^2 + 9$ (c) $f(x) = x^3 - 5x^2 - 8x + 3$

(d) $f(x) = \dfrac{x^2}{x-1}$ (e) $f(x) = \dfrac{x^2}{x^2+1}$

23.6 Sketch the graphs of the following functions, showing extrema (relative or absolute), inflection points, asymptotes, and behavior at infinity.

(a) $f(x) = (x^2 - 1)^3$ (b) $f(x) = x^3 - 2x^2 - 4x + 3$ (c) $f(x) = x(x-2)^2$

(d) $f(x) = x^4 + 4x^3$ (e) $f(x) = 3x^5 - 20x^3$ (f) $f(x) = \sqrt[3]{x-1}$

(g) $f(x) = x^2 + \dfrac{2}{x}$ (h) $f(x) = \dfrac{x^2-3}{x^3}$ (i) $f(x) = \dfrac{(x-1)^3}{x^2}$

23.7 If, for all x, $f'(x) > 0$ and $f''(x) < 0$, which of the curves in Fig. 23-13 could be part of the graph of f?

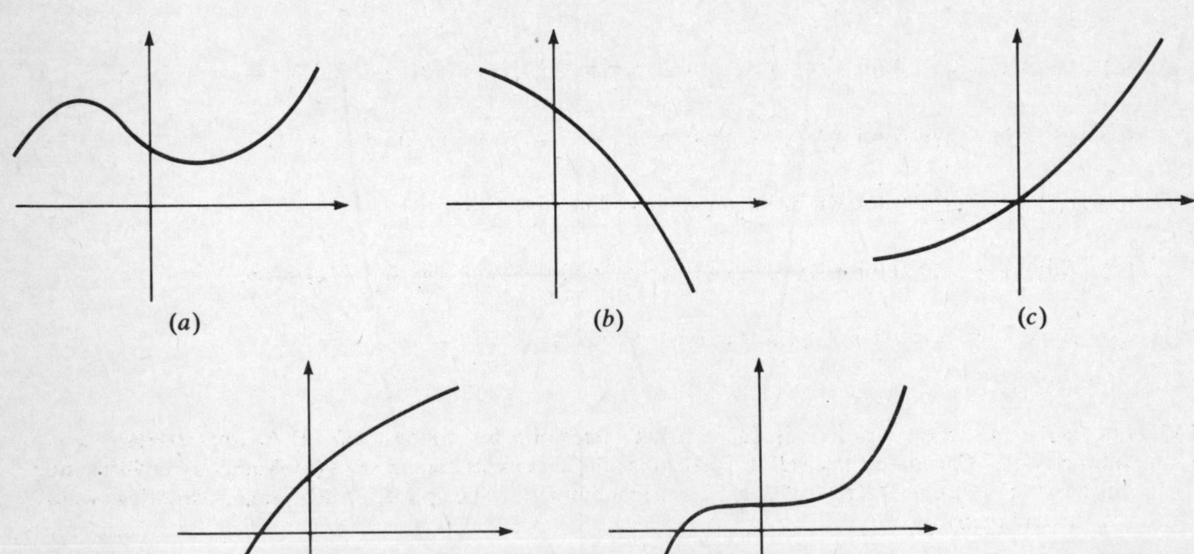

Fig. 23-13

23.8 At which of the five indicated points of the graph in Fig. 23-14 do y' and y'' have the same sign?

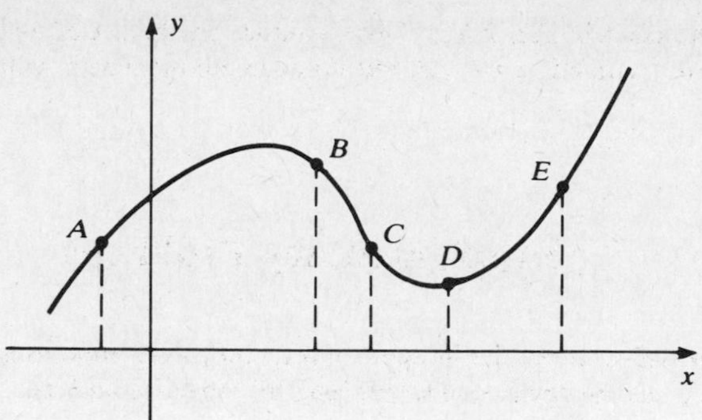

Fig. 23-14

23.9 Let $f(x) = ax^2 + bx + c$, with $a \neq 0$. (*a*) How many relative extrema does f have? (*b*) How many points of inflection does the graph of f have? (*c*) What kind of curve is the graph of f?

23.10 Let f be continuous for all x, with a relative maximum at $(-1, 4)$ and a relative minimum at $(3, -2)$. Which of the following *must* be true? (*a*) The graph of f has a point of inflection for some x in $(-1, 3)$. (*b*) The graph of f has a vertical asymptote. (*c*) The graph of f has a horizontal asymptote. (*d*) $f'(3) = 0$. (*e*) The graph of f has a horizontal tangent line at $x = -1$. (*f*) The graph of f intersects both the x-axis and the y-axis. (*g*) f has an absolute maximum on the set of all real numbers.

23.11 If $f(x) = x^3 + 3x^2 + k$ has three distinct real roots, what are the bounds on k? (*Hint*: Sketch the graph of f, using f' and f''. At how many points does the graph cross the x-axis?)

23.12 Sketch the graph of a continuous function f such that:

(*a*) $f(1) = -2$, $f'(1) = 0$, $f''(x) > 0$ for all x;

(*b*) $f(2) = 3$, $f'(2) = 0$, $f''(x) < 0$ for all x;

(*c*) $f(1) = 1$, $f''(x) < 0$ for $x > 1$, $f''(x) > 0$ for $x < 1$, $\lim_{x \to +\infty} f(x) = +\infty$, $\lim_{x \to -\infty} f(x) = -\infty$;

(*d*) $f(0) = 0$, $f''(x) < 0$ for $x > 0$, $f''(x) > 0$ for $x < 0$, $\lim_{x \to +\infty} f(x) = 1$, $\lim_{x \to -\infty} f(x) = -1$;

(*e*) $f(0) = 1$, $f''(x) < 0$ for $x \neq 0$, $\lim_{x \to 0^+} f'(x) = +\infty$, $\lim_{x \to 0^-} f'(x) = -\infty$;

(*f*) $f(0) = 0$, $f''(x) > 0$ for $x < 0$, $f''(x) < 0$ for $x > 0$, $\lim_{x \to 0^-} f'(x) = +\infty$, $\lim_{x \to 0^+} f'(x) = +\infty$;

(*g*) $f(0) = 1$, $f''(x) < 0$ if $x \neq 0$, $\lim_{x \to 0^+} f'(x) = 0$, $\lim_{x \to 0^-} f'(x) = -\infty$.

23.13 Let $f(x) = x|x - 1|$ for x in $[-1, 2]$. (*a*) At what values of x is f continuous? (*b*) At what values of x is f differentiable? Calculate $f'(x)$. (*Hint*: Distinguish the cases $x > 1$ and $x < 1$.) (*c*) Where is f an increasing function? (*d*) Calculate $f''(x)$. (*e*) Where is the graph of f concave upward, and where concave downward? (*f*) Sketch the graph of f.

23.14 Given functions f and g such that, for all x, (i) $(g(x))^2 - (f(x))^2 = 1$; (ii) $f'(x) = (g(x))^2$; (iii) $f''(x)$ and $g''(x)$ exist; (iv) $g(x) < 0$; (v) $f(0) = 0$. Show that (*a*) $g'(x) = f(x)g(x)$, (*b*) g has a relative maximum at $x = 0$, (*c*) f has a point of inflection at $x = 0$.

23.15 For what value of k will $x - kx^{-1}$ have a relative maximum at $x = -2$?

23.16 Let $f(x) = x^4 + Ax^3 + Bx^2 + Cx + D$. Assume that the graph of $y = f(x)$ is symmetric with respect to the y-axis, has a relative maximum at (0, 1), and has an absolute minimum at $(k, -3)$. Find A, B, C, and D, as well as the possible value(s) of k.

23.17 Prove Theorem 23.1. (*Hint*: Assume that $f''(x) > 0$ on (a, b), and let c be in (a, b). The equation of the tangent line at $x = c$ is

$$y = f'(c)(x - c) + f(c)$$

It must be shown that $f(x) > f'(c)(x - c) + f(c)$. But the Mean-Value Theorem gives

$$f(x) = f'(x^*)(x - c) + f(c)$$

where x^* is between x and c, and, since $f''(x) > 0$ on (a, b), f' is increasing.)

Chapter 24

More Maximum and Minimum Problems

Until now, we have been able to find the absolute maxima and minima of differentiable functions only on *closed* intervals (see Section 14.2). The following result often enables us to handle cases where the function is defined on a half-open interval, open interval, infinite interval, or the set of all real numbers. Remember that, in general, there is no guarantee that a function has an absolute maximum or an absolute minimum on such intervals.

Theorem 24.1: Let f be a continuous function on an interval $\mathscr{I}$, with a *single* relative extremum within $\mathscr{I}$. Then this relative extremum is also an absolute extremum on $\mathscr{I}$.

Intuitive Argument: Refer to Fig. 24-1. Suppose that f has a relative maximum at c, and no other relative extremum inside $\mathscr{I}$. Take any other number d in $\mathscr{I}$. The curve moves downward on both sides of c. Hence, if the value $f(d)$ were greater than $f(c)$, then, at some point u between c and d, the curve would have to change direction and start moving upward. But then f would have a relative minimum at u—a contradiction.

For a rigorous proof, see Problem 24.20.

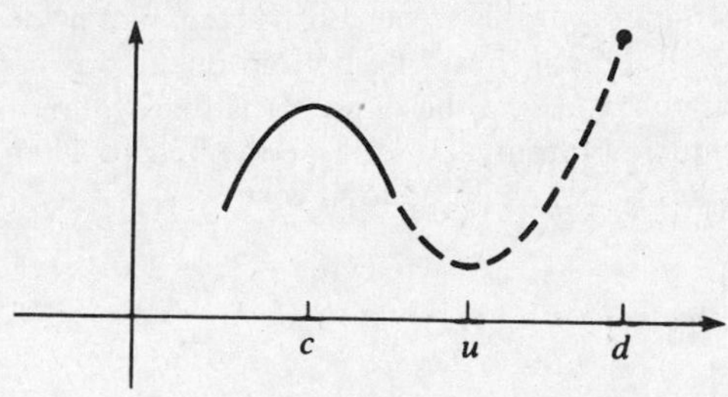

Fig. 24-1

EXAMPLES

(*a*) Find the shortest distance from the point $P(1, 0)$ to the parabola $x = y^2$ (see Fig. 24-2(*a*)).

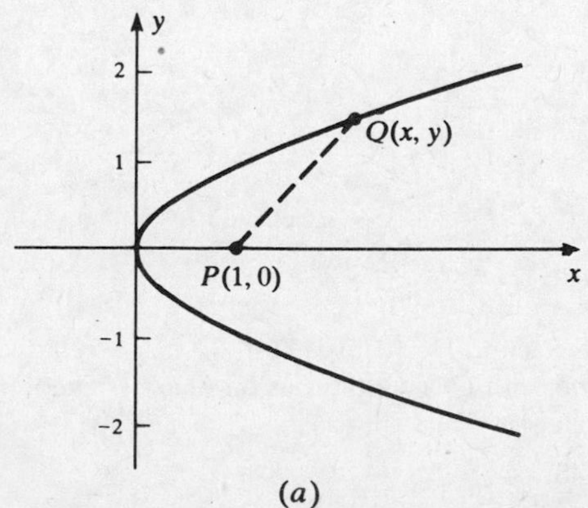

(*a*)

(*b*)

Fig. 24-2

The distance from an arbitrary point $Q(x, y)$ on the parabola to the point $P(1, 0)$ is, by *(2.1)*,

$$\begin{aligned} u &= \sqrt{(x-1)^2 + y^2} \\ &= \sqrt{(x-1)^2 + x} \quad [y^2 = x \text{ at } Q] \\ &= \sqrt{x^2 - 2x + 1 + x} = \sqrt{x^2 - x + 1} \end{aligned}$$

But, minimizing u is equivalent to minimizing

$$u^2 \equiv F(x) = x^2 - x + 1$$

on the interval $[0, +\infty)$ (the value of x is restricted by the fact that $x = y^2 \geq 0$).

$$F'(x) = 2x - 1 \qquad F''(x) = 2$$

The only critical number is the solution of

$$F'(x) = 2x - 1 = 0 \qquad \text{or} \qquad x = \frac{1}{2}$$

Now, $F''(\frac{1}{2}) = 2 > 0$; so, by the Second-Derivative Test, the function F has a relative minimum at $x = \frac{1}{2}$. Theorem 24.1 implies that this is an absolute minimum. When $x = \frac{1}{2}$,

$$y^2 = x = \frac{1}{2} \qquad \text{and} \qquad y = \pm\frac{1}{\sqrt{2}} = \pm\frac{\sqrt{2}}{2}$$

Thus, the points on the parabola closest to $(1, 0)$ are

$$\left(\frac{1}{2}, \frac{\sqrt{2}}{2}\right) \qquad \text{and} \qquad \left(\frac{1}{2}, -\frac{\sqrt{2}}{2}\right)$$

(*b*) An open box (that is, a box without a top) is to be constructed with a square base (see Fig. 24-2(*b*)), and is required to have a volume of 48 cubic inches. The bottom of the box costs 3 cents per square inch, while the sides costs 2 cents per square inch. Find the dimensions that will minimize the cost of the box.

Let x be the side of the square bottom, and let h be the height. Then the cost of the bottom is $3x^2$ and the cost of each of the four sides is $2xh$, giving a total cost

$$C = 3x^2 + 4(2xh) = 3x^2 + 8xh$$

The volume is $V = 48 = x^2h$. Hence, $h = 48/x^2$ and

$$C = 3x^2 + 8x\left(\frac{48}{x^2}\right) = 3x^2 + \frac{384}{x} = 3x^2 + 384x^{-1}$$

which is to be minimized on $(0, +\infty)$. Now,

$$\frac{dC}{dx} = 6x - 384x^{-2} = 6x - \frac{384}{x^2}$$

and so the critical numbers are the solutions of:

$$\begin{aligned} 6x - \frac{384}{x^2} &= 0 \\ 6x &= \frac{384}{x^2} \\ x^3 &= 64 \\ x &= 4 \end{aligned}$$

Now $$\frac{d^2C}{dx^2} = 6 - (-2)384x^{-3} = 6 + \frac{768}{x^3} > 0$$

for all positive x; in particular, for $x = 4$. By the Second-Derivative Test, C has a relative minimum at $x = 4$. But, since 4 is the only positive critical number and C is continuous on $(0, +\infty)$, Theorem 24.1 tells us that C has an absolute minimum at $x = 4$. When $x = 4$,

$$h = \frac{48}{x^2} = \frac{48}{16} = 3$$

So, the side of the base should be 4 inches, and the height 3 inches.

Solved Problems

24.1 A farmer must fence in a rectangular field with one side along a stream; no fence is needed on that side. If the area must be 1800 square meters, and the fencing costs \$2 per meter, what dimensions will minimize the cost?

Let x and y be the lengths of the sides respectively parallel and perpendicular to the stream. Then the cost C is:

$$C = 2(x + 2y) = 2x + 4y$$

But, $1800 = xy$, or $x = 1800/y$, so that

$$C = 2\left(\frac{1800}{y}\right) + 4y = \frac{3600}{y} + 4y = 3600y^{-1} + 4y$$

and
$$\frac{dC}{dy} = -3600y^{-2} + 4 = -\frac{3600}{y^2} + 4$$

We wish to minimize $C(y)$ for $y > 0$; so we look for positive critical numbers:

$$-\frac{3600}{y^2} + 4 = 0$$

$$4 = \frac{3600}{y^2}$$

$$y^2 = \frac{3600}{4} = 900$$

$$y = +30$$

Now
$$\frac{d^2C}{dy^2} = \frac{7200}{y^3}$$

which is positive at $y = +30$. Thus, by the Second-Derivative Test, C has a relative minimum at $y = 30$. Since $y = 30$ is the only positive critical number, there can be no other relative extremum in the interval $(0, +\infty)$. Therefore C has an absolute minimum at $y = 30$, by Theorem 24.1. When $y = 30$ meters,

$$x = \frac{1800}{y} = \frac{1800}{30} = 60 \text{ meters}$$

24.2 If $c_1, c_2, \ldots, c_n$ are the results of n measurements of an unknown quantity, a method for estimating the value of that quantity is to find the number x which minimizes the function

$$f(x) = (x - c_1)^2 + (x - c_2)^2 + \cdots + (x - c_n)^2$$

This method is called the *Least-Squares Principle.* Find the value of x determined by the Least-Squares Principle.

$$f'(x) = 2(x - c_1) + 2(x - c_2) + \cdots + 2(x - c_n)$$

To find the critical numbers:

$$2(x - c_1) + 2(x - c_2) + \cdots + 2(x - c_n) = 0$$

$$(x - c_1) + (x - c_2) + \cdots + (x - c_n) = 0$$

$$nx - (c_1 + c_2 + \cdots + c_n) = 0$$

$$nx = c_1 + c_2 + \cdots + c_n$$

$$x = \frac{c_1 + c_2 + \cdots + c_n}{n}$$

As $f''(x) = 2 + 2 + \cdots + 2 = 2n > 0$, we have, by the Second-Derivative Test, a relative minimum of f at the unique critical number. By Theorem 24.1, this relative minimum is also an absolute minimum on the set of all real x. Thus, the Least-Squares Principle prescribes the *average of the n measurements.*

24.3 Let

$$f(x) = \frac{4x^2 - 3}{x - 1}$$

for $0 \le x < 1$. Find the absolute extrema, if any, of f on $[0, 1)$.

It will simplify the differentiations if we write

$$f(x) = \frac{4x^2 - 4 + 1}{x - 1} = \frac{4(x^2 - 1) + 1}{x - 1} = \frac{4(x - 1)(x + 1) + 1}{x - 1}$$

$$= 4(x + 1) + \frac{1}{x - 1}$$

Then, by the Chain Rule,

$$f'(x) = 4 - \frac{1}{(x - 1)^2} \qquad (1)$$

$$f''(x) = \frac{+2}{(x - 1)^3} \qquad (2)$$

From (*1*), the critical numbers are the roots of:

$$4 - \frac{1}{(x - 1)^2} = 0$$

$$4 = \frac{1}{(x - 1)^2}$$

$$(x - 1)^2 = \frac{1}{4}$$

$$x - 1 = \pm\frac{1}{2}$$

$$x = 1 \pm \frac{1}{2} = \frac{3}{2}, \frac{1}{2}$$

Thus, the only critical number in $[0, 1)$ is $\frac{1}{2}$; and, from (*2*),

$$f''(\tfrac{1}{2}) = \frac{2}{(\frac{1}{2} - 1)^3} = \frac{2}{(-\frac{1}{2})^3} = \frac{2}{-\frac{1}{8}} < 0$$

It follows that there is a single relative extremum—a relative maximum—in $[0, 1)$ at $x = \frac{1}{2}$. By Theorem 24.1, this extremum is also an absolute maximum. The function f has no absolute minimum on $[0, 1)$; rather, it has the line $x = 1$ as a vertical asymptote (see Fig. 24.3).

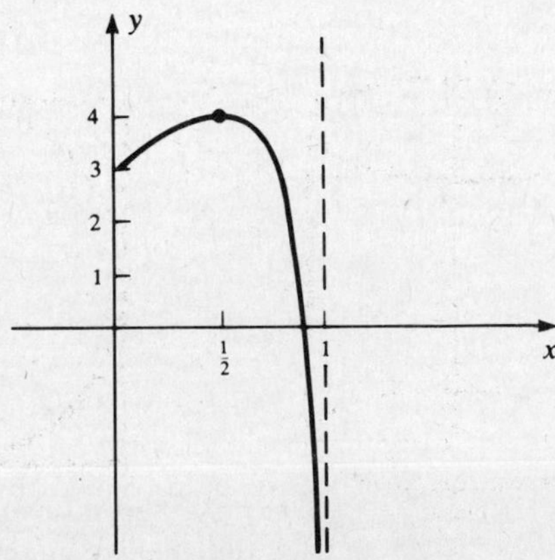

Fig. 24-3

Supplementary Problems

24.4 A rectangular field is to be fenced in so that the resulting area is 100 square meters. Find the dimensions of that field (if any) for which the perimeter is (*a*) a maximum, (*b*) a minimum.

24.5 Find the point(s) on the parabola $2x = y^2$ closest to the point (1, 0).

24.6 Find the point(s) on the hyperbola $x^2 - y^2 = 2$ closest to the point (0, 1).

24.7 A closed box with a square base is to contain 252 cubic feet. The bottom costs \$5 per square foot, the top costs \$2 per square foot, and the sides cost \$3 per square foot. Find the dimensions that will minimize the cost.

24.8 Find the absolute maxima and minima (if any) of

$$f(x) = \frac{x^2 + 4}{x - 2}$$

on the interval [0, 2).

24.9 A printed page must contain 60 cm^2 of printed material. There are to be margins of 5 cm on either side, and margins of 3 cm each on the top and the bottom. How long should the printed lines be in order to minimize the amount of paper used?

24.10 A farmer wishes to fence in a rectangular field of 10 000 square feet. The north-south fences will cost \$1.50 per foot, while the east-west fences will cost \$6.00 per foot. Find the dimensions of the field that will minimize the cost.

24.11 (*a*) Sketch the graph of

$$y = \frac{1}{1 + x^2}$$

and (*b*) find the point on the graph where the tangent line has the greatest slope.

24.12 (*a*) Find the dimensions of the closed cylindrical can (Fig. 24-4(*a*)) that will have a capacity of k volume units and use the minimum amount of material. Find the ratio of the height h to the radius r of the top and bottom. (The volume is $V = \pi r^2 h$, and the lateral surface area is $S = 2\pi rh$.) (*b*) If the bottom and top of the can have to be cut from square pieces of metal and the rest of these squares is wasted (Fig. 24-4(*b*)), find the dimensions that will minimize the amount of material used, and find the ratio of the height to the radius.

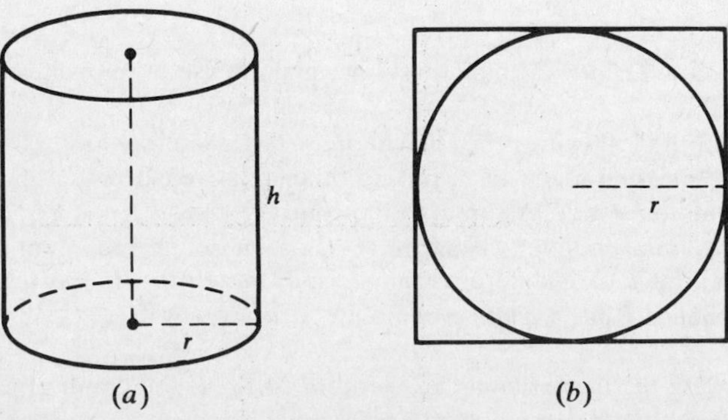

Fig. 24-4

24.13 A thin-walled cone-shaped cup is to hold 36π cubic inches of water when full. What dimensions will minimize the amount of material needed for the cup? (The volume is $V = \frac{1}{3}\pi r^2 h$ and the surface area is $A = \pi rs$; see Fig. 24-5.)

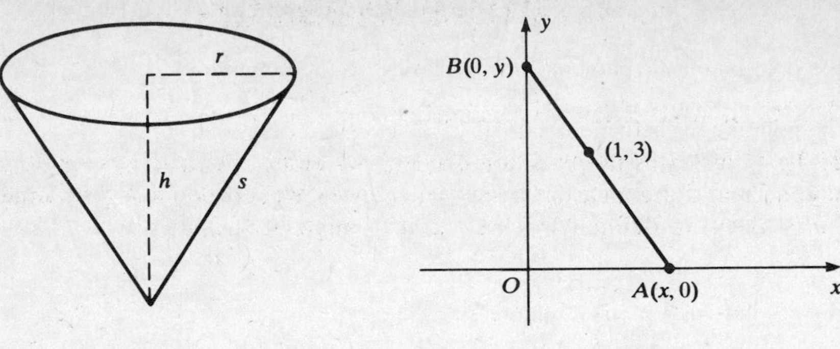

Fig. 24-5 **Fig. 24-6**

24.14 (*a*) Find the absolute extrema on $[0, +\infty)$ (if they exist) of

$$f(x) = \frac{x}{(x^2+1)^{3/2}}$$

(*b*) Sketch the graph of f.

24.15 A rectangular bin, open at the top, is required to contain a volume of 128 cubic meters. If the bottom is to be a square, at a cost of \$2 per square meter, while the sides cost \$0.50 per square meter, what dimensions will minimize the cost?

24.16 The selling price P of an item is $100 - 0.02x$ dollars, where x is the number of items produced per day. If the cost C of producing and selling x items is $40x + 15\,000$ dollars per day, how many items should be produced and sold every day in order to maximize the profit?

24.17 Consider all lines through the point (1, 3) and intersecting the positive x-axis at $A(x, 0)$ and the positive y-axis at $B(0, y)$ (see Fig. 24-6). Find the line which makes the area of $\triangle BOA$ a minimum.

24.18 Consider the function

$$f(x) = \frac{1}{2}x^2 + \frac{k}{x}$$

(*a*) For what value of k will f have a relative minimum at $x = -2$? (*b*) For the value of k found in (*a*), sketch the graph of f. (*c*) For what value(s) of k will f have an absolute minimum?

24.19 Find the point(s) on the graph of

$$3x^2 + 10xy + 3y^2 = 9$$

closest to the origin. (*Hint*: Minimize $x^2 + y^2$, making use of implicit differentiation.)

24.20 Fill in the gaps in the following proof of Theorem 24.1. Assume that f is continuous on an interval $\mathscr{I}$. Let f have a relative maximum at c in $\mathscr{I}$, but no other relative extremum in $\mathscr{I}$. We must show that f has an absolute maximum on $\mathscr{I}$ at c. Assume, to the contrary, that $d \neq c$ is a point in $\mathscr{I}$ with $f(c) < f(d)$. On the closed interval $\mathscr{I}$ with endpoints c and d. f has an absolute minimum at some point u. Since f has a relative maximum at c, u is different from c, and, therefore, $f(u) < f(c)$. Hence, $u \neq d$. So, u is in the interior of $\mathscr{I}$, whence f has a *relative* minimum at $u \neq c$.

24.21 Prove the following theorem, similar to Theorem 24.1: If the graph of f is concave upward (downward) over an interval $\mathscr{I}$, then any relative minimum (maximum) of f in $\mathscr{I}$ is an absolute minimum (maximum) on $\mathscr{I}$. (*Hint*: Consider the relationship of the graph of f to the tangent line at the relative extremum.)

Chapter 25

Angle Measure

25.1 ARC LENGTH AND RADIAN MEASURE

Figure 25-1(a) illustrates the traditional system of angle measure. A complete rotation is divided into 360 equal parts, and the measure assigned to each part is called a *degree*. In modern mathematics and science, it is useful to define a different unit of angle measure.

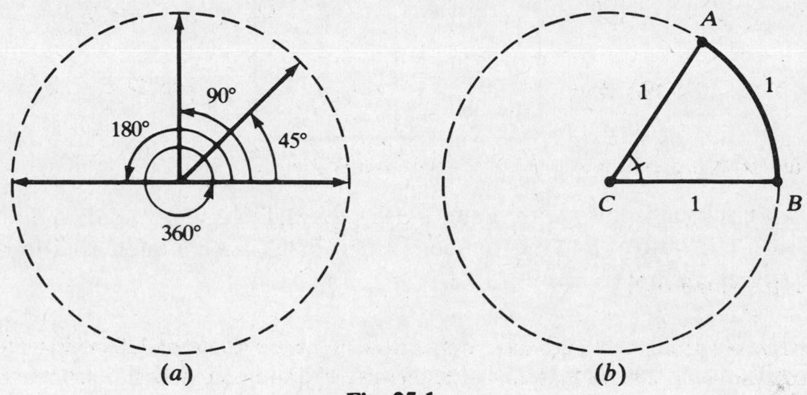

Fig. 25-1

Definition: Consider a circle with a radius of one unit (Fig. 25-1(b)). Let the center be C, and let CA and CB be two radii for which the intercepted arc $\widehat{AB}$ of the circle has length 1. Then the central angle ACB is taken to be the unit of measure, *one radian*.

Let X be the number of degrees in $\measuredangle ACB$ of radian measure 1. Then the ratio of X to 360° (a complete rotation) is the same as the ratio of $\widehat{AB}$ to the entire circumference, 2π. Since $\widehat{AB} = 1$,

$$\frac{X}{360} = \frac{1}{2\pi} \qquad \text{or} \qquad X = \frac{360}{2\pi} = \frac{180}{\pi}$$

Thus,

$$1 \text{ radian} = \frac{180}{\pi} \text{ degrees} \tag{25.1}$$

If we approximate π as 3.14, then 1 radian is about 57.3 degrees. If we multiply (*25.1*) by $\pi/180$, we obtain:

$$1 \text{ degree} = \frac{\pi}{180} \text{ radians} \tag{25.2}$$

EXAMPLE Let us find the radian measures of some "common" angles given in degrees. Clearly, the null angle is 0 in either measure. For an angle of 30°, (*25.2*) gives:

$$30° = 30\left(\frac{\pi}{180} \text{ radians}\right) = \frac{\pi}{6} \text{ radians}$$

for an angle of 45°,

$$45° = 45\left(\frac{\pi}{180} \text{ radians}\right) = \frac{\pi}{4} \text{ radians}$$

and so on, generating Table 25-1. This table should be memorized by the student, who will often be going back and forth between degrees and radians.

Table 25-1

Degrees	Radians
0	0
30	$\frac{\pi}{6}$
45	$\frac{\pi}{4}$
60	$\frac{\pi}{3}$
90	$\frac{\pi}{2}$
180	π
270	$\frac{3\pi}{2}$
360	2π

Consider now a circle of radius r with center O (Fig. 25-2). Let $\measuredangle DOE$ contain θ radians and let s be the length of arc $\widehat{DE}$. The ratio of θ to the number 2π of radians in a complete rotation is the same as the ratio of s to the entire circumference $2\pi r$: $\theta/2\pi = s/2\pi r$. Hence,

$$s = r\theta \tag{25.3}$$

gives the basic relationship between the arc length, the radius, and the radian measure of the central angle.

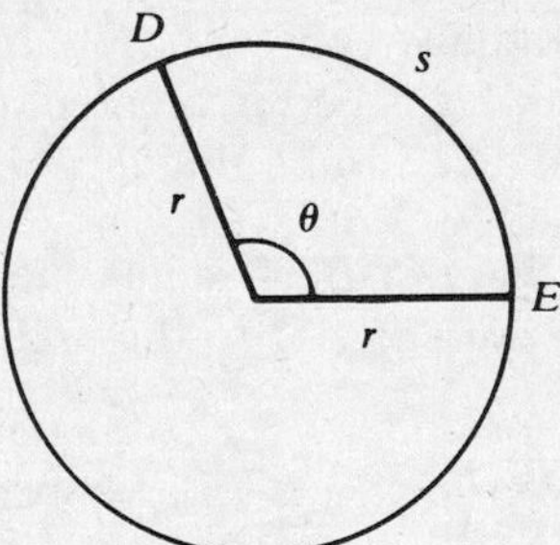

Fig. 25-2

25.2 DIRECTED ANGLES

Angles can be classified as positive or negative according to the direction of the rotation that generates them. In Fig. 25-3(a), the directed angle AOB is *positive* because it is obtained by rotating the arrow OA *counterclockwise* toward arrow OB. On the other hand, in Fig. 25-3(b), the directed angle AOB is *negative* because it is obtained by rotating arrow OA *clockwise* toward arrow OB. Some examples of directed angles and their radian measures are shown in Fig. 25-4.

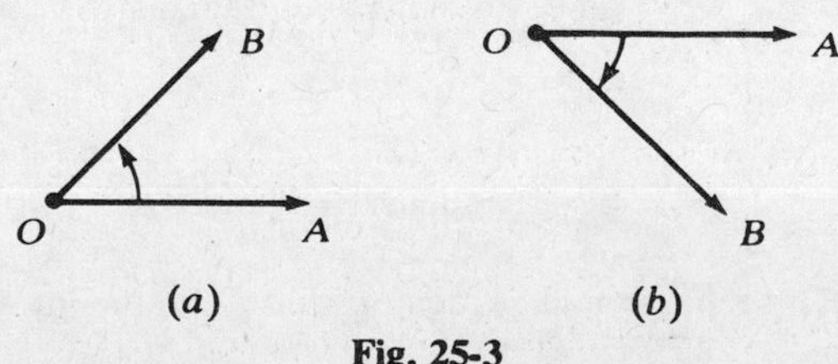

Fig. 25-3

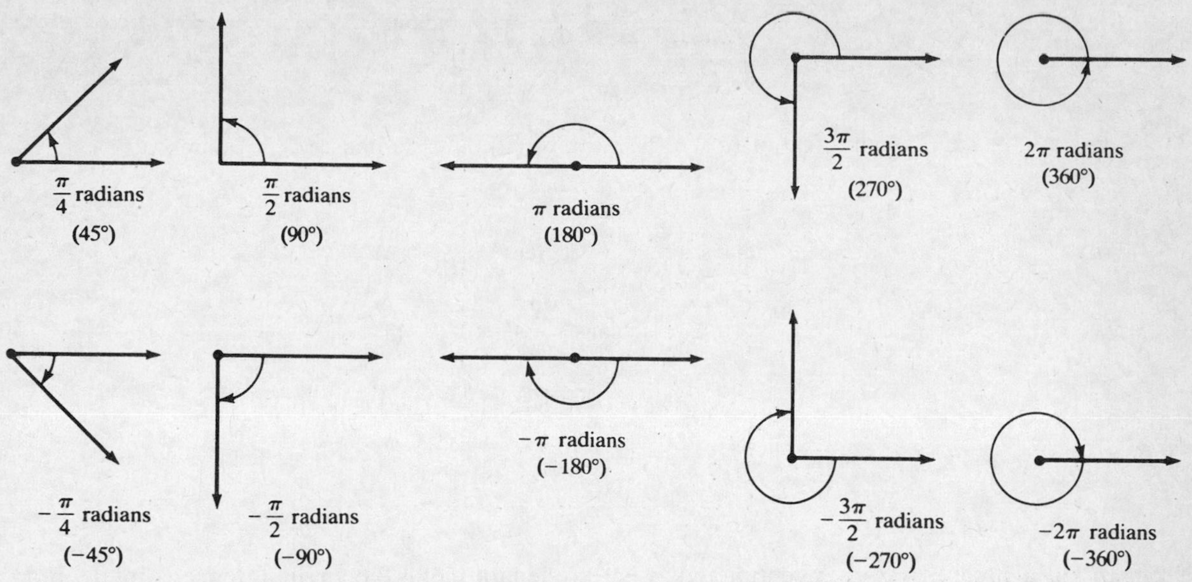

Fig. 25-4

Some directed angles corresponding to more than one complete rotation are shown in Fig. 25-5. It is apparent that directed angles whose radian measures differ by an integral multiple of 2π (e.g., the first and the last angles in Fig. 25-5) represent identical configurations of the two arrows; we shall say that such angles "have the same sides."

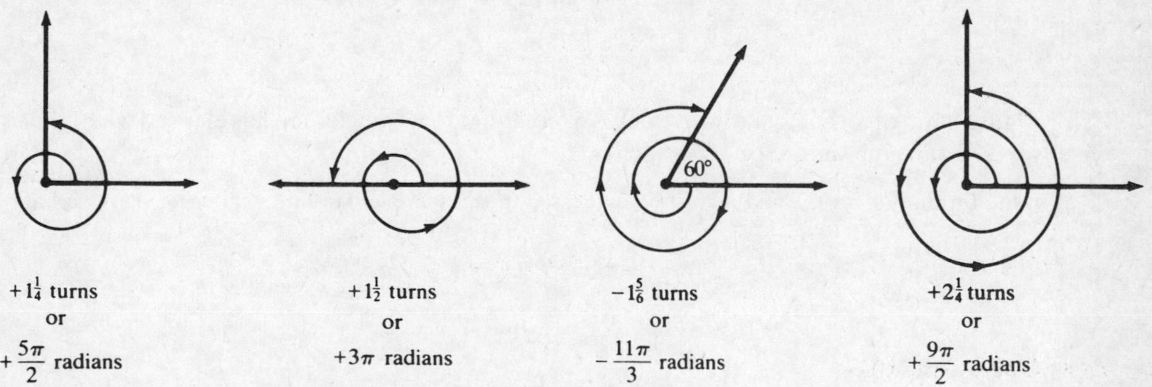

Fig. 25-5

Solved Problems

25.1 Express in radians an angle of (*a*) 72°, (*b*) 150°.

Use (*25.2*).

(*a*)
$$72° = 72\left(\frac{\pi}{180}\text{ radians}\right) = \frac{2\cdot 36}{5\cdot 36}(\pi\text{ radians}) = \frac{2\pi}{5}\text{ radians}$$

(*b*) $$150° = \frac{150}{180}\pi = \frac{5(30)}{6(30)}\pi = \frac{5\pi}{6} \text{ radians}$$

25.2 Express in degrees an angle of (*a*) $5\pi/12$ radians, (*b*) 0.3π radians, (*c*) 3 radians.

Use (*25.1*).

(*a*) $$\frac{5\pi}{12} \text{ radians} = \frac{5\pi}{12}\left(\frac{180}{\pi} \text{ degrees}\right) = \frac{5}{12} \times 180° = 75°$$

(*b*) $$0.3\pi \text{ radians} = \frac{0.3\pi}{\pi} \times 180° = 54°$$

(*c*) $$3 \text{ radians} = \frac{3}{\pi} \times 180° = \left(\frac{540}{\pi}\right)^{\circ} \approx 172°$$

since $\pi \approx 3.14$.

25.3 (*a*) In a circle of radius 5 centimeters, what arc length along the circumference is intercepted by a central angle of $\pi/3$ radians? (*b*) In a circle of radius 12 feet, what arc length along the circumference is intercepted by a central angle of 30°?

Use (*25.3*), $s = r\theta$.

(*a*) $$s = 5 \times \frac{\pi}{3} = \frac{5\pi}{3} \text{ centimeters}$$

(*b*) The central angle must be changed to radian measure. By Table 25-1,

$$30° = \frac{\pi}{6} \text{ radians} \qquad \text{and so} \qquad s = 12 \times \frac{\pi}{6} = 2\pi \text{ feet}$$

25.4 The minute hand of an ancient tower-clock is 5 feet long. How much time has elapsed when the tip has traveled through an arc of 188.4 inches?

In the formula $\theta = s/r$, s and r must be expressed in the same length unit. Choosing feet, we have

$$s = \frac{188.4}{12} = 15.7 \text{ feet} \qquad r = 5 \text{ feet}$$

whence $$\theta = \frac{15.7}{5} = 3.14 \text{ radians}$$

This is very nearly π radians, which is a half-revolution, or 30 minutes of time.

25.5 What (positive) angles between 0 and 2π radians have the same sides as angles with the following measures?

(*a*) $\frac{9\pi}{4}$ radians (*b*) 390° (*c*) $-\frac{\pi}{2}$ radians (*d*) -3π radians

(*a*) $$\frac{9\pi}{4} = \left(2 + \frac{1}{4}\right)\pi = 2\pi + \frac{\pi}{4}$$

Hence, $9\pi/4$ radians determines a counterclockwise complete rotation (2π radians), plus a counterclockwise rotation of $\pi/4$ radians (45°). See Fig. 25-6(*a*). The "reduced angle"—i.e., the angle with measure in $[0, 2\pi)$ and having the same sides as the given angle—therefore is $\pi/4$ radians.

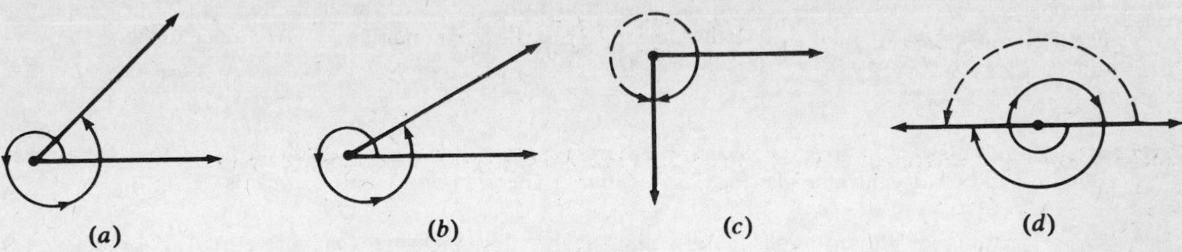

Fig. 25-6

(*b*)

$$390° = 360° + 30° = (2\pi \text{ radians}) + \left(\frac{\pi}{6} \text{ radians}\right)$$

See Fig. 25-6(*b*). The reduced angle is $\pi/6$ radians (or 30°).

(*c*) A clockwise rotation of $\pi/2$ radians (90°) is equivalent to a counterclockwise rotation of

$$2\pi - \frac{\pi}{2} = \frac{3\pi}{2} \text{ radians}$$

(see Fig. 25-6(*c*)). Thus the reduced angle is $3\pi/2$ radians.

(*d*) Adding a suitable multiple of 2π to the given angle, we have:

$$-3\pi + 4\pi = +\pi \text{ radians}$$

that is, a clockwise rotation of 3π radians is equivalent to a counterclockwise rotation of π radians (see Fig. 25-6(*d*)). The reduced angle is π radians.

Supplementary Problems

25.6 Convert the following degree measures of angles into radian measures: (*a*) 36°, (*b*) 15°, (*c*) 2°, (*d*) $\left(\frac{90}{\pi}\right)^{\circ}$, (*e*) 144°.

25.7 Convert the following radian measures of angles into degree measures: (*a*) 2 radians, (*b*) $\pi/5$ radians, (*c*) $7\pi/12$ radians, (*d*) $5\pi/4$ radians, (*e*) $7\pi/6$ radians.

25.8 If a bug moves a distance of 3π centimeters along a circular arc and if this arc subtends a central angle of 45°, what is the radius of the circle?

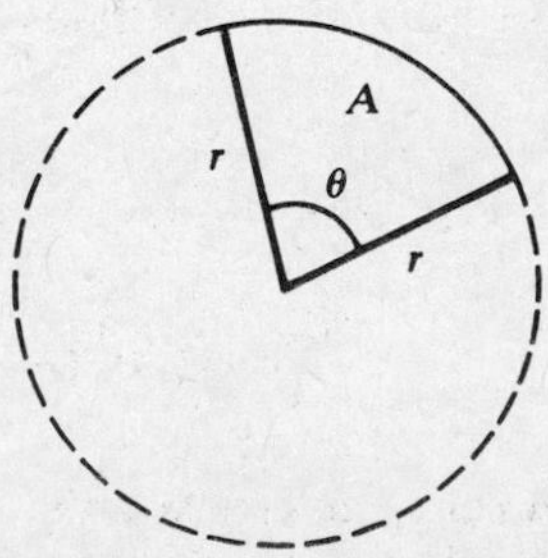

Fig. 25-7

25.9 In each of the following cases, from the information about two of the quantities s (intercepted arc), r (radius), and θ (central angle), find the third quantity. (If only a number is given for θ, assume that it is the number of radians.) (*a*) $r = 10$, $\theta = \pi/5$; (*b*) $\theta = 60°$, $s = 11/21$; (*c*) $r = 1$, $s = \pi/4$; (*d*) $r = 2$, $s = 3$; (*e*) $r = 3$, $\theta = 90°$; (*f*) $\theta = 180°$, $s = 6.28318$; (*g*) $r = 10$, $\theta = 120°$.

25.10 If a central angle of a circle of radius r measures θ radians, find the area A of the sector of the circle determined by the central angle (Fig. 25-7). (*Hint*: The area of the entire circle is πr^2.)

25.11 Draw pictures of the rotations determining angles that measure (*a*) 405°, (*b*) $11\pi/4$ radians, (*c*) $7\pi/2$ radians, (*d*) $-60°$, (*e*) $-\pi/6$ radians, (*f*) $-5\pi/2$ radians.

25.12 Reduce each angle in Problem 25.11 to the range 0 to 2π radians.

Chapter 26

The Sine and Cosine Functions

26.1 GENERAL DEFINITION

The fundamental trigonometric functions, the *sine* and the *cosine*, will play an important role in the calculus. These functions will now be defined for all real numbers.

Definition: Place an arrow $\overrightarrow{OA}$ of unit length so that its initial point O is the origin of a coordinate system and its endpoint A is the point $(1, 0)$ of the x-axis. See Fig. 26-1. For any given number θ, rotate $\overrightarrow{OA}$ about the point O through an angle with radian measure θ. Let $\overrightarrow{OB}$ be the final position of the arrow after the rotation. Then: (i) the x-coordinate of B is defined to be the *cosine* of θ, denoted $\cos\theta$; (ii) the y-coordinate of B is defined to be the *sine* of θ, denoted $\sin\theta$. Thus, $B = (\cos\theta, \sin\theta)$.

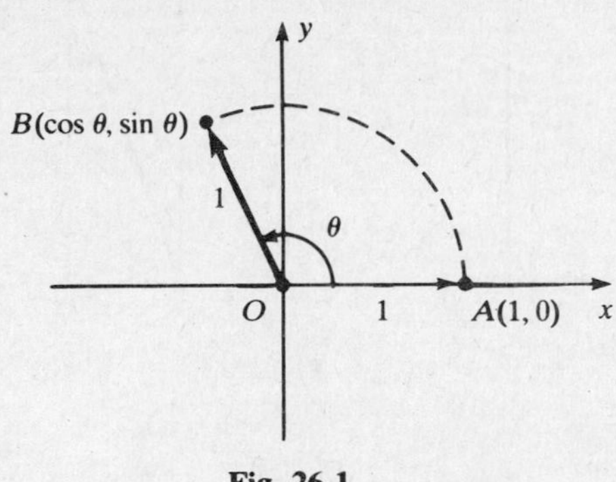

Fig. 26-1

EXAMPLES

(*a*) Let $\theta = \pi/2$. If we rotate $\overrightarrow{OA}$ $\pi/2$ radians in the counterclockwise direction, the final position B is $(0, 1)$; see Fig. 26-2(*a*). Hence,

$$\cos\frac{\pi}{2} = 0 \qquad \sin\frac{\pi}{2} = 1$$

(*b*) Let $\theta = \pi$. If we rotate $\overrightarrow{OA}$ through π radians in the counterclockwise direction, the final position B is $(-1, 0)$. See Fig. 26-2(*b*). So,

$$\cos\pi = -1 \qquad \sin\pi = 0$$

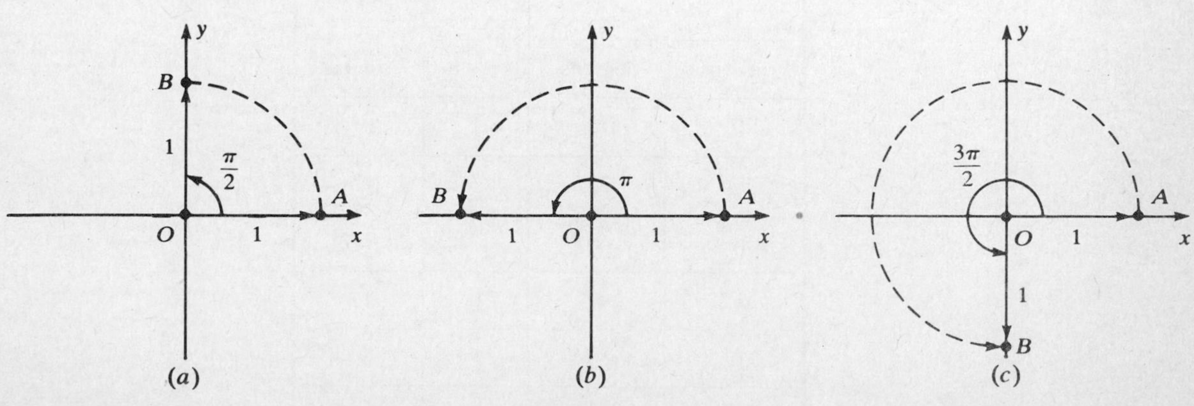

Fig. 26-2

(*c*) Let $\theta = 3\pi/2$. Then the final position B, after a rotation through $3\pi/2$ radians in the counterclockwise direction, is $(0, -1)$. See Fig. 26-2(*c*). Hence,

$$\cos\frac{3\pi}{2} = 0 \qquad \sin\frac{3\pi}{2} = -1$$

(*d*) Let $\theta = 0$. If $\overrightarrow{OA}$ is rotated through 0 radians, the final position is still $(1, 0)$. Therefore,

$$\cos 0 = 1 \qquad \sin 0 = 0$$

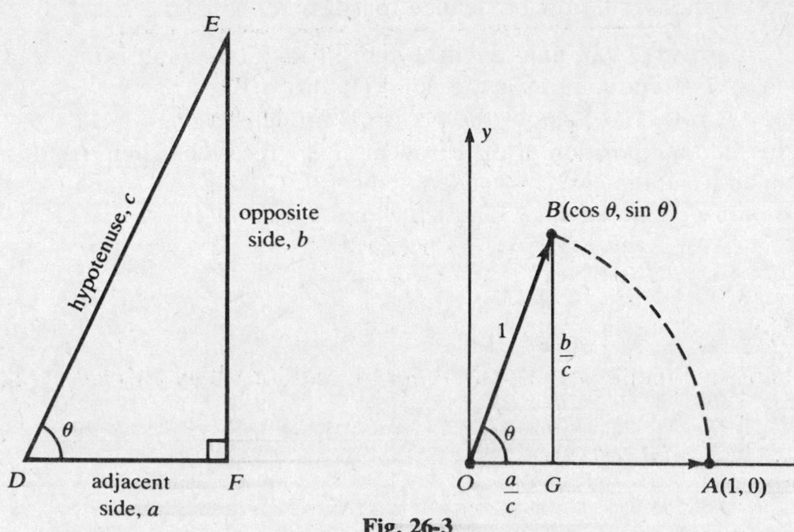

Fig. 26-3

(*e*) Let θ be an *acute angle* $(0 < \theta < \pi/2)$ of right triangle *DEF*, and let $\triangle OBG$ be a similar triangle with hypotenuse 1. See Fig. 26-3. By proportionality, $\overline{BG} = b/c$ and $\overline{OG} = a/c$. So, by definition, $\cos\theta = a/c$, $\sin\theta = b/c$. This agrees with the traditional definitions:

$$\cos\theta = \frac{\text{adjacent side}}{\text{hypotenuse}} \qquad \sin\theta = \frac{\text{opposite side}}{\text{hypotenuse}}$$

Consequently, we can appropriate the values of the functions for $\theta = \pi/6$, $\pi/4$, $\pi/3$ from high-school trigonometry. The results are collected in Table 26-1, which ought to be memorized.

Table 26-1

θ			
radians	degrees	$\cos\theta$	$\sin\theta$
0	0	1	0
$\pi/6$	30	$\sqrt{3}/2$	1/2
$\pi/4$	45	$\sqrt{2}/2$	$\sqrt{2}/2$
$\pi/3$	60	1/2	$\sqrt{3}/2$
$\pi/2$	90	0	1
π	180	-1	0
$3\pi/2$	270	0	-1

The above Definition implies that the signs of $\cos\theta$ and $\sin\theta$ are determined by the quadrant in which point B lies. In the first quadrant, $\cos\theta > 0$ and $\sin\theta > 0$. In the second quadrant, $\cos\theta < 0$ and $\sin\theta > 0$. In the third quadrant, $\cos\theta < 0$ and $\sin\theta < 0$. In the fourth quadrant, $\cos\theta > 0$ and $\sin\theta < 0$. See Fig. 26-4.

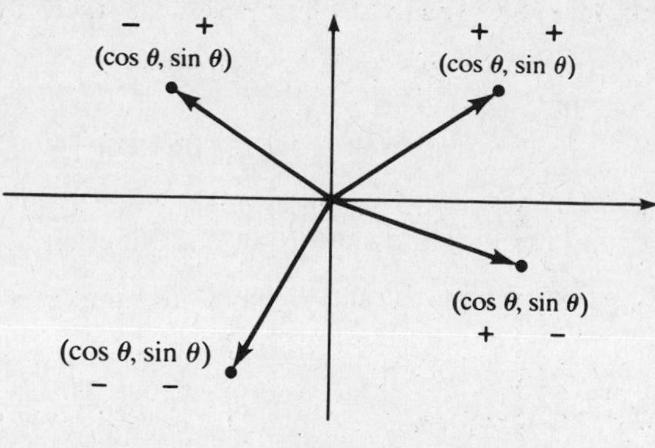

Fig. 26-4

26.2 PROPERTIES

We list, as theorems, the important properties of, and identities satisfied by, the sine and cosine functions.

Theorem 26.1 (*Fundamental Relation*):

$$\sin^2\theta + \cos^2\theta = 1$$

Proof. In Fig. 26-1, the length of $\overrightarrow{OB}$ is given by (*2.1*) as

$$1 = \sqrt{(\cos\theta - 0)^2 + (\sin\theta - 0)^2} = \sqrt{(\cos\theta)^2 + (\sin\theta)^2}$$

Squaring both sides, and using the conventional notations

$$(\sin\theta)^2 \equiv \sin^2\theta \qquad (\cos\theta)^2 \equiv \cos^2\theta$$

gives Theorem 26.1.

EXAMPLE Let θ be the radian measure of an acute angle such that $\sin\theta = 3/5$. By Theorem 26.1,

$$\left(\frac{3}{5}\right)^2 + \cos^2\theta = 1$$

$$\frac{9}{25} + \cos^2\theta = 1$$

$$\cos^2\theta = 1 - \frac{9}{25} = \frac{16}{25}$$

$$\cos\theta = \pm\sqrt{\frac{16}{25}} = \pm\frac{4}{5}$$

Since θ is an acute angle, $\cos\theta$ is positive; $\cos\theta = 4/5$.

We have already seen (Chapter 25) that two angles that differ by a multiple of 2π radians (360°) have the same sides. This establishes:

Theorem 26.2: The cosine and sine functions are *periodic*, of *period* 2π; that is, for all θ,

$$\cos(\theta + 2\pi) = \cos\theta$$
$$\sin(\theta + 2\pi) = \sin\theta$$

(Moreover, 2π is the smallest positive number with this property.)

In view of Theorem 26.2, it is sufficient to know the values of $\cos\theta$ and $\sin\theta$ for $0 \le \theta < 2\pi$.

EXAMPLES

(a) $$\sin\frac{7\pi}{3} = \sin\left(2\pi + \frac{\pi}{3}\right) = \sin\frac{\pi}{3} = \frac{\sqrt{3}}{2}$$

(b) $$\cos 5\pi = \cos(3\pi + 2\pi) = \cos 3\pi = \cos(\pi + 2\pi) = \cos \pi = -1$$

(c) $$\cos 390° = \cos(30° + 360°) = \cos 30° = \frac{\sqrt{3}}{2}$$

(d) $$\sin 405° = \sin(45° + 360°) = \sin 45° = \frac{\sqrt{2}}{2}$$

Theorem 26.3: $\cos\theta$ is an even function, and $\sin\theta$ is an odd function.

The proof is obvious from Fig. 26-5: points B' and B have the same x-coordinate,

$$\cos(-\theta) = \cos\theta$$

but their y-coordinates differ in sign,

$$\sin(-\theta) = -\sin\theta$$

Because of Theorem 26.3, we now need to know the values of $\cos\theta$ and $\sin\theta$ only for $0 \le \theta \le \pi$.

Consider any point $A(x, y)$ different from the origin O, as in Fig. 26-6. Let r be its distance from the origin and let θ be the radian measure of the angle that line OA makes with the positive x-axis. We call r and θ *polar coordinates* of point A.

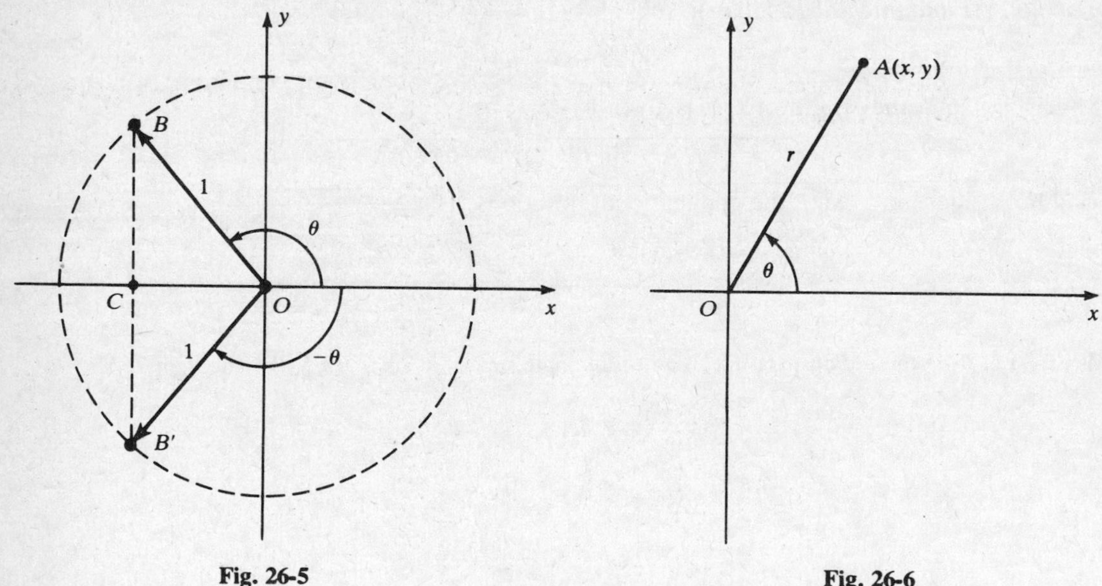

Fig. 26-5 **Fig. 26-6**

Theorem 26.4: The polar coordinates of a point, and its x- and y-coordinates, are related by

$$x = r\cos\theta \qquad y = r\sin\theta$$

For the proof, see Problem 26.2.

Theorem 26.5 (*Law of Cosines*): In any triangle ABC (Fig. 26-7),

$$c^2 = a^2 + b^2 - 2ab\cos\theta$$

For a proof, see Problem 26.3. Note that the Pythagorean Theorem is obtained as the special case $\theta = \pi/2$.

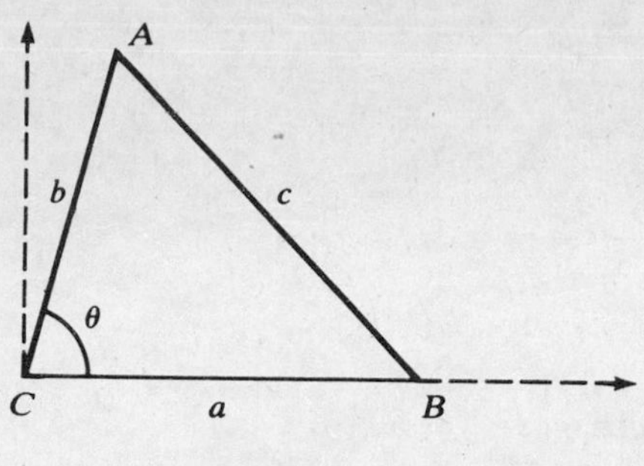

Fig. 26-7

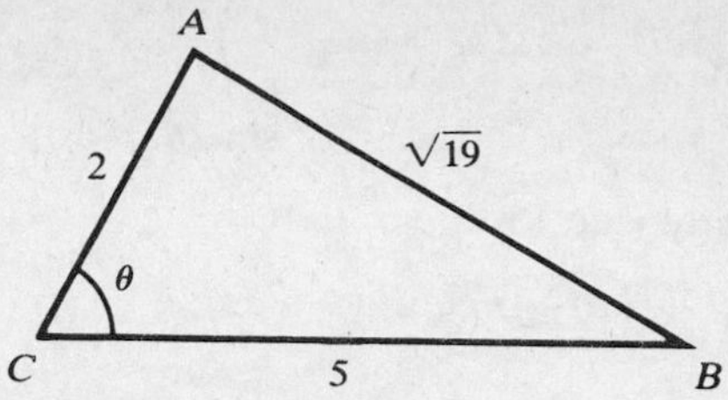

Fig. 26-8

EXAMPLE Find angle θ in the triangle of Fig. 26-8.
Solving the Law of Cosines for $\cos\theta$,

$$\cos\theta = \frac{a^2+b^2-c^2}{2ab} = \frac{(5)^2+(2)^2-(\sqrt{19})^2}{2(5)(2)}$$
$$= \frac{25+4-19}{20} = \frac{1}{2}$$

Then, from Table 26-1, $\theta = \pi/3$.

Theorem 26.6 (*Sum and Difference Formulas*):

$$\cos(u \pm v) = \cos u \cos v \mp \sin u \sin v$$
$$\sin(u \pm v) = \sin u \cos v \pm \cos u \sin v$$

Once any one of the four formulas has been proved, the other three follow easily. A proof of the difference formula for the cosine is sketched in Problem 26.14.

EXAMPLES

(*a*)
$$\cos\frac{\pi}{12} = \cos\left(\frac{\pi}{3}-\frac{\pi}{4}\right) = \cos\frac{\pi}{3}\cos\frac{\pi}{4} + \sin\frac{\pi}{3}\sin\frac{\pi}{4}$$
$$= \left(\frac{1}{2}\right)\left(\frac{\sqrt{2}}{2}\right) + \left(\frac{\sqrt{3}}{2}\right)\left(\frac{\sqrt{2}}{2}\right) = \frac{\sqrt{2}}{4} + \frac{\sqrt{6}}{4} = \frac{\sqrt{2}+\sqrt{6}}{4}$$

(*b*)
$$\cos 135^\circ = \cos(90^\circ + 45^\circ) = \cos 90^\circ \cos 45^\circ - \sin 90^\circ \sin 45^\circ$$
$$= (0)\left(\frac{\sqrt{2}}{2}\right) - (1)\left(\frac{\sqrt{2}}{2}\right) = -\frac{\sqrt{2}}{2}$$

(*c*)
$$\sin\frac{7\pi}{12} = \sin\left(\frac{\pi}{2}+\frac{\pi}{12}\right) = \sin\frac{\pi}{2}\cos\frac{\pi}{12} + \cos\frac{\pi}{2}\sin\frac{\pi}{12}$$
$$= (1)\left(\frac{\sqrt{2}+\sqrt{6}}{4}\right) + (0)\left(\sin\frac{\pi}{12}\right) \quad \text{[by Example (a)]}$$
$$= \frac{\sqrt{2}+\sqrt{6}}{4}$$

Substituting $u = \pi/2$, $v = \theta$ in the difference formulas of Theorem 26.6 gives

$$\cos\left(\frac{\pi}{2}-\theta\right) = \sin\theta \qquad \sin\left(\frac{\pi}{2}-\theta\right) = \cos\theta$$

or

Theorem 26.7: The sine of an angle is the cosine of its complement.

Substituting $u = v = \theta$ in the sum formulas of Theorem 26.6 gives, with the help of Theorem 26.1,

Theorem 26.8 (*Double-Angle Formulas*):

$$\cos 2\theta = \cos^2\theta - \sin^2\theta$$
$$= 2\cos^2\theta - 1 = 1 - 2\sin^2\theta$$
$$\sin 2\theta = 2\sin\theta\cos\theta$$

or, equivalently (see Problem 26.16),

Theorem 26.9 (*Half-Angle Formulas*):

$$\cos^2\frac{\theta}{2} = \frac{1+\cos\theta}{2}$$
$$\sin^2\frac{\theta}{2} = \frac{1-\cos\theta}{2}$$

EXAMPLES

(*a*)
$$\sin 120° = \sin(2\times 60°) = 2\sin 60°\cos 60°$$
$$= 2\left(\frac{\sqrt{3}}{2}\right)\left(\frac{1}{2}\right) = \frac{\sqrt{3}}{2}$$

(*b*)
$$\cos\frac{2\pi}{3} = \cos\left(2\times\frac{\pi}{3}\right) = \cos^2\frac{\pi}{3} - \sin^2\frac{\pi}{3}$$
$$= \left(\frac{1}{2}\right)^2 - \left(\frac{\sqrt{3}}{2}\right)^2 = \frac{1}{4} - \frac{3}{4} = -\frac{1}{2}$$

(*c*)
$$\cos^2\frac{\pi}{12} = \cos^2\left(\frac{1}{2}\times\frac{\pi}{6}\right) = \frac{1+\cos\frac{\pi}{6}}{2} = \frac{1+\frac{\sqrt{3}}{2}}{2}$$
$$= \frac{2+\sqrt{3}}{4} \quad \text{[multiply numerator and denominator by 2]}$$

CHECK Earlier, we deduced from Theorem 26.6 that

$$\cos\frac{\pi}{12} = \frac{\sqrt{2}+\sqrt{6}}{4}$$

Hence, in view of the identity $(u+v)^2 = u^2 + 2uv + v^2$,

$$\cos^2\frac{\pi}{12} = \frac{(\sqrt{2})^2 + 2(\sqrt{2})(\sqrt{6}) + (\sqrt{6})^2}{(4)^2} = \frac{2 + 2(\sqrt{2})(\sqrt{2})(\sqrt{3}) + 6}{16}$$
$$= \frac{8+4\sqrt{3}}{16} = \frac{2+\sqrt{3}}{4}$$

(*d*)
$$\sin^2\frac{\pi}{8} = \sin^2\left(\frac{1}{2}\times\frac{\pi}{4}\right) = \frac{1-\cos\frac{\pi}{4}}{2} = \frac{1-\frac{\sqrt{2}}{2}}{2} = \frac{2-\sqrt{2}}{4}$$

Hence, since $\pi/8$ is in the first quadrant,

$$\sin\frac{\pi}{8} = +\frac{\sqrt{2-\sqrt{2}}}{2}$$

Solved Problems

26.1 Evaluate (*a*) $\cos(\pi/8)$, (*b*) $\cos 210°$, (*c*) $\sin 135°$, (*d*) $\cos 17°$.

(*a*) By Theorem 26.9,

$$\cos^2\frac{\pi}{8} = \cos^2\left(\frac{1}{2}\times\frac{\pi}{4}\right) = \frac{1+\cos\frac{\pi}{4}}{2} = \frac{1+\frac{\sqrt{2}}{2}}{2} = \frac{2+\sqrt{2}}{4}$$

Hence, since $\pi/8$ is an acute angle,

$$\cos\frac{\pi}{8} = \frac{\sqrt{2+\sqrt{2}}}{2}$$

(*b*) Writing $210° = 180° + 30°$, we have, by Theorem 26.6,

$$\cos 210° = \cos(180° + 30°) = \cos 180° \cos 30° - \sin 180° \sin 30°$$
$$= (-1)\left(\frac{\sqrt{3}}{2}\right) - (0)\left(\frac{1}{2}\right) = -\frac{\sqrt{3}}{2}$$

(*c*) Using the previously derived value $\cos 135° = -\sqrt{2}/2$, we have, from Theorem 26.1,

$$\sin^2 135° = 1 - \cos^2 135° = 1 - \left(-\frac{\sqrt{2}}{2}\right)^2 = 1 - \frac{2}{4} = \frac{2}{4}$$

Hence, since 135° is in the second quadrant,

$$\sin 135° = +\sqrt{\frac{2}{4}} = \frac{\sqrt{2}}{2}$$

(*d*) 17° cannot be expressed in terms of more familiar angles (such as 30°, 45°, 60°) in such a way as to allow the application of any of our formulas. We must then use the cosine table in Appendix D, which gives 0.9563 as the value of cos 17°. This is an approximation, correct to four decimal places.

26.2 Prove Theorem 26.4.

Refer to Fig. 26-9. Let D be the foot of the perpendicular from A to the x-axis. Let F be the point on the ray OA at a unit distance from the origin; then, $F = (\cos\theta, \sin\theta)$. If E is the foot of the perpendicular from F to the x-axis, we have

$$\overline{OE} = \cos\theta \qquad \overline{FE} = \sin\theta$$

Now, $\triangle ADO$ is similar to $\triangle FEO$ (by the AA-criterion), whence

$$\frac{\overline{OD}}{\overline{OE}} = \frac{\overline{OA}}{\overline{OF}} = \frac{\overline{AD}}{\overline{FE}} \qquad \text{or} \qquad \frac{x}{\cos\theta} = \frac{r}{1} = \frac{y}{\sin\theta}$$

Therefore, $x = r\cos\theta$ and $y = r\sin\theta$. A similar proof holds when $A(x, y)$ is in one of the other quadrants. The proof is immediate when $A(x, y)$ is on the x-axis or y-axis.

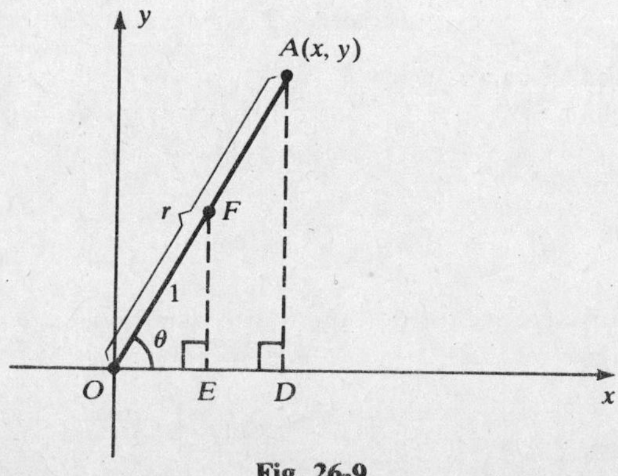

Fig. 26-9

26.3 Prove Theorem 26.5.

In Fig. 26-7, set up a coordinate system with C as origin and B on the positive x-axis. Then B has coordinates $(a, 0)$. Let (x, y) be the coordinates of A. By Theorem 26.4,

$$x = b\cos\theta \qquad y = b\sin\theta$$

By the distance formula, (*2.1*),

$$c = \sqrt{(x-a)^2 + (y-0)^2} = \sqrt{(x-a)^2 + y^2}$$

Hence,

$$c^2 = (x-a)^2 + y^2 = (b\cos\theta - a)^2 + (b\sin\theta)^2$$

ALGEBRA $\quad (P-Q)^2 = P^2 - 2PQ + Q^2$

$$= b^2\cos^2\theta - 2ab\cos\theta + a^2 + b^2\sin^2\theta$$
$$= a^2 + b^2(\cos^2\theta + \sin^2\theta) - 2ab\cos\theta$$
$$= a^2 + b^2 - 2ab\cos\theta$$

26.4 Derive the following identities:

$$(a)\quad (\sin\theta - \cos\theta)^2 = 1 - \sin 2\theta \qquad (b)\quad \frac{\sin\theta}{\cos\theta} = \frac{1-\cos 2\theta}{\sin 2\theta}$$

(*a*) By the ALGEBRA of Problem 26.3,

$$\begin{aligned}(\sin\theta - \cos\theta)^2 &= \sin^2\theta - 2\sin\theta\cos\theta + \cos^2\theta\\ &= 1 - 2\sin\theta\cos\theta \quad \text{[Theorem 26.1]}\\ &= 1 - \sin 2\theta \quad \text{[Theorem 26.8]}\end{aligned}$$

(*b*)

$$\begin{aligned}\frac{1-\cos 2\theta}{\sin 2\theta} &= \frac{1-(\cos^2\theta - \sin^2\theta)}{2\sin\theta\cos\theta} \quad \text{[Theorem 26.8]}\\ &= \frac{\sin^2\theta + \sin^2\theta}{2\sin\theta\cos\theta} \quad \text{[Theorem 26.1]}\\ &= \frac{(2\sin\theta)(\sin\theta)}{(2\sin\theta)(\cos\theta)} = \frac{\sin\theta}{\cos\theta}\end{aligned}$$

26.5 Given that θ is in the fourth quadrant and that $\cos\theta = 2/3$, find (*a*) $\cos 2\theta$, (*b*) $\sin\theta$, (*c*) $\cos(\theta/2)$, (*d*) $\sin(\theta/2)$.

(*a*) By Theorem 26.8,

$$\begin{aligned}\cos 2\theta &= 2\cos^2\theta - 1 = 2\left(\frac{2}{3}\right)^2 - 1\\ &= 2\left(\frac{4}{9}\right) - 1 = \frac{8}{9} - \frac{9}{9} = -\frac{1}{9}\end{aligned}$$

(*b*) By Theorem 26.1,

$$\sin^2\theta = 1 - \cos^2\theta = 1 - \left(\frac{2}{3}\right)^2 = 1 - \frac{4}{9} = \frac{5}{9}$$

Then, since $\sin\theta$ is negative for θ in the fourth quadrant, $\sin\theta = -\sqrt{5}/3$.

(*c*) By Theorem 26.9,

$$\cos^2\frac{\theta}{2} = \frac{1+\cos\theta}{2} = \frac{1+(2/3)}{2} = \frac{3+2}{6} = \frac{5}{6}$$

Since θ is in the fourth quadrant, $3\pi/2 < \theta < 2\pi$; therefore,

$$\frac{3\pi}{4} < \frac{\theta}{2} < \pi$$

Thus, $\theta/2$ is in the second quadrant, and $\cos(\theta/2)$ is negative. Hence,

$$\cos\frac{\theta}{2} = -\sqrt{\frac{5}{6}}$$

(*d*) By Theorem 26.9,

$$\sin^2\frac{\theta}{2} = \frac{1-\cos\theta}{2} = \frac{1-(2/3)}{2} = \frac{3-2}{6} = \frac{1}{6}$$

and, from (*c*), $\sin(\theta/2)$ is positive. Thus,

$$\sin\frac{\theta}{2} = \frac{1}{\sqrt{6}} = \frac{\sqrt{6}}{6}$$

26.6 Let $\triangle ABC$ be any triangle. Using the notation in Fig. 26-10, prove

$$\frac{\sin A}{a} = \frac{\sin B}{b} = \frac{\sin C}{c} \qquad \textbf{law of sines}$$

(Here, $\sin A$ is the sine of $\measuredangle CAB$, and similarly for $\sin B$ and $\sin C$.)

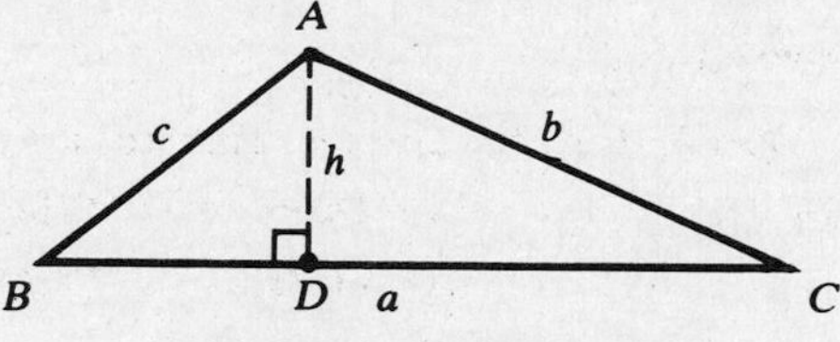

Fig. 26-10

Let D be the foot of the perpendicular from A to side BC. Let $h = \overline{AD}$. Then

$$\sin B = \frac{\overline{AD}}{\overline{AB}} = \frac{h}{c} \qquad \text{or} \qquad h = c \sin B$$

and so

$$\text{area of } \triangle ABC = \frac{1}{2}(\text{base} \times \text{height}) = \frac{1}{2}ah = \frac{1}{2}ac \sin B$$

(Check that this is also correct when $\measuredangle B$ is obtuse.) Similarly,

$$\text{area} = \frac{1}{2}bc \sin A \qquad \text{area} = \frac{1}{2}ab \sin C$$

Hence

$$\frac{1}{2}ac \sin B = \frac{1}{2}bc \sin A = \frac{1}{2}ab \sin C$$

and division by $\frac{1}{2}abc$ gives the Law of Sines.

Supplementary Problems

26.7 Evaluate (*a*) $\sin(4\pi/3)$, (*b*) $\cos(11\pi/6)$, (*c*) $\sin 240°$, (*d*) $\cos 315°$, (*e*) $\sin 75°$, (*f*) $\sin 15°$, (*g*) $\sin(11\pi/12)$, (*h*) $\cos 71°$, (*i*) $\sin 12°$.

26.8 If θ is acute and $\cos\theta = 1/9$, evaluate (*a*) $\sin\theta$, (*b*) $\sin 2\theta$, (*c*) $\cos 2\theta$, (*d*) $\cos(\theta/2)$, (*e*) $\sin(\theta/2)$.

26.9 If θ is in the third quadrant and $\sin\theta = -2/5$, evaluate (*a*) $\cos\theta$, (*b*) $\sin 2\theta$, (*c*) $\cos 2\theta$, (*d*) $\cos(\theta/2)$, (*e*) $\sin(\theta/2)$.

26.10 In $\triangle ABC$, $\overline{AB} = 5$, $\overline{AC} = 7$, and $\overline{BC} = 6$. Find $\cos B$.

26.11 In Fig. 26-11, D is a point on side BC of $\triangle ABC$ such that $\overline{AB} = 2$, $\overline{AD} = 5$, $\overline{BD} = 4$, and $\overline{DC} = 3$. Find $\overline{AC}$. (*Hint*: Use the Law of Cosines twice.)

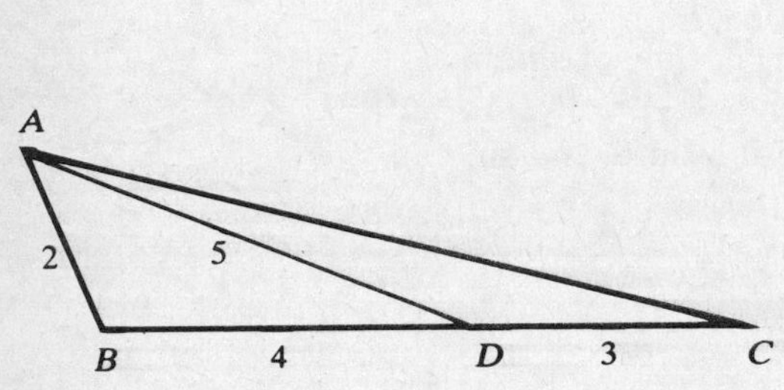

Fig. 26-11

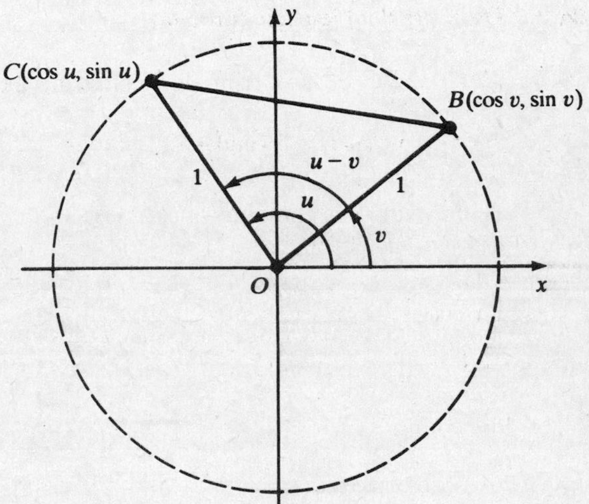

Fig. 26-12

26.12 Prove the following identities:

(*a*) $\cos^2 5\theta = (1 - \sin 5\theta)(1 + \sin 5\theta)$ (*b*) $\dfrac{1 + \cos 2\theta}{\sin 2\theta} = \dfrac{\cos\theta}{\sin\theta}$

(*c*) $(\sin\theta + \cos\theta)^2 = 1 + \sin 2\theta$ (*d*) $\cos^4\theta - \sin^4\theta = \cos^2\theta - \sin^2\theta$

26.13 For each of the following, either prove that it is an identity or give an example in which it is false.

(*a*) $\sin 4\theta = 4\sin\theta\cos\theta$ (*b*) $\dfrac{\sin\theta}{1 + \cos\theta} + \dfrac{\cos\theta}{\sin\theta} = \dfrac{1}{\sin\theta}$

(*c*) $\sin^4\theta + \cos^4\theta = 1$ (*d*) $\dfrac{\cos\theta\sin^2\theta}{1 + \cos\theta} = \cos\theta - \cos^2\theta$

(*e*) $\dfrac{\cos\theta}{\sin\theta} + \dfrac{\sin\theta}{\cos\theta} = \dfrac{2}{\sin 2\theta}$ (*f*) $2\sin\dfrac{3}{2}\theta\cos\dfrac{3}{2}\theta = \sin 3\theta$

26.14 Fill in all details of the following proof of the identity

$$\cos(u - v) = \cos u\cos v + \sin u\sin v$$

Consider the case where $0 \le v < u < v + \pi$ (see Fig. 26-12). By the Law of Cosines,

$$\overline{BC}^2 = 1^2 + 1^2 - 2(1)(1)\cos\measuredangle BOC$$

$$(\cos u - \cos v)^2 + (\sin u - \sin v)^2 = 2 - 2\cos(u - v)$$

$$\cos^2 u - 2\cos u \cos v + \cos^2 v$$
$$+ \sin^2 u - 2\sin u \sin v + \sin^2 v = 2 - 2\cos(u - v)$$
$$(\cos^2 u + \sin^2 u) + (\cos^2 v + \sin^2 v)$$
$$-2(\cos u \cos v + \sin u \sin v) = 2 - 2\cos(u - v)$$
$$1 + 1 - 2(\cos u \cos v + \sin u \sin v) = 2 - 2\cos(u - v)$$
$$\cos u \cos v + \sin u \sin v = \cos(u - v)$$

Verify that all other cases can be derived from the case just considered.

26.15 Prove the following identities:

$$(a)\quad \cos\left(\theta + \frac{\pi}{2}\right) = -\sin\theta \qquad (b)\quad \sin\left(\theta + \frac{\pi}{2}\right) = \cos\theta$$

$$(c)\quad \cos(\pi + \theta) = -\cos\theta \qquad (d)\quad \sin(\pi + \theta) = -\sin\theta$$

26.16 From the double-angle formulas

$$\cos\theta = 2\cos^2\frac{\theta}{2} - 1 = 1 - 2\sin^2\frac{\theta}{2}$$

infer the half-angle formulas.

Chapter 27

Graphs and Derivatives of the Sine and Cosine Functions

27.1 GRAPHS

Let us first observe that $\cos x$ and $\sin x$ are continuous functions. This means that, for any fixed $x = \theta$,

$$\lim_{h\to 0} \cos(\theta + h) = \cos\theta \qquad \lim_{h\to 0} \sin(\theta + h) = \sin\theta$$

as is obvious from Fig. 27-1. Indeed, as h approaches 0, point C approaches point B; therefore, the x-coordinate (the cosine) of C approaches the x-coordinate of B, and the y-coordinate (the sine) of C approaches the y-coordinate of B.

Table 27-1

x	$\cos x$	$\sin x$
0	1	0
$\frac{\pi}{6}$	$\frac{\sqrt{3}}{2} \approx 0.87$	$\frac{1}{2}$
$\frac{\pi}{4}$	$\frac{\sqrt{2}}{2} \approx 0.71$	$\frac{\sqrt{2}}{2} \approx 0.71$
$\frac{\pi}{3}$	$\frac{1}{2}$	$\frac{\sqrt{3}}{2} \approx 0.87$
$\frac{\pi}{2}$	0	1
$\frac{2\pi}{3}$	$-\frac{1}{2}$	$\frac{\sqrt{3}}{2} \approx 0.87$
$\frac{3\pi}{4}$	$-\frac{\sqrt{2}}{2} \approx -0.71$	$\frac{\sqrt{2}}{2} \approx 0.71$
$\frac{5\pi}{6}$	$-\frac{\sqrt{3}}{2} \approx -0.87$	$\frac{1}{2}$
π	-1	0

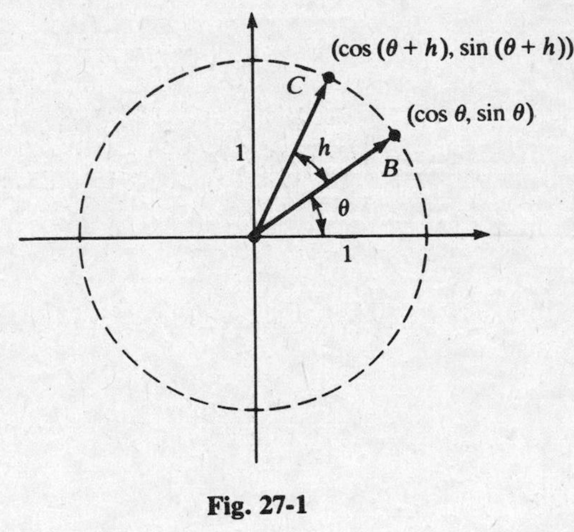

Fig. 27-1

Now we can sketch the graphs of $y = \cos x$ and $y = \sin x$. Table 27-1 contains the values of $\cos x$ and $\sin x$ for the standard values of x between 0 and $\pi/2$; these values are taken from Table 26-1. Also listed are the values for $2\pi/3$ (120°), $3\pi/4$ (135°), and $5\pi/6$ (150°). These are obtained from the formulas (see Problem 26.15)

$$\cos\left(\theta + \frac{\pi}{2}\right) = -\sin\theta \qquad \sin\left(\theta + \frac{\pi}{2}\right) = \cos\theta$$

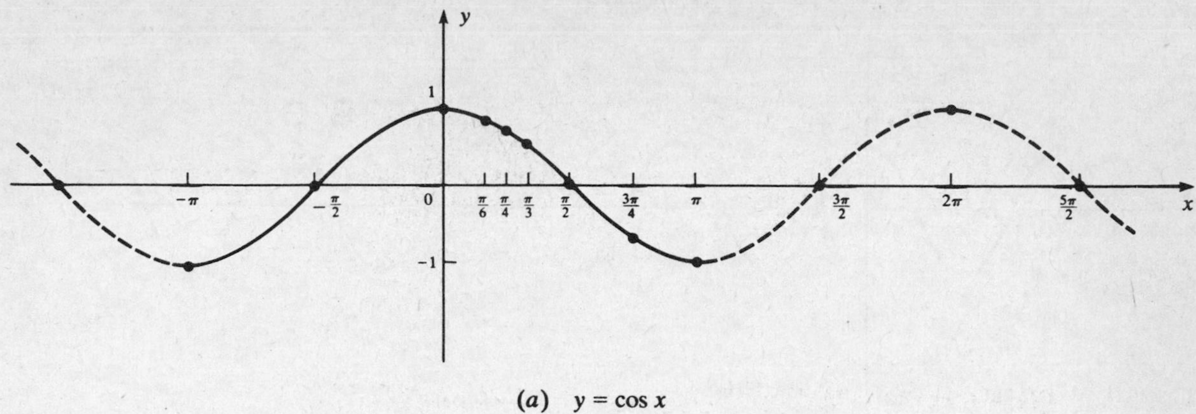

(*a*) $y = \cos x$

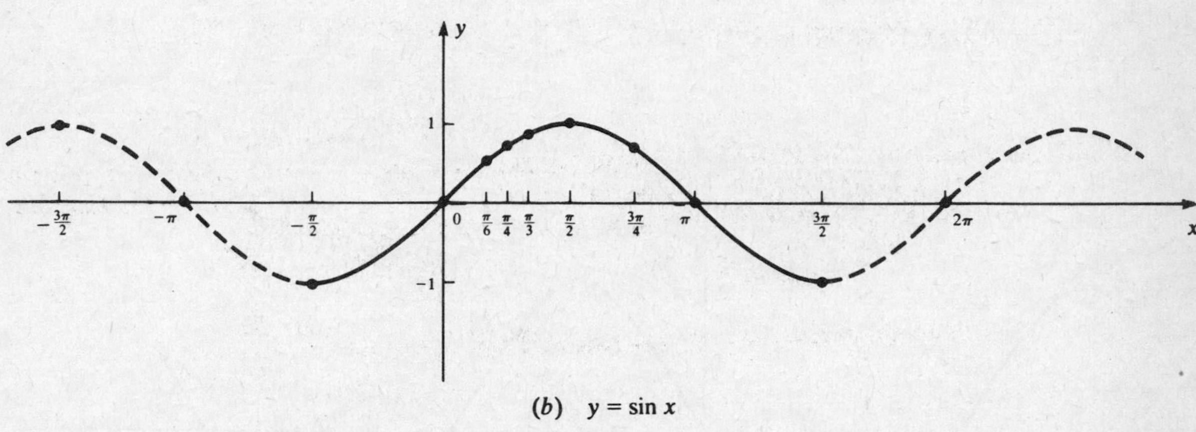

(*b*) $y = \sin x$

Fig. 27-2

The graph of $y = \cos x$ is sketched in Fig. 27-2(*a*). For arguments between $-\pi$ and 0, we have used the identity $\cos(-x) = \cos x$ (Theorem 26.3). Outside the interval $[-\pi, \pi]$, the curve repeats itself in accordance with Theorem 26.2.

The graph of $y = \sin x$, Fig. 27-2(*b*), is obtained in the same way. Notice that this graph is the result of moving the graph of $y = \cos x$ to the right by $\pi/2$ units. This can be verified by observing that

$$\cos\left(x - \frac{\pi}{2}\right) = \sin x$$

The graphs of $y = \cos x$ and $y = \sin x$ have the shape of repeated waves, with each complete wave extending over an interval of length 2π (the period). The length and height of the waves can be changed by multiplying the argument and functional value, respectively, by constants.

EXAMPLES

(*a*) $y = \cos 3x$. The graph of this function is sketched in Fig. 27-3. Because

$$\cos 3\left(x + \frac{2\pi}{3}\right) = \cos(3x + 2\pi) = \cos 3x$$

the function is of period $p = 2\pi/3$. Hence, the length of each wave is $2\pi/3$. The number of waves over an interval of length 2π (corresponding to one complete revolution of the ray determining angle x) is 3. This number is called the *frequency f* of $\cos 3x$. In general,

$$pf = (\text{length of each wave}) \times (\text{number of waves in an interval of length } 2\pi) = 2\pi$$

and so

$$f = \frac{2\pi}{p}$$

For an arbitrary constant $b > 0$, the function $\cos bx$ (or $\sin bx$) has frequency b and period $2\pi/b$.

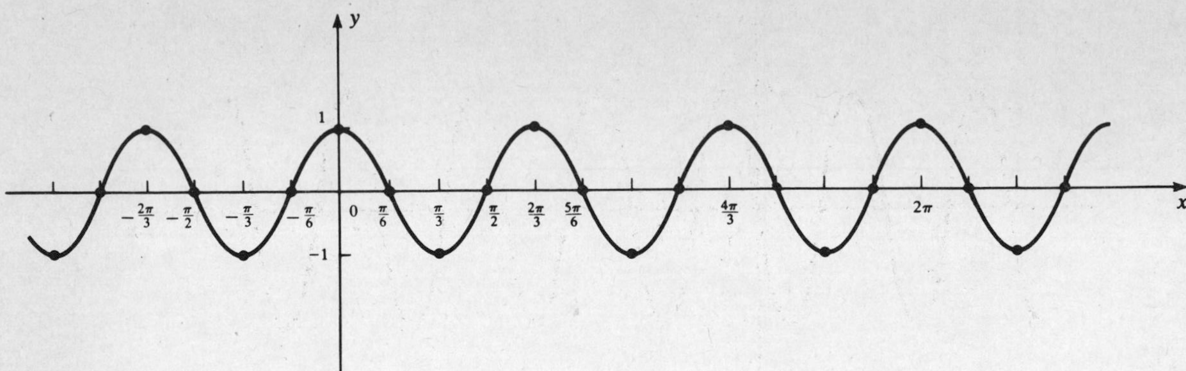

Fig. 27-3

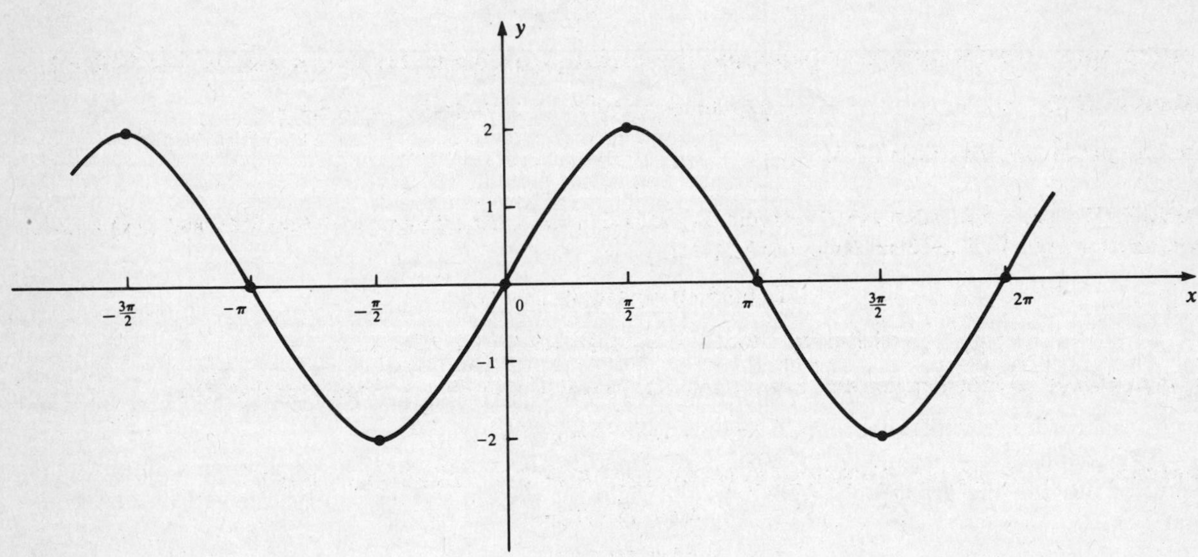

Fig. 27-4

(*b*) $y = 2 \sin x$. The graph of this function, Fig. 27-4, is obtained from the graph of $y = \sin x$, Fig. 27-2, by multiplying each ordinate by 2. The period (wavelength) and frequency of the function $2 \sin x$ are the same as those of the function $\sin x$: $p = 2\pi, f = 1$. But the ***amplitude***, the maximum height of each wave, of $2 \sin x$ is 2, or 2 times the amplitude of $\sin x$.

NOTE The *total oscillation* of a sine or cosine function, which is the vertical distance from crest to trough, is *twice* the amplitude.

For an arbitrary constant A, the function $A \sin x$ (or $A \cos x$) has amplitude $|A|$.

(*c*) Putting together Examples (*a*) and (*b*) above, we see that the functions

$$A \sin bx \qquad \text{and} \qquad A \cos bx$$

($b > 0$) have period $2\pi/b$, frequency b, and amplitude $|A|$. Figure 27-5 gives the graph of $y = 1.5 \sin 4x$.

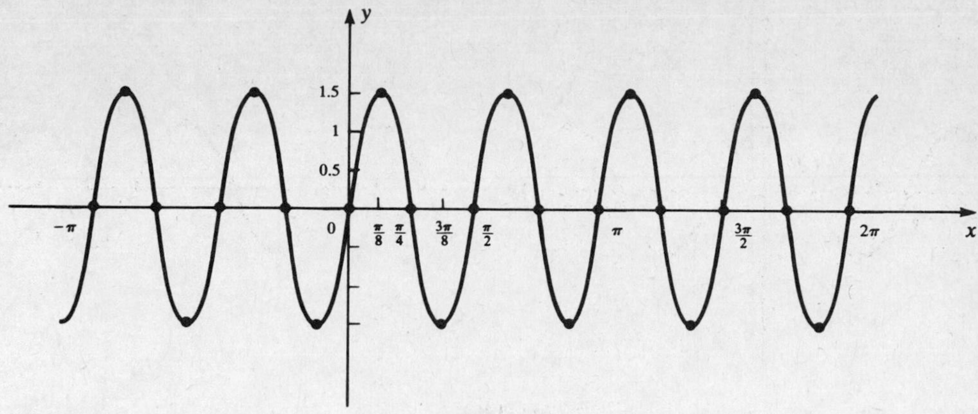

Fig. 27-5

27.2 DERIVATIVES

Lemma 27.1:

$$\lim_{\theta \to 0} \frac{\sin \theta}{\theta} = 1$$

For the proof, see Problem 27.4. Note that if we write $f(\theta) = \sin \theta$, the lemma asserts that $f'(0) = 1$.

Theorem 27.2:

$$D_x(\sin x) = \cos x \qquad D_x(\cos x) = -\sin x$$

For the proof, see Problem 27.5.

EXAMPLES (*a*) Find $D_x(\sin 2x)$. The function $\sin 2x$ is a composite function: if $f(x) = 2x$ and $g(x) = \sin x$, then $\sin 2x = g(f(x))$. The Chain Rule, (*15.2*), gives

$$D_x(g(f(x)) = g'(f(x)) \cdot D_x(f(x))$$

or

$$D_x(\sin 2x) = (\cos 2x) \cdot D_x(2x) = (\cos 2x) \cdot 2 = 2 \cos 2x$$

(*b*) Find $D_x(\cos^4 x)$. The function $\cos^4 x$ is a composite function: if $f(x) = \cos x$ and $g(x) = x^4$, then

$$\cos^4 x = (\cos x)^4 = g(f(x))$$

Therefore

$$\begin{aligned} D_x(\cos^4 x) = D_x((\cos x)^4) &= 4(\cos x)^3 \cdot D_x(\cos x) \quad \text{[Power Chain Rule]} \\ &= 4(\cos x)^3 \cdot (-\sin x) \\ &= -4 \cos^3 x \sin x \end{aligned}$$

Solved Problems

27.1 Find the period, frequency, and amplitude of

$$(a)\ \ 4 \sin \frac{x}{2} \qquad (b)\ \ \frac{1}{2} \cos 3x$$

and sketch their graphs.

(*a*) For $4 \sin \frac{1}{2}x \equiv A \sin bx$, the frequency is $f = b = \frac{1}{2}$, the period is

$$p = \frac{2\pi}{f} = \frac{2\pi}{1/2} = 4\pi$$

and the amplitude is $A = 4$. The graph is sketched in Fig. 27-6(*a*).

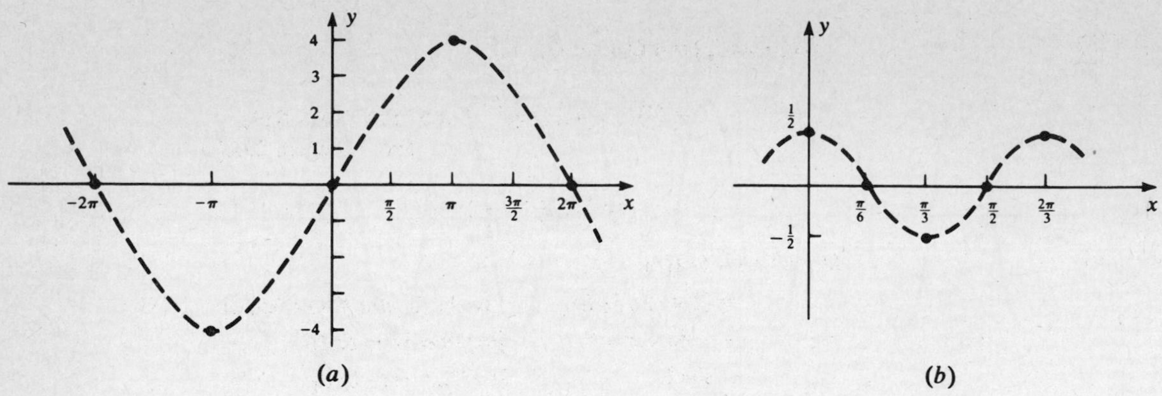

Fig. 27-6

(*b*) For $\frac{1}{2}\cos 3x \equiv A\cos bx$, the frequency f is $b = 3$, the period is

$$p = \frac{2\pi}{b} = \frac{2\pi}{3}$$

and the amplitude is $A = \frac{1}{2}$. The graph is sketched in Fig. 27-6(*b*).

27.2 Find all solutions of the equations

$$(a)\quad \cos x = 0 \qquad (b)\quad \sin x = \frac{1}{2}$$

(*a*) For $-\pi \leq x < \pi$, inspection of Fig. 27-2(*a*) shows that the only solutions of $\cos x = 0$ are

$$x = -\frac{\pi}{2} \qquad \text{and} \qquad x = \frac{\pi}{2}$$

Since 2π is the period of $\cos x$, the solutions of $\cos x = 0$ are obtained by adding arbitrary integral multiples of 2π to $-\pi/2$ and $\pi/2$:

$$-\frac{\pi}{2} + 2\pi n = \frac{\pi}{2}(4n-1) \qquad \text{and} \qquad \frac{\pi}{2} + 2\pi n = \frac{\pi}{2}(4n+1)$$

Together, $4n - 1$ and $4n + 1$ range over all odd integers, and so the solutions of $\cos x = 0$ are all odd multiples of $\pi/2$:

$$x = (2k+1)\frac{\pi}{2} \qquad (k = 0, \pm 1, \pm 2, \ldots)$$

(*b*) Figure 27-2(*b*) shows that, for $-\pi/2 \leq x < 3\pi/2$, the only solutions are $x = \pi/6$ and $x = 5\pi/6$. Hence, all the solutions are

$$\frac{\pi}{6} + 2\pi n \qquad \text{and} \qquad \frac{5\pi}{6} + 2\pi n$$

where $n = 0, \pm 1, \pm 2, \ldots$.

27.3 Find the derivatives of

$$(a)\quad 3\sin 5x \qquad (b)\quad 4\cos\frac{x}{2} \qquad (c)\quad \sin^2 x \qquad (d)\quad \cos^3\frac{x}{2}$$

(*a*)

$$\begin{aligned} D_x(3\sin 5x) &= 3\,D_x(\sin 5x) \\ &= 3(\cos 5x)\,D_x(5x) \quad \text{[by the Chain Rule and Theorem 27.2]} \\ &= 3(\cos 5x)(5) = 15\cos 5x \end{aligned}$$

(b)
$$D_x\left(4\cos\frac{x}{2}\right) = 4\,D_x\left(\cos\frac{x}{2}\right)$$
$$= (4)\left(-\sin\frac{x}{2}\right)D_x\left(\frac{x}{2}\right) \quad \text{[by the Chain Rule and Theorem 27.2]}$$
$$= -4\left(\sin\frac{x}{2}\right)\left(\frac{1}{2}\right) = -2\sin\frac{x}{2}$$

(c)
$$D_x(\sin^2 x) = D_x((\sin x)^2)$$
$$= (2\sin x)\,D_x(\sin x) \quad \text{[by the Power Chain Rule]}$$
$$= 2\sin x\cos x \quad \text{[by Theorem 27.2]}$$
$$= \sin 2x \quad \text{[by Theorem 26.8]}$$

(d)
$$D_x\left(\cos^3\frac{x}{2}\right) = D_x\left(\left(\cos\frac{x}{2}\right)^3\right)$$
$$= 3\left(\cos\frac{x}{2}\right)^2 D_x\left(\cos\frac{x}{2}\right) \quad \text{[by the Power Chain Rule]}$$
$$= \left(3\cos^2\frac{x}{2}\right)\left(-\left(\sin\frac{x}{2}\right)D_x\left(\frac{x}{2}\right)\right) \quad \text{[by the Chain Rule and Theorem 27.2]}$$
$$= \left(3\cos^2\frac{x}{2}\right)\left(-\frac{1}{2}\sin\frac{x}{2}\right) = -\frac{3}{2}\cos^2\frac{x}{2}\sin\frac{x}{2}$$

27.4 Prove Lemma 27.1.

Consider the case when $\theta > 0$. Using the notation of Fig. 27-7,

$$\text{area}(\triangle OBC) < \text{area (sector } OBA) < \text{area}(\triangle ODA) \tag{1}$$

Now,

$$\text{area}(\triangle OBC) = \frac{1}{2}(\overline{OC})(\overline{BC}) = \frac{1}{2}\cos\theta\sin\theta$$

$$\text{area (sector } OBA) = \frac{\theta}{2\pi}\times(\text{area of circle})$$
$$= \frac{\theta}{2\pi}\times(\pi 1^2) = \frac{\theta}{2}$$

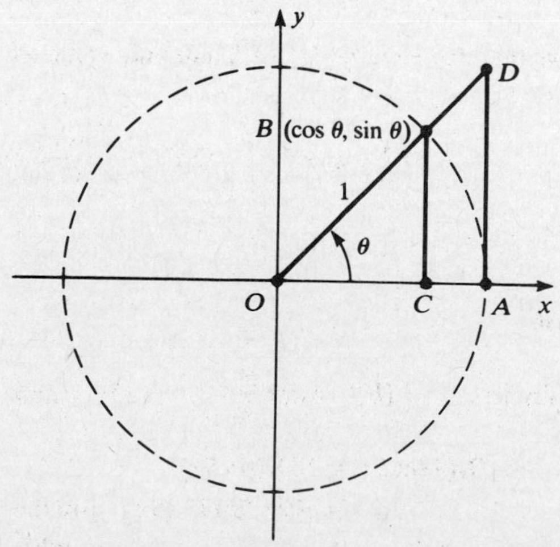

Fig. 27-7

Furthermore, since $\triangle OBC$ and $\triangle ODA$ are similar (by the AA-criterion),

$$\frac{\overline{DA}}{\overline{BC}} = \frac{\overline{OA}}{\overline{OC}} \quad \text{or} \quad \frac{\overline{DA}}{\sin\theta} = \frac{1}{\cos\theta} \quad \text{or} \quad \overline{DA} = \frac{\sin\theta}{\cos\theta}$$

$$\text{area}\,(\triangle ODA) = \frac{1}{2}(\overline{DA})(\overline{OA}) = \frac{1}{2}\left(\frac{\sin\theta}{\cos\theta}\right)(1)$$

and (*1*) becomes, after dividing through by the positive quantity $\frac{1}{2}\sin\theta$,

$$\cos\theta < \frac{\theta}{\sin\theta} < \frac{1}{\cos\theta}$$

which is equivalent to

$$\cos\theta < \frac{\sin\theta}{\theta} < \frac{1}{\cos\theta} \tag{2}$$

ALGEBRA If a and b are positive numbers, then

$$a < b \quad \text{if and only if} \quad \frac{1}{a} > \frac{1}{b}$$

In (*2*), let θ approach 0 from the right. Both $\cos\theta$ and $1/(\cos\theta)$ approach 1; therefore, by Problem 8.8(*f*),

$$\lim_{\theta\to 0^+} \frac{\sin\theta}{\theta} = 1 \tag{3}$$

Now, the function $(\sin\theta)/\theta$ being even (verify this), (*3*) implies that

$$\lim_{\theta\to 0^-} \frac{\sin\theta}{\theta} = 1 \tag{4}$$

and (*3*) and (*4*) together mean simply

$$\lim_{\theta\to 0} \frac{\sin\theta}{\theta} = 1$$

which is Lemma 27.1.

27.5 Prove Theorem 27.2.

It suffices to prove either half of the theorem; say,

$$D_x(\sin x) = \cos x$$

The other half will follow at once from Theorem 26.7.

By Theorem 26.6,

$$\sin(x+h) = \sin x \cos h + \cos x \sin h$$

and so

$$\begin{aligned}
\frac{\sin(x+h) - \sin x}{h} &= \frac{\sin x \cos h + \cos x \sin h - \sin x}{h} \\
&= \frac{\cos x \sin h - (\sin x)(1 - \cos h)}{h} \\
&= (\cos x)\frac{\sin h}{h} - (\sin x)\frac{1 - \cos h}{h} \\
&= (\cos x)\frac{\sin h}{h} - (\sin x)\frac{\sin^2(h/2)}{h/2} \qquad \text{[by Theorem 26.9]} \\
&= (\cos x)\frac{\sin h}{h} - \left(\sin\frac{h}{2}\right)(\sin x)\frac{\sin(h/2)}{h/2}
\end{aligned}$$

Letting $h \to 0$, which also makes $h/2 \to 0$, and applying Lemma 27.1,

$$D_x(\sin x) = \lim_{h \to 0} \frac{\sin(x+h) - \sin x}{h}$$
$$= (\cos x)1 - (0)(\sin x)1 = \cos x$$

27.6 Evaluate:

$$\lim_{\theta \to 0} \frac{\theta^2}{1 - \cos \theta}$$

$$\frac{\theta^2}{1-\cos\theta} = \frac{\theta^2}{1-\cos\theta}\left(\frac{1+\cos\theta}{1+\cos\theta}\right) = \frac{\theta^2(1+\cos\theta)}{1-\cos^2\theta} = \frac{\theta^2(1+\cos\theta)}{\sin^2\theta}$$

Hence

$$\lim_{\theta\to 0}\frac{\theta^2}{1-\cos\theta} = \lim_{\theta\to 0}\frac{\theta^2(1+\cos\theta)}{\sin^2\theta}$$

$$= \lim_{\theta\to 0}\frac{\theta^2}{\sin^2\theta}\cdot\lim_{\theta\to 0}(1+\cos\theta) = \left(\lim_{\theta\to 0}\frac{\theta}{\sin\theta}\right)^2\cdot 2$$

$$= 2\left(\frac{1}{\lim\limits_{\theta\to 0}\dfrac{\sin\theta}{\theta}}\right)^2 = 2\left(\frac{1}{1}\right)^2 = 2$$

27.7 Sketch the graph of the function $f(x) = \sin x - \sin^2 x$.

Along with $\sin x$, $f(x)$ has 2π as a period; therefore, we need only sketch the graph for $0 \le x < 2\pi$.

$$f'(x) = \cos x - 2\sin x\cos x = \cos x - \sin 2x$$
$$f''(x) = -\sin x - (\cos 2x)\,D_x(2x) = -\sin x - 2\cos 2x$$

To find the critical numbers, solve $f'(x) = 0$:

$$\cos x - 2\sin x\cos x = 0$$
$$(\cos x)(1 - 2\sin x) = 0$$
$$\cos x = 0 \quad\text{or}\quad 1 - 2\sin x = 0$$
$$\cos x = 0 \quad\text{or}\quad \sin x = \frac{1}{2}$$
$$x = \frac{\pi}{2}, \frac{3\pi}{2} \quad\text{or}\quad x = \frac{\pi}{6}, \frac{5\pi}{6}$$

where Problem 27.2(*b*) has been used. Application of the Second-Derivative Test (Theorem 23.2) at each critical number gives the results shown in Table 27-2. The graph is sketched in Fig. 27-8.

Table 27-2

x	$f(x)$	$f''(x)$
$\frac{\pi}{6}$	$\frac{1}{4}$	$-\frac{3}{2}$ *rel. max.*
$\frac{\pi}{2}$	0	1 *rel. min.*
$\frac{5\pi}{6}$	$\frac{1}{4}$	$-\frac{1}{2}$ *rel. max.*
$\frac{3\pi}{2}$	-2	3 *rel. min.*

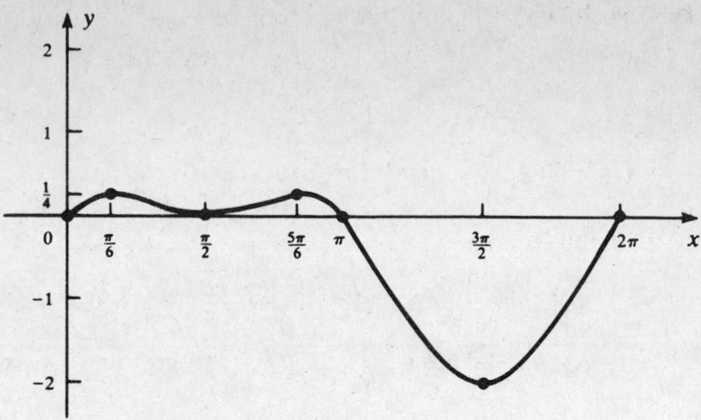

Fig. 27-8

27.8 Find the absolute extrema of the function

$$f(x) = 2 \sin x - \cos 2x$$

on the closed interval $[0, \pi]$.

For this simple function it is unnecessary to resort to the tabular method of Section 14.2. It suffices to observe that the sine and cosine functions have maximum absolute value 1. Thus, if there is an argument x in $[0, \pi]$ such that

$$f(x) = 2(+1) - (-1) = 3$$

that argument must represent an absolute maximum (on any interval whatever). Clearly, there is one and only one such argument, $x = \pi/2$.

On the other hand, since $\sin x \geq 0$ on $[0, \pi]$, the arguments $x = 0$ and $x = \pi$ minimize $\sin x$, and, at the same time, *maximize* $\cos 2x$. Hence,

$$f(0) = f(\pi) = 2(0) - (+1) = -1$$

is the absolute minimum value on $[0, \pi]$.

Supplementary Problems

27.9 Find the period, frequency, and amplitude of each function, and sketch its graph.

$$(a)\ \cos \frac{4x}{3} \qquad (b)\ \frac{1}{2} \sin 3x$$

27.10 Find the period, frequency, and amplitude of

$$(a)\ 2 \sin \frac{x}{3} \qquad (b)\ -\cos 2x \qquad (c)\ 5 \sin \frac{5x}{2} \qquad (d)\ \cos(-4x)$$

(*Hint*: In (*d*), the function is even.)

27.11 Find all solutions of the following equations:

(*a*) $\sin x = 0$ (*b*) $\cos x = 1$ (*c*) $\sin x = 1$ (*d*) $\cos x = -1$

(*e*) $\sin x = -1$ (*f*) $\cos x = \frac{1}{2}$ (*g*) $\sin x = \frac{\sqrt{2}}{2}$ (*h*) $\cos x = \frac{\sqrt{3}}{2}$

(*Hint*: By Theorem 26.1, only two of (*a*), (*b*), and (*d*) need be solved.)

27.12 Find the derivatives of the following functions:

(a) $4 \sin^3 x$ (b) $\sin x + 2 \cos x$ (c) $x \sin x$ (d) $x^2 \cos 2x$

(e) $\dfrac{\sin x}{x}$ (f) $\dfrac{1 - \cos x}{x^2}$ (g) $5 \sin 3x \cos x$ (h) $\cos^2 2x$

(i) $\cos (2x^2 - 3)$ (j) $\sin^3 (5x + 4)$ (k) $\sqrt{\cos 2x}$

27.13 Calculate the following limits:

(a) $\lim_{x \to 0} \dfrac{\sin x}{3x}$ (b) $\lim_{x \to 0} \dfrac{\cos x - 1}{2x}$ (c) $\lim_{x \to 0} \dfrac{\sin 2x}{\sin 3x}$

(d) $\lim_{\theta \to 0} \dfrac{\theta + 1}{\cos \theta}$ (e) $\lim_{x \to 0} \dfrac{\sin 4x}{5x}$ (f) $\lim_{x \to 0} \dfrac{\sin^3 x}{4x^3}$

(g) $\lim_{x \to 0} \dfrac{\sin x^2}{x}$ (h) $\lim_{x \to \pi/2} \dfrac{\cos 3x}{\cos x}$

27.14 Sketch the graphs of the following functions, making use of oddness or evenness when possible.

(a) $f(x) = \sin^2 x$ (b) $g(x) = \sin x + \cos x$ (c) $f(x) = 3 \sin x - \sin^3 x$

(d) $h(x) = \cos x - \cos^2 x$ (e) $g(x) = |\sin x|$ (f) $f(x) = \sin x + x$

27.15 Find the absolute extrema of each function on the given interval.

(a) $f(x) = \sin x + x$, on $[0, 2\pi]$ (b) $f(x) = \sin x - \cos x$, on $[0, \pi]$

(c) $f(x) = \cos^2 x + \sin x$, on $[0, \pi]$ (d) $f(x) = 2 \sin x + \sin 2x$, on $[0, 2\pi]$

(e) $f(x) = \left| (\cos x) - \dfrac{1}{4} \right|$, on $[0, 2\pi]$ (f) $f(x) = \dfrac{1}{2}x - \sin x$, on $[0, 2\pi]$

(g) $f(x) = 3 \sin x - 4 \cos x$, on $[0, 2\pi]$

27.16 (a) Show that for $f(x) = A \cos x + B \sin x$,

$$\text{period} = 2\pi \qquad \text{amplitude} = \sqrt{A^2 + B^2}$$

(*Hint*: If ϕ be such that

$$\sin \phi = \frac{A}{\sqrt{A^2 + B^2}} \qquad \cos \phi = \frac{B}{\sqrt{A^2 + B^2}}$$

show that $f(x) = \sqrt{A^2 + B^2} \sin (x + \phi)$.)

(b) Find the amplitude and period of

(i) $3 \cos x - 4 \sin x$ (ii) $5 \sin 2x + 12 \cos 2x$ (iii) $2 \cos x + \sqrt{2} \sin x$

27.17 Evaluate:

(a) $\lim_{h \to 0} \dfrac{\sin \left(\dfrac{\pi}{4} + h \right) - \sin \dfrac{\pi}{4}}{h}$ (b) $\lim_{h \to 0} \dfrac{\cos \left(\dfrac{\pi}{3} + h \right) - \dfrac{1}{2}}{h}$

(*Hint*: Recall the definition of the derivative.)

27.18 For what value of A does $2 \sin Ax$ have period 2?

27.19 Find an equation of the tangent line to the graph of $y = \sin^2 x$ at the point where $x = \pi/3$.

27.20 Find an equation of the normal line to the curve $y = 1 + \cos x$ at the point $(\pi/3, 3/2)$.

27.21 Find the smallest positive integer n such that $D_x^n(\cos x) = \cos x$.

27.22 Let $f(x) = \sin x + \sin |x|$. (*a*) Sketch the graph of f. (*b*) Is f continuous at $x = 0$? (*c*) Is f differentiable at $x = 0$?

27.23 Find the slope of the tangent line to the graph of (*a*) $y = x + \cos(xy)$, at $(0, 1)$; (*b*) $\sin(xy) = y$, at $(\pi/2, 1)$. (*Hint*: Use implicit differentiation.)

27.24 Use implicit differentiation to find y':

(*a*) $\cos y = x$ (*b*) $\sin(xy) = y^2$

27.25 (*a*) A ladder 26 feet long is leaning against a vertical wall (Fig. 27-9). If the bottom of the ladder, A, is slipping away from the base of the wall at the rate of 3 feet per second, how fast is the angle θ between the ladder and the ground changing when the bottom of the ladder is 10 feet from the base of the wall? (*b*) An airplane is ascending at a speed of 400 kilometers per hour along a line making an angle of 60° with the ground. How fast is the altitude of the plane changing?

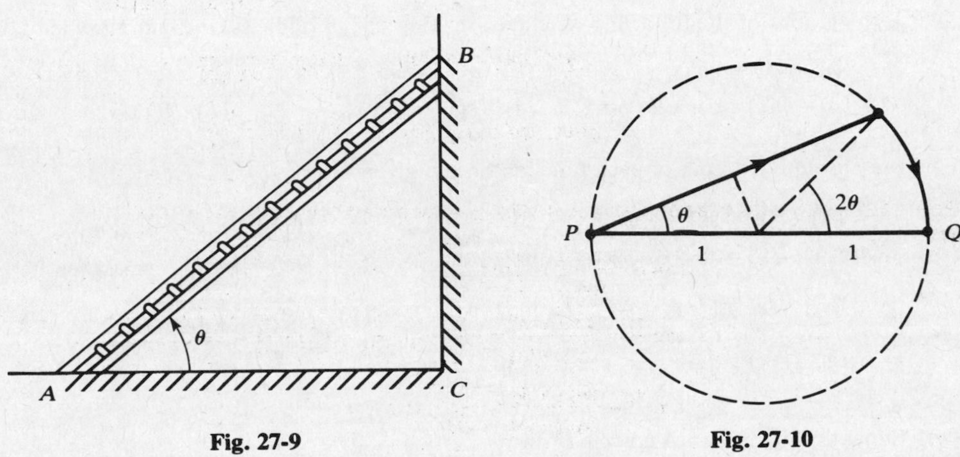

Fig. 27-9 **Fig. 27-10**

27.26 (*a*) A man at a point P on the shore of a circular lake of radius one mile (Fig. 27-10) wants to reach the point Q on the shore diametrically opposite P. He can row 1.5 miles per hour and walk 3 miles per hour. At what angle θ $(0 \le \theta \le \pi/2)$ to the diameter PQ should he row in order to minimize the time required to reach Q? (When $\theta = 0$, he rows all the way; when $\theta = \pi/2$, he walks all the way.) (*b*) Rework (*a*) if, instead of rowing, the man can paddle a canoe at 4 miles per hour.

27.27 (*a*) Find the absolute extrema of $f(x) = x - \sin x$ on $[0, \pi/2]$. (*b*) From (*a*), infer that $\sin x < x$ holds for all positive x.

Chapter 28

The Tangent and Other Trigonometric Functions

Besides the sine and cosine functions, there are four other important trigonometric functions, all of them expressible in terms of $\sin x$ and $\cos x$.

Definitions:

tangent function $\tan x = \dfrac{\sin x}{\cos x}$

cotangent function $\cot x = \dfrac{\cos x}{\sin x} = \dfrac{1}{\tan x}$

secant function $\sec x = \dfrac{1}{\cos x}$

cosecant function $\csc x = \dfrac{1}{\sin x}$

EXAMPLE Let us calculate, and collect in Table 28-1, some of the values of $\tan x$.

$$\tan 0 = \frac{\sin 0}{\cos 0} = \frac{0}{1} = 0$$

$$\tan \frac{\pi}{6} = \frac{\sin(\pi/6)}{\cos(\pi/6)} = \frac{1/2}{\sqrt{3}/2} = \frac{1}{\sqrt{3}} = \frac{1}{\sqrt{3}}\frac{\sqrt{3}}{\sqrt{3}} = \frac{\sqrt{3}}{3}$$

$$\tan \frac{\pi}{4} = \frac{\sin(\pi/4)}{\cos(\pi/4)} = \frac{\sqrt{2}/2}{\sqrt{2}/2} = 1$$

$$\tan \frac{\pi}{3} = \frac{\sin(\pi/3)}{\cos(\pi/3)} = \frac{\sqrt{3}/2}{1/2} = \sqrt{3}$$

Table 28-1

x	$\tan x$
0	0
$\dfrac{\pi}{6}$	$\dfrac{\sqrt{3}}{3} \approx 0.58$
$\dfrac{\pi}{4}$	1
$\dfrac{\pi}{3}$	$\sqrt{3} \approx 1.73$

Notice that $\tan(\pi/2)$ is not defined, since $\sin(\pi/2) = 1$ and $\cos(\pi/2) = 0$. Because $\cos x$ changes sign at $x = \pi/2$, we have:

$$\lim_{x \to (\pi/2)^-} (\tan x) = +\infty \qquad \lim_{x \to (\pi/2)^+} (\tan x) = -\infty$$

The principal properties of the supplementary trigonometric functions will be summarized in the form of theorems.

Theorem 28.1: The tangent and cotangent functions are periodic, of period π. Moreover, they are odd functions.

The proof is immediate from Problem 26.4(*b*). As a consequence of Theorem 28.1, it is necessary to evaluate $\tan x$ or $\cot x$ only in the interval $(0, \pi/2)$.

Theorem 28.2 (*Derivatives*):

$$D_x(\tan x) = \sec^2 x$$
$$D_x(\cot x) = -\csc^2 x$$
$$D_x(\sec x) = \tan x \sec x$$
$$D_x(\csc x) = -\cot x \csc x$$

For the proofs, see Problem 28.1.

EXAMPLE From Theorem 28.2 and the Power Chain Rule,

$$D_x^2(\tan x) = D_x((\sec x)^2) = 2(\sec x)\, D_x(\sec x)$$
$$= 2(\sec x)(\tan x \sec x) = 2 \tan x \sec^2 x$$

Now, in $(0, \pi/2)$, $\tan x > 0$ (since both $\sin x$ and $\cos x$ are positive), making $D_x^2(\tan x) > 0$. Thus (Theorem 23.1) the graph of $y = \tan x$ is concave upward on $(0, \pi/2)$. Knowing this, we can easily sketch the graph on $(0, \pi/2)$, and hence everywhere. See Fig. 28-1.

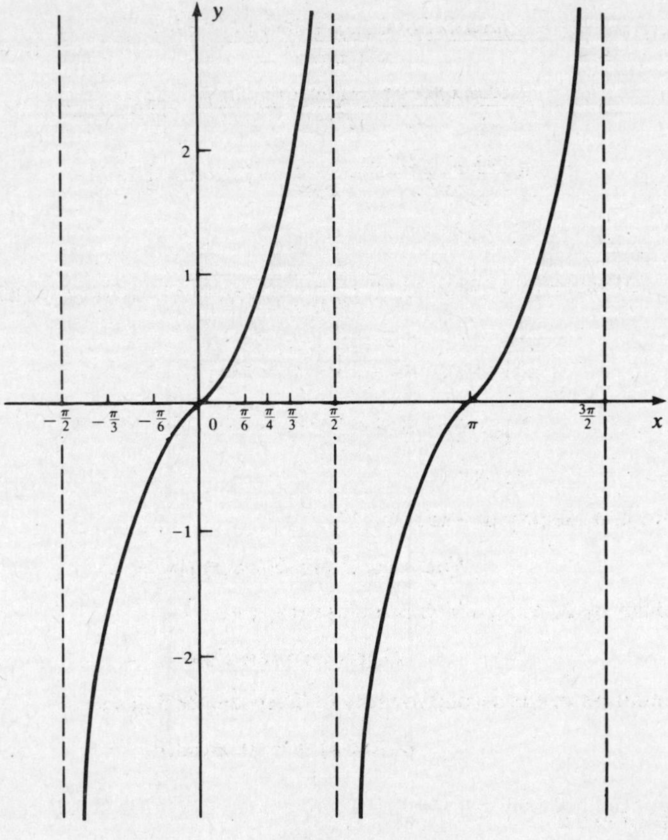

Fig. 28-1

Theorem 28.3 (*Identities*):

$$\tan^2 x + 1 = \sec^2 x \qquad \cot^2 x + 1 = \csc^2 x$$

Proof: Divide $\sin^2 x + \cos^2 x = 1$ by $\cos^2 x$ or $\sin^2 x$.

Traditional Definitions

As was the case with $\sin\theta$ and $\cos\theta$, the supplementary trigonometric functions were originally defined only for an acute angle of a right triangle. Referring to Fig. 26-3, we have:

$$\tan\theta \equiv \frac{\text{opposite side}}{\text{adjacent side}}$$

$$\cot\theta \equiv \frac{\text{adjacent side}}{\text{opposite side}}$$

$$\sec\theta \equiv \frac{\text{hypotenuse}}{\text{adjacent side}}$$

$$\csc\theta \equiv \frac{\text{hypotenuse}}{\text{opposite side}}$$

Solved Problems

28.1 Prove Theorem 28.2.

$$\begin{aligned} D_x(\tan x) &= D_x\left(\frac{\sin x}{\cos x}\right) \\ &= \frac{(\cos x)\,D_x(\sin x) - (\sin x)\,D_x(\cos x)}{(\cos x)^2} \qquad \text{[Quotient Rule]} \\ &= \frac{(\cos x)(\cos x) - (\sin x)(-\sin x)}{\cos^2 x} \qquad \text{[Theorem 27.2]} \\ &= \frac{\cos^2 x + \sin^2 x}{\cos^2 x} = \frac{1}{\cos^2 x} \qquad \text{[Theorem 26.1]} \\ &= \sec^2 x \end{aligned}$$

$$\begin{aligned} D_x(\cot x) &= D_x((\tan x)^{-1}) = \frac{-1}{(\tan x)^2}\,D_x(\tan x) \qquad \text{[Power Chain Rule]} \\ &= \frac{-1}{(\tan x)^2}\,\frac{1}{(\cos x)^2} = \frac{-1}{(\tan x \cos x)^2} \\ &= \frac{-1}{(\sin x)^2} \qquad [\tan x = (\sin x)/(\cos x)] \\ &= -\csc^2 x \end{aligned}$$

Differentiating the first identity of Theorem 28.3,

$$2(\tan x)(\sec^2 x) = 2(\sec x)\,D_x(\sec x)$$

and dividing through by $2(\sec x)$, which is never zero, gives

$$D_x(\sec x) = \tan x \sec x$$

Similarly, differentiation of the second identity of Theorem 28.3 gives

$$D_x(\csc x) = -\cot x \csc x$$

28.2 Draw the graph of $\csc x$.

Because $\csc x = 1/(\sin x)$, it is, along with $\sin x$, periodic, of period 2π, and odd. Therefore, we need only find the graph for $0 < x < \pi$. Now,

$$\csc\frac{\pi}{2} = \frac{1}{\sin\frac{\pi}{2}} = \frac{1}{1} = 1$$

As x decreases from $\pi/2$ toward 0, $\sin x$ decreases from 1 toward 0; therefore, $\csc x$ increases from 1 to $+\infty$. Likewise, as x increases from $\pi/2$ toward π, $\sin x$ decreases from 1 toward 0, and $\csc x$ increases from 1 to $+\infty$. In fact, since

$$\sin\left(\frac{\pi}{2}+u\right)=\sin\frac{\pi}{2}-u$$

the graph will be symmetric about the line $x=\pi/2$. See Fig. 28-2.

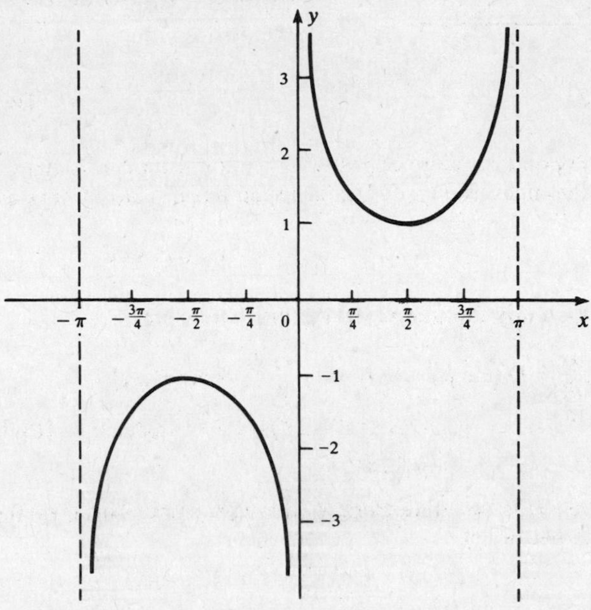

Fig. 28-2

28.3 Prove the identity

$$\tan(u-v)=\frac{\tan u-\tan v}{1+\tan u\tan v}$$

$$\tan(u-v)=\frac{\sin(u-v)}{\cos(u-v)}=\frac{\sin u\cos v-\cos u\sin v}{\cos u\cos v+\sin u\sin v}\quad\text{[Theorem 26.6]}$$

$$=\frac{\dfrac{\sin u}{\cos u}-\dfrac{\sin v}{\cos v}}{1+\dfrac{\sin u}{\cos u}\dfrac{\sin v}{\cos v}}\quad\text{[divide numerator and denominator by } \cos u\cos v\text{]}$$

$$=\frac{\tan u-\tan v}{1+\tan u\tan v}$$

28.4 Calculate (*a*) $D_x(3\tan^2 x)$, (*b*) $D_x(\sec x\tan x)$.

(*a*)
$$D_x(3\tan^2 x)=3\,D_x(\tan^2 x)=3(2\tan x)\,D_x(\tan x)\quad\text{[Power Chain Rule]}$$
$$=6\tan x\sec^2 x$$

(*b*)
$$D_x(\sec x\tan x)=(\sec x)\,D_x(\tan x)+(\tan x)\,D_x(\sec x)\quad\text{[Product Rule]}$$
$$=(\sec x)(\sec^2 x)+(\tan x)(\tan x\sec x)\quad\text{[Theorem 28.2]}$$
$$=(\sec x)(\sec^2 x+\tan^2 x)=(\sec x)(\tan^2 x+1+\tan^2 x)\quad\text{[Theorem 28.3]}$$
$$=(\sec x)(2\tan^2 x+1)$$

28.5 A lighthouse, H, one mile from a straight shore, has a beam that rotates counterclockwise (as seen from above) at the rate of six revolutions per minute. How fast is point P, where the beam of light hits the shore, moving?

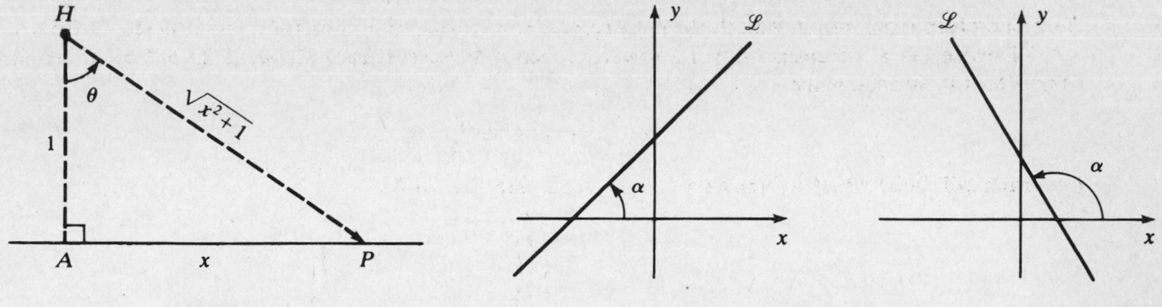

Fig. 28-3

Fig. 28-4

Let A be the point on shore opposite H, and let x be the distance $\overline{AP}$ (see Fig. 28-3). We must find dx/dt. Let θ be the measure of $\measuredangle PHA$. Since the beam makes 6 revolutions (of 2π radians) per minute,

$$\frac{d\theta}{dt} = 12\pi \text{ rad/min}$$

Now, $\tan\theta = x/1 = x$ (by the traditional definition); hence,

$$\begin{aligned}\frac{dx}{dt} &= D_t(\tan\theta) = D_\theta(\tan\theta)\frac{d\theta}{dt} \quad \text{[Chain Rule]}\\ &= (\sec^2\theta)(12\pi) = (\sqrt{x^2+1})^2(12\pi) \quad \text{[trad. defn. of secant]}\\ &= 12\pi(x^2+1)\end{aligned}$$

For instance, when P is one mile from A, the light is moving along the shore at 24π miles per minute (about 4522 miles per hour).

28.6 The *angle of inclination* of a nonvertical line $\mathscr{L}$ is defined to be the smaller counterclockwise angle α from the positive x-axis to the line. (See Fig. 28-4.) Show that

$$\tan\alpha = m$$

where m is the slope of $\mathscr{L}$.

By taking a parallel line, we may always assume that $\mathscr{L}$ goes through the origin (Fig. 28-5). $\mathscr{L}$ intersects the unit circle with center at the origin at the point $P(\cos\alpha, \sin\alpha)$. By definition of slope,

$$m = \frac{\sin\alpha - 0}{\cos\alpha - 0} = \frac{\sin\alpha}{\cos\alpha} = \tan\alpha$$

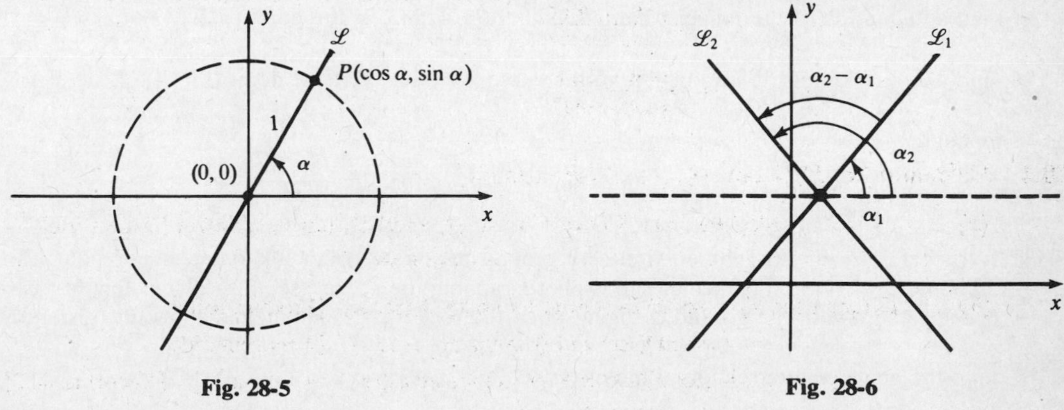

Fig. 28-5

Fig. 28-6

28.7 Find the angle at which the lines

$$\mathscr{L}_1:\ y = 2x + 1 \qquad \text{and} \qquad \mathscr{L}_2:\ y = -3x + 5$$

intersect.

Let α_1 and α_2 be the angles of inclination of $\mathscr{L}_1$ and $\mathscr{L}_2$ (see Fig. 28-6).

$$\tan(\alpha_2-\alpha_1)=\frac{\tan\alpha_2-\tan\alpha_1}{1+\tan\alpha_1\tan\alpha_2}\quad\text{[by Problem 28.3]}$$

$$=\frac{m_2-m_1}{1+m_1m_2}\quad\text{[by Problem 28.6]}$$

$$=\frac{-3-2}{1+(-3)(2)}=\frac{-5}{1-6}=\frac{-5}{-5}=1$$

Since $\tan(\alpha_2-\alpha_1)=1$,

$$\alpha_2-\alpha_1=\frac{\pi}{4}\text{ radians}=45^\circ$$

In general, given $\tan(\alpha_2-\alpha_1)$, the value of $\alpha_2-\alpha_1$ can be estimated from the table in Appendix D.

It should be noted that, in certain cases, the above method will yield the *larger* angle of intersection.

Supplementary Problems

28.8 Sketch the graphs of (*a*) $\sec x$, (*b*) $\cot x$.

28.9 Prove the identity

$$\tan(u+v)=\frac{\tan u+\tan v}{1-\tan u\tan v}$$

28.10 Calculate:

(*a*) $D_x\left(2\tan\frac{x}{2}-5\right)$ (*b*) $D_x(\tan x-\sec x)$ (*c*) $D_x(\cot^2 x)$ (*d*) $D_x(\cot 4x+3x)$

(*e*) $D_x(\sec^2 3x)$ (*f*) $D_x(\csc(3x-5))$ (*g*) $D_x(\csc\sqrt{x})$

28.11 Find y' by implicit differentiation:

(*a*) $\tan(xy)=y$ (*b*) $\sec^2 y+\csc^2 x=3$ (*c*) $\tan^2(y+1)=3\sin x$ (*d*) $y=\tan^2(x+y)$

28.12 Find an equation of the tangent line to the curve $y=\tan x$ at the point $(\pi/3, \sqrt{3})$.

28.13 Find an equation of the normal line to the curve $y=3\sec^2 x$ at the point $(\pi/6, 4)$.

28.14 Evaluate

(*a*) $\lim_{x\to 0}\frac{\tan x}{x}$ (*b*) $\lim_{x\to 0}\frac{\tan^3 2x}{x^3}$

28.15 A rocket is rising straight up from the ground at a rate of 1000 kilometers per hour. An observer 2 kilometers from the launching site is photographing the rocket (see Fig. 28-7). How fast is the angle θ of the camera with the ground changing when the rocket is 1.5 kilometers above the ground?

28.16 Find the angle of intersection of the lines $\mathscr{L}_1$: $y=x-3$ and $\mathscr{L}_2$: $y=-5x+4$.

28.17 Find the angle of intersection of the tangent lines to the curves $xy=1$ and $y=x^3$ at the common point (1, 1).

28.18 Find the angle of intersection of the tangent lines to the curves $y=\cos x$ and $y=\sin 2x$ at the common point $(\pi/6, \sqrt{3}/2)$.

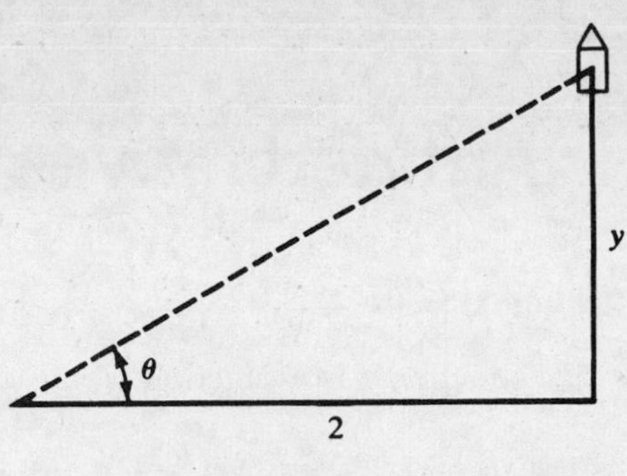

Fig. 28-7

28.19 Find an equation of the tangent line to the curve

$$1 + 16x^2y = \tan(x - 2y)$$

at the point $(\pi/4, 0)$.

28.20 Find the relative maxima and minima, inflection points, and vertical asymptotes of the graphs of the following functions, on $[0, \pi]$:

(*a*) $f(x) = 2x - \tan x$ (*b*) $f(x) = \tan x - 4x$

28.21 Find the intervals where the function $f(x) = \tan x - \sin x$ is increasing.

Chapter 29

Antiderivatives

29.1 DEFINITION AND NOTATION

Definition: An *antiderivative* of a function f is a function whose derivative is f.

EXAMPLES (*a*) x^2 is an antiderivative of $2x$, since $D_x(x^2) = 2x$. (*b*) $x^4/4$ is an antiderivative of x^3, since $D_x(x^4/4) = x^3$. (*c*) $3x^3 - 4x^2 + 5$ is an antiderivative of $9x^2 - 8x$, since

$$D_x(3x^3 - 4x^2 + 5) = 9x^2 - 8x$$

(*d*) $x^2 + 3$ is an antiderivative of $2x$, since $D_x(x^2 + 3) = 2x$. (*e*) $\sin x$ is an antiderivative of $\cos x$, since $D_x(\sin x) = \cos x$.

Examples (*a*) and (*d*) show that a function can have more than one antiderivative. This is true for all functions: If $g(x)$ is an antiderivative of $f(x)$, then $g(x) + C$ is also an antiderivative of $f(x)$, where C is any constant. The reason is that $D_x(C) = 0$, whence

$$D_x(g(x) + C) = D_x(g(x))$$

Let us find the relationship between any two antiderivatives of a function.

Theorem 29.1: If $F'(x) = 0$ for all x in an interval $\mathcal{I}$, then $F(x)$ is a constant on $\mathcal{I}$.

The assumption $F'(x) = 0$ tells us that the graph of F always has a horizontal tangent. It is then obvious that the graph of F must be a horizontal straight line; that is, $F(x)$ is constant. For a rigorous proof, see Problem 29.4.

Corollary 29.2: If $g'(x) = h'(x)$ for all x in an interval $\mathcal{I}$, then there is a constant C such that $g(x) = h(x) + C$ for all x in $\mathcal{I}$.

Indeed,

$$D_x(g(x) - h(x)) = g'(x) - h'(x) = 0$$

whence, by Theorem 29.1, $g(x) - h(x) = C$, or $g(x) = h(x) + C$.

According to Corollary 29.2, any two antiderivatives of a given function differ only by a constant. Thus, if we know one antiderivative of a function, we know them all.

Notation. $\int f(x)\,dx$ stands for *any* antiderivative of f. Thus,

$$D_x\left(\int f(x)\,dx\right) = f(x)$$

OTHER TERMINOLOGY Sometimes the term *indefinite integral* is used instead of antiderivative, and the process of finding antiderivatives is termed *integration*. In an expression $\int f(x)\,dx$, $f(x)$ is called the *integrand*. The motive for this nomenclature will become clear in Chapter 31.

EXAMPLES

(*a*)
$$\int x^2\,dx = \frac{x^3}{3} + C$$

Since $D_x(x^3/3) = x^2$, we know that $x^3/3$ is an antiderivative of x^2. By Corollary 29.2, any other antiderivative of x^2 is of the form $(x^3/3) + C$, where C is a constant

(*b*)
$$\int \cos x\,dx = \sin x + C$$

(c) $$\int \sin x \, dx = -\cos x + C$$

(d) $$\int \sec^2 x \, dx = \tan x + C$$

(e) $$\int 0 \, dx = C$$

(f) $$\int 1 \, dx = x + C$$

29.2 RULES FOR ANTIDERIVATIVES

The rules for derivatives—in particular, the sum-or-difference rule and the Chain Rule—yield corresponding rules for antiderivatives.

RULE 1. $$\int a \, dx = ax + C$$

for any constant a.

EXAMPLE $$\int 3 \, dx = 3x + C$$

RULE 2. $$\int x^r \, dx = \frac{x^{r+1}}{r+1} + C$$

for any rational number r other than $r = -1$.

NOTE The antiderivative of x^{-1} will be dealt with in Chapter 34.

Rule 2 follows from Theorem 15.4, according to which

$$D_x(x^{r+1}) = (r+1)x^r \qquad \text{or} \qquad D_x\left(\frac{x^{r+1}}{r+1}\right) = x^r$$

EXAMPLES

(a) $$\int \sqrt{x} \, dx = \int x^{1/2} \, dx = \frac{x^{3/2}}{3/2} + C = \frac{2}{3}x^{3/2} + C$$

(b) $$\int \frac{1}{x^3} \, dx = \int x^{-3} \, dx = \frac{x^{-2}}{-2} + C = -\frac{1}{2}x^{-2} + C = -\frac{1}{2x^2} + C$$

RULE 3. $$\int af(x) \, dx = a \int f(x) \, dx$$

for any constant a.

This follows from:

$$D_x\left(a \cdot \int f(x) \, dx\right) = a \cdot D_x\left(\int f(x) \, dx\right) = af(x)$$

EXAMPLE $$\int 5x^2 \, dx = 5 \int x^2 \, dx = 5\left(\frac{x^3}{3}\right) + C = \frac{5x^3}{3} + C$$

RULE 4. $$\int (f(x) \pm g(x)) \, dx = \int f(x) \, dx \pm \int g(x) \, dx$$

For $$D_x\left(\int f(x)\,dx \pm \int g(x)\,dx\right) = D_x\left(\int f(x)\,dx\right) \pm D_x\left(\int g(x)\,dx\right) = f(x) \pm g(x)$$

EXAMPLE $$\int (x^2 + x^3)\,dx = \int x^2\,dx + \int x^3\,dx = \frac{x^3}{3} + \frac{x^4}{4} + C$$

Notice that we find a specific antiderivative,

$$\frac{x^3}{3} + \frac{x^4}{4}$$

and then add the "arbitrary" constant C.

Rules 1 through 4 enable us to compute the antiderivative of any polynomial.

EXAMPLE

$$\int \left(3x^5 - \frac{1}{2}x^4 + 7x^2 + x - 3\right) dx = 3\left(\frac{x^6}{6}\right) - \frac{1}{2}\left(\frac{x^5}{5}\right) + 7\left(\frac{x^3}{3}\right) + \frac{x^2}{2} - 3x + C$$

$$= \frac{x^6}{2} - \frac{x^5}{10} + \frac{7}{3}x^3 + \frac{x^2}{2} - 3x + C$$

RULE 5. $$\int (g(x))^r g'(x)\,dx = \frac{(g(x))^{r+1}}{r+1} + C$$

The Power Chain Rule implies that

$$D_x\left(\frac{(g(x))^{r+1}}{r+1}\right) = (g(x))^r \cdot g'(x)$$

which yields Rule 5.

EXAMPLES

(*a*) $$\int \left(\frac{1}{2}x^2 + 5\right)^7 x\,dx = \frac{1}{8}\left(\frac{1}{2}x^2 + 5\right)^8 + C$$

(*b*) $$\int \sqrt{2x - 5}\,dx = \frac{1}{2}\int (2x - 5)^{1/2}(2)\,dx = \frac{1}{2}\frac{(2x - 5)^{3/2}}{3/2} + C = \frac{1}{3}(2x - 5)^{3/2} + C$$

RULE 6 (Substitution Method). Deferring the general formulation and justification to Problem 29.8, we illustrate the method by three examples.

(i) Find $\int x^2 \cos x^3\,dx$. Let $x^3 = u$. Then, by Section 21.3, the differential of u is given by

$$du = D_x(x^3)\,dx = 3x^2\,dx \qquad \text{or} \qquad x^2\,dx = \frac{1}{3}\,du$$

Now substitute u for x^3 and $\frac{1}{3}du$ for $x^2\,dx$:

$$\int x^2 \cos x^3\,dx = \int \frac{1}{3}\cos u\,du = \frac{1}{3}\int \cos u\,du = \frac{1}{3}\sin u + C$$

$$= \frac{1}{3}\sin x^3 + C$$

(ii) Find $\int (x^2 + 3x - 5)^3(2x + 3)\,dx$. Let $u = x^2 + 3x - 5$, $du = (2x + 3)\,dx$. Then

$$\int (x^2 + 3x - 5)^3(2x + 3)\,dx = \int u^3\,du = \frac{u^4}{4} + C = \frac{1}{4}(x^2 + 3x - 5)^4 + C$$

(iii) Find $\int \sin^2 x \cos x\,dx$. Let $u = \sin x$. Then $du = \cos x\,dx$, and

$$\int \sin^2 x \cos x\,dx = \int u^2\,du = \frac{u^3}{3} + C = \frac{\sin^3 x}{3} + C$$

Notice that Rule 5 is a special case of Rule 6, corresponding to the substitution $u = g(x)$.

Solved Problems

29.1 Find the following antiderivatives:

$$(a)\ \int (\sqrt[3]{x} - 5x^2)\,dx \qquad (b)\ \int (4x + \sqrt{x^5} - 2)\,dx$$

$$(c)\ \int (x^2 - \sec^2 x)\,dx \qquad (d)\ \int \frac{2\sqrt{x} + 3x^2}{x}\,dx$$

(*a*)

$$\int (\sqrt[3]{x} - 5x^2)\,dx = \int (x^{1/3} - 5x^2)\,dx$$

$$= \frac{x^{4/3}}{4/3} - 5\left(\frac{x^3}{3}\right) + C \quad \text{[by Rules 2 and 4]}$$

$$= \frac{3}{4}x^{4/3} - \frac{5}{3}x^3 + C$$

(*b*)

$$\int (4x + \sqrt{x^5} - 2)\,dx = \int (4x + x^{5/2} - 2)\,dx = 4\left(\frac{x^2}{2}\right) + \frac{x^{7/2}}{7/2} - 2x + C$$

$$= 2x^2 + \frac{2}{7}x^{7/2} - 2x + C$$

(*c*)

$$\int (x^2 - \sec^2 x)\,dx = \int x^2\,dx - \int \sec^2 x\,dx = \frac{x^3}{3} - \tan x + C$$

(*d*)

$$\int \frac{2\sqrt{x} + 3x^2}{x}\,dx = \int \left(\frac{2}{\sqrt{x}} + 3x\right)dx = 2\int x^{-1/2}\,dx + 3\int x\,dx \quad \text{[Rules 1 and 4]}$$

$$= 2\frac{x^{1/2}}{1/2} + 3\frac{x^2}{2} + C = 4\sqrt{x} + \frac{3}{2}x^2 + C$$

29.2 Find the following antiderivatives:

$$(a)\ \int (2x^3 - x)^4(6x^2 - 1)\,dx \qquad (b)\ \int \sqrt[3]{5x^2 - 1}\,x\,dx$$

(*a*) Notice that $D_x(2x^3 - x) = 6x^2 - 1$. So, by Rule 5,

$$\int (2x^3 - x)^4(6x^2 - 1)\,dx = \frac{1}{5}(2x^3 - x)^5 + C$$

(*b*) Observe that $D_x(5x^2 - 1) = 10x$. By Rules 5 and 1,

$$\int \sqrt[3]{5x^2 - 1}\,x\,dx = \int (5x^2 - 1)^{1/3}\,x\,dx = \frac{1}{10}\int (5x^2 - 1)^{1/3}\,10x\,dx$$

$$= \frac{1}{10}\frac{(5x^2 - 1)^{4/3}}{4/3} + C = \frac{3}{40}(5x^2 - 1)^{4/3} + C$$

$$= \frac{3}{40}(\sqrt[3]{5x^2 - 1})^4 + C = \frac{3}{40}\sqrt[3]{(5x^2 - 1)^4} + C$$

(For manipulations of rational powers, review Chapter 15.)

29.3 Use the substitution method to evaluate

$$(a)\ \int \frac{\sin\sqrt{x}}{\sqrt{x}}\,dx \qquad (b)\ \int x\sec^2(3x^2 - 1)\,dx \qquad (c)\ \int x^2\sqrt{x + 2}\,dx$$

(*a*) Let $u = \sqrt{x}$ Then

$$du = D_x(\sqrt{x})\,dx = D_x(x^{1/2})\,dx = \frac{1}{2}x^{-1/2}\,dx = \frac{1}{2\sqrt{x}}\,dx$$

Hence,

$$\int \frac{\sin \sqrt{x}}{\sqrt{x}}\, dx = 2\int \sin u\, du = -2\cos u + C = -2\cos \sqrt{x} + C$$

(*b*) Let $u = 3x^2 - 1$. Then $du = 6x\, dx$, and

$$\int x \sec^2 (3x^2 - 1)\, dx = \frac{1}{6}\int \sec^2 u\, du = \frac{1}{6}\tan u + C = \frac{1}{6}\tan (3x^2 - 1) + C$$

(*c*) Let $u = x + 2$. Then $du = dx$ and $x = u - 2$. Hence,

$$\int x^2\sqrt{x+2}\, dx = \int (u-2)^2\sqrt{u}\, du = \int (u^2 - 4u + 4)u^{1/2}\, du$$

$$= \int (u^{5/2} - 4u^{3/2} + 4u^{1/2})\, du \quad [\text{by } u^r u^s = u^{r+s}]$$

$$= \frac{2}{7}u^{7/2} - \frac{8}{5}u^{5/2} + \frac{8}{3}u^{3/2} + C = \frac{2}{7}(x+2)^{7/2} - \frac{8}{5}(x+2)^{5/2} + \frac{8}{3}(x+2)^{3/2} + C$$

The substitution $u = \sqrt{x+2}$ also would work.

29.4 Prove Theorem 29.1.

Let a and b be any two numbers in $\mathscr{I}$. By the Mean-Value Theorem (Theorem 17.2), there is a number c between a and b, and therefore in $\mathscr{I}$, such that

$$F'(c) = \frac{F(b) - F(a)}{b - a}$$

But, by hypothesis, $F'(c) = 0$; hence, $F(b) - F(a) = 0$, or $F(b) = F(a)$.

29.5 Rework Problem 18.7 by means of antiderivatives.

In free-fall problems,

$$v = \int a\, dt \qquad \text{and} \qquad s = \int v\, dt$$

since, by definition, $a = dv/dt$ and $v = ds/dt$. Since $a = -32\ \text{ft/sec}^2$ (when *up* is positive),

$$v = \int -32\, dt = -32t + C_1$$

$$s = \int (-32t + C_1)\, dt = (-32)\frac{t^2}{2} + C_1 t + C_2$$

$$= -16t^2 + C_1 t + C_2$$

in which the values of C_1 and C_2 are determined by the problem at hand. In the present case, it is given that $v(0) = 128$ ft/sec and $s(0) = 0$ ft; hence,

$$128 = 0 + C_1 \qquad \text{and} \qquad 0 = 0 + 0 + C_2$$

so that

$$v = -32t + 128 \tag{1}$$

$$s = -16t^2 + 128t \tag{2}$$

(*a*) By (*2*),

$$s(1) = -16(1)^2 + 128(1) = 112 \text{ ft} \qquad s(2) = -16(2)^2 + 128(2) = 192 \text{ ft}$$

(*b*) For maximum height,

$$\frac{ds}{dt} = v = -32t + 128 = 0 \qquad \text{or} \qquad t = \frac{128}{32} = 4 \text{ sec}$$

(*c*) Substituting $t = 4$ in (*2*),

$$s(4) = -16(4)^2 + 128(4) = -256 + 512 = 256 \text{ ft}$$

(*d*) Setting $s = 0$ in (*2*),

$$-16t^2 + 128t = 0$$
$$-16t(t-8) = 0$$
$$t = 0 \text{ or } 8 \text{ sec}$$

The rocket leaves the ground at $t = 0$ and returns at $t = 8$ sec.

(*e*) Substituting $t = 8$ in (*1*), $v(8) = -32(8) + 128 = -128$ ft/sec. The speed is the absolute value of the velocity, or 128 ft/sec.

Supplementary Problems

29.6 Find the following antiderivatives:

(*a*) $\int (2x^3 - 5x^2 + 3x + 1)\, dx$ (*b*) $\int \left(5 - \frac{1}{\sqrt{x}}\right) dx$ (*c*) $\int 2\sqrt[4]{x}\, dx$

(*d*) $\int 5\sqrt[3]{x^2}\, dx$ (*e*) $\int \frac{3}{x^4}\, dx$ (*f*) $\int (x^2 - 1)\sqrt{x}\, dx$

(*g*) $\int \left(\frac{1}{x^3} - \frac{1}{x^5}\right) dx$ (*h*) $\int \frac{3x^2 - 2x + 1}{\sqrt{x}}\, dx$ (*i*) $\int (3 \sin x + 5 \cos x)\, dx$

(*j*) $\int (7 \sec^2 x - \sec x \tan x)\, dx$ (*k*) $\int (\csc^2 x + 3x^2)\, dx$ (*l*) $\int x\sqrt{3x}\, dx$

(*m*) $\int \frac{1}{\sec x}\, dx$ (*n*) $\int \tan^2 x\, dx$

(*Hint*: Use Theorem 28.3 in (*n*).)

29.7 Evaluate the following antiderivatives by using Rule 5 or Rule 6. (In (*m*), $a \neq 0$.)

(*a*) $\int \sqrt{7x + 4}\, dx$ (*b*) $\int \frac{1}{\sqrt{x - 1}}\, dx$ (*c*) $\int (3x - 5)^{12}\, dx$

(*d*) $\int \sin(3x - 1)\, dx$ (*e*) $\int \sec^2 \frac{x}{2}\, dx$ (*f*) $\int \frac{\cos \sqrt{x}}{\sqrt{x}}\, dx$

(*g*) $\int (4 - 2t^2)^7 t\, dt$ (*h*) $\int x^2 \sqrt[3]{x^3 + 5}\, dx$ (*i*) $\int \frac{x}{\sqrt{x + 1}}\, dx$

(*j*) $\int \sqrt[3]{x^2 - 2x + 1}\, dx$ (*k*) $\int (x^4 + 1)^{1/3} x^7\, dx$ (*l*) $\int \frac{x}{\sqrt{1 + 5x^2}}\, dx$

(*m*) $\int x\sqrt{ax + b}\, dx$ (*n*) $\int \frac{\cos 3x}{\sin^2 3x}\, dx$ (*o*) $\int \sqrt{1 - x}\, x^2\, dx$

29.8 Justify the following form of the Substitution Method (Rule 6):

$$\int f(g(x))\, g'(x)\, dx = \int f(u)\, du$$

where u is replaced by $g(x)$ after the integration on the right side. (*Hint*: By the Chain Rule,

$$D_x\left(\int f(u)\, du\right) = f(u)\frac{du}{dx}$$

and $u = g(x)$.)

29.9 A rocket is shot vertically upward from a tower 240 feet above the ground, with an initial velocity of 224 feet per second. (*a*) When will it attain its maximum height? (*b*) What will be its maximum height? (*c*) When will it strike the ground? (*d*) With what speed will it hit the ground?

29.10 (Rectilinear motion: Chapter 18). A particle moves along the x-axis with acceleration $a = 2t - 3$ (ft/sec²). At time $t = 0$, it is at the origin and moving with a speed of 4 ft/sec in the positive direction. (*a*) Find a formula for its velocity v in terms of t. (*b*) Find a formula for its position x in terms of t. (*c*) When and where does the particle change direction? (*d*) At what times is the particle moving toward the left?

29.11 Rework Problem 29.10 if

$$a = t^2 - \frac{13}{3} \quad (\text{ft/sec}^2)$$

29.12 A rocket shot straight up from ground level hits the ground ten seconds later. (*a*) What was its initial velocity? (*b*) How high did it go?

29.13 A motorist applies the brakes on a car moving at 45 miles per hour on a straight road, and the brakes cause a constant deceleration of 22 ft/sec². (*a*) In how many seconds will the car stop? (*b*) How many feet will the car have traveled after the time the brakes were applied? (*Hint*: Put the origin at the point where the brakes were initially applied, and let $t = 0$ at that time. Note that the speed and deceleration involve different units of distance and time; change the speed to ft/sec.)

29.14 A particle moving on a straight line has acceleration $a = 5 - 3t$, and its velocity is 7 at time $t = 2$. If $s(t)$ is the distance from the origin at time t, find $s(2) - s(1)$.

29.15 (*a*) Find an equation of the curve passing through the point (3, 2) and having slope $2x^2 - 5$ at point (x, y). (*b*) Find an equation of the curve passing through the point (0, 1) and having slope $12x + 1$ at point (x, y).

Chapter 30

The Definite Integral

30.1 SIGMA NOTATION

The Greek capital letter Σ is used in mathematics to indicate repeated addition.

EXAMPLES (Note the rule for a constant factor illustrated in (*c*).)

(*a*)
$$\sum_{i=1}^{99} i = 1+2+3+\cdots+99$$

i.e., the sum of the first ninety-nine positive integers.

(*b*)
$$\sum_{i=1}^{6} (2i-1) = 1+3+5+7+9+11$$

i.e., the sum of the first six odd positive integers.

(*c*)
$$\sum_{i=2}^{5} 3i = 6+9+12+15 = 3(2+3+4+5) = 3\sum_{i=2}^{5} i$$

(*d*)
$$\sum_{j=1}^{15} j^2 = 1^2+2^2+3^2+\cdots+15^2 = 1+4+9+\cdots+225$$

(*e*)
$$\sum_{j=1}^{5} \sin j\pi = \sin\pi + \sin 2\pi + \sin 3\pi + \sin 4\pi + \sin 5\pi$$

In general, given a function f defined on the integers, and given integers k and $n \geq k$,

$$\sum_{i=k}^{n} f(i) = f(k) + f(k+1) + \cdots + f(n)$$

30.2 AREA UNDER A CURVE

Let f be a function such that $f(x) \geq 0$ for all x in the closed interval $[a, b]$. Then its graph is a curve lying on or above the x-axis (see Fig. 30-1). We have an intuitive idea of the *area A* of the

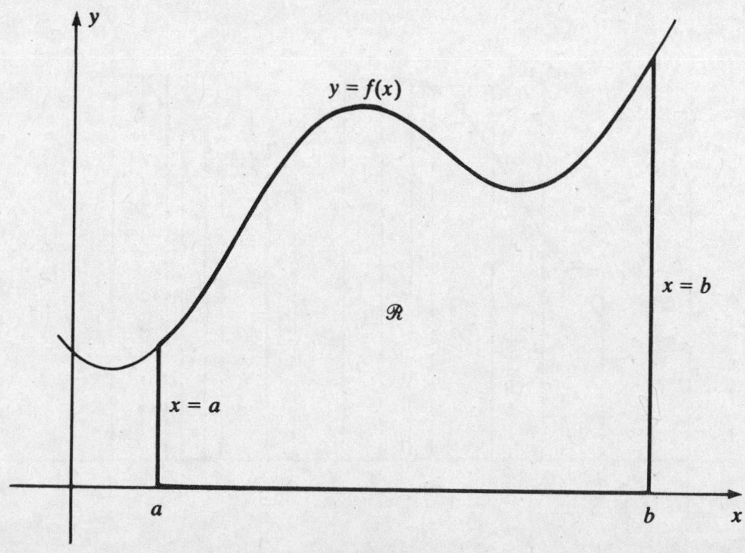

Fig. 30-1

region $\mathcal{R}$ lying under the curve, above the x-axis, and between the vertical lines $x = a$ and $x = b$. Let us set up a procedure for finding the value of the area A.

Select points $x_1, x_2, \ldots, x_{n-1}$ inside $[a, b]$, as in Fig. 30-2; let $x_0 = a$ and $x_n = b$:

$$a = x_0 < x_1 < x_2 < \cdots < x_{n-1} < x_n = b$$

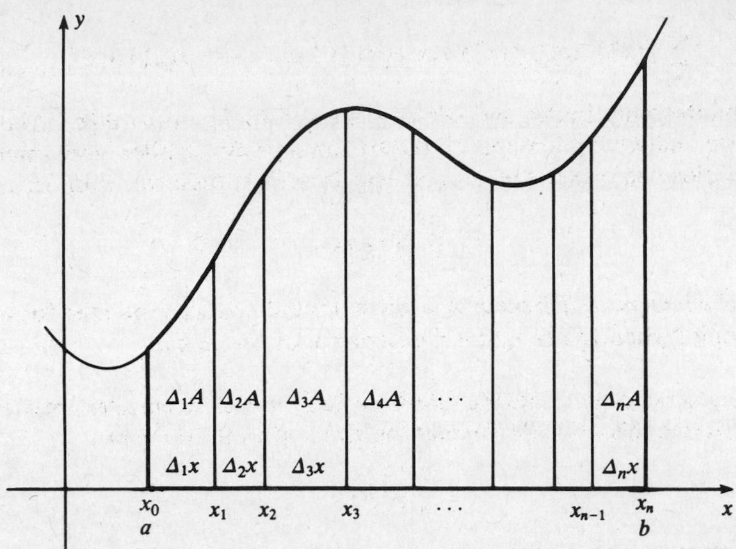

Fig. 30-2

These divide $[a, b]$ into the n subintervals $[x_0, x_1], [x_1, x_2], \ldots, [x_{n-1}, x_n]$. Let the lengths of these subintervals be $\Delta_1 x, \Delta_2 x, \ldots, \Delta_n x$, where

$$\Delta_i x \equiv x_i - x_{i-1}$$

Draw vertical lines $x = x_i$ from the x-axis up to the graph, thereby dividing the region $\mathcal{R}$ into n strips. If $\Delta_i A$ is the area of the ith strip, then

$$A = \sum_{i=1}^{n} \Delta_i A$$

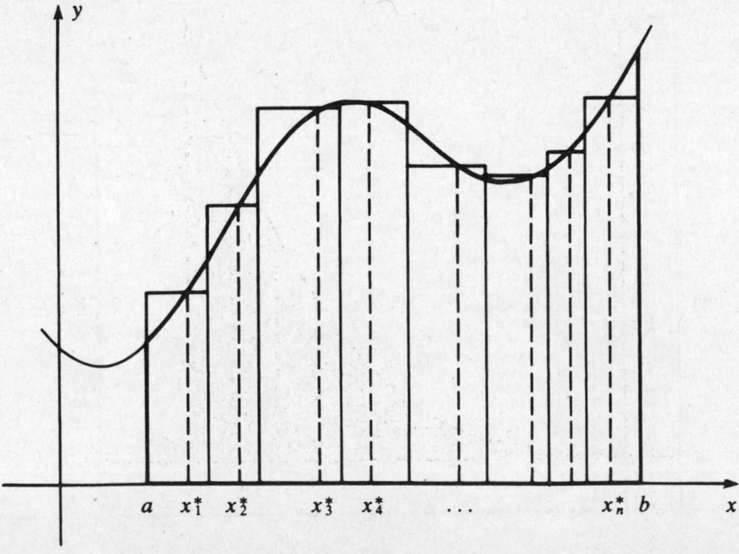

Fig. 30-3

Approximate the area $\Delta_i A$ as follows. Choose any point x_i^* in the ith subinterval, $[x_{i-1}, x_i]$, and draw the vertical line segment from the point x_i^* up to the graph (see the dashed lines in Fig. 30-3); the length of this segment is $f(x_i^*)$. The rectangle with base $\Delta_i x$ and height $f(x_i^*)$ has area $f(x_i^*)\,\Delta_i x$, which is approximately the area $\Delta_i A$ of the ith strip. So, the total area A under the curve is approximately the sum

$$\sum_{i=1}^{n} f(x_i^*)\,\Delta_i x = f(x_1^*)\,\Delta_1 x + f(x_2^*)\,\Delta_2 x + \cdots + f(x_n^*)\,\Delta_n x \tag{30.1}$$

The approximation becomes better and better as we divide the interval $[a, b]$ into more and more subintervals and as we make the lengths of these subintervals smaller and smaller. If successive approximations get as close as one wishes to a specific number, then this number is denoted

$$\int_a^b f(x)\,dx$$

and is called the *definite integral of f from a to b*. Such a number does not exist for all functions f, but it does exist, for example, when the function f is continuous on $[a, b]$.

EXAMPLE Approximating the definite integral by a *finite* number n of rectangular areas does not usually give good numerical results. To see this, consider the function $f(x) = x^2$ on $[0, 1]$. Then,

$$\int_0^1 x^2\,dx$$

is the area under the parabola $y = x^2$, above the x-axis, between $x = 0$ and $x = 1$. Divide $[0, 1]$ into $n = 10$ equal subintervals by the points 0.1, 0.2, . . . , 0.9 (see Fig. 30-4). Thus, each $\Delta_i x$ equals 1/10. In the ith subinterval, choose x_i^* to be the left-hand endpoint, $(i-1)/10$. Then:

$$\begin{aligned}\int_0^1 x^2\,dx \approx \sum_{i=1}^{n} f(x_i^*)\,\Delta_i x &= \sum_{i=1}^{10}\left(\frac{i-1}{10}\right)^2\left(\frac{1}{10}\right) = \sum_{i=1}^{10}\frac{(i-1)^2}{100}\left(\frac{1}{10}\right)\\ &= \frac{1}{1000}\sum_{i=1}^{10}(i-1)^2 \quad \text{[by Example (c) above]}\\ &= \frac{1}{1000}(0+1+4+\cdots+81) = \frac{1}{1000}(285) = 0.285\end{aligned}$$

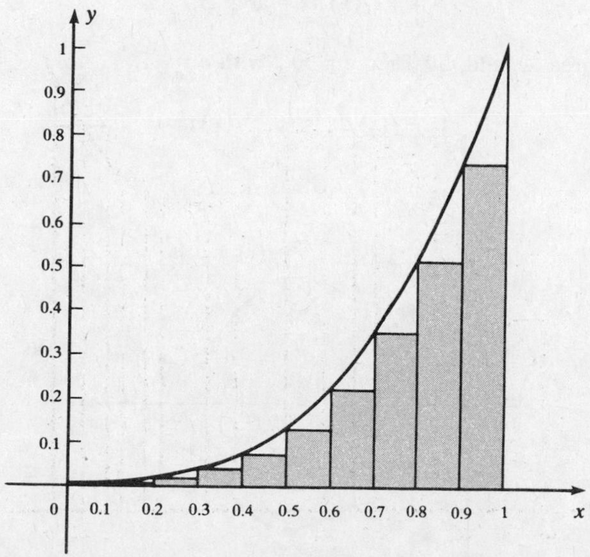

Fig. 30-4

As will be shown in Problem 30.2, the exact value is

$$\int_0^1 x^2\,dx = \frac{1}{3} = 0.333\cdots$$

so the above approximation is not too good. In terms of Fig. 30-4, there is too much unfilled space between the curve and the tops of the rectangles.

Now, for an arbitrary (not necessarily nonnegative) function f on $[a, b]$, a sum of the form *(30.1)* can be defined, without any reference to the graph of f or to the notion of area. The precise epsilon-delta procedure of Problem 8.4(a) can be used to determine whether this sum, considered as a function of n, has a limit as n approaches ∞ (in such a way that max $\Delta_i x \to 0$). If it does, the function f is said to be *integrable* on $[a, b]$, and the limit is called the *definite integral*

$$\int_a^b f(x)\,dx$$

In the following Section, we shall state several properties of the definite integral, omitting any proof that depends on the precise definition in favor of the intuitive picture of the definite integral as an area (when $f(x) \geq 0$).

30.3 PROPERTIES OF THE DEFINITE INTEGRAL

Theorem 30.1: If f is continuous on $[a, b]$, then f is integrable on $[a, b]$.

Theorem 30.2:

$$\int_a^b c\,f(x)\,dx = c\int_a^b f(x)\,dx$$

for any constant c.

Obviously, since the respective approximating sums enjoy this relationship [Example (c) above], the limits enjoy it as well.

EXAMPLE Suppose that $f(x) \leq 0$ for all x in $[a, b]$. The graph of f—along with its mirror image, the graph of $-f$—is shown in Fig. 30-5. Since $-f(x) \geq 0$,

$$\int_a^b -f(x)\,dx = \text{area } B$$

But, by symmetry, area B = area A; and, by Theorem 30.2 with $c = -1$,

$$\int_a^b -f(x)\,dx = -\int_a^b f(x)\,dx$$

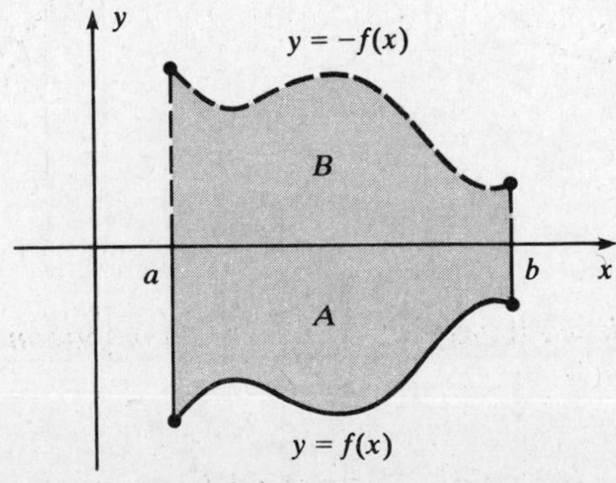

Fig. 30-5

It follows that

$$\int_a^b f(x)\,dx = -(\text{area } A)$$

In other words, the definite integral of a nonpositive function is the *negative* of the area *above* the graph of the function and *below* the x-axis.

Theorem 30.3: If f and g are integrable on $[a, b]$, then so are $f+g$ and $f-g$, and

$$\int_a^b (f(x) \pm g(x))\,dx = \int_a^b f(x)\,dx \pm \int_a^b g(x)\,dx$$

Again, this property is implied by the corresponding property of the approximating sums:

$$\sum_{i=1}^{n} (P(i) \pm Q(i)) = \sum_{i=1}^{n} P(i) \pm \sum_{i=1}^{n} Q(i)$$

Theorem 30.4: If $a < c < b$ and if f is integrable on $[a, c]$ and on $[c, b]$, then f is integrable on $[a, b]$, and

$$\int_a^b f(x)\,dx = \int_a^c f(x)\,dx + \int_c^b f(x)\,dx$$

For $f(x) \geq 0$, the theorem is obvious: the area under the graph from a to b must be the sum of the areas from a to c and from c to b.

EXAMPLE Theorem 30.4 yields a geometric interpretation for the definite integral when the graph of f has the appearance shown in Fig. 30-6. Here,

$$\begin{aligned}\int_a^b f(x)\,dx &= \int_a^{c_1} f(x)\,dx + \int_{c_1}^{c_2} f(x)\,dx + \int_{c_2}^{c_3} f(x)\,dx + \int_{c_3}^{c_4} f(x)\,dx + \int_{c_4}^{b} f(x)\,dx \\ &= A_1 - A_2 + A_3 - A_4 + A_5\end{aligned}$$

That is, the definite integral may be considered a total area, in which areas *above* the x-axis are counted as *positive*, and areas *below* the x-axis are counted as *negative*. Thus, we can infer from Fig. 27-2(*b*) that

$$\int_0^{2\pi} \sin x\,dx = 0$$

because the positive area from 0 to π is just canceled by the negative area from π to 2π.

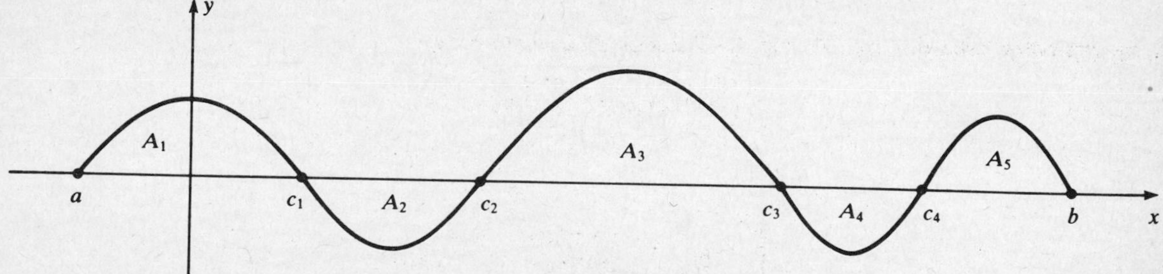

Fig. 30-6

Arbitrary Limits of Integration

In defining $\int_a^b f(x)\,dx$, we have assumed that the *limits of integration*, a and b, are such that $a < b$. Extend the definition as follows:

(1) $\int_a^a f(x)\,dx = 0$.
(2) If $a > b$, let $\int_a^b f(x)\,dx = -\int_b^a f(x)\,dx$ (with the definite integral on the right falling under the original definition).

Under this extended definition, *interchanging the limits of integration in any definite integral reverses the algebraic sign of the integral.* Moreover, the equations of Theorems 30.2, 30.3, and 30.4 now hold for arbitrary limits of integration a, b, and c.

Solved Problems

30.1 Show that

$$\int_a^b 1\,dx = b - a$$

For any subdivision

$$a = x_0 < x_1 < x_2 < \cdots < x_{n-1} < x_n = b$$

of $[a, b]$, the approximating sum *(30.1)* is

$$\begin{aligned}\sum_{i=1}^{n} f(x_i^*)\,\Delta_i x &= \sum_{i=1}^{n} \Delta_i x \quad [\text{since } f(x) = 1 \text{ for all } x]\\ &= (x_1 - x_0) + (x_2 - x_1) + (x_3 - x_2) + \cdots + (x_n - x_{n-1}) = x_n - x_0 = b - a\end{aligned}$$

Since every approximating sum is equal to $b - a$,

$$\int_a^b 1\,dx = b - a$$

30.2 Calculate $\int_0^1 x^2\,dx$. (You may assume the formula

$$1^2 + 2^2 + \cdots + n^2 = \frac{n(n+1)(2n+1)}{6}$$

which is established in Problem 30.12(*b*).)

Divide the interval [0, 1] into n equal parts, as indicated in Fig. 30-7, making each $\Delta_i x = 1/n$. In the ith subinterval, $[(i-1)/n, i/n]$, let x_i^* be the right endpoint, i/n. Then *(30.1)* becomes

$$\begin{aligned}\sum_{i=1}^{n} f(x_i^*)\,\Delta_i x &= \sum_{i=1}^{n} \left(\frac{i}{n}\right)^2 \frac{1}{n} = \frac{1}{n^3}\sum_{i=1}^{n} i^2\\ &= \frac{1}{n^3}\,\frac{n(n+1)(2n+1)}{6} = \frac{1}{6}\left(\frac{n+1}{n}\right)\left(\frac{2n+1}{n}\right)\\ &= \frac{1}{6}\left(1+\frac{1}{n}\right)\left(2+\frac{1}{n}\right)\end{aligned}$$

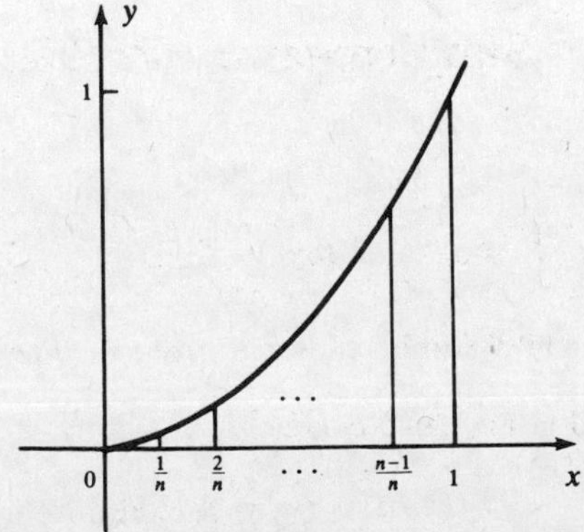

Fig. 30-7

We can make the subdivision finer and finer by letting n approach infinity. Then,

$$\int_0^1 x^2\,dx = \lim_{n\to\infty} \frac{1}{6}\left(1+\frac{1}{n}\right)\left(2+\frac{1}{n}\right) = \frac{1}{6}(1)(2) = \frac{1}{3}$$

This kind of direct calculation of a definite integral is possible only for the very simplest functions $f(x)$. A much more powerful method will be explained in Chapter 31.

30.3 Let $f(x)$ and $g(x)$ be integrable on $[a, b]$. (*a*) If $f(x) \geq 0$ on $[a, b]$, show that

$$\int_a^b f(x)\,dx \geq 0$$

(*b*) If $f(x) \leq g(x)$ on $[a, b]$, show that

$$\int_a^b f(x)\,dx \leq \int_a^b g(x)\,dx$$

(*c*) If $m \leq f(x) \leq M$ on $[a, b]$, show that

$$m(b-a) \leq \int_a^b f(x)\,dx \leq M(b-a)$$

(*a*) The definite integral, being the area under the graph of f, cannot be negative. More fundamentally, every approximating sum (*30.1*) is nonnegative, since $f(x_i^*) \geq 0$ and $\Delta_i x > 0$. Hence (as shown in Problem 9.10), the limiting value of the approximating sums is also nonnegative.

(*b*) Because $g(x) - f(x) \geq 0$ on $[a, b]$,

$$\int_a^b (g(x) - f(x))\,dx \geq 0 \quad \text{[by (a)]}$$

$$\int_a^b g(x)\,dx - \int_a^b f(x)\,dx \geq 0 \quad \text{[by Theorem 30.3]}$$

$$\int_a^b g(x)\,dx \geq \int_a^b f(x)\,dx$$

(*c*)

$$\int_a^b m\,dx \leq \int_a^b f(x)\,dx \leq \int_a^b M\,dx \quad \text{[by (b)]}$$

$$m\int_a^b 1\,dx \leq \int_a^b f(x)\,dx \leq M\int_a^b 1\,dx \quad \text{[by Theorem 30.2]}$$

$$m(b-a) \leq \int_a^b f(x)\,dx \leq M(b-a) \quad \text{[by Problem 30.1]}$$

Supplementary Problems

30.4 Evaluate

$$(a)\ \int_2^5 8\,dx \qquad (b)\ \int_0^1 5x^2\,dx \qquad (c)\ \int_0^1 (x^2+4)\,dx$$

(*Hint*: Use Problems 30.1 and 30.2.)

30.5 For the function f graphed in Fig. 30-8, express

$$\int_0^5 f(x)\,dx$$

in terms of the areas A_1, A_2, A_3.

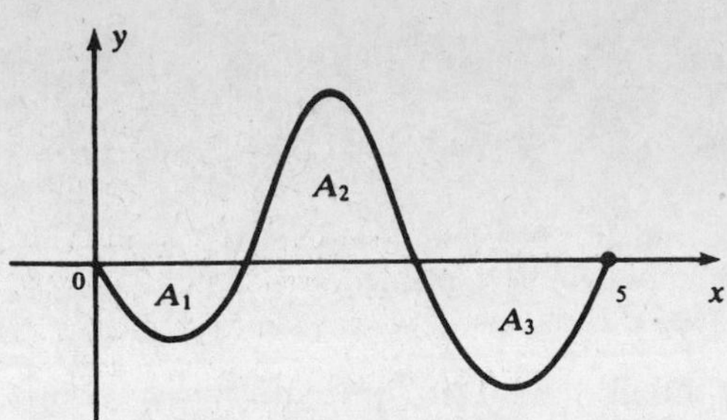

Fig. 30-8

30.6 (*a*) Show that

$$\int_0^b x\,dx = \frac{b^2}{2}$$

You may assume the formula

$$1+2+\cdots+n = \frac{n(n+1)}{2}$$

proved in Problem 30.12(*a*). Check your result by using the standard formula for the area of a triangle. (*Hint*: Divide the interval $[0, b]$ into n equal subintervals, and choose $x_i^* = ib/n$, the right endpoint of the ith subinterval.) (*b*) Show that

$$\int_a^b x\,dx = \frac{b^2-a^2}{2}$$

(*Hint*: Use (*a*) and Theorem 30.4.) (*c*) Evaluate

$$\int_1^3 5x\,dx$$

(*Hint*: Use Theorem 30.2 and (*b*).)

30.7 Show that the equation of Theorem 30.4,

$$\int_a^c f(x)\,dx + \int_c^b f(x)\,dx = \int_a^b f(x)\,dx$$

holds for any numbers a, b, c, such that the two definite integrals on the left can be defined in the extended sense. (*Hint*: Consider all six arrangements of distinct a, b, c: $a<b<c$, $a<c<b$, $b<a<c$, $b<c<a$, $c<a<b$, $c<b<a$. Also consider the cases where two of the numbers are equal or all three are equal.)

30.8 Show that

$$1 \le \int_1^2 x^3\,dx \le 8$$

(*Hint*: Use Problem 30.3(*c*).)

30.9 (*a*) Find

$$\int_0^2 \sqrt{4-x^2}\,dx$$

by using a formula of geometry. (*Hint*: What curve is the graph of $y=\sqrt{4-x^2}$?) (*b*) From (*a*) infer that $0 \le \pi \le 4$. (Much closer estimates of π are obtainable this way.)

30.10 Evaluate:

$$(a)\ \sum_{i=1}^{3}(3i-1) \qquad (b)\ \sum_{k=0}^{4}(3k^2+4)$$

$$(c)\ \sum_{j=0}^{3}\sin(j\pi/6) \qquad (d)\ \sum_{n=1}^{5}f\left(\frac{1}{n}\right)\ \text{ if } f(x)=\frac{1}{x}$$

30.11 If f is continuous on $[a, b]$, $f(x) \geq 0$ on $[a, b]$, and $f(x) > 0$ for some x in $[a, b]$, show that

$$\int_a^b f(x)\,dx > 0$$

(*Hint*: By continuity, $f(x) > K > 0$ on some closed interval inside $[a, b]$. Use Theorem 30.4 and Problem 30.3(*c*).)

30.12 Use mathematical induction (Problem 12.2) to prove

$$(a)\ 1+2+\cdots+n=\frac{n(n+1)}{2} \qquad (b)\ 1^2+2^2+\cdots+n^2=\frac{n(n+1)(2n+1)}{6}$$

Chapter 31

The Fundamental Theorem of Calculus

31.1 CALCULATION OF THE DEFINITE INTEGRAL

We shall develop a simple method for calculating

$$\int_a^b f(x)\,dx$$

—a method based on a profound and surprising connection between differentiation and integration. This connection, discovered by Isaac Newton and Gottfried von Leibniz, the co-inventors of the calculus, is expressed in the following

Theorem 31.1: Let f be continuous on $[a, b]$. Then, for x in $[a, b]$,

$$\int_a^x f(t)\,dt$$

is a function of x such that

$$D_x\left(\int_a^x f(t)\,dt\right) = f(x)$$

A proof may be found in Problem 31.5.

Now for the computation of the definite integral. Let

$$F(x) = \int f(x)\,dx$$

denote some known antiderivative of $f(x)$ (for x in $[a, b]$). According to Theorem 31.1, the function

$$\int_a^x f(t)\,dt$$

is also an antiderivative of $f(x)$; hence, by Corollary 29.2,

$$\int_a^x f(t)\,dt = F(x) + C$$

for some constant C. When $x = a$,

$$0 = \int_a^a f(t)\,dt = F(a) + C \qquad \text{or} \qquad C = -F(a)$$

Thus, when $x = b$,

$$\int_a^b f(t)\,dt = F(b) - F(a)$$

and we have proved

Theorem 31.2 (*Fundamental Theorem of Calculus*): Let f be continuous on $[a, b]$, and let $F(x) = \int f(x)\,dx$. Then,

$$\int_a^b f(x)\,dx = F(b) - F(a)$$

NOTATION The difference $F(b) - F(a)$ will often be denoted $F(x)\Big]_a^b$, and the Fundamental Theorem notated as

$$\int_a^b f(x)\,dx = \int f(x)\,dx\,\Big]_a^b$$

EXAMPLES

(*a*) Recall the complicated evaluation

$$\int_0^1 x^2\,dx = \frac{1}{3}$$

in Problem 30.2. If, instead, we choose the antiderivative $x^3/3$ and apply the Fundamental Theorem,

$$\int_0^1 x^2\,dx = \frac{x^3}{3}\bigg]_0^1 = \frac{1^3}{3} - \frac{0^3}{3} = \frac{1}{3}$$

(*b*) Let us find the area A under one arch of the curve $y = \sin x$; say, the arch from $x = 0$ to $x = \pi$. With

$$\int \sin x\,dx = -\cos x + \sqrt{5}$$

the Fundamental Theorem gives

$$\begin{aligned} A = \int_0^\pi \sin x\,dx &= (-\cos x + \sqrt{5})\bigg]_0^\pi = (-\cos \pi + \sqrt{5}) - (-\cos 0 + \sqrt{5}) \\ &= (-(-1) + \sqrt{5}) - (-1 + \sqrt{5}) = 1 + 1 + \sqrt{5} - \sqrt{5} = 2 \end{aligned}$$

Observe that the $\sqrt{5}$-terms canceled out in the calculation of A. Ordinarily, we pick the "simplest" antiderivative (here, $-\cos x$) for use in the Fundamental Theorem.

31.2 THE AVERAGE VALUE OF A FUNCTION

The *average* or *mean* of two numbers a_1 and a_2 is

$$\frac{a_1 + a_2}{2}$$

For n numbers $a_1, a_2, \ldots, a_n$, the average is

$$\frac{a_1 + a_2 + \cdots + a_n}{n}$$

Now, consider a function f defined on an interval $[a, b]$. Since f may assume infinitely many values, we cannot directly use the above definition to talk about the average of all the values of f. However, let us divide the interval $[a, b]$ into n equal subintervals, each of length

$$\Delta x = \frac{b - a}{n}$$

Choose an arbitrary point x_i^* in the ith subinterval; then the average of the n numbers $f(x_1^*), f(x_2^*), \ldots, f(x_n^*)$ is

$$\frac{f(x_1^*) + f(x_2^*) + \cdots + f(x_n^*)}{n} = \frac{1}{n}\sum_{i=1}^n f(x_i^*)$$

If n is large, this value should be a good estimate of the intuitive idea of the "average value of f on $[a, b]$." But,

$$\frac{1}{n}\sum_{i=1}^n f(x_i^*) = \frac{1}{b - a}\sum_{i=1}^n f(x_i^*)\,\Delta x \qquad \left[\text{since } \frac{1}{n} = \frac{1}{b - a}\Delta x\right]$$

As n approaches infinity, the sum on the right approaches

$$\int_a^b f(x)\,dx$$

(by definition of the definite integral), and we are led to the

Definition: The average value of f on $[a, b]$ is

$$\frac{1}{b-a}\int_a^b f(x)\,dx$$

EXAMPLES (*a*) The average value V of $\sin x$ on $[0, \pi]$ is

$$V = \frac{1}{\pi - 0}\int_0^\pi \sin x\,dx = \frac{1}{\pi}(2) \quad \text{[by Example (}b\text{) above]}$$

$$= \frac{2}{\pi} \approx 0.64$$

(*b*) The average value V of x^3 on $[0, 1]$ is

$$V = \frac{1}{1-0}\int_0^1 x^3\,dx = \int_0^1 x^3\,dx$$

Now,

$$\int x^3\,dx = \frac{x^4}{4}$$

Hence, by the Fundamental Theorem,

$$V = \int_0^1 x^3\,dx = \frac{x^4}{4}\Big]_0^1 = \frac{1^4}{4} - \frac{0^4}{4} = \frac{1}{4}$$

With the mean value of a function defined in this fashion, we have the following useful

Theorem 31.3 (*Mean-Value Theorem for Integrals*): If a function f is continuous on $[a, b]$, it assumes its mean value in $[a, b]$. That is,

$$\frac{1}{b-a}\int_a^b f(x)\,dx = f(c)$$

for some $a \le c \le b$.

For the proof, see Problem 31.4. Note that, by contrast, the average of a finite set of numbers $a_1, a_2, \ldots, a_n$ in general does not coincide with any of the a_i.

31.3 CHANGE OF VARIABLE IN A DEFINITE INTEGRAL

To evaluate a definite integral by the Fundamental Theorem, an antiderivative $\int f(x)\,dx$ is required. It was seen in Chapter 29 that the substitution of a new variable, u, may be useful in finding $\int f(x)\,dx$. When the substitution is made in the definite integral too, the limits of integration, a and b, must be replaced by the corresponding values of u.

EXAMPLE Let us compute

$$\int_0^1 \sqrt{5x+4}\,dx$$

Let $u = 5x + 4$; then $du = 5\,dx$. Consider the limits of integration: when $x = 0$, $u = 4$; when $x = 1$, $u = 9$. Therefore,

$$\int_0^1 \sqrt{5x+4}\,dx = \int_4^9 \sqrt{u}\,\frac{1}{5}\,du = \frac{1}{5}\int_4^9 u^{1/2}\,du = \frac{1}{5}\left(\frac{2}{3}u^{3/2}\right)\Big]_4^9$$

$$= \frac{2}{15}(9^{3/2} - 4^{3/2}) = \frac{2}{15}((\sqrt{9})^3 - (\sqrt{4})^3)$$

$$= \frac{2}{15}(3^3 - 2^3) = \frac{2}{15}(27-8) = \frac{2}{15}(19) = \frac{38}{15}$$

See Problem 31.6 for a justification of this procedure.

Solved Problems

31.1 Calculate the area A under the parabola $y = x^2 + 2x$ and above the x-axis, between $x = 0$ and $x = 1$.

Since $x^2 + 2x \geq 0$ for $x \geq 0$, we know that the graph of $y = x^2 + 2x$ is on or above the x-axis between $x = 0$ and $x = 1$. Hence the area A is given by the definite integral

$$\int_0^1 (x^2 + 2x)\,dx$$

Evaluating by the Fundamental Theorem,

$$A = \int_0^1 (x^2 + 2x)\,dx = \left(\frac{x^3}{3} + x^2\right)\Bigg]_0^1 = \left(\frac{1^3}{3} + 1^2\right) - \left(\frac{0^3}{3} + 0^2\right) = \frac{1}{3} + 1 = \frac{4}{3}$$

31.2 Compute

$$\int_a^{a+2\pi} \sin x\,dx$$

(Compare the Example following Theorem 30.4, where $a = 0$.)

By the Fundamental Theorem,

$$\int_a^{a+2\pi} \sin x\,dx = -\cos x\,\Bigg]_a^{a+2\pi} = 0$$

since the cosine function has period 2π.

31.3 Compute the mean value V of $\sqrt{x}$ on $[0, 4]$. For what x in $[0, 4]$ does this value occur (as guaranteed by Theorem 31.3)?

$$V = \frac{1}{4-0}\int_0^4 \sqrt{x}\,dx = \frac{1}{4}\int_0^4 x^{1/2}\,dx = \frac{1}{4}\left(\frac{2}{3}x^{3/2}\right)\Bigg]_0^4$$

$$= \frac{1}{6}(4^{3/2} - 0^{3/2}) = \frac{1}{6}((\sqrt{4})^3 - 0) = \frac{1}{6}(2^3) = \frac{8}{6} = \frac{4}{3}$$

This average value, 4/3, is the value of $\sqrt{x}$ when

$$x = \left(\frac{4}{3}\right)^2 = \frac{16}{9}$$

31.4 Prove the Mean-Value Theorem for Integrals (Theorem 31.3).

Write

$$V \equiv \frac{1}{b-a}\int_a^b f(x)\,dx$$

Let m and M be the minimum and maximum values of f on $[a, b]$. (The existence of m and M is guaranteed by Theorem 14.2.) Thus, $m \leq f(x) \leq M$ for all x in $[a, b]$, so that Problem 30.3(*c*) gives

$$m(b-a) \leq \int_a^b f(x)\,dx \leq M(b-a) \qquad \text{or} \qquad m \leq V \leq M$$

But then, by the Intermediate-Value Theorem (Theorem 17.4), the value V is assumed by f somewhere in $[a, b]$.

31.5 Prove Theorem 31.1.

Write

$$g(x) \equiv \int_a^x f(t)\,dt$$

Then,

$$\begin{aligned} g(x+h) - g(x) &= \int_a^{x+h} f(t)\,dt - \int_a^x f(t)\,dt \\ &= \int_a^x f(t)\,dt + \int_x^{x+h} f(t)\,dt - \int_a^x f(t)\,dt \quad \text{[Theorem 30.4]} \\ &= \int_x^{x+h} f(t)\,dt \end{aligned}$$

By the Mean-Value Theorem for Integrals, the last integral is equal to $hf(x^*)$, for some x^* between x and $x+h$. Hence,

$$\frac{g(x+h)-g(x)}{h} = f(x^*)$$

and

$$D_x\left(\int_a^x f(t)\,dt\right) = D_x(g(x)) = \lim_{h\to 0} \frac{g(x+h)-g(x)}{h} = \lim_{h\to 0} f(x^*)$$

Now, as $h \to 0$, $x+h \to x$, and so $x^* \to x$ (since x^* lies between x and $x+h$). Since f is continuous,

$$\lim_{h\to 0} f(x^*) = f(x)$$

and the proof is complete.

31.6 (*Change of Variables*) Consider

$$\int_a^b f(x)\,dx$$

Let $x = g(u)$, where, as x varies from a to b, u increases or decreases from c to d. (See Fig. 31-1; in effect, we rule out $g'(u) = 0$ in $[c, d]$.) Show that

$$\int_a^b f(x)\,dx = \int_c^d f(g(u))\,g'(u)\,du$$

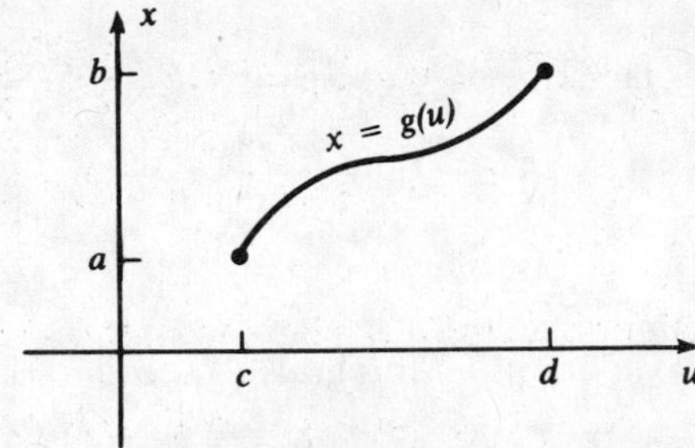

Fig. 31-1

(The right-hand side is obtained by substituting $g(u)$ for x, $g'(u)\,du$ for dx, and changing the limits of integration from a and b to c and d.)

Let

$$F(x) = \int f(x)\,dx \qquad \text{or} \qquad F'(x) = f(x)$$

The Chain Rule gives

$$D_u(F(g(u)) = F'(g(u))\, g'(u) = f(g(u))\, g'(u)$$

Hence
$$\int f(g(u))\, g'(u)\, du = F(g(u))$$

By the Fundamental Theorem,

$$\int_c^d f(g(u))\, g'(u)\, du = F(g(u))\Big]_c^d = F(g(d)) - F(g(c))$$
$$= F(b) - F(a) = \int_a^b f(x)\, dx$$

31.7 Calculate

$$\int_0^1 \sqrt{x^2+1}\, x\, dx$$

Let us find the antiderivative of $\sqrt{x^2+1}\,x$ by making the substitution $u = x^2+1$. Then $du = 2x\,dx$, and

$$\int \sqrt{x^2+1}\, x\, dx = \int \sqrt{u}\,\frac{1}{2}\, du = \frac{1}{2}\int u^{1/2}\, du = \frac{1}{2}\frac{u^{3/2}}{3/2}$$
$$= \frac{1}{3}u^{3/2} = \frac{1}{3}(x^2+1)^{3/2} = \frac{1}{3}(\sqrt{x^2+1})^3$$

Hence, by the Fundamental Theorem,

$$\int_0^1 \sqrt{x^2+1}\, x\, dx = \frac{1}{3}(\sqrt{x^2+1})^3\Big]_0^1 = \frac{1}{3}((\sqrt{1^2+1})^3 - (\sqrt{0^2+1})^3)$$
$$= \frac{1}{3}((\sqrt{2})^3 - (\sqrt{1})^3) = \frac{1}{3}(2\sqrt{2}-1)$$

ALGEBRA $\quad (\sqrt{2})^3 = (\sqrt{2})^2 \cdot \sqrt{2} = 2\sqrt{2} \quad$ and $\quad (\sqrt{1})^3 = 1^3 = 1$

Alternate Method

Make the same substitution as above, but directly in the definite integral, changing the limits of integration accordingly. When $x = 0$, $u = 0^2+1 = 1$; when $x = 1$, $u = 1^2+1 = 2$. Thus, the first line of the computation above yields:

$$\int_0^1 \sqrt{x^2+1}\, x\, dx = \frac{1}{2}\int_1^2 u^{1/2}\, du = \frac{1}{2}\frac{u^{3/2}}{3/2}\Big]_1^2 = \frac{1}{3}u^{3/2}\Big]_1^2$$
$$= \frac{1}{3}((\sqrt{2})^3 - (\sqrt{1})^3) = \frac{1}{3}(2\sqrt{2}-1)$$

31.8 (*a*) If f is an even function (Section 7.3), show that, for any $a > 0$,

$$\int_{-a}^a f(x)\, dx = 2\int_0^a f(x)\, dx$$

(*b*) If f is an odd function (Section 7.3), show that, for any $a > 0$,

$$\int_{-a}^a f(x)\, dx = 0$$

If $u = -x$, then $du = -dx$. Hence, for any integrable function $f(x)$,

$$\int_{-a}^0 f(x)\, dx = \int_a^0 f(-u)(-du) = -\int_a^0 f(-u)\, du = \int_0^a f(-u)\, du$$

NOTATION Renaming the variable in a definite integral does not affect the value of the integral:

$$\int_a^b g(x)\,dx = \int_a^b g(t)\,dt = \int_a^b g(\theta)\,d\theta = \cdots$$

Thus, changing u to x,

$$\int_{-a}^{0} f(x)\,dx = \int_0^a f(-x)\,dx \qquad (1)$$

and so

$$\int_{-a}^{a} f(x)\,dx = \int_{-a}^{0} f(x)\,dx + \int_0^a f(x)\,dx \quad \text{[by Theorem 30.4]}$$

$$= \int_0^a f(-x)\,dx + \int_0^a f(x)\,dx \quad \text{[by (1)]}$$

$$= \int_0^a (f(x)+f(-x))\,dx \quad \text{[by Theorem 30.3]}$$

(*a*) For an even function, $f(x)+f(-x)=2f(x)$, whence

$$\int_{-a}^{a} f(x)\,dx = \int_0^a 2f(x)\,dx = 2\int_0^a f(x)\,dx$$

(*b*) For an odd function, $f(x)+f(-x)=0$, whence

$$\int_{-a}^{a} f(x)\,dx = \int_0^a 0\,dx = 0\int_0^a dx = 0$$

NOTATION One usually writes

$$\int_a^b dx \qquad \text{instead of} \qquad \int_a^b 1\,dx$$

31.9 (*a*) Let $f(x) \geq 0$ on $[a, b]$, and let $[a, b]$ be divided into n equal parts, of length $\Delta x = (b-a)/n$, by means of points $x_1, x_2, \ldots, x_{n-1}$ (Fig. 31-2(*a*)). Show that

$$\int_a^b f(x)\,dx \approx \frac{\Delta x}{2}\left(f(a) + 2\sum_{i=1}^{n-1} f(x_i) + f(b)\right) \qquad \textbf{trapezoidal rule}$$

(*b*) Use the Trapezoidal Rule, with $n = 10$, to approximate

$$\int_0^1 x^2\,dx \;(= 0.333\cdots)$$

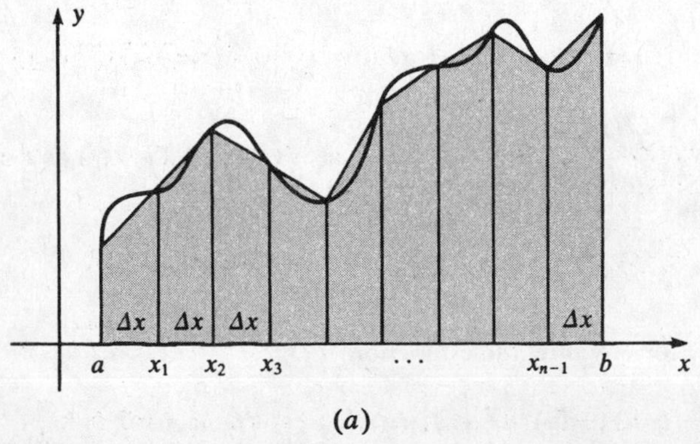

(*a*)

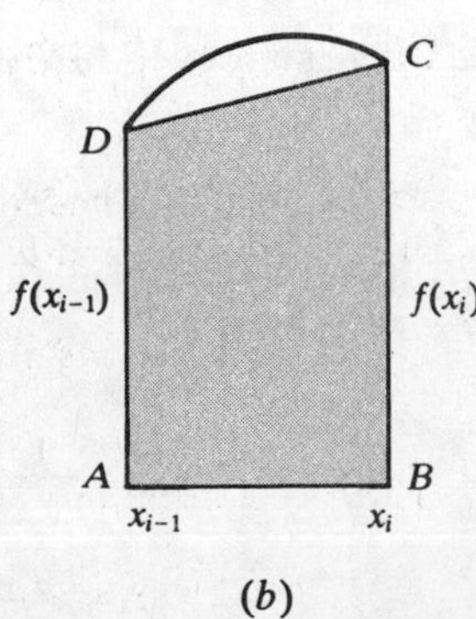

(*b*)

Fig. 31-2

(a) The area in the strip over the interval $[x_{i-1}, x_i]$ is approximately the area of trapezoid $ABCD$ in Fig. 31-2(b), which is

$$\frac{\Delta x}{2}(f(x_{i-1})+f(x_i))$$

GEOMETRY The area of a trapezoid of height h and bases b_1 and b_2 is

$$\frac{1}{2}h(b_1+b_2)$$

---

where we understand $x_0 = a$, $x_n = b$. The area under the curve is then approximated by the sum of the trapezoidal areas:

$$\int_a^b f(x)\,dx \approx \frac{\Delta x}{2}[(f(x_0)+f(x_1))+(f(x_1)+f(x_2))+\cdots+(f(x_{n-1})+f(x_n))]$$
$$=\frac{\Delta x}{2}\left(f(a)+2\sum_{i=1}^{n-1} f(x_i)+f(b)\right)$$

(b) By the Trapezoidal Rule, with $n = 10$, $a = 0$, $b = 1$, $\Delta x = 1/10$, $x_i = i/10$,

$$\int_0^1 x^2\,dx \approx \frac{1}{20}\left(0^2+2\sum_{i=1}^{9}\frac{i^2}{100}+1^2\right)=\frac{1}{20}\left(\frac{2}{100}\sum_{i=1}^{9} i^2+1\right)$$
$$=\frac{1}{20}\left(\frac{2}{100}(285)+1\right) \quad \text{[by arithmetic or Problem 30.12(b)]}$$
$$=\frac{285}{1000}+\frac{1}{20}=0.285+0.050=0.335$$

an approximation correct to two decimal places.

Supplementary Problems

31.10 Use the Fundamental Theorem to compute the following definite integrals:

(a) $\int_{-1}^{3}(3x^2-2x+1)\,dx$ (b) $\int_0^{\pi/4}\cos x\,dx$ (c) $\int_0^{\pi/3}\sec^2 x\,dx$

(d) $\int_1^{16} x^{3/2}\,dx$ (e) $\int_4^5\left(\frac{2}{\sqrt{x}}-x\right)dx$ (f) $\int_0^1 \sqrt{x^2-6x+9}\,dx$

31.11 Calculate the areas under the graphs of the following functions, above the x-axis, and between the two indicated values, a and b, of x. (In (g), area *below* the x-axis is counted negative.)

(a) $f(x)=\sin x \quad \left(a=\frac{\pi}{6}, b=\frac{\pi}{3}\right)$ (b) $f(x)=x^2+4x \quad (a=0, b=3)$

(c) $f(x)=\frac{1}{\sqrt[3]{x}} \quad (a=1, b=8)$ (d) $f(x)=\sqrt{4x+1} \quad (a=0, b=2)$

(e) $f(x)=x^2-3x \quad (a=3, b=5)$ (f) $f(x)=\sin^2 x\cos x \quad \left(a=0, b=\frac{\pi}{2}\right)$

(g) $f(x)=x^2(x^3-2) \quad (a=1, b=2)$

31.12 Compute the following definite integrals:

(*a*) $\int_0^{\pi/2} \cos x \sin x\, dx$ (*b*) $\int_0^{\pi/4} \tan x \sec^2 x\, dx$ (*c*) $\int_{-1}^{1} \sqrt{3x^2-2x+3}\,(3x-1)\, dx$

(*d*) $\int_0^{\pi/2} \sqrt{\sin x+1} \cos x\, dx$ (*e*) $\int_{-1}^{2} \sqrt{x+2}\, x^2\, dx$ (*f*) $\int_2^5 \sqrt{x^3-4}\, x^5\, dx$

(*g*) $\int_3^{15} \sqrt[3]{x^2-9}\, x^3\, dx$ (*h*) $\int_0^1 \frac{x}{(2x^2+1)^3}\, dx$ (*i*) $\int_0^8 \frac{x}{(x+1)^{3/2}}\, dx$

(*j*) $\int_{-1}^{2} |x-1|\, dx$ (*k*) $\int_1^2 \frac{x^7-2x+1}{4x^3}\, dx$

(*Hint*: Apply Theorem 30.4 to (*j*).)

31.13 Compute the average value of each of the following functions on the given interval.

(*a*) $f(x)=\sqrt[3]{x}$, on $[0,1]$ (*b*) $f(x)=\sec^2 x$, on $\left[0, \frac{\pi}{4}\right]$ (*c*) $f(x)=x^2-2x-1$, on $[-2,3]$

31.14 Verify the Mean-Value Theorem for Integrals in the following cases:

(*a*) $f(x)=x+2$, on $[1,2]$ (*b*) $f(x)=x^3$, on $[0,1]$ (*c*) $f(x)=x^2+5$, on $[0,3]$

31.15 Evaluate by the change-of-variable technique:

(*a*) $\int_{1/2}^{3} \sqrt{2x+3}\, x^2\, dx$ (*b*) $\int_0^{\pi/2} \sin^5 x \cos x\, dx$

31.16 Using only geometric reasoning, calculate the average value of

$$f(x)=\sqrt{2x-x^2}$$

on $[0,2]$. (*Hint*: If $y=f(x)$, then $(x-1)^2+y^2=1$. Draw the graph.)

31.17 If, in a period of time T, an object moves along the x-axis from x_1 to x_2, calculate its average velocity. (*Hint*: $\int v\, dt = x$.)

31.18 Find

(*a*) $D_x\left(\int_2^x \sqrt{5+7t^2}\, dt\right)$ (*b*) $D_x\left(\int_x^1 \sin^3 t\, dt\right)$ (*c*) $D_x\left(\int_{-x}^{x} \sqrt[3]{t^4+1}\, dt\right)$

(*Hint*: In (*c*), use Problem 31.8(*a*).)

31.19 Evaluate

$$\int_{-3}^{3} x^2 \sin x\, dx$$

31.20 (*a*) Find

$$D_x\left(\int_1^{3x^2} \sqrt{t^5+1}\, dt\right)$$

(*Hint*: With $u=3x^2$, the Chain Rule yields

$$D_x\left(\int_1^u \sqrt{t^5+1}\, dt\right) = D_u\left(\int_1^u \sqrt{t^5+1}\, dt\right)\cdot D_x(u)$$

and Theorem 31.1 applies on the right-hand side.) (*b*) Find a formula for

$$D_x\left(\int_a^{h(x)} f(t)\, dt\right)$$

31.21 Solve for b: $$\int_1^b x^{n-1}\,dx = \frac{2}{n}$$

31.22 If
$$\int_3^5 f(x-k)\,dx = 1$$
compute
$$\int_{3-k}^{5-k} f(x)\,dx$$
(*Hint*: Let $x = u - k$.)

31.23 If
$$f(x) = \begin{cases} \sin x & \text{for } x < 0 \\ 3x^2 & \text{for } x \ge 0 \end{cases}$$
find $\int_{-\pi/2}^{1} f(x)\,dx$.

31.24 Given that
$$2x^2 - 8 = \int_a^x f(t)\,dt$$
(*a*) find a formula for $f(x)$, (*b*) find the value of a.

31.25 Define
$$H(x) \equiv \int_1^x \frac{1}{1+t^2}\,dt$$
(*a*) Find $H(1)$. (*b*) Find $H'(1)$. (*c*) Show that $H(4) - H(2) < 2/5$.

31.26 If the average value of $f(x) = x^3 + bx - 2$ on $[0, 2]$ is 4, find b.

31.27 Find
$$\lim_{h \to 0} \left(\frac{1}{h} \int_2^{2+h} \sqrt[3]{x^2 + 2}\,dx \right)$$

31.28 If g is continuous, which of the following integrals are equal?
$$(a)\ \int_a^b g(x)\,dx \qquad (b)\ \int_{a+1}^{b+1} g(x-1)\,dx \qquad (c)\ \int_0^{b-a} g(x+a)\,dx$$

31.29 The region above the x-axis and under the curve $y = \sin x$, between $x = 0$ and $x = \pi$, is divided into two parts by the line $x = c$. The area of the left part is 1/3 the area of the right part. Find c.

31.30 Find the value(s) of k for which
$$\int_0^2 x^k\,dx = \int_0^2 (2-x)^k\,dx$$

31.31 The velocity v of an object moving on the x-axis is $\cos 3t$. It is at the origin at $t = 0$. (*a*) Find a formula for the position x at any time t. (*b*) Find the average value of the position x over the interval $0 \le t \le \pi/3$. (*c*) For what values of t in $[0, \pi/3]$ is the object moving to the right? (*d*) What are the maximum and minimum x-coordinates of the object?

31.32 Evaluate
$$\lim_{n \to +\infty} \frac{1}{n} \left(\sin\frac{\pi}{n} + \sin\frac{2\pi}{n} + \cdots + \sin\frac{n\pi}{n} \right)$$
(*Hint*: Calculate the average value of $\sin x$ on $[0, \pi]$.)

31.33 An object moves on a straight line with velocity $v = 3t - 1$, where v is measured in meters per second. How far does the object move in the period $0 \le t \le 2$ seconds? (*Hint*: Apply the Fundamental Theorem.)

31.34 Use the Trapezoidal Rule (Problem 31.9(*a*)), with $n = 10$, to approximate

$$\int_0^1 x^3\,dx$$

Compare the answer with the exact value obtained by the Fundamental Theorem of Calculus. (You may assume the formula

$$1^3 + 2^3 + \cdots + n^3 = \left[\frac{n(n+1)}{2}\right]^2$$

for use in the Trapezoidal Rule.)

Chapter 32

Applications of Integration I: Area and Arc Length

32.1 AREA BETWEEN A CURVE AND THE y-AXIS

We have learned how to find the area of a region like that shown in Fig. 32-1. Now let us consider what happens when x and y are interchanged.

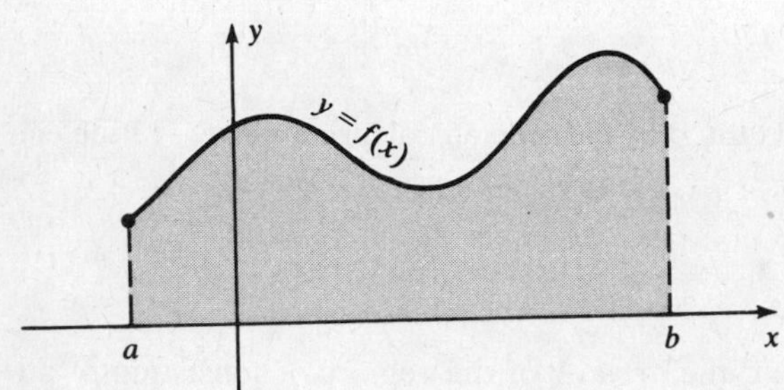

Fig. 32-1

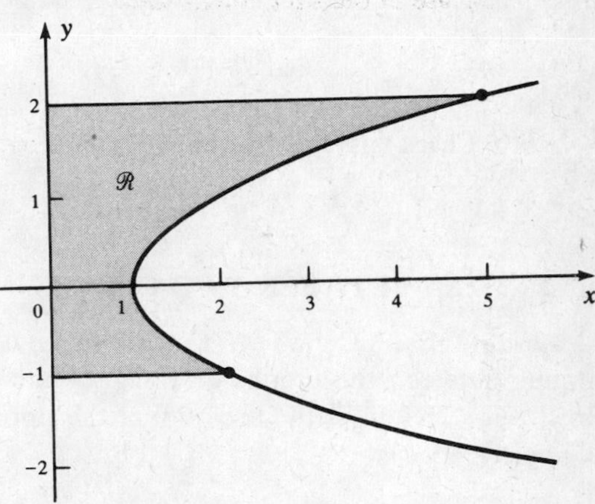

Fig. 32-2

EXAMPLES

(a) The graph of $x = y^2 + 1$ is a parabola, with its "nose" at (1, 0) and the positive x-axis as its axis of symmetry (see Fig. 32-2). Consider the region $\mathscr{R}$ consisting of all points to the left of this graph, to the right of the y-axis, and between $y = -1$ and $y = 2$. If we apply the reasoning used to calculate the area of a region like that shown in Fig. 32-1, but with x and y interchanged, we must integrate "along the y-axis." Thus, the area of $\mathscr{R}$ is given by the definite integral

$$\int_{-1}^{2} (y^2 + 1)\, dy$$

The Fundamental Theorem gives:

$$\int_{-1}^{2} (y^2 + 1)\, dy = \left(\frac{y^3}{3} + y\right)\Big]_{-1}^{2} = \left(\frac{2^3}{3} + 2\right) - \left(\frac{(-1)^3}{3} + (-1)\right)$$
$$= \left(\frac{8}{3} + 2\right) - \left(-\frac{1}{3} - 1\right) = \frac{9}{3} + 3 = 3 + 3 = 6$$

(b) Find the area of the region above the line $y = x - 3$ in the first quadrant, and below the line $y = 4$ (the shaded region of Fig. 32-3). Thinking of x as a function of y, namely,

$$x = y + 3$$

we can express the area as

$$\int_{0}^{4} (y + 3)\, dy = \left(\frac{y^2}{2} + 3y\right)\Big]_{0}^{4}$$
$$= \left(\frac{4^2}{2} + 3(4)\right) - \left(\frac{0^2}{2} + 3(0)\right) = \frac{16}{2} + 12 = 20$$

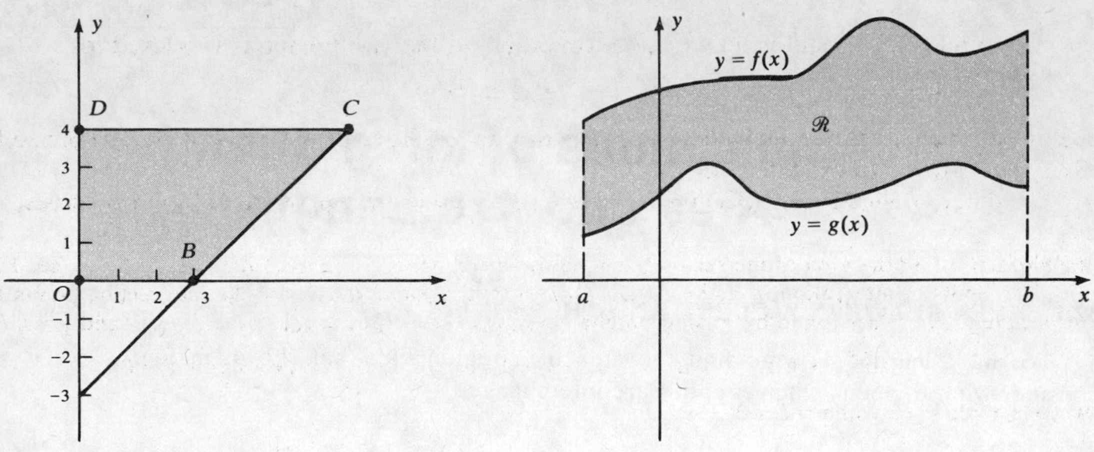

Fig. 32-3 Fig. 32-4

Check this result by computing the area of trapezoid *OBCD* by the geometrical formula given in Problem 31.9.

32.2 AREA BETWEEN TWO CURVES

Assume that $0 \le g(x) \le f(x)$ for x in $[a, b]$. Let us find the area A of the region $\mathcal{R}$ consisting of all points between the graphs of $y = g(x)$ and $y = f(x)$, and between $x = a$ and $x = b$. As may be seen from Fig. 32-4, A is the area under the upper curve, $y = f(x)$, minus the area under the lower curve, $y = g(x)$; that is,

$$A = \int_a^b f(x)\,dx - \int_a^b g(x)\,dx = \int_a^b (f(x) - g(x))\,dx \tag{32.1}$$

EXAMPLE Figure 32-5 shows the region $\mathcal{R}$ under the line

$$y = \frac{1}{2}x + 2$$

above the parabola $y = x^2$, and between the y-axis and $x = 1$. Its area is:

$$\begin{aligned}
\int_0^1 \left(\left(\frac{1}{2}x + 2\right) - x^2\right) dx &= \left(\frac{x^2}{4} + 2x - \frac{x^3}{3}\right)\Bigg]_0^1 \\
&= \left(\frac{1^2}{4} + 2(1) - \frac{1^3}{3}\right) - \left(\frac{0^2}{4} + 2(0) - \frac{0^3}{3}\right) \\
&= \frac{1}{4} + 2 - \frac{1}{3} = \frac{9}{4} - \frac{1}{3} = \frac{27 - 4}{12} = \frac{23}{12}
\end{aligned}$$

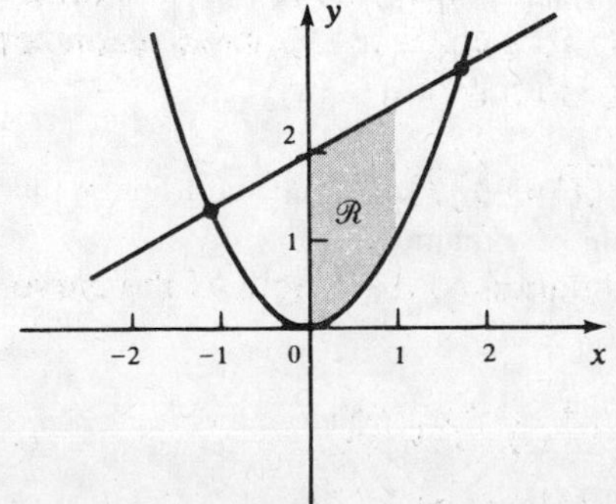

Fig. 32-5

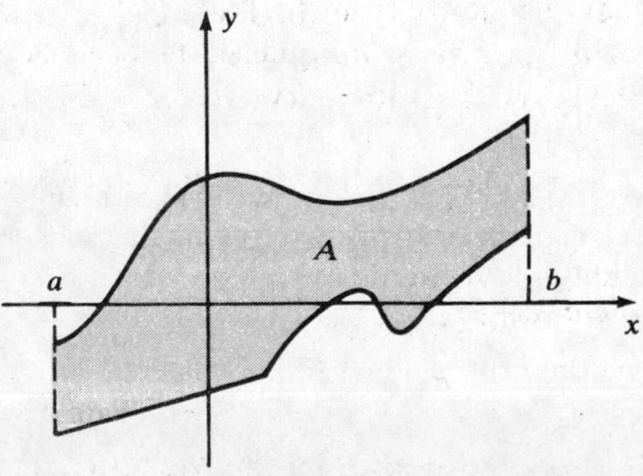

Fig. 32-6

Formula (*32.1*) is still valid when the condition on the two functions is relaxed to

$$g(x) \leq f(x)$$

i.e., when the curves are allowed to lie partly or totally below the x-axis, as in Fig. 32-6. See Problem 32.3 for a proof of this statement.

Another application of (*32.1*) is in finding the area of a region enclosed by two curves.

EXAMPLE Find the area of the region bounded by the parabola $y = x^2$ and the line $y = x + 2$ (see Fig. 32-7).

The limits of integration, a and b, in (*32.1*) must be the x-coordinates of the intersection points P and Q, respectively. These are found by solving simultaneously the equations of the curves, $y = x^2$ and $y = x + 2$. Thus,

$$x^2 = x + 2 \qquad \text{or} \qquad x^2 - x - 2 = 0 \qquad \text{or} \qquad (x-2)(x+1) = 0$$

whence $x = a = -1$ and $x = b = 2$.

$$A = \int_{-1}^{2} ((x+2) - x^2)\,dx = \left(\frac{x^2}{2} + 2x - \frac{x^3}{3}\right)\Big]_{-1}^{2}$$

$$= \left(\frac{2^2}{2} + 2(2) - \frac{2^3}{3}\right) - \left(\frac{(-1)^2}{2} + 2(-1) - \frac{(-1)^3}{3}\right)$$

$$= \left(\frac{4}{2} + 4 - \frac{8}{3}\right) - \left(\frac{1}{2} - 2 + \frac{1}{3}\right) = \frac{3}{2} + 6 - \frac{9}{3}$$

$$= \frac{3}{2} + 6 - 3 = \frac{3}{2} + 3 = \frac{3+6}{2} = \frac{9}{2}$$

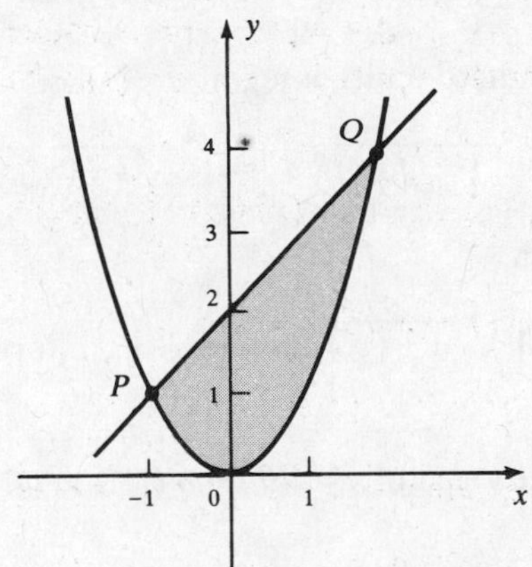

Fig. 32-7

32.3 ARC LENGTH

Consider a differentiable (not just continuous) function f on a closed interval $[a, b]$. The graph of f is a curve running from $(a, f(a))$ to $(b, f(b))$. We shall find a formula for the length L of this curve.

Divide $[a, b]$ into n equal parts, each of length Δx. To each x_i in this subdivision corresponds the point $P_i(x_i, f(x_i))$ on the curve (see Fig. 32-8). For large n, the sum

$$\overline{P_0P_1} + \overline{P_1P_2} + \cdots + \overline{P_{n-1}P_n} \equiv \sum_{i=1}^{n} \overline{P_{i-1}P_i}$$

of the lengths of the line segments $P_{i-1}P_i$ is an approximation to the length of the curve. Now, by the Distance Formula, (*2.1*),

$$\overline{P_{i-1}P_i} = \sqrt{(x_i - x_{i-1})^2 + (f(x_i) - f(x_{i-1}))^2}$$

But, $x_i - x_{i-1} = \Delta x$; also, by the Mean-Value Theorem (Theorem 17.2),

$$f(x_i) - f(x_{i-1}) = (x_i - x_{i-1})f'(x_i^*) = (\Delta x)f'(x_i^*)$$

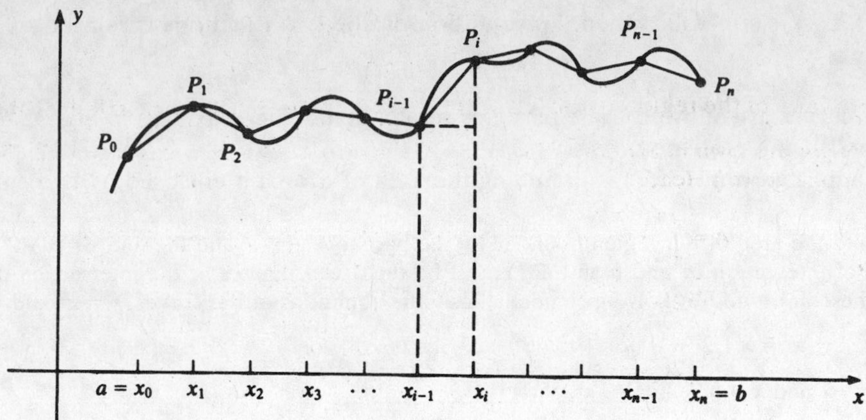

Fig. 32-8

for some x_i^* in $[x_{i-1}, x_i]$. Hence,

$$\overline{P_{i-1}P_i} = \sqrt{(\Delta x)^2 + (\Delta x)^2(f'(x_i^*))^2} = \sqrt{(1 + (f'(x_i^*))^2)(\Delta x)^2}$$
$$= \sqrt{1 + (f'(x_i^*))^2}\sqrt{(\Delta x)^2} = \sqrt{1 + (f'(x_i^*))^2}\,\Delta x$$

and

$$\sum_{i=1}^{n} \overline{P_{i-1}P_i} = \sum_{i=1}^{n} \sqrt{1 + (f'(x_i^*))^2}\,\Delta x$$

The right-hand sum approximates the definite integral

$$\int_a^b \sqrt{1 + (f'(x))^2}\,dx = \int_a^b \sqrt{1 + (y')^2}\,dx$$

Therefore, letting $n \to \infty$, we obtain

$$L = \int_a^b \sqrt{1 + (y')^2}\,dx \quad \textbf{arc-length formula} \tag{32.2}$$

EXAMPLE Find the arc length of the graph of $y = x^{3/2}$ from (1, 1) to (4, 8).

We have:

$$y' = \frac{3}{2}x^{1/2} \qquad \text{and} \qquad (y')^2 = \frac{9}{4}x$$

Hence, by the Arc-Length Formula,

$$L = \int_1^4 \sqrt{1 + \frac{9}{4}x}\,dx$$

Let

$$u = 1 + \frac{9}{4}x \qquad du = \frac{9}{4}dx \qquad dx = \frac{4}{9}du$$

When $x = 1$, $u = 13/4$; when $x = 4$, $u = 10$. Thus,

$$L = \int_{13/4}^{10} \sqrt{u}\,\frac{4}{9}\,du = \frac{4}{9}\int_{13/4}^{10} u^{1/2}\,du = \frac{4}{9}\left(\frac{2}{3}u^{3/2}\right)\Bigg]_{13/4}^{10}$$
$$= \frac{8}{27}\left(10^{3/2} - \left(\frac{13}{4}\right)^{3/2}\right) = \frac{8}{27}\left((\sqrt{10})^3 - \left(\frac{\sqrt{13}}{2}\right)^3\right)$$
$$= \frac{8}{27}\left(10\sqrt{10} - \frac{13\sqrt{13}}{8}\right) = \frac{1}{27}(80\sqrt{10} - 13\sqrt{13})$$

where, in the next-to-last step, we have used the identity $(\sqrt{c})^3 = (\sqrt{c})^2(\sqrt{c}) = c\sqrt{c}$.

Solved Problems

32.1 Find the area A of the region to the left of the parabola $x = -y^2 + 4$ and to the right of the y-axis.

The region is shown in Fig. 32-9. Notice that the parabola cuts the y-axis at $y = \pm 2$. (Set $x = 0$ in the equation of the curve.) Hence,

$$A = \int_{-2}^{2} (-y^2 + 4)\,dy = 2\int_0^2 (-y^2 + 4)\,dy \quad \text{[by Problem 31.8(a)]}$$

$$= 2\left(-\frac{y^3}{3} + 4y\right)\Big]_0^2 = 2\left[\left(-\frac{2^3}{3} + 4(2)\right) - \left(\frac{0^3}{3} + 4(0)\right)\right]$$

$$= 2\left(-\frac{8}{3} + 8\right) = 2\left(-\frac{8}{3} + \frac{24}{3}\right) = 2\left(\frac{16}{3}\right) = \frac{32}{3}$$

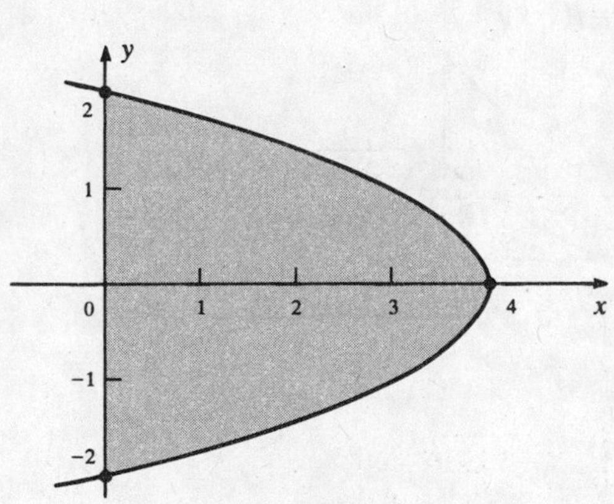

Fig. 32-9

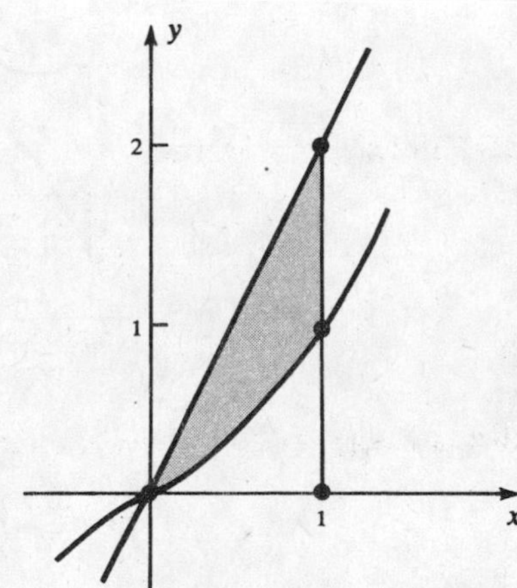

Fig. 32-10

32.2 Find the area of the region between the curves $y = x^3$ and $y = 2x$, between $x = 0$ and $x = 1$ (see Fig. 32-10).

For $0 \le x \le 1$,

$$2x - x^3 = x(2 - x^2) = x(\sqrt{2} + x)(\sqrt{2} - x) \ge 0$$

since all three factors are nonnegative. Thus, $y = x^3$ is the lower curve, and $y = 2x$ is the upper curve. By (*32.1*),

$$A = \int_0^1 (2x - x^3)\,dx = \left(x^2 - \frac{x^4}{4}\right)\Big]_0^1 = \left(1^2 - \frac{1^4}{4}\right) - \left(0^2 - \frac{0^4}{4}\right) = 1 - \frac{1}{4} = \frac{3}{4}$$

32.3 Prove that the formula for the area,

$$A = \int_a^b (f(x) - g(x))\,dx$$

holds whenever $g(x) \le f(x)$ on $[a, b]$.

Let $m < 0$ be the absolute minimum of g on $[a, b]$. (See Fig. 32-11(*a*); if $m \ge 0$, both curves lie above or on the x-axis, and this case is already known.) "Raise" both curves by $|m|$ units; the new graphs, shown in Fig. 32-11(*b*), are on or above the x-axis and include the same area A as the original graphs. Thus, by (*32.1*),

$$A = \int_a^b ((f(x) + |m|) - (g(x) + |m|))\,dx = \int_a^b (f(x) - g(x) + 0)\,dx = \int_a^b (f(x) - g(x))\,dx$$

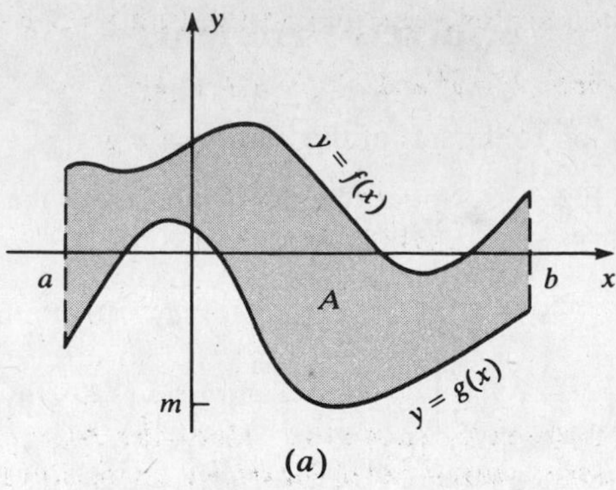

(a)

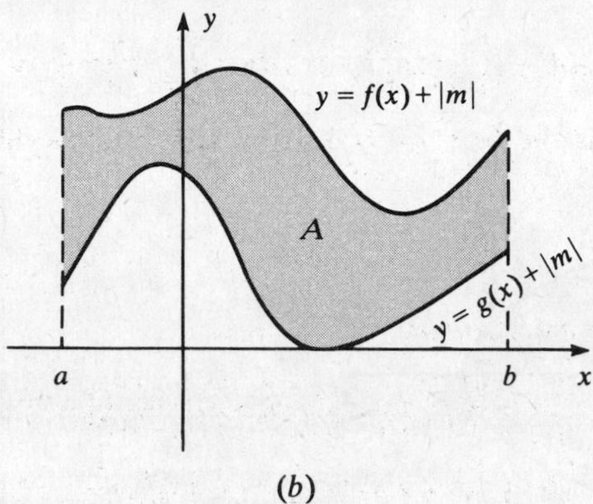

(b)

Fig. 32-11

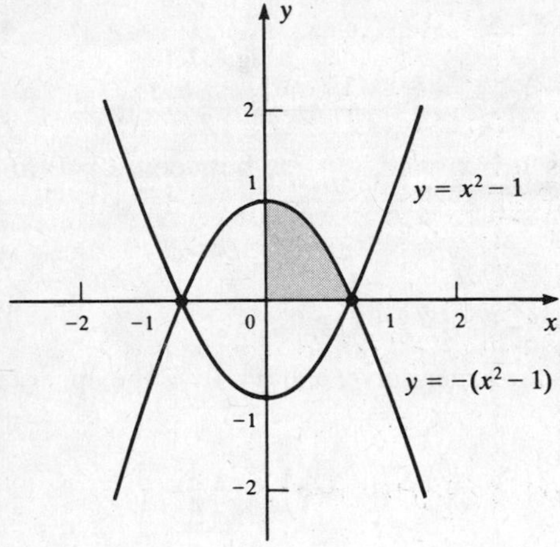

Fig. 32-12

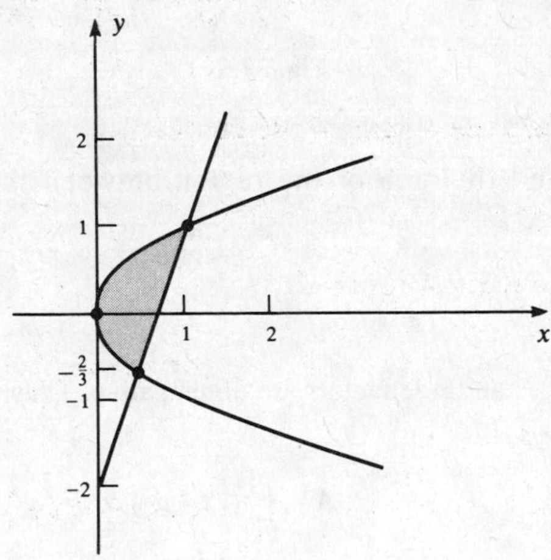

Fig. 32-13

32.4 Find the area A between the parabolas

$$y = x^2 - 1 \qquad \text{and} \qquad y = -(x^2 - 1)$$

From the symmetry of Fig. 32-12, it is clear that A will be equal to four times the area of the shaded region:

$$A = 4\int_0^1 -(x^2 - 1)\,dx = 4\int_0^1 (1 - x^2)\,dx = 4\left(x - \frac{x^3}{3}\right)\bigg]_0^1$$

$$= 4\left(\left(1 - \frac{1^3}{3}\right) - \left(0 - \frac{0^3}{3}\right)\right) = 4\left(\frac{2}{3}\right) = \frac{8}{3}$$

32.5 Find the area between the parabola $x = y^2$ and the line $y = 3x - 2$. (See Fig. 32-13.)

Find the intersection points: $x = y^2$ and $y = 3x - 2$ imply

$$y = 3y^2 - 2$$
$$3y^2 - y - 2 = 0$$
$$(3y + 2)(y - 1) = 0$$
$$3y + 2 = 0 \quad \text{or} \quad y - 1 = 0$$
$$y = -\frac{2}{3} \quad \text{or} \quad y = 1$$

Notice that we cannot find the area by integrating "along the x-axis" (unless we break the region into two parts). Integration along the y-axis is called for (which requires only the ordinates of the intersection points).

$$A = \int_{-2/3}^{1} \left(\frac{y+2}{3} - y^2\right) dy = \int_{-2/3}^{1} \left(\frac{y}{3} + \frac{2}{3} - y^2\right) dy$$

Here, the "upper" curve is the line $y = 3x - 2$; we had to solve this equation for x in terms of y, obtaining

$$x = \frac{y+2}{3}$$

Evaluating by the Fundamental Theorem,

$$A = \left(\frac{y^2}{6} + \frac{2}{3}y - \frac{y^3}{3}\right)\Big]_{-2/3}^{1} = \left(\frac{1}{6} + \frac{2}{3} - \frac{1}{3}\right) - \left(\frac{1}{6}\left(\frac{4}{9}\right) + \frac{2}{3}\left(-\frac{2}{3}\right) - \frac{1}{3}\left(-\frac{2}{3}\right)^3\right)$$
$$= \left(\frac{1}{6} + \frac{1}{3}\right) - \left(\frac{2}{27} - \frac{4}{9} + \frac{8}{81}\right) = \frac{1}{2} - \left(\frac{6}{81} - \frac{36}{81} + \frac{8}{81}\right) = \frac{1}{2} - \left(\frac{-22}{81}\right)$$
$$= \frac{1}{2} + \frac{22}{81} = \frac{81 + 44}{162} = \frac{125}{162}$$

32.6 Find the length of the curve

$$y = \frac{x^3}{6} + \frac{1}{2x}$$

from $x = 1$ to $x = 2$.

$$y = \frac{x^3}{6} + \frac{1}{2}x^{-1} \qquad y' = \frac{x^2}{2} - \frac{1}{2}x^{-2} = \frac{1}{2}(x^2 - x^{-2})$$

Hence,

$$1 + (y')^2 = 1 + \frac{1}{4}(x^2 - x^{-2})^2 = 1 + \frac{1}{4}(x^4 - 2 + x^{-4})$$
$$= \frac{1}{4}(4 + (x^4 - 2 + x^{-4})) = \frac{1}{4}(x^4 + 2 + x^{-4}) = \frac{1}{4}(x^2 + x^{-2})^2$$

Therefore

$$\sqrt{1 + (y')^2} = \frac{1}{2}(x^2 + x^{-2})$$

and the Arc-Length Formula gives

$$L = \frac{1}{2}\int_1^2 (x^2 + x^{-2})\, dx = \frac{1}{2}\left(\frac{x^3}{3} - x^{-1}\right)\Big]_1^2$$
$$= \frac{1}{2}\left[\left(\frac{8}{3} - \frac{1}{2}\right) - \left(\frac{1}{3} - 1\right)\right] = \frac{1}{2}\left(\frac{7}{3} + \frac{1}{2}\right) = \frac{1}{2}\left(\frac{17}{6}\right) = \frac{17}{12}$$

Supplementary Problems

32.7 Sketch, and find the area of, (*a*) the region to the left of the parabola $x = 2y^2$, to the right of the y-axis, and between $y = 1$ and $y = 3$; (*b*) the region above the line $y = 3x - 2$, in the first quadrant, and below the line $y = 4$; (*c*) the region between the curve $y = x^3$ and the lines $y = -x$ and $y = 1$.

32.8 Sketch the following regions and find their areas.

(*a*) The region between the curves $y = x^2$ and $y = x^3$.

(*b*) The region between the parabola $y = 4x^2$ and the line $y = 6x - 2$.

(*c*) The region between the curves $y = \sqrt{x}$, $y = 1$, and $x = 4$.

(*d*) The region under the curve $\sqrt{x} + \sqrt{y} = 1$ and in the first quadrant.

(*e*) The region between the curves $y = \sin x$, $y = \cos x$, $x = 0$, and $x = \pi/4$.

(*f*) The region between the parabola $x = -y^2$ and the line $y = x + 6$.

(*g*) The region between the parabola $y = x^2 - x - 6$ and the line $y = -4$.

(*h*) The region between the curves $y = \sqrt{x}$ and $y = x^3$.

(*i*) The region in the first quadrant between the curves $4y + 3x = 7$ and $y = x^{-2}$.

(*j*) The region bounded by the parabolas $y = x^2$ and $y = -x^2 + 6x$.

(*k*) The region bounded by the parabola $x = y^2 + 2$ and the line $y = x - 8$.

(*l*) The region bounded by the parabolas $y = x^2 - x$ and $y = x - x^2$.

(*m*) The region in the first quadrant bounded by the curves $y = x^2$ and $y = x^4$.

32.9 Find the lengths of the following curves:

(*a*) $y = \frac{x^4}{8} + \frac{1}{4x^2}$, from $x = 1$ to $x = 2$

(*b*) $y = 3x - 2$, from $x = 0$ to $x = 1$

(*c*) $y = x^{2/3}$, from $x = 1$ to $x = 8$

(*d*) $x^{2/3} + y^{2/3} = 4$, from $x = 1$ to $x = 8$

(*e*) $y = \frac{x^5}{15} + \frac{1}{4x^3}$, from $x = 1$ to $x = 2$

Chapter 33

Applications of Integration II: Volume

The volumes of certain kinds of solids can be calculated by means of definite integrals.

33.1 SOLIDS OF REVOLUTION

Disk and Ring Methods

Let f be a continuous function such that $f(x) \geq 0$ for $a \leq x \leq b$. Consider the region $\mathcal{R}$ under the graph of $y = f(x)$, above the x-axis, and between $x = a$ and $x = b$ (Fig. 33-1). If $\mathcal{R}$ is revolved about the x-axis, the resulting solid is called a *solid of revolution*. The generating regions $\mathcal{R}$ for some familiar solids of revolution are shown in Fig. 33-2.

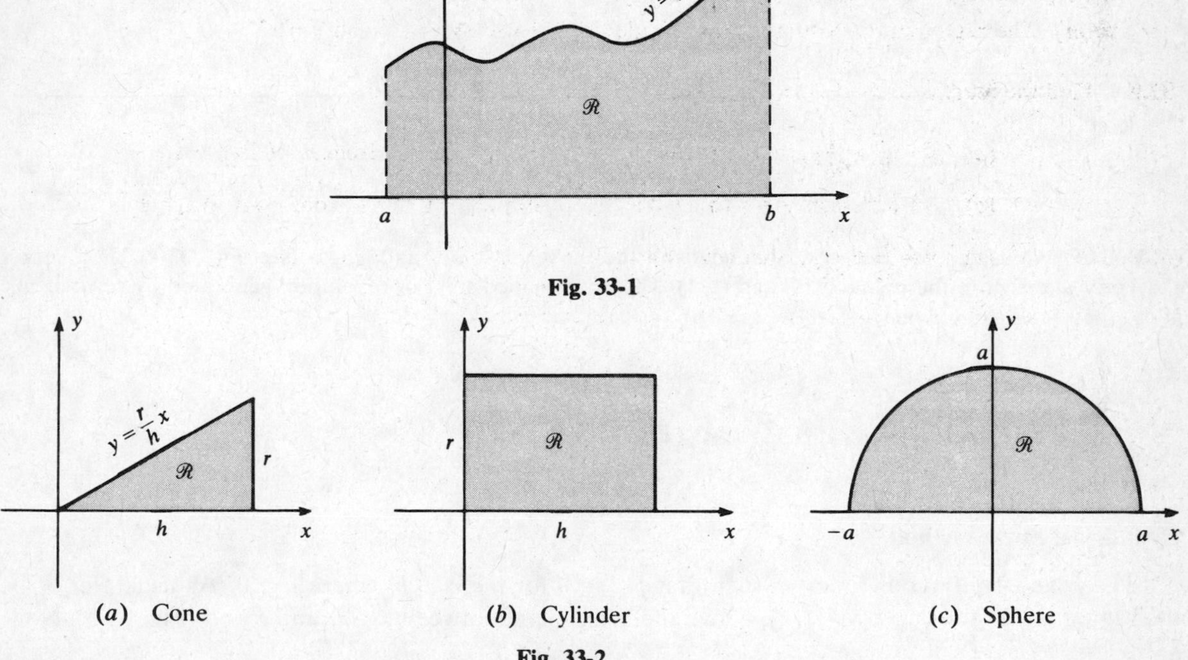

y
y = f(x)
ℛ
a
b
x

Fig. 33-1

y
y = (r/h) x
ℛ
r
h
x
(a) Cone
y
r
ℛ
h
x
(b) Cylinder
y
a
ℛ
−a
a
x
(c) Sphere

Fig. 33-2

Theorem 33.1: The volume V of the solid of revolution obtained by revolving the region of Fig. 33-1 about the x-axis is given by

$$V = \pi \int_a^b (f(x))^2\, dx = \pi \int_a^b y^2\, dx \quad \textbf{disk formula}$$

An argument for the disk formula is sketched in Problem 33.4.

If we interchange the roles of x and y, and revolve the area "under" the graph of $x = g(y)$ about the y-axis, then the same reasoning leads to the disk formula

$$V = \pi \int_c^d (g(y))^2\, dy = \pi \int_c^d x^2\, dy$$

EXAMPLE Applying the disk formula to Fig. 33-2(*a*), we obtain

$$V = \pi \int_0^h \left(\frac{r}{h}x\right)^2 dx = \pi \int_0^h \frac{r^2}{h^2}x^2\,dx$$
$$= \frac{\pi r^2}{h^2}\frac{x^3}{3}\bigg]_0^h = \frac{\pi r^2}{h^2}\left(\frac{h^3}{3} - 0\right) = \frac{\pi r^2 h}{3}$$

which is the standard formula for the volume of a cone with height h and radius of base r.

Now let f and g be two functions such that $0 \le g(x) \le f(x)$ for $a \le x \le b$, and revolve the region $\mathcal{R}$ between the curves $y = f(x)$ and $y = g(x)$ about the x-axis (see Fig. 33-3). The resulting solid of revolution has a volume V which is the difference between the volume of the solid of revolution generated by the region under $y = f(x)$ and the volume of the solid of revolution generated by the region under $y = g(x)$. Hence, by Theorem 33.1,

$$V = \pi \int_a^b ((f(x))^2 - (g(x))^2)\,dx \quad \textbf{circular ring formula}$$

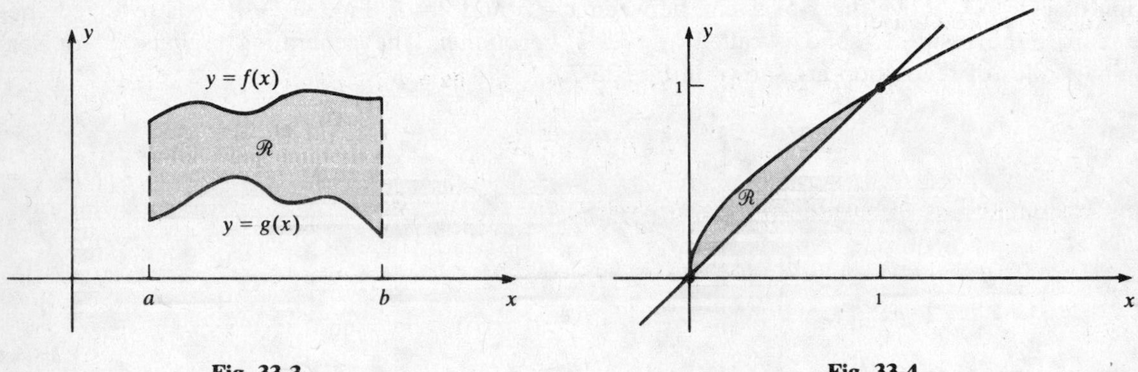

Fig. 33-3 Fig. 33-4

EXAMPLE Consider the region $\mathcal{R}$ bounded by the curves $y = \sqrt{x}$ and $y = x$ (see Fig. 33-4). The curves obviously intersect in the points (0, 0) and (1, 1). The bowl-shaped solid of revolution generated by revolving $\mathcal{R}$ about the x-axis has volume

$$V = \pi \int_0^1 ((\sqrt{x})^2 - x^2)\,dx = \pi \int_0^1 (x - x^2)\,dx = \pi\left(\frac{x^2}{2} - \frac{x^3}{3}\right)\bigg]_0^1$$
$$= \pi\left(\left(\frac{1}{2} - \frac{1}{3}\right) - 0\right) = \frac{\pi}{6}$$

Cylindrical Shell Method

Let f be a continuous function such that $f(x) \ge 0$ for $a \le x \le b$, where $a \ge 0$. As usual, let $\mathcal{R}$ be the region under the curve $y = f(x)$, above the x-axis, and between $x = a$ and $x = b$ (Fig. 33-5). Now,

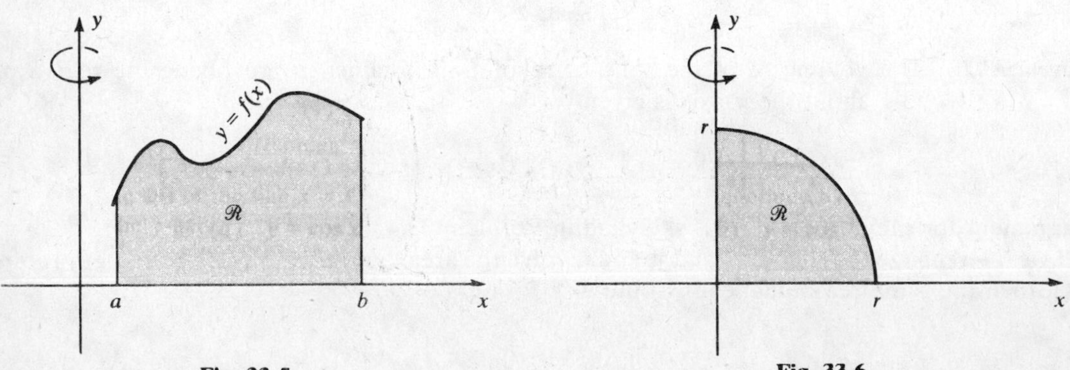

Fig. 33-5 Fig. 33-6

however, revolve $\mathscr{R}$ about the y-axis. The resulting solid of revolution has volume

$$V = 2\pi \int_a^b xf(x)\,dx = 2\pi \int_a^b xy\,dx \quad \textbf{cylindrical shell formula}$$

For the basic idea behind this formula and its name, see Problem 33.5.

EXAMPLE Consider the function

$$f(x) = \sqrt{r^2 - x^2}$$

for $0 \le x \le r$. The graph of f is the part of the circle $x^2 + y^2 = r^2$ that lies in the first quadrant. Revolution about the y-axis of the region $\mathscr{R}$ under the graph of f (see Fig. 33-6) produces a solid hemisphere of radius r. By the cylindrical shell formula,

$$V = 2\pi \int_0^r x\sqrt{r^2 - x^2}\,dx$$

To evaluate V substitute $u = r^2 - x^2$. Then $du = -2x\,dx$, and the limits of integration $x = 0$ and $x = r$ become $u = r^2$ and $u = 0$, respectively.

$$\begin{aligned} V &= 2\pi \int_{r^2}^0 u^{1/2}\left(-\frac{1}{2}\,du\right) = -\pi \int_{r^2}^0 u^{1/2}\,du = \pi \int_0^{r^2} u^{1/2}\,du \\ &= \frac{2}{3}\pi u^{3/2}\Big]_0^{r^2} = \frac{2}{3}\pi (r^2)^{3/2} = \frac{2}{3}\pi r^3 \end{aligned}$$

(This result is more easily obtained by the disk formula,

$$V = \pi \int_0^r x^2\,dy$$

Try it.)

33.2 VOLUME BASED ON CROSS SECTIONS

Assume that a solid (not necessarily a solid of revolution) lies entirely between the plane perpendicular to the x-axis at $x = a$ and the plane perpendicular to the x-axis at $x = b$. For $a \le x \le b$, let the plane perpendicular to the x-axis at that value of x intersect the solid in a region of area $A(x)$, as indicated in Fig. 33-7. Then the volume V of the solid is given by

$$V = \int_a^b A(x)\,dx \quad \textbf{cross-section formula}$$

For a derivation, see Problem 33.6.

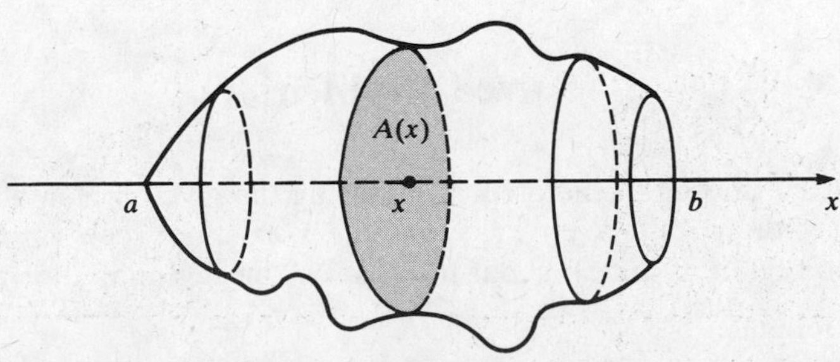

Fig. 33-7

EXAMPLES

(*a*) Assume that half of a salami, of length h, is such that a cross section perpendicular to the axis of the salami at a distance x from the end O is a circle of radius $\sqrt{kx}$. (See Fig. 33-8.) Thus,

$$A(x) = \pi(\sqrt{kx})^2 = \pi kx$$

and the cross-section formula gives

$$V = \int_0^h \pi kx\, dx = \pi k \int_0^h x\, dx = \pi k \frac{x^2}{2}\bigg]_0^h = \frac{\pi kh^2}{2}$$

Note that for this solid of revolution the disk formula would give the same expression for V.

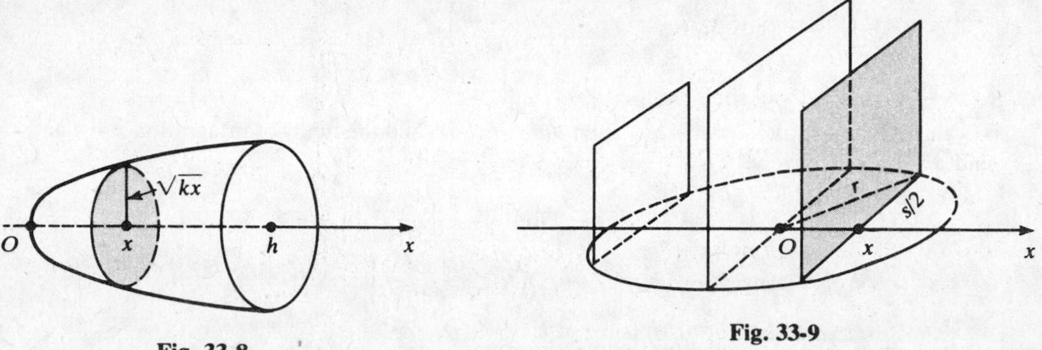

Fig. 33-8

Fig. 33-9

(*b*) Assume that a solid has a base which is a circle of radius r. Assume that there is a diameter D such that all plane sections of the solid perpendicular to diameter D are squares (see Fig. 33-9). Find the volume.

Let the origin be the center of the circle and let the x-axis be the special diameter D. For a given value of x, with $-r \le x \le r$, the side $s(x)$ of the square cross section is given by the Pythagorean Theorem (see Fig. 33-9):

$$x^2 + \left(\frac{s}{2}\right)^2 = r^2$$

$$x^2 + \frac{s^2}{4} = r^2$$

$$s^2 = 4(r^2 - x^2) = A(x)$$

Then, by the cross-section formula, and the fact that $A(x)$ is an even function,

$$V = \int_{-r}^{r} 4(r^2 - x^2)\, dx = 8\int_0^r (r^2 - x^2)\, dx = 8\left(r^2x - \frac{x^3}{3}\right)\bigg]_0^r$$

$$= 8\left[\left(r^2(r) - \frac{r^3}{3}\right) - (0 - 0)\right] = 8\left(\frac{2}{3}r^3\right) = \frac{16}{3}r^3$$

Solved Problems

33.1 Find the volume of the solid generated by revolving the given region about the given axis. (*a*) The region under the parabola $y = x^2$, above the x-axis, between $x = 0$ and $x = 1$; about the x-axis. (*b*) Same region as in (*a*); about the y-axis. The region is shown in Fig. 33-10.

(*a*) Use the disk formula:

$$V = \pi \int_0^1 (x^2)^2\, dx = \pi \int_0^1 x^4\, dx = \pi\left(\frac{x^5}{5}\right)\bigg]_0^1 = \pi\left(\frac{1}{5}\right) = \frac{\pi}{5}$$

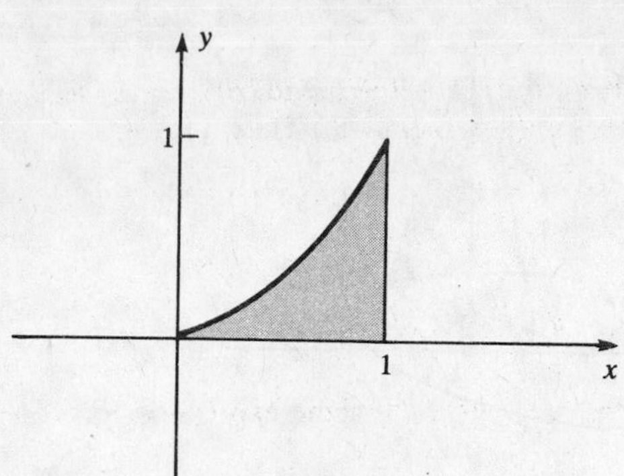

Fig. 33-10

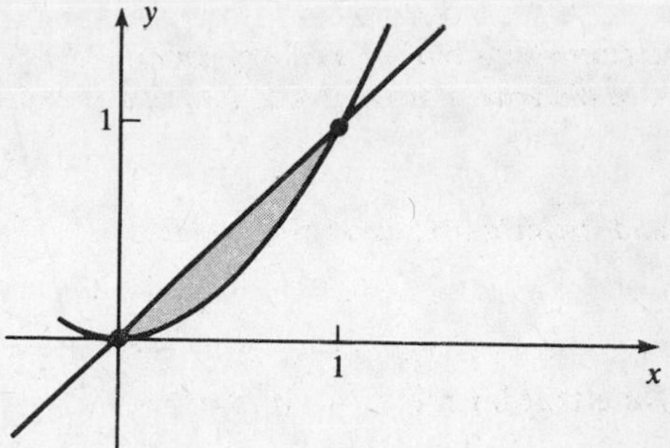

Fig. 33-11

(*b*) Use the cylindrical shell formula:

$$V = 2\pi \int_0^1 x(x^2)\,dx = 2\pi \int_0^1 x^3\,dx = 2\pi\left(\frac{x^4}{4}\right)\Big]_0^1 = 2\pi\left(\frac{1}{4}\right) = \frac{\pi}{2}$$

33.2 Let $\mathcal{R}$ be the region between $y = x^2$ and $y = x$ (Fig. 33-11). Find the volume of the solid obtained by revolving $\mathcal{R}$ around (*a*) the *x*-axis, (*b*) the *y*-axis.

The curves intersect at (0, 0) and (1, 1).

(*a*) By the circular ring formula,

$$V = \pi \int_0^1 (x^2 - (x^2)^2)\,dx = \pi \int_0^1 (x^2 - x^4)\,dx = \pi\left(\frac{x^3}{3} - \frac{x^5}{5}\right)\Big]_0^1$$
$$= \pi\left(\frac{1}{3} - \frac{1}{5}\right) = \frac{2\pi}{15}$$

(*b*) Use the circular ring formula, along the *y*-axis.

$$V = \pi \int_0^1 ((\sqrt{y})^2 - y^2)\,dy = \pi \int_0^1 (y - y^2)\,dy = \pi\left(\frac{y^2}{2} - \frac{y^3}{3}\right)\Big]_0^1$$
$$= \pi\left(\frac{1}{2} - \frac{1}{3}\right) = \frac{\pi}{6}$$

33.3 Find the volume of the solid whose base is a circle of radius *r* and such that every cross section perpendicular to a particular fixed diameter *D* is an equilateral triangle.

Let the center of the circular base be the origin, and let the *x*-axis be the diameter *D*. The area of the cross section at *x* is

$$A(x) = h\frac{s}{2}$$

(see Fig. 33-12). Now, in the horizontal right triangle,

$$x^2 + \left(\frac{s}{2}\right)^2 = r^2 \qquad \text{or} \qquad \frac{s}{2} = \sqrt{r^2 - x^2}$$

and in the vertical right triangle,

$$h^2 + \left(\frac{s}{2}\right)^2 = s^2$$
$$h^2 + \frac{s^2}{4} = s^2$$
$$h^2 = 3\frac{s^2}{4}$$
$$h = \sqrt{3}\frac{s}{2} = \sqrt{3}\sqrt{r^2 - x^2}$$

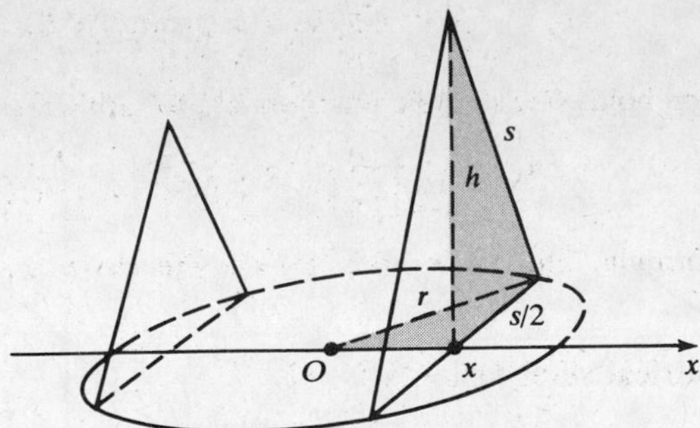

Fig. 33-12

Hence, $A(x) = \sqrt{3}(r^2 - x^2)$—an even function—and the cross-section formula gives

$$V = \sqrt{3}\int_{-r}^{r}(r^2 - x^2)\,dx = 2\sqrt{3}\int_{0}^{r}(r^2 - x^2)\,dx = 2\sqrt{3}\left(r^2x - \frac{x^3}{3}\right)\Big]_0^r$$

$$= 2\sqrt{3}\left(r^3 - \frac{r^3}{3}\right) = 2\sqrt{3}\left(\frac{2}{3}r^3\right) = \frac{4\sqrt{3}}{3}r^3$$

33.4 Establish the disk formula,

$$V = \pi\int_2^b (f(x))^2\,dx$$

We assume as valid the expression $\pi r^2 h$ for the volume of a cylinder of radius r and height h. Divide the interval $[a, b]$ into n equal subintervals, each of length $\Delta x = (b - a)/n$ (see Fig. 33-13). Consider the volume V_i obtained by revolving the region $\mathcal{R}_i$ above the ith subinterval about the x-axis. If m_i and M_i denote the absolute minimum and absolute maximum of f on the ith subinterval, it is plain that V_i must lie between the volume of a cylinder of radius m_i and height Δx and the volume of a cylinder of radius M_i and height Δx:

$$\pi m_i^2\,\Delta x \le V_i \le \pi M_i^2\,\Delta x \qquad \text{or} \qquad m_i^2 \le \frac{V_i}{\pi\,\Delta x} \le M_i^2$$

The Intermediate-Value Theorem for the continuous function $(f(x))^2$ guarantees the existence of some point x_i^* in the ith subinterval such that

$$\frac{V_i}{\pi\,\Delta x} = (f(x_i^*))^2 \qquad \text{or} \qquad V_i = \pi(f(x_i^*))^2\,\Delta x$$

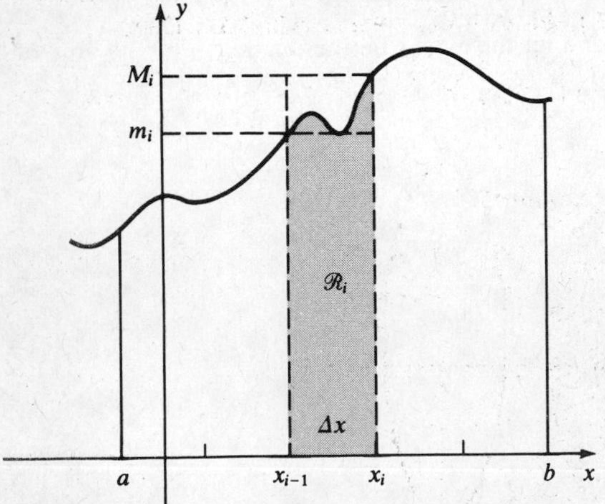

Fig. 33-13

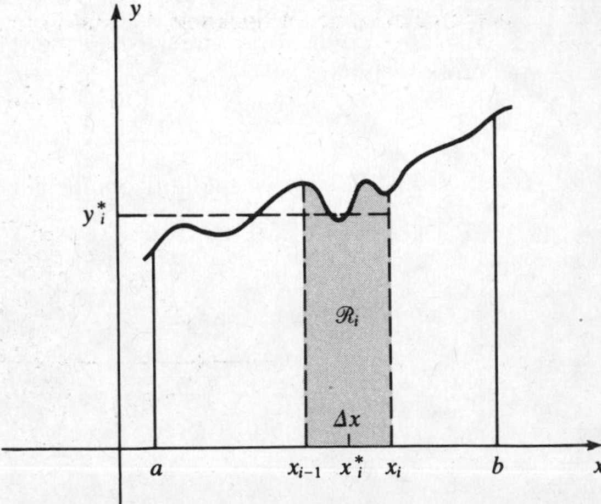

Fig. 33-14

Hence
$$V = \sum_{i=1}^{n} V_i = \pi \sum_{i=1}^{n} (f(x_i^*))^2 \, \Delta x$$

Since this relation holds (for suitable numbers x_i^*) for arbitrary n, it must hold in the limit as $n \to \infty$:

$$V = \pi \lim_{n\to\infty} \left[\sum_{i=1}^{n} (f(x_i^*))^2 \, \Delta x \right] = \pi \int_a^b (f(x))^2 \, dx$$

which is the disk formula. The name derives from the use of cylindrical *disks* (of thickness Δx) to approximate the V_i.

33.5 Establish the cylindrical shell formula,

$$V = 2\pi \int_a^b xf(x) \, dx$$

Divide the interval $[a, b]$ into n equal subintervals, each of length Δx. Let $\mathcal{R}_i$ be the region above the ith subinterval (see Fig. 33-14). Let x_i^* be the midpoint of the ith subinterval,

$$x_i^* = \frac{x_{i-1} + x_i}{2}$$

Now, the solid obtained by revolving the region $\mathcal{R}_i$ about the y-axis is approximately the solid obtained by revolving the rectangle with base Δx and height $y_i^* = f(x_i^*)$. The latter solid is a *cylindrical shell*; that is, it is the difference between the cylinders obtained by rotating the rectangles with the same height $f(x_i^*)$ and with bases $[0, x_{i-1}]$ and $[0, x_i]$. Hence, it has volume

$$\pi x_i^2 f(x_i^*) - \pi x_{i-1}^2 f(x_i^*) = \pi f(x_i^*)\,(x_i^2 - x_{i-1}^2) = \pi f(x_i^*)\,(x_i + x_{i-1})(x_i - x_{i-1})$$
$$= \pi f(x_i^*)(2x_i^*)(\Delta x) = 2\pi x_i^* f(x_i^*)\,\Delta x$$

Thus, the total V is approximated by

$$2\pi \sum_{i=1}^{n} x_i^* f(x_i^*)\,\Delta x$$

which in turn approximates the definite integral

$$2\pi \int_a^b xf(x)\,dx$$

33.6 Establish the cross-section formula,

$$V = \int_a^b A(x)\,dx$$

Divide the interval $[a, b]$ into n equal subintervals $[x_{i-1}, x_i]$, each of length Δx. Choose a point x_i^* in $[x_{i-1}, x_i]$. If n is large, making Δx small, the piece of the solid between x_{i-1} and x_i will be very nearly a (noncircular) disk, of thickness Δx and base area $A(x_i^*)$; this disk has volume $A(x_i^*)\,\Delta x$. Thus,

$$V \approx \sum_{i=1}^{n} A(x_i^*)\,\Delta x \to \int_a^b A(x)\,dx$$

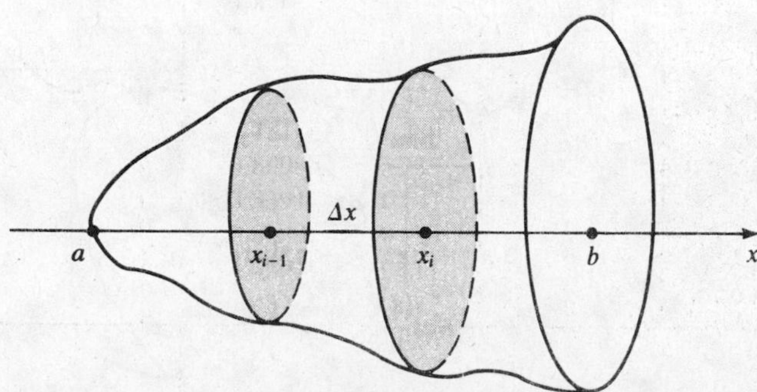

Fig. 33-15

Supplementary Problems

33.7 Find the volume of the solid generated by revolving the given region about the given axis. (*a*) The region above the curve $y = x^3$, under the line $y = 1$, and between $x = 0$ and $x = 1$; about the x-axis. (*b*) The region of (*a*); about the y-axis. (*c*) The region below the line $y = 2x$, above the x-axis, and between $x = 0$ and $x = 1$; about the y-axis. (*d*) The region between the parabolas $y = x^2$ and $x = y^2$; about either the x-axis or the y-axis. (*e*) The region (Fig. 33-16) inside the semicircle

$$x^2 + y^2 = r^2$$

with $0 \le x \le a < r$; about the y-axis. (This gives the volume cut from a sphere of radius r by a pipe of radius a whose axis is a diameter of the sphere.) (*f*) The region (Fig. 33-17) below the quarter-circle

$$x^2 + y^2 = r^2 \quad (x \ge 0, y \ge 0)$$

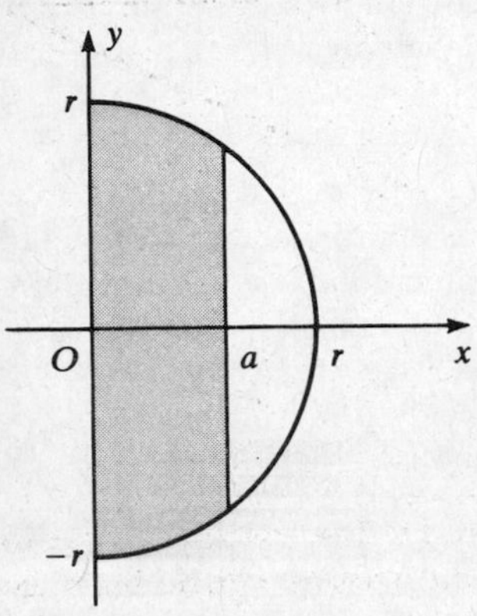

Fig. 33-16

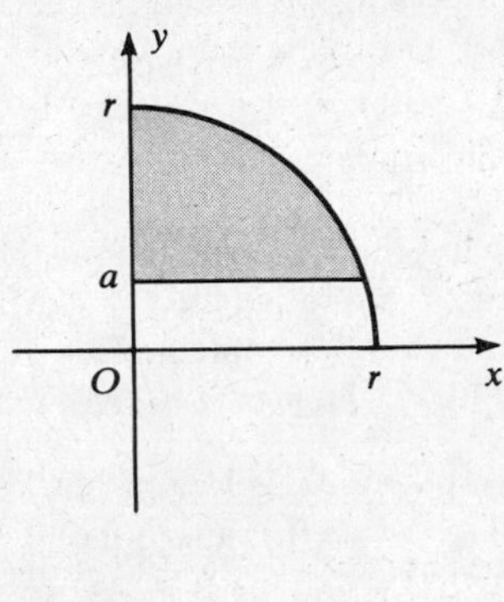

Fig. 33-17

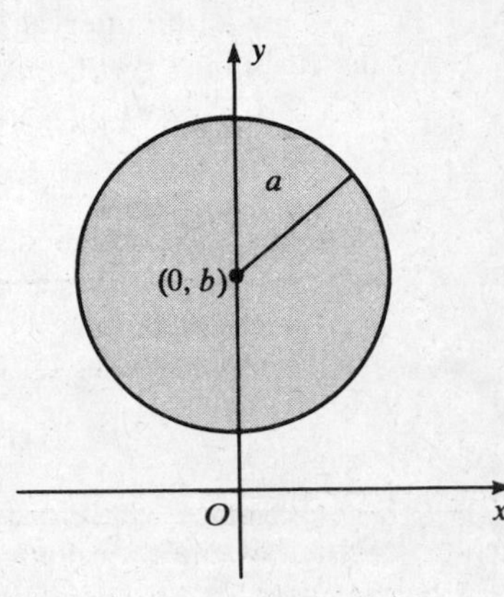

Fig. 33-18

and above the line $y = a$, where $0 \le a < r$; about the y-axis. (This gives the volume of a polar cap of a sphere.) (*g*) The region bounded by $y = 1 + x^2$ and $y = 5$; about the x-axis. (*h*) The region inside the circle

$$x^2 + (y - b)^2 = a^2 \quad (0 < a < b)$$

about the x-axis. (*Hint*: When you obtain an integral of the form

$$\int_{-a}^{a} \sqrt{a^2 - x^2}\, dx$$

notice that this is the area of a semicircle of radius a.) This problem gives the volume of a doughnut-shaped solid. (*i*) The region bounded by $x^2 = 4y$ and $y = x/2$; about the y-axis. (*j*) The region bounded by $y = 4/x$ and $y = (x - 3)^2$; about the x-axis. (Notice that the curves intersect when $x = 1$ and $x = 4$. What is special about the intersection at $x = 1$?) (*k*) The region of (*j*); about the y-axis. (*l*) The region bounded by $xy = 1$, $x = 1$, $x = 3$, $y = 0$; about the x-axis. (*m*) The region of (*l*); about the y-axis.

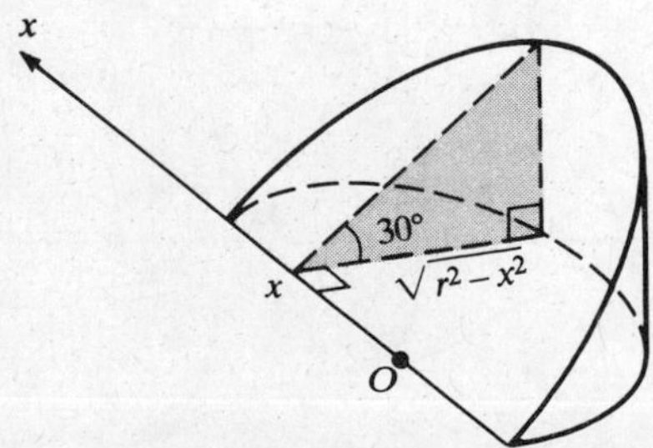

Fig. 33-19

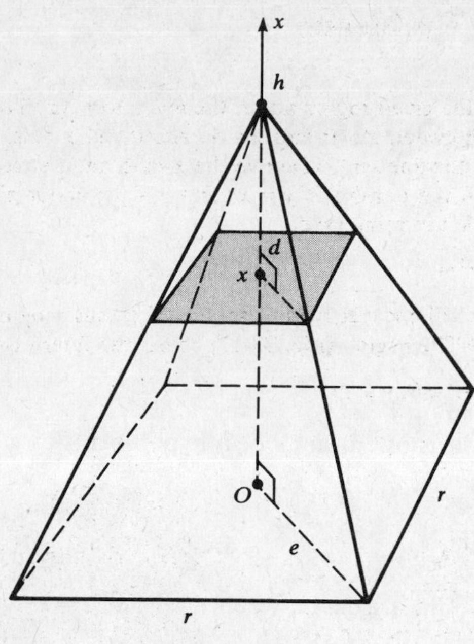

Fig. 33-20

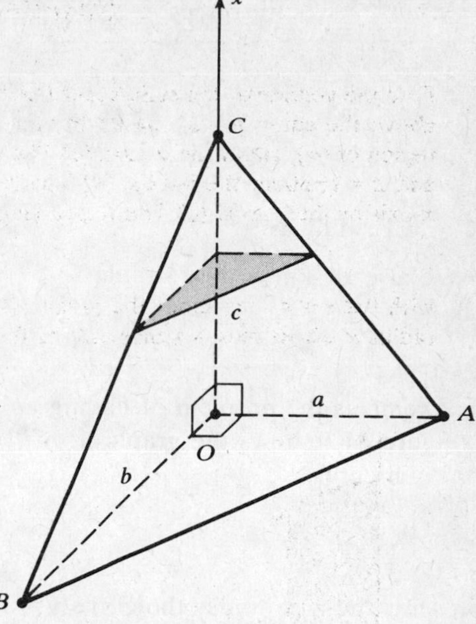

Fig. 33-21

33.8 Use the cross-section formula to find the volume of the following solids. (*a*) The solid has a base which is a circle of radius *r*. Each cross section perpendicular to a fixed diameter of the circle is an isosceles triangle with altitude equal to one half of its base. (*b*) The solid is a wedge, cut from a perfectly round tree of radius *r* by two planes, one perpendicular to the axis of the tree and the other intersecting the first plane at an angle of 30° along a diameter (see Fig. 33-19). (*c*) A square pyramid with a height of *h* units and a base of side *r* units. (*Hint*: Locate the *x*-axis as in Fig. 33-20. By the similar right triangles,

$$\frac{d}{e} = \frac{h-x}{h}$$

and

$$\frac{A(x)}{r^2} = \left(\frac{d}{e}\right)^2$$

which determine $A(x)$.) (*d*) The tetrahedron (Fig. 33-21) formed by three mutually perpendicular faces and three mutually perpendicular edges of lengths *a*, *b*, *c*. (*Hint*: Another pyramid; proceed as in (*c*).)

Chapter 34

The Natural Logarithm

34.1 DEFINITION

We already know the formula

$$\int x^r\,dx = \frac{x^{r+1}}{r+1} + C \qquad \text{for } r \neq -1$$

There remains the problem of finding an antiderivative of x^{-1}.

Figure 34-1 shows the graph of $y = 1/t$ for $t > 0$; it is one branch of a hyperbola. For $x > 1$, the definite integral

$$\int_1^x \frac{1}{t}\,dt$$

represents the area under the curve $y = 1/t$ and above the t-axis, between $t = 1$ and $t = x$.

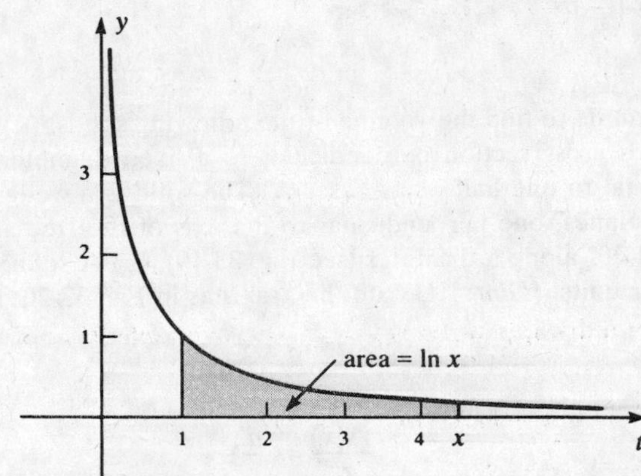

Fig. 34-1

Definition: $$\ln x = \int_1^x \frac{1}{t}\,dt \qquad \text{for } x > 0$$

The function $\ln x$ is called the *natural logarithm*. By Theorem 31.1,

$$D_x(\ln x) = \frac{1}{x} \qquad \text{for } x > 0 \tag{34.1}$$

and so the natural logarithm is the desired antiderivative of x^{-1}, *but only on* $(0, \infty)$. An antiderivative for all $x \neq 0$ will be constructed in the following Section.

34.2 PROPERTIES

PROPERTY 1. $\ln 1 = 0$

as follows at once from the definition of $\ln x$.

PROPERTY 2. If $x > 1$, $\ln x > 0$.

This is apparent from the area interpretation (Fig. 34-1) or, more rigorously, from Problem 30.11.

PROPERTY 3. If $0 < x < 1$, $\ln x < 0$.

In fact,
$$\ln x = \int_1^x \frac{1}{t}\,dt = -\int_x^1 \frac{1}{t}\,dt \quad \text{[reversing limits of integration]}$$
and, for $0 < x < 1$,
$$\int_x^1 \frac{1}{t}\,dt > 0$$
by Problem 30.11.

PROPERTY 4.
$$\int \frac{1}{x}\,dx = \ln|x| + C \qquad (x \neq 0)$$

In other words, $\ln|x|$ is an antiderivative of x^{-1} for all nonzero x. The proof is simple: When $x > 0$, then $|x| = x$, and so
$$D_x(\ln|x|) = D_x(\ln x) = \frac{1}{x} \quad \text{[by (34.1)]}$$
When $x < 0$, then $|x| = -x$, and so
$$\begin{aligned} D_x(\ln|x|) = D_x(\ln(-x)) &= D_u(\ln u)D_x(u) \quad \text{[Chain Rule; } u = -x > 0] \\ &= \left(\frac{1}{u}\right)(-1) \quad \text{[by (34.1)]} \\ &= \frac{1}{-u} = \frac{1}{x} \end{aligned}$$

PROPERTY 5.
$$\ln uv = \ln u + \ln v$$

For a proof, see Problem 34.2.

EXAMPLE Property 5 is basic; indeed, it can serve as an alternate definition of the natural logarithm. To see this, choose $v = 1$, obtaining
$$\ln u = \ln u + \ln 1 \qquad \text{or} \qquad \ln 1 = 0$$
Now choose $u = x$ and $v = 1 + hx^{-1}$, obtaining
$$\ln(x+h) = \ln x + \ln(1 + hx^{-1})$$
from which
$$\begin{aligned} \frac{\ln(x+h) - \ln x}{h} = \frac{\ln(1+hx^{-1})}{h} &= x^{-1}\frac{\ln(1+hx^{-1})}{hx^{-1}} \\ &= x^{-1}\frac{\ln(1+hx^{-1}) - \ln 1}{hx^{-1}} \quad \text{[since } \ln 1 = 0] \end{aligned}$$
If $\ln x$ is assumed differentiable at $x = 1$, then we may let $h \to 0$ in the last equation, obtaining in the limit
$$D_x(\ln x) = x^{-1} \cdot D_x(\ln x)|_{x=1}$$
Assigning the constant on the right the value 1 (the *natural* choice), we recover (*34.1*). The Fundamental Theorem of Calculus then gives
$$\int_1^x \frac{1}{t}\,dt = \ln t\,\Big]_1^x = \ln x$$
which is our original definition of the *natural* logarithm.

PROPERTY 6.
$$\ln\frac{u}{v} = \ln u - \ln v$$

Proof. In Property 5, replace u by u/v.

PROPERTY 7.
$$\ln \frac{1}{v} = -\ln v$$

Proof. Let $u = 1$ in Property 6.

PROPERTY 8. For any rational number r,
$$\ln x^r = r \ln x$$

See Problem 34.3 for a proof. Observe that the special case $r = -1$ gives Property 7.

PROPERTY 9. $\ln x$ is an increasing function.

This follows from the fact that its derivative is positive. By Property 9, $\ln u$ and $\ln v$ are unequal if u and v are unequal; i.e.,

PROPERTY 10. $\ln u = \ln v$ implies $u = v$.

PROPERTY 11.
$$\lim_{x \to +\infty} \ln x = +\infty$$

Proof. In view of Property 9, we need only show that $\ln x$ eventually exceeds any given positive integer k. Now, for $x > 2^{2k}$,
$$\ln x > \ln 2^{2k} = 2k \ln 2 \quad \text{[Property 8]}$$

But, by Problem 30.3(*c*),
$$\ln 2 = \int_1^2 \frac{1}{t}\, dt \geq \frac{1}{2}(2-1) = \frac{1}{2}$$

and so $\ln x > k$.

PROPERTY 12.
$$\lim_{x \to 0^+} \ln x = -\infty$$

This follows from Properties 11 and 7.

Solved Problems

34.1 Sketch the graph of $y = \ln x$.

We know that $\ln x$ is increasing (Property 9). We also know that $\ln 1 = 0$ and that
$$\frac{1}{2} < \ln 2 < 1$$

(see the proof of Property 11; it can be shown that $\ln 2 = 0.693\cdots$). From the ordinate at $x = 2$ we can obtain the ordinates at $x = 4, 8, 16, \ldots,$ and at $x = 1/2, 1/4, 1/8, \ldots,$ by Property 8:
$$\ln 4 = 2 \ln 2 \qquad \ln 8 = 3 \ln 2 \qquad \ln 16 = 4 \ln 2 \qquad \cdots$$

and
$$\ln \frac{1}{2} = -1 \ln 2 \qquad \ln \frac{1}{4} = -2 \ln 2 \qquad \ln \frac{1}{8} = -3 \ln 2 \qquad \cdots$$

Because $D_x(\ln x) = x^{-1}$,

$$D_x^2(\ln x) = D_x(x^{-1}) = -x^{-2} < 0$$

and the graph is concave downward (Theorem 23.1). There is no horizontal asymptote (by Property 11), but the negative y-axis is a vertical asymptote (by Property 12). The graph is sketched in Fig. 34-2. Notice that $\ln x$ assumes all real numbers as values.

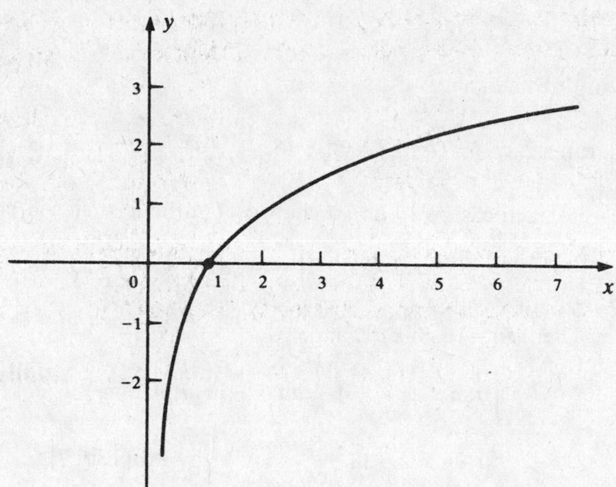

Fig. 34-2

34.2 Prove: $\ln uv = \ln u + \ln v$.

In

$$\ln v = \int_1^v \frac{1}{t}\,dt$$

make the change of variable $w = ut$ (u fixed). Then $dw = u\,dt$, and the limits of integration, $t = 1$ and $t = v$, go over into $w = u$ and $w = uv$, respectively.

$$\ln v = \int_u^{uv} \frac{u}{w}\frac{1}{u}\,dw = \int_u^{uv} \frac{1}{w}\,dw = \int_u^{uv} \frac{1}{t}\,dt$$

Then, by Theorem 30.4,

$$\ln u + \ln v = \int_1^u \frac{1}{t}\,dt + \int_u^{uv} \frac{1}{t}\,dt = \int_1^{uv} \frac{1}{t}\,dt = \ln uv$$

34.3 Prove: $\ln x^r = r \ln x$, for rational r.

By the Chain Rule,

$$D_x(\ln x^r) = \frac{1}{x^r} D_x(x^r) = \frac{1}{x^r}(rx^{r-1})$$

$$= r\frac{1}{x} = r D_x(\ln x) = D_x(r \ln x)$$

Then, by Corollary 29.2, $\ln x^r = r \ln x + C$, for some constant C. Substituting $x = 1$, we find that $C = 0$, and the proof is complete.

34.4 Evaluate

(a) $D_x(\ln(x^3 - 2x))$ (b) $D_x(\ln(\sin x))$ (c) $D_x(\cos(\ln x))$

Use the Chain Rule.

(a) $$D_x(\ln(x^3-2x)) = \frac{1}{x^3-2x}D_x(x^3-2x) = \frac{1}{x^3-2x}(3x^2-2) = \frac{3x^2-2}{x^3-2x}$$

(b) $$D_x(\ln(\sin x)) = \frac{1}{\sin x}D_x(\sin x) = \frac{1}{\sin x}(\cos x) = \frac{\cos x}{\sin x} = \cot x$$

(c) $$D_x(\cos(\ln x)) = -\sin(\ln x)\cdot D_x(\ln x) = -\sin(\ln x)\cdot\frac{1}{x} = -\frac{\sin(\ln x)}{x}$$

34.5 Find the following antiderivatives:

$$(a)\ \int \tan x\,dx \qquad (b)\ \int \sec x\,dx \qquad (c)\ \int \frac{1}{2x-1}\,dx \qquad (d)\ \int \frac{x}{x^2-5}\,dx$$

The technique in each case is to find a change of variable that will enable use of Property 4.

(a) $$\int \tan x\,dx = \int \frac{\sin x}{\cos x}\,dx$$

Let $u = \cos x$. Then $du = -\sin x\,dx$, and

$$\int \tan x\,dx = -\int \frac{1}{u}\,du = -\ln|u| + C = -\ln|\cos x| + C$$

$$= \ln\frac{1}{|\cos x|} + C \quad \text{[by Property 7]}$$

$$= \ln\left|\frac{1}{\cos x}\right| + C = \ln|\sec x| + C$$

(b) First we employ a trick:

$$\int \sec x\,dx = \int (\sec x)\left(\frac{\sec x + \tan x}{\sec x + \tan x}\right)dx = \int \frac{\sec^2 x + \sec x \tan x}{\sec x + \tan x}\,dx$$

Let $u = \sec x + \tan x$. Then $du = (\sec x \tan x + \sec^2 x)\,dx$, and

$$\int \sec x\,dx = \int \frac{\sec^2 x + \sec x \tan x}{\sec x + \tan x}\,dx = \int \frac{1}{u}\,du = \ln|u| + C$$

$$= \ln|\sec x + \tan x| + C$$

(c) Let $u = 2x-1$. Then $du = 2\,dx$, and

$$\int \frac{1}{2x-1}\,dx = \frac{1}{2}\int \frac{1}{u}\,du = \frac{1}{2}\ln|u| + C = \frac{1}{2}\ln|2x-1| + C$$

(d) Let $u = x^2-5$. Then $du = 2x\,dx$, and

$$\int \frac{x}{x^2-5}\,dx = \frac{1}{2}\int \frac{1}{u}\,du = \frac{1}{2}\ln|u| + C = \frac{1}{2}\ln|x^2-5| + C$$

34.6 Show that

$$\int \frac{f'(x)}{f(x)}\,dx = \ln|f(x)| + C$$

Let $u = f(x)$. Then $du = f'(x)\,dx$, and

$$\int \frac{f'(x)}{f(x)}\,dx = \int \frac{1}{u}\,du = \ln|u| + C = \ln|f(x)| + C$$

34.7 (*Logarithmic Differentiation*) Find the derivative of

$$y = \frac{x^2\sqrt{8x+5}}{(2x-1)^3}$$

Instead of using the Product and Quotient Rules for differentiation, it is easier to find the logarithms of both sides,

$$\begin{aligned}\ln y &= \ln \frac{x^2\sqrt{8x+5}}{(2x-1)^3}\\ &= \ln (x^2\sqrt{8x+5}) - \ln ((2x-1)^3) \quad \text{[by Property 6]}\\ &= \ln x^2 + \ln ((8x+5)^{1/2}) - \ln ((2x-1)^3) \quad \text{[by Property 5]}\\ &= 2\ln x + \frac{1}{2}\ln (8x+5) - 3\ln (2x-1) \quad \text{[by Property 8]}\end{aligned}$$

and then to differentiate:

$$\begin{aligned}\frac{1}{y}D_x y &= 2\left(\frac{1}{x}\right) + \frac{1}{2}\left(\frac{1}{8x+5}\cdot 8\right) - 3\left(\frac{1}{2x-1}\cdot 2\right)\\ &= \frac{2}{x} + \frac{4}{8x+5} - \frac{6}{2x-1}\end{aligned}$$

Therefore,

$$\begin{aligned}D_x y &= y\left(\frac{2}{x} + \frac{4}{8x+5} - \frac{6}{2x-1}\right)\\ &= \frac{x^2\sqrt{8x+5}}{(2x-1)^3}\left(\frac{2}{x} + \frac{4}{8x+5} - \frac{6}{2x-1}\right)\end{aligned}$$

This procedure of first taking the logarithm and then differentiating is called *logarithmic differentiation.*

34.8 Show that

$$1 - \frac{1}{x} \le \ln x \le x - 1$$

for $x > 0$.

For $x \ge 1$, apply Problem 30.3(*c*) to

$$\ln x = \int_1^x \frac{1}{t}\,dt$$

noting that $1/t$ is a decreasing function on $[1, x]$. Thus,

$$\frac{1}{x}(x-1) \le \ln x \le \frac{1}{1}(x-1) \qquad \text{or} \qquad 1 - \frac{1}{x} \le \ln x \le x - 1$$

For $0 < x < 1$, do the same with

$$\ln x = -\int_x^1 \frac{1}{t}\,dt = \int_x^1 -\frac{1}{t}\,dt$$

where $-1/t$ is an increasing function on $[x, 1]$. Thus,

$$\left(-\frac{1}{x}\right)(1-x) \le \ln x \le \left(-\frac{1}{1}\right)(1-x) \qquad \text{or} \qquad 1 - \frac{1}{x} \le \ln x \le x - 1$$

Supplementary Problems

34.9 Find the derivatives of the following functions:

(*a*) $\ln(4x-1)$ (*b*) $(\ln x)^3$ (*c*) $\sqrt{\ln x}$ (*d*) $\ln(\ln x)$

(*e*) $x^2 \ln x$ (*f*) $\ln\left(\frac{x-1}{x+1}\right)$ (*g*) $\ln|5x-2|$

34.10 Find the following antiderivatives:

(*a*) $\int \frac{1}{3x}\,dx$ (*b*) $\int \frac{1}{7x-2}\,dx$ (*c*) $\int \frac{x^3}{x^4-1}\,dx$ (*d*) $\int \cot x\,dx$

(*e*) $\int \frac{dx}{x\ln x}$ (*f*) $\int \frac{\cos 2x}{1-\sin 2x}\,dx$ (*g*) $\int \frac{3x^5+2x^2-3}{x^3}\,dx$

(*h*) $\int \frac{\sec^2 x}{\tan x}\,dx$ (*i*) $\int \frac{1}{\sqrt{x}\,(1-\sqrt{x})}\,dx$ (*j*) $\int \frac{\ln x}{x}\,dx$

34.11 Use logarithmic differentiation to find y':

(*a*) $y = x^3\sqrt{4-x^2}$ (*b*) $y = \dfrac{(x-2)^4\sqrt[3]{x+5}}{\sqrt{x^2+4}}$

(*c*) $y = \dfrac{\sqrt{x^2-1}\sin x}{(2x+3)^4}$ (*d*) $y = \sqrt[3]{\dfrac{x+2}{x-2}}$

34.12 (*a*) Show that $\ln x < x$. (*Hint*: Use Problem 34.8.) (*b*) Show that

$$\frac{\ln x}{x} < \frac{2}{\sqrt{x}}$$

(*Hint*: Put $\sqrt{x}$ for x in (*a*).) (*c*) Prove:

$$\lim_{x\to+\infty} \frac{\ln x}{x} = 0$$

(*Hint*: Use (*b*).) (*d*) Prove:

$$\lim_{x\to 0^+} (x \ln x) = 0$$

(*Hint*: Let $x = 1/y$ in (*c*).) (*e*) Show that

$$\lim_{x\to+\infty} (x - \ln x) = +\infty$$

34.13 Calculate the following in terms of ln 2 and ln 5:

(*a*) $\ln 10$ (*b*) $\ln\frac{1}{2}$ (*c*) $\ln\frac{1}{5}$ (*d*) $\ln 25$

(*e*) $\ln\sqrt{2}$ (*f*) $\ln\sqrt[3]{5}$ (*g*) $\ln\frac{1}{20}$ (*h*) $\ln 2^{12}$

34.14 Find an equation of the tangent line to the curve $y = \ln x$ at the point $(1, 0)$.

34.15 Find the area of the region bounded by the curves $y = x^2$, $y = 1/x$, and $x = 1/2$.

34.16 Find the average value of $1/x$ on $[1, 4]$.

34.17 Find the volume of the solid obtained by revolving about the x-axis the region in the first quadrant under $y = x^{-1/2}$, between $x = \frac{1}{4}$ and $x = 1$.

34.18 Sketch the graphs of

$$(a)\quad y = \ln (x+1) \qquad (b)\quad y = \ln \frac{1}{x} \qquad (c)\quad y = x - \ln x \qquad (d)\quad y = \ln (\cos x)$$

34.19 An object moves along the x-axis with acceleration

$$a = t - 1 + \frac{6}{t}$$

(*a*) Find a formula for the velocity $v(t)$, if $v(1) = 1.5$. (*b*) What is the maximum value of v in the interval $[1, 9]$?

34.20 Use implicit differentiation to find y':

$$(a)\quad y^2 = \ln (x^2 + y^2) \qquad (b)\quad \ln xy + 2x - y = 1 \qquad (c)\quad \ln (x + y^2) = y^3$$

34.21 Find

$$\lim_{h \to 0} \left(\frac{1}{h} \ln \frac{3+h}{3} \right)$$

(*Hint*: Recall the definition of the derivative.)

34.22 Derive the formula $\int \csc x \, dx = \ln |\csc x - \cot x| + C$. (*Hint*: Similar to Problem 34.5(*b*).)

34.23 Use the Trapezoidal Rule (Problem 31.9(*a*)), with $n = 10$, to approximate

$$\ln 2 = \int_1^2 \frac{1}{x} \, dx$$

(You may wish to use a calculator to obtain the reciprocals of 1.1, 1.2, . . . , 1.9.)

Chapter 35

Exponential Functions

35.1 INTRODUCTION

Let a be any positive real number. We wish to define a function a^x which has the usual meaning when x is rational. For example, we want to obtain the usual results

$$4^3 = 4 \cdot 4 \cdot 4 = 64 \qquad 5^{-2} = \frac{1}{5^2} = \frac{1}{25} \qquad 8^{2/3} = (\sqrt[3]{8})^2 = 2^2 = 4$$

In addition, an expression like $5^{\sqrt{2}}$, which does not yet have any meaning, will be assigned a value by the new definition.

Definition: a^x is the unique positive real number such that

$$\ln a^x = x \ln a \tag{35.1}$$

To see that $y = a^x$ is a proper function, recall that the equation

$$\ln y = x \ln a$$

must have at least one positive solution for each real number x, because $\ln y$ takes on all real values on $0 < y < \infty$. Furthermore, the equation has at most one solution, in view of Section 34.2, Property 10.

35.2 PROPERTIES OF a^x

It is shown in Problem 35.4 that the function a^x possesses all the standard properties of (rational) powers:

(I) $a^0 = 1$.

(II) $a^1 = a$.

(III) $a^{u+v} = a^u a^v$.

(IV) $a^{u-v} = \dfrac{a^u}{a^v}$.

(V) $a^{-v} = \dfrac{1}{a^v}$.

(VI) $(ab)^x = a^x b^x$.

(VII) $\left(\dfrac{a}{b}\right)^x = \dfrac{a^x}{b^x}$.

(VIII) $(a^u)^v = a^{uv}$.

35.3 THE FUNCTION e^x

For a particular choice of the real number a, the function a^x becomes the *inverse* of the function $\ln x$.

NOMENCLATURE Two functions f and g are *inverses* of each other if f undoes the effect of g, and g undoes the effect of f. In terms of compositions (Section 15.1), this means:

$$f(g(x)) = x \qquad \text{and} \qquad g(f(x)) = x$$

Definition: Let e denote the unique real number such that

$$\ln e = 1 \tag{35.2}$$

Theorem 35.1: The functions e^x and $\ln x$ are inverses of each other:

$$\ln e^x = x \qquad e^{\ln x} = x$$

Indeed, with $a = e$ in (*35.1*),

$$\ln e^x = x \ln e = x \cdot 1 = x$$

and if in this relation x is replaced by $\ln x$,

$$\ln (e^{\ln x}) = \ln x \quad \Rightarrow \quad e^{\ln x} = x$$

(by Section 34.2, Property 10).

Theorem 35.1 shows that the natural logarithm, $\ln x$, is what one ordinarily calls in high school the "logarithm to the base e": "$e^{\ln x} = x$" means that $\ln x$ is the power to which e has to be raised to obtain x.

Theorem 35.2:

$$a^x = e^{x \ln a}$$

Thus, every exponential function a^x is definable in terms of the particular exponential function e^x, which for this reason is often referred to as *the* exponential function. To see why Theorem 35.2 is true, notice that, by Theorem 35.1,

$$\ln (e^{x \ln a}) = x \ln a$$

But $y = a^x$ is the ***unique*** solution of the equation $\ln y = x \ln a$; therefore,

$$e^{x \ln a} = a^x$$

In Problem 35.9, it is shown that e^x is differentiable; in fact, that

Theorem 35.3:

$$D_x(e^x) = e^x$$

Thus, e^x has the property of being its own derivative. Problem 35.27 shows how special this property is. In terms of antiderivatives,

$$\int e^x \, dx = e^x + C \tag{35.3}$$

Knowing the derivative of e^x, we have from Theorem 35.2:

$$\begin{aligned} D_x(a^x) = D_x(e^{x \ln a}) &= e^{x \ln a} D_x(x \ln a) \quad \text{[by the Chain Rule]} \\ &= e^{x \ln a} \ln a = a^x \ln a \end{aligned}$$

This proves:

Theorem 35.4:

$$D_x(a^x) = (\ln a) a^x$$

or, in terms of antiderivatives,

$$\int a^x \, dx = \frac{a^x}{\ln a} + C \qquad (a \neq 1) \tag{35.4}$$

We know that $D_x(x^r) = rx^{r-1}$ for any rational number r. Now this formula can be extended to arbitrary exponents. By Theorem 35.2, replacing a by x (and thereby forcing $x > 0$) and x by r, $x^r = e^{r \ln x}$. Hence,

$$\begin{aligned} D_x(x^r) = D_x(e^{r \ln x}) &= e^{r \ln x} D_x(r \ln x) \quad \text{[by the Chain Rule]} \\ &= e^{r \ln x} \left(r \frac{1}{x} \right) = x^r \left(r \frac{1}{x} \right) = rx^{r-1} \end{aligned}$$

Thus, we have the following:

Theorem 35.5: For any real number r and for all positive x,

$$D_x(x^r) = rx^{r-1}$$

Solved Problems

35.1 Evaluate (*a*) $e^{2\ln x}$, (*b*) $\ln e^2$, (*c*) $e^{(\ln u)-1}$, (*d*) 1^x.

(*a*)
$$e^{2\ln x} = (e^{\ln x})^2 \quad \text{[by Property (VIII)]}$$
$$= x^2 \quad \text{[by Theorem 35.1]}$$

(*b*)
$$\ln e^2 = 2 \quad \text{[by Theorem 35.1]}$$

(*c*)
$$e^{(\ln u)-1} = \frac{e^{\ln u}}{e^1} \quad \text{[by Property (IV)]}$$
$$= \frac{u}{e} \quad \text{[by Theorem 35.1 and Property (II)]}$$

(*d*) By definition,

$$\ln 1^x = x(\ln 1) = x(0) = 0 = \ln 1$$

Hence, $1^x = 1$.

35.2 Find the derivatives of

(*a*) $e^{\sqrt{x}}$ (*b*) 3^{2x} (*c*) xe^x (*d*) $3x^{\sqrt{2}}$

(*a*)
$$D_x(e^{\sqrt{x}}) = e^{\sqrt{x}}\,D_x(\sqrt{x}) \quad \text{[by the Chain Rule]}$$
$$= e^{\sqrt{x}}\,D_x(x^{1/2}) = e^{\sqrt{x}}\frac{1}{2}x^{-1/2} = \frac{e^{\sqrt{x}}}{2\sqrt{x}}$$

(*b*)
$$D_x(3^{2x}) = (\ln 3)3^{2x}\,D_x(2x) \quad \text{[Chain Rule and Theorem 35.4]}$$
$$= (\ln 3)3^{2x}(2) = (2\ln 3)3^{2x}$$

(*c*)
$$D_x(xe^x) = x\,D_x(e^x) + D_x(x)\,e^x \quad \text{[Product Rule]}$$
$$= xe^x + (1)e^x = e^x(x+1)$$

(*d*)
$$D_x(3x^{\sqrt{2}}) = 3\,D_x(x^{\sqrt{2}}) = 3(\sqrt{2}\,x^{\sqrt{2}-1}) = 3\sqrt{2}\,x^{\sqrt{2}-1}$$

35.3 Find the following antiderivatives (e^{x^2} stands for $e^{(x^2)}$):

(*a*) $\int 10^x\,dx$ (*b*) $\int xe^{x^2}\,dx$

(*a*) By Theorem 35.4,

$$\int 10^x\,dx = \frac{1}{\ln 10}10^x + C$$

(*b*) Let $u = x^2$. Then $du = 2x\,dx$, and

$$\int xe^{x^2}\,dx = \frac{1}{2}\int e^u\,du = \frac{1}{2}e^u + C = \frac{1}{2}e^{x^2} + C$$

35.4 Prove Properties (I)–(VIII) of a^x (Section 35.2).

By definition, a^x has a given value y if and only if $\ln y = x \ln a$.

(I) $a^0 = 1$.

$$\ln 1 = 0 = 0(\ln a)$$

(II) $a^1 = a$.

$$\ln a = 1(\ln a)$$

(III) $a^{u+v} = a^u a^v$.

$$\ln (a^u a^v) = \ln a^u + \ln a^v = u \ln a + v \ln a = (u+v) \ln a$$

(IV) $a^{u-v} = \dfrac{a^u}{a^v}$.

By (III), $a^{u-v}a^v = a^{(u-v)+v} = a^u$. Now divide by a^v.

(V) $a^{-v} = \dfrac{1}{a^v}$.

Let $u = 0$ in (IV), and use (I).

(VI) $(ab)^x = a^x b^x$.

$$\ln a^x b^x = \ln a^x + \ln b^x = x \ln a + x \ln b = x(\ln a + \ln b) = x \ln ab$$

(VII) $\left(\dfrac{a}{b}\right)^x = \dfrac{a^x}{b^x}$.

By (VI),

$$\left(\frac{a}{b}\right)^x b^x = \left(\frac{a}{b} b\right)^x = a^x$$

Now divide by b^x.

(VIII) $(a^u)^v = a^{uv}$.

$$\ln (a^u)^v = v \ln a^u = v(u \ln a) = uv \ln a$$

35.5 Show that

$$\int g'(x)\, e^{g(x)}\, dx = e^{g(x)} + C$$

By Theorem 35.3 and the Chain Rule,

$$\frac{d}{dx} e^{g(x)} = e^{g(x)}\, g'(x)$$

Thus, $e^{g(x)}$ is a particular antiderivative of $g'(x)\, e^{g(x)}$, and so $e^{g(x)} + C$ is the general antiderivative.

35.6 Use logarithmic differentiation to find $\dfrac{dy}{dx}$:

$$(a)\quad y = x^x \qquad (b)\quad y = \frac{\sqrt{x+1}\, e^{5x}}{2\sqrt{x}}$$

(*a*)

$$\ln y = \ln x^x = x \ln x$$

Now differentiate:

$$\frac{1}{y}\frac{dy}{dx} = x\, D_x(\ln x) + D_x(x)(\ln x)$$

$$= x\frac{1}{x} + (1)(\ln x) = 1 + \ln x$$

$$\frac{dy}{dx} = y(1 + \ln x) = x^x(1 + \ln x)$$

(b)
$$\ln y = \ln(\sqrt{x+1}\, e^{5x}) - \ln 2^{\sqrt{x}} = \ln \sqrt{x+1} + \ln e^{5x} - \sqrt{x}\ln 2$$
$$= \ln (x+1)^{1/2} + 5x - (\ln 2)\sqrt{x} = \frac{1}{2}\ln(x+1) + 5x - (\ln 2)x^{1/2}$$

Differentiate:

$$\frac{1}{y}\frac{dy}{dx} = \frac{1}{2}\frac{1}{x+1} + 5 - \frac{\ln 2}{2}x^{-1/2} = \frac{1}{2(x+1)} + 5 - \frac{\ln 2}{2\sqrt{x}}$$
$$\frac{dy}{dx} = y\left(\frac{1}{2(x+1)} + 5 - \frac{\ln 2}{2\sqrt{x}}\right) = \frac{\sqrt{x+1}\, e^{5x}}{2^{\sqrt{x}}}\left(\frac{1}{2(x+1)} + 5 - \frac{\ln 2}{2\sqrt{x}}\right)$$

35.7 Prove the following facts about e^x: (a) $e^x > 0$; (b) e^x is increasing; (c) $e^x > x$;

$$(d)\ \lim_{x\to+\infty} e^x = +\infty \qquad (e)\ \lim_{x\to-\infty} e^x = 0$$

(a) $a^x > 0$, by definition (Section 35.1).

(b) $D_x(e^x) = e^x > 0$, and a function with positive derivative is increasing. More generally, the definition of a^x, together with the fact that $\ln y$ is increasing, implies that a^x is increasing if $a > 1$.

(c) We know (Problem 34.8) that $\ln u < u$. Hence, $x = \ln e^x < e^x$.

(d) This is a direct consequence of (c).

(e) Let $u = -x$; then, by Property (V),

$$e^x = e^{-u} = \frac{1}{e^u}$$

As $x \to -\infty$, $u \to +\infty$, and the denominator on the right becomes arbitrarily large (by (d)). Hence, the fraction becomes arbitrarily small.

35.8 Sketch the graph of $y = e^x$, and show that it is the reflection in the diagonal line $y = x$ of the graph of $y = \ln x$.

From Problem 35.7 we know that e^x is positive and increasing, and that it approaches $+\infty$ on the right and approaches 0 on the left. Moreover, since $D_x^2(e^x) = e^x > 0$, the graph will be concave upward for all x. The graph is shown in Fig. 35-1(a).

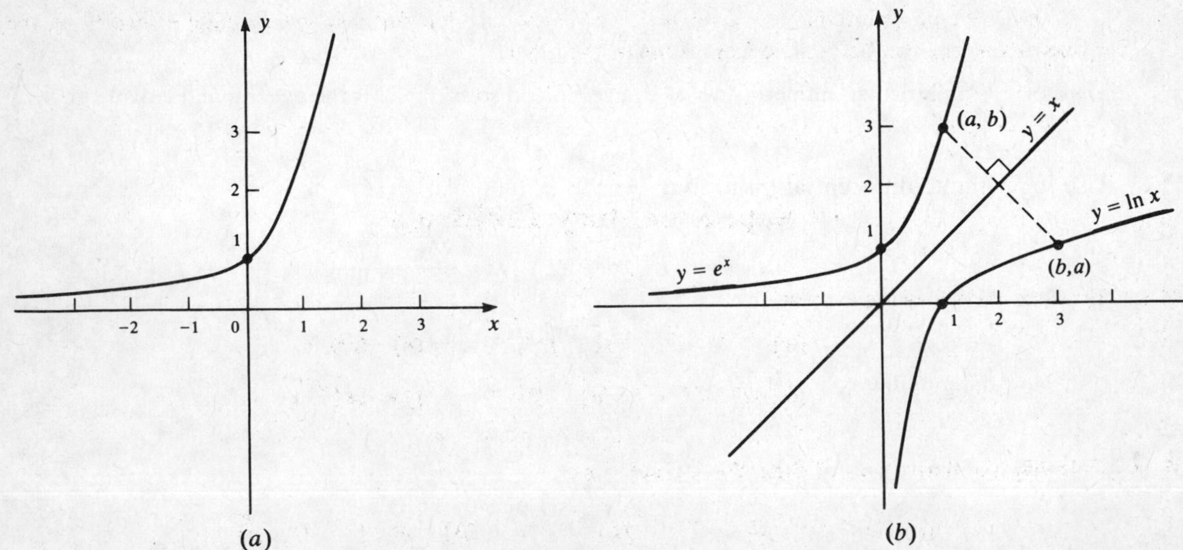

Fig. 35-1

By Theorem 35.1, $y = e^x$ is equivalent to $x = \ln y$. So, a point (a, b) is on the graph of $y = e^x$ if and only if (b, a) is on the graph of $y = \ln x$. But, the points (a, b) and (b, a) are symmetric with respect to the line $y = x$ (see Fig. 35-1(*b*)).

In general, the graphs of any pair of inverse functions are mirror images of each other in the 45° line.

35.9 Prove that e^x is differentiable.

The proof will consist in finding

$$\lim_{h\to 0} \frac{e^{x+h} - e^x}{h}$$

thereby exhibiting the derivative of e^x.

First, because

$$\frac{e^{x+h} - e^x}{h} = \frac{e^x e^h - e^x}{h} = e^x \frac{e^h - 1}{h}$$

it suffices to investigate

$$\lim_{h\to 0} \frac{e^h - 1}{h}$$

Let $k = e^h - 1$; then $e^h = 1 + k$, or $h = \ln(1+k)$ [by Theorem 35.1]. As $h \to 0$, $k \to 0$ [since $e^0 = 1$], and

$$\lim_{h\to 0} \frac{e^h - 1}{h} = \lim_{k\to 0} \frac{k}{\ln(1+k)}$$

$$= \lim_{k\to 0} \frac{1}{\dfrac{\ln(1+k) - \ln 1}{k}} \qquad \text{[since } \ln 1 = 0\text{]}$$

$$= \frac{1}{\lim\limits_{k\to 0} \dfrac{\ln(1+k) - \ln 1}{k}} \qquad \text{[by Section 8.2, Property VI]}$$

The denominator on the right is recognized as the derivative of $\ln x$ at $x = 1$, which we know to have the value $1/1 = 1$. Thus,

$$\lim_{h\to 0} \frac{e^h - 1}{h} = 1 \qquad \text{and} \qquad D_x(e^x) = e^x \cdot 1 = e^x$$

The above procedure is quite general: If two functions are mutually inverse (like e^x and $\ln x$), the properties of one can be derived from those of the other.

Supplementary Problems

35.10 Evaluate the following expressions:

(*a*) $e^{-\ln x}$ (*b*) $\ln e^{-x}$ (*c*) $(e^2)^{\ln x}$

(*d*) $(3e)^{\ln x}$ (*e*) $e^{\ln(x-1)}$ (*f*) $\ln\left(\dfrac{e^x}{x}\right)$

35.11 Calculate the derivatives of the following functions:

(*a*) e^{-x} (*b*) $e^{1/x}$ (*c*) $e^{\cos x}$ (*d*) $\tan e^x$ (*e*) $\dfrac{e^x}{x}$

(*f*) $e^x \ln x$ (*g*) x^{π} (*h*) π^x (*i*) $\ln e^{2x}$ (*j*) $e^x - e^{-x}$

35.12 Evaluate the following antiderivatives:

(a) $\int e^{3x}\,dx$ (b) $\int e^{-x}\,dx$ (c) $\int e^x\sqrt{e^x-2}\,dx$

(d) $\int e^{\cos x}\sin x\,dx$ (e) $\int 3^{2x}\,dx$ (f) $\int \sqrt{e^x}\,dx$

(g) $\int x^{\pi}\,dx$ (h) $\int e^x e^{2x}\,dx$ (i) $\int \frac{e^{2x}}{e^x+1}\,dx$

(j) $\int x^2 2^{x^3}\,dx$ (k) $\int x^3 e^{-x^4}\,dx$

35.13 Use implicit differentiation to find y':

(a) $e^y = y + \ln x$ (b) $\tan e^{y-x} = x^2$ (c) $e^{1/y} + e^y = 2x$

(d) $x^2 + e^{xy} + y^2 = 1$ (e) $\sin x = e^y$

35.14 Use logarithmic differentiation to find y':

(a) $y = 3^{\sin x}$ (b) $y = (\sqrt{2})^{e^x}$ (c) $y = x^{\ln x}$

(d) $y = (\ln x)^{\ln x}$ (e) $y^2 = (x+1)(x+2)$

35.15 Solve the following equations for x:

(a) $e^{3x} = 2$ (b) $\ln x^3 = -1$ (c) $e^x - 2e^{-x} = 1$ (d) $\ln(\ln x) = 1$ (e) $\ln(x-1) = 0$

35.16 Consider the region $\mathcal{R}$ under the curve $y = e^x$, above the x-axis, and between $x = 0$ and $x = 1$. Find (a) the area of $\mathcal{R}$, (b) the volume of the solid generated by revolving $\mathcal{R}$ about the x-axis.

35.17 Consider the region $\mathcal{R}$ bounded by the curve $y = e^{x/2}$, the y-axis, and the line $y = e$. Find (a) the area of $\mathcal{R}$, (b) the volume of the solid generated by revolving $\mathcal{R}$ about the x-axis.

35.18 Let $\mathcal{R}$ be the region bounded by the curve $y = e^{x^2}$, the x-axis, the y-axis, and the line $x = 1$. Find the volume of the solid generated by revolving $\mathcal{R}$ about the y-axis.

35.19 Find the absolute extrema of $y = e^{\sin x}$ on the interval $[-\pi, \pi]$. (*Hint*: e^u is an increasing function of u.)

35.20 If $y = e^{nx}$, where n is a positive integer, find the nth derivative $y^{(n)}$.

35.21 Let $y = 2e^{\sin x}$. (a) Find y' and y''. (b) Assume that x and y vary with time and that y increases at a constant rate of 4 units per second. How fast is x changing when $x = \pi$?

35.22 The acceleration of an object moving on the x-axis is $9e^{3t}$. (a) If the velocity at time $t = 0$ is 4 units per second, find a formula for the velocity $v(t)$. (b) How far does the object move while its velocity increases from 4 to 10 units per second? (c) If the object is at the origin when $t = 0$, find a formula for its position $x(t)$.

35.23 Find an equation of the tangent line to the curve $y = 2e^x$ at the point $(0, 2)$.

35.24 Sketch the graphs of the following functions, indicating relative extrema, inflection points, and asymptotes:

(a) $y = e^{-x^2}$ (b) $y = x\ln x$ (c) $y = \frac{\ln x}{x}$

(d) $y = e^{-x}$ (e) $y = (1-\ln x)^2$ (f) $y = \frac{1}{x} + \ln x$

(*Hint*: For (b) and (c), you will need the results of Problem 34.12(d) and (c).)

35.25 Sketch the graphs of $y = 2^x$ and $y = 2^{-x}$ (*Hint*: $a^x = e^{x \ln a}$.)

35.26 For $a > 0$ and $a \neq 1$, define

$$\log_a x \equiv \frac{\ln x}{\ln a}$$

This function is called the *logarithm of x to the base a* ($\log_{10} x$ is called the *common logarithm* of x). Prove the following properties:

(*a*) $D_x(\log_a x) = \dfrac{1}{x \ln a}$ (*b*) $a^{\log_a x} = x$

(*c*) $\log_a a^x = x$ (*d*) $\log_e x = \ln x$

(*e*) $\log_a uv = \log_a u + \log_a v$ (*f*) $\log_a \dfrac{u}{v} = \log_a u - \log_a v$

(*g*) $\log_a u^r = r \log_a u$ (*h*) $\ln x = \dfrac{\log_a x}{\log_a e}$

35.27 Show that the only functions $f(x)$ such that $f'(x) = f(x)$ are the functions Ce^x, where C is a constant. (*Hint*: Let

$$F(x) = \frac{f(x)}{e^x}$$

and find $F'(x)$.)

35.28 Find the absolute extrema of $f(x) = (\ln x)^2/x$ on $[1, e]$.

35.29 Prove:

$$e^x = \lim_{u \to +\infty} \left(1 + \frac{x}{u}\right)^u$$

(*Hint*: Let $y = \left(1 + \dfrac{x}{u}\right)^u$; then

$$\ln y = u \ln \left(1 + \frac{x}{u}\right) = u[\ln (u + x) - \ln u] = u \cdot x \cdot \frac{1}{u^*} \quad \text{[Theorem 17.2]}$$

where $u < u^* < u + x$. Show that

$$\lim_{u \to +\infty} \frac{u}{u^*} = 1$$

since $1 < u^*/u < 1 + (x/u)$.)

35.30 Show that, for any positive n,

$$\lim_{x \to +\infty} \frac{x^n}{e^x} = 0$$

(*Hint*:

$$\frac{x^n}{e^x} = \frac{e^{n \ln x}}{e^x} = \frac{1}{e^{x - n \ln x}} = \frac{1}{e^{x(1 - n \ln x/x)}}$$

Now apply Problems 34.12(*c*) and 35.7(*d*).)

Chapter 36

L'Hôpital's Rule. Exponential Growth and Decay

36.1 L'HÔPITAL'S RULE

The following theorem allows us to compute limits of the form

$$\lim \frac{f(x)}{g(x)}$$

in the "indeterminate case" when the numerator $f(x)$ and denominator $g(x)$ both approach 0 or both approach $\pm\infty$.

Theorem 36.1 (*L'Hôpital's Rule*): If $f(x)$ and $g(x)$ either both approach 0 or both approach $\pm\infty$, then

$$\lim \frac{f(x)}{g(x)} = \lim \frac{f'(x)}{g'(x)}$$

In L'Hôpital's Rule, the operation "lim" stands for any of

$$\lim_{x\to+\infty} \qquad \lim_{x\to-\infty} \qquad \lim_{x\to a} \qquad \lim_{x\to a^+} \qquad \lim_{x\to a^-}$$

The rule is proved in more advanced courses.

EXAMPLES Let us verify L'Hôpital's Rule for four limits found earlier by other means.

(*a*) See Problem 9.4(*a*).

$$\lim_{x\to-\infty} \frac{3x^2-5}{4x^2+5} = \lim_{x\to-\infty} \frac{6x}{8x} = \lim_{x\to-\infty} \frac{6}{8} = \frac{6}{8} = \frac{3}{4}$$

(*b*) See Problem 27.4.

$$\lim_{\theta\to 0} \frac{\sin\theta}{\theta} = \lim_{\theta\to 0} \frac{\cos\theta}{1} = \cos 0 = 1$$

(*c*) See Problem 34.12(*d*). To apply Theorem 36.1, we must rewrite the function as a fraction that becomes indeterminate:

$$\lim_{x\to 0^+} (x \ln x) = \lim_{x\to 0^+} \frac{\ln x}{x^{-1}} = \lim_{x\to 0^+} \frac{x^{-1}}{-x^{-2}} \quad [D_x(\ln x) = x^{-1}]$$

$$= \lim_{x\to 0^+} -x \quad \left[\text{since } \frac{x^{-1}}{x^{-2}} = x^{-1-(-2)} = x^1 = x\right]$$

$$= 0$$

(*d*) See Problem 35.30, where now n is supposed to be an integer. By L'Hôpital's Rule,

$$\lim_{x\to+\infty} \frac{x^n}{e^x} = \lim_{x\to+\infty} \frac{nx^{n-1}}{e^x}$$

As the right side is indeterminate, L'Hôpital's Rule can be applied again:

$$\lim_{x\to+\infty} \frac{nx^{n-1}}{e^x} = \lim_{x\to+\infty} \frac{n(n-1)x^{n-2}}{e^x}$$

Continuing in this fashion, we reach after n steps:

$$\lim_{x\to+\infty} \frac{n(n-1)\cdots(2)(1)}{e^x} = n!\lim_{x\to+\infty}\frac{1}{e^x} = n!(0) = 0$$

This example illustrates the importance of having the function expressed as a fraction "in the right way." Suppose we had chosen "the wrong way," and tried to evaluate the same limit as

$$\lim_{x\to+\infty} \frac{e^{-x}}{x^{-n}}$$

Then repeated application of L'Hôpital's Rule would have given

$$\lim \frac{e^{-x}}{x^{-n}} = \lim \frac{(-1)e^{-x}}{(-n)x^{-(n+1)}} = \lim \frac{(-1)^2e^{-x}}{(-1)^2(n)(n+1)x^{-(n+2)}} = \cdots$$

and we should never have arrived at a definite value.

Warning: When the conditions for L'Hôpital's Rule do not hold, use of the rule usually leads to false results.

EXAMPLE

$$\lim_{x\to2} \frac{x^2+1}{x^2-1} = \frac{2^2+1}{2^2-1} = \frac{5}{3}$$

If we used L'Hôpital's Rule, we would conclude mistakenly that

$$\lim_{x\to2} \frac{x^2+1}{x^2-1} = \lim_{x\to2} \frac{2x}{2x} = \lim_{x\to2} 1 = 1$$

36.2 EXPONENTIAL GROWTH AND DECAY

Example (*d*) above shows that e^x grows much faster than any power of x. There are many natural processes, like bacterial growth or radioactive decay, in which quantities increase or decrease at an "exponential rate."

Definition: Assume that a quantity y varies with time t. Then y is said to *grow or decay exponentially* if its instantaneous rate of change (Chapter 19) is proportional to its instantaneous value; i.e.,

$$\frac{dy}{dt} = Ky \tag{36.1}$$

where K is a constant.

Suppose that y satisfies (*36.1*). Let us make the change of variable $u = Kt$. Then, by the Chain Rule,

$$Ky = \frac{dy}{dt} = \frac{dy}{du}\frac{du}{dt} = \frac{dy}{du}K \qquad \text{or} \qquad \frac{dy}{du} = y$$

and so, by Problem 35.27,

$$y = Ce^u = Ce^{Kt} \tag{36.2}$$

where C is another constant. We can now see why the process y is called "exponential": if $K > 0$ (the *growth constant*), y increases exponentially with time; if $K < 0$ (the *decay constant*), y decreases exponentially with time.

Let y_0 be the value of y at $t = 0$. Substituting 0 for t in (*36.2*), we obtain

$$y_0 = Ce^0 = C(1) = C$$

so that (*36.2*) can be rewritten as

$$y = y_0e^{Kt} \tag{36.3}$$

EXAMPLES

(*a*) Assume that a culture consisting of 1000 bacteria has a growth constant of 2% per hour. Let us find a formula for the number y of bacteria present after t hours, and let us compute how long it will take until 100 000 bacteria are present in the culture.

To say that the culture has a growth constant of 2% per hour means that $K = 0.02$ in (*36.1*). Since $y_0 = 1000$, the desired formula for y is given by (*36.3*) as

$$y = 1000e^{0.02t}$$

Now set $y = 100\,000$ and solve for t:

$$\begin{aligned} 100\,000 &= 1000e^{0.02t} \\ 100 &= e^{0.02t} \\ \ln 100 &= \ln e^{0.02t} \\ 2 \ln 10 &= 0.02t \quad [\ln 10^2 = 2\ln 10;\ \ln e^u = u] \\ t &= 100 \ln 10 \end{aligned}$$

Appendix E gives the approximate value 2.3026 for ln 10. Thus,

$$t \approx 230.26 \text{ hours} \approx 9\tfrac{1}{2} \text{ days}$$

(*b*) If the decay constant of a given radioactive element is $K < 0$, compute the time T after which only half of any original quantity remains.

At $t = T$, (*36.3*) gives

$$\begin{aligned} \frac{1}{2} y_0 &= y_0 e^{KT} \\ \frac{1}{2} &= e^{KT} \\ \ln \frac{1}{2} &= KT \\ -\ln 2 &= KT \quad \left[\ln \frac{1}{x} = -\ln x\right] \\ -\frac{\ln 2}{K} &= T \end{aligned} \tag{36.4}$$

The number T is called the *half-life* of the given element. Knowing either the half-life or the decay constant, we can find the other from (*36.4*).

Solved Problems

36.1 Show that

$$\lim_{x\to+\infty} \frac{(\ln x)^n}{x} = 0$$

for any positive integer n.

The proof will be by mathematical induction (see Problem 12.2). The assertion is true for $n = 1$, by Problem 34.12(*c*). Assuming that it is true for $n = k$, we have, for $n = k + 1$,

$$\begin{aligned} \lim_{x\to+\infty} \frac{(\ln x)^{k+1}}{x} &= \lim_{x\to+\infty} \frac{(k+1)(\ln x)^k x^{-1}}{1} \quad \text{[by Theorem 36.1]} \\ &= (k+1) \lim_{x\to+\infty} \frac{(\ln x)^k}{x} = (k+1)(0) = 0 \end{aligned}$$

Thus, the assertion is also true for $n = k + 1$, and the proof is complete.

A direct proof is supplied in Problem 36.19.

36.2 Find

$$\lim_{x \to 0} \frac{e^x - 1}{\sin x}$$

The numerator and denominator are continuous functions, each 0 at $x = 0$. Therefore, L'Hôpital's Rule applies:

$$\lim_{x \to 0} \frac{e^x - 1}{\sin x} = \lim_{x \to 0} \frac{e^x}{\cos x} = \frac{e^0}{\cos 0} = \frac{1}{1} = 1$$

36.3 Find

$$\lim_{x \to +\infty} \left(x \sin \frac{\pi}{x} \right)$$

Since $\displaystyle\lim_{x \to +\infty} \frac{\pi}{x} = 0$ whence $\displaystyle\lim_{x \to +\infty} \sin \frac{\pi}{x} = \sin 0 = 0$

there is no obvious way of solving the problem. However, if we put the function in the form

$$\frac{\sin \pi x^{-1}}{x^{-1}}$$

then L'Hôpital's Rule is applicable.

$$\lim_{x \to +\infty} \frac{\sin \pi x^{-1}}{x^{-1}} = \lim_{x \to +\infty} \frac{(\cos \pi x^{-1})(-\pi x^{-2})}{-x^{-2}} = \pi \lim_{x \to +\infty} \cos \pi x^{-1}$$

$$= \pi \cos 0 = \pi(1) = \pi$$

36.4 Find $\displaystyle\lim_{x \to 0^+} x^x$.

We know that $x \ln x \to 0$ as $x \to 0^+$; hence,

$$\lim_{x \to 0^+} x^x = \lim_{x \to 0^+} e^{x \ln x} = e^0 = 1$$

36.5 Sketch the graph of $y = xe^x$.

By the Product Rule, $y' = xe^x + e^x = e^x(x+1)$. Since e^x is always positive, the only critical number occurs when $x + 1 = 0$; that is, when $x = -1$. When $x = -1$,

$$y = (-1)e^{-1} = -\frac{1}{e}$$

Again by the Product Rule, $y'' = e^x(1) + e^x(x+1) = e^x(x+2)$. When $x = -1$,

$$y'' = e^{-1}((-1)+2) = \frac{1}{e} > 0$$

So, by the Second-Derivative Test, there is a relative minimum at the point $P(-1, -1/e)$. From the expression for y'', the curve is concave upward (that is, $y'' > 0$) when $x > -2$, and concave downward ($y'' < 0$) when $x < -2$. Thus, the point $I(-2, -2/e^2)$ is an inflection point.

It is clear that the curve rises without bound as $x \to +\infty$. To see what happens as $x \to -\infty$, let $u = -x$; then,

$$\lim_{x \to -\infty} xe^x = \lim_{u \to +\infty} -ue^{-u} = -\lim_{u \to +\infty} \frac{u}{e^u} = 0$$

by Problem 35.30. Hence, the negative x-axis is an asymptote. This enables us to complete the sketch of the graph in Fig. 36-1.

36.6 If $y(t)$ defines an exponential growth or decay process, find a formula for the average value of y over the time interval $[0, \tau]$.

By definition (see Section 31.2), the desired average value is given by

$$y_{\text{avg}} = \frac{1}{\tau} \int_0^\tau y\,dt = \frac{1}{K\tau} \int_0^\tau Ky\,dt$$

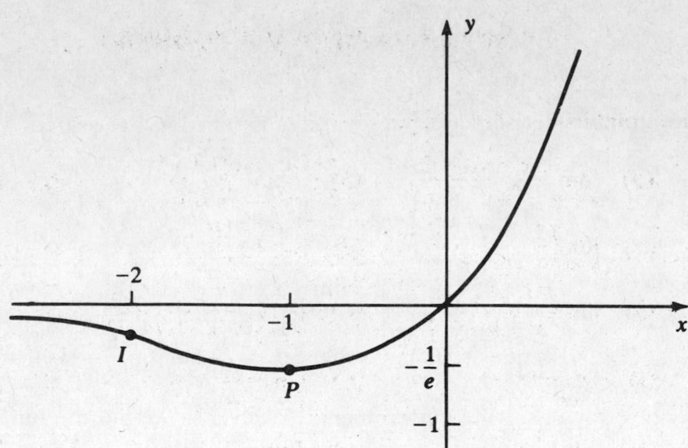

Fig. 36-1

But (*36.1*) states that an antiderivative of Ky is y itself. Hence, by the Fundamental Theorem of Calculus,

$$y_{\text{avg}} = \frac{1}{K\tau} y \Big]_0^\tau = \frac{y(\tau) - y_0}{K\tau}$$

Note that this relation was obtained without reference to the explicit form of $y(t)$. Rewritten as

$$y_{\text{avg}} = \frac{\Delta y}{K\,\Delta t} \qquad \text{or} \qquad \Delta y = K y_{\text{avg}}\,\Delta t$$

it provides a useful description of exponential processes: *the change in the quantity over any time interval is proportional to the size of the interval and to the average value of the quantity over that interval.*

36.7 If the bacteria in a culture grow exponentially and if their number y doubles in one hour, how long will it take before ten times the original number is present?

We know that $y = y_0 e^{Kt}$. Since the number after one hour is $2y_0$,

$$2y_0 = y_0 e^{K(1)} = y_0 e^K \qquad \text{or} \qquad 2 = e^K \qquad \text{or} \qquad K = \ln 2$$

So, $y = y_0 e^{(\ln 2)t}$; and when y is $10y_0$,

$$\begin{aligned} 10y_0 &= y_0 e^{(\ln 2)t} \\ 10 &= e^{(\ln 2)t} \\ \ln 10 &= (\ln 2)t \\ t &= \frac{\ln 10}{\ln 2} \approx \frac{2.3026}{0.6931} \approx 3.32 \end{aligned}$$

Thus, it takes a little less than $3\frac{1}{3}$ hours for the number of bacteria to increase tenfold.

36.8 Given that the half-life T of radium is 1690 years, how much will remain of 1 gram of radium after 10 000 years?

$$K = -\frac{\ln 2}{T} = -\frac{\ln 2}{1690}$$

by (*36.4*), and the quantity y of radium is given by

$$y = y_0 e^{Kt} = (1\text{ g})e^{-(\ln 2)t/1690}$$

After 10 000 years,

$$y = (1\text{ g})e^{-(\ln 2)(10\,000)/1690} \approx e^{-6931/1690} \approx e^{-4.1} \approx 0.0166\text{ g}$$

where Appendix F was used to evaluate $e^{-4.1}$. Hence, about 16.6 milligrams of radium is left after 10 000 years.

Supplementary Problems

36.9 Find the indicated limits.

(a) $\lim_{x\to+\infty} \frac{5x^3-4x+3}{2x^2-1}$ (b) $\lim_{x\to+\infty} \frac{\ln(1+e^x)}{1+x}$ (c) $\lim_{x\to 0} \frac{1-\cos x}{x}$

(d) $\lim_{x\to 0} \left(\frac{1}{x}-\frac{1}{\sin x}\right)$ (e) $\lim_{x\to 0} \frac{1-e^x}{x}$ (f) $\lim_{x\to+\infty} \frac{x^2}{(\ln x)^3}$

(g) $\lim_{x\to 1} \frac{x^3-x^2+x-1}{x+\ln x-1}$ (h) $\lim_{x\to 0} \left(\frac{1}{\ln(x+1)}-\frac{1}{x}\right)$ (i) $\lim_{x\to 0} \frac{\tan x}{x}$

(j) $\lim_{x\to+\infty} x^{1/x}$ (k) $\lim_{x\to 0} \frac{3^x-2^x}{x}$ (l) $\lim_{x\to 0} \frac{e^x}{x^2}$

(m) $\lim_{x\to\pi/2} \frac{1-\sin x}{x-(\pi/2)}$ (n) $\lim_{x\to 0^+} x^{\sin x}$ (o) $\lim_{x\to 1} \frac{x^3-1}{x+1}$

(p) $\lim_{x\to+\infty} \left(1+\frac{3}{x}\right)^x$ (q) $\lim_{x\to 0^+} \frac{\tan x}{x^2}$ (r) $\lim_{x\to 0^+} \frac{\sin x}{\sqrt{x}}$

(s) $\lim_{x\to 1} \frac{\ln x}{\tan \pi x}$ (t) $\lim_{x\to 0} \frac{\sin 3x}{\sin 7x}$ (u) $\lim_{x\to 0} \frac{e^{3x}-1}{\tan x}$

(v) $\lim_{x\to 0} \frac{1-\cos^2 2x}{x^2}$ (w) $\lim_{x\to 0} \frac{\tan x-\sin x}{x^3}$

36.10 Sketch the graphs of the following functions, indicating relative extrema, concavity, inflection points, and asymptotes:

(a) $y=\frac{x}{e^x}$ (b) $y=x^2e^{-x}$ (c) $y=x^3e^{-x}$ (d) $y=x^2\ln x$ (e) $y=e^{-x}\sin x$

36.11 A bacteria culture grows exponentially so that the initial number has doubled in three hours. How many times the initial number will be present after nine hours?

36.12 Under *continuous compounding of interest,* an invested amount will grow exponentially. What is the initial sum that will be multiplied by five in 8 years and will amount to \$10 000 after 24 years?

36.13 The half-life of radium is 1690 years. If 10% of an original quantity of radium remains, how long ago was the radium created?

36.14 If radioactive carbon-14 has a half-life of 5750 years, what will remain of one gram after 3000 years?

36.15 If 20% of a radioactive element disappears in one year, compute its half-life.

36.16 A quantity y is said to *grow exponentially at the rate of r percent per year* if

$$y=y_0\left(1+\frac{r}{100}\right)^t$$

with t in years. Find the exponential growth constant K for such a quantity.

36.17 Fruit flies are being bred in an enclosure that can hold a maximum of 640 flies. If the flies grow exponentially with growth constant $K=0.05$ per day, how long will it take an initial population of 20 to fill the enclosure? (Approximate $\ln 2$ by 0.6931.)

36.18 A certain chemical decomposes exponentially. Assume that 200 grams becomes 50 grams in one hour. How much will remain after three hours?

36.19 Prove:

$$\lim_{x\to+\infty} \frac{(\ln x)^n}{x}=0$$

by making the substitution $u=\ln x$ and utilizing Problem 35.30.

Chapter 37

Inverse Trigonometric Functions

37.1 ONE-ONE FUNCTIONS

In Section 35.3 we introduced the notion of the *inverse* of a function, and showed that the inverse of $\ln x$ is e^x, and vice versa. Not all functions, however, have inverses.

EXAMPLES (*a*) Consider the function f such that $f(x) = x^2$ for all x. Then $f(1) = 1$ and $f(-1) = 1$. If there were an inverse g of f, then $g(f(x)) = x$. Therefore, $g(1) = g(f(1)) = 1$ and $g(1) = g(f(-1)) = -1$, implying that $1 = -1$, which is impossible. (*b*) Let f be any periodic function:

$$f(x + p) = f(x)$$

for all x (see Section 26.2). The argument of Example (*a*), for two points x_0 and $x_0 + p$, shows that f cannot have an inverse. Now, all the trigonometric functions are periodic (with either $p = 2\pi$ or $p = \pi$). Hence, the trigonometric functions do not have inverses!

The functions that have inverses turn out to be the *one-one* functions.

Definition: A function f is one-one if, whenever $u \neq v$,

$$f(u) \neq f(v)$$

Thus, a one-one function takes different numbers into different numbers. A function is one-one if and only if its graph intersects any horizontal line in *at most one* point. Figure 37-1(*a*) is the graph of a one-one function; Fig. 37-1(*b*) graphs a function that is not one-one, because $f(u) = f(v) = c$.

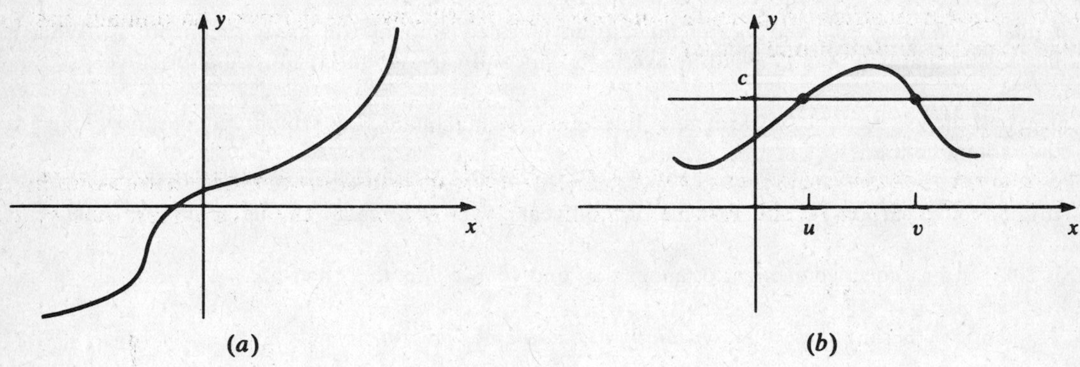

Fig. 37-1

NOTATION The inverse of a one-one function f will be denoted f^{-1} (instead of g):

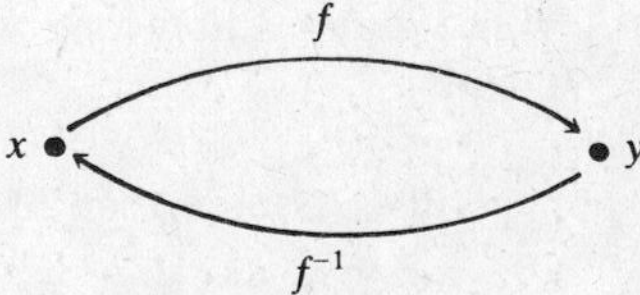

Warning: Do not confuse the inverse f^{-1} with the reciprocal $1/f$.

If a one-one function f is defined by means of a formula, $y = f(x)$, we can *sometimes* solve this equation for x in terms of y. This solution constitutes the formula for the inverse function, $x = f^{-1}(y)$.

EXAMPLES

(*a*) Let $f(x) = 3x + 1$ (a one-one function). Let y stand for $f(x)$; then $y = 3x + 1$. Solve this equation for x in terms of y:

$$y - 1 = 3x$$
$$\frac{y-1}{3} = x$$

Therefore, the inverse f^{-1} is given by the formula

$$f^{-1}(y) = \frac{y-1}{3}$$

(*b*) Consider the one-one function $f(x) = 2e^x - 5$.

$$y = 2e^x - 5$$
$$y + 5 = 2e^x$$
$$\frac{y+5}{2} = e^x$$
$$\ln \frac{y+5}{2} = \ln e^x = x$$

Thus $$f^{-1}(y) = \ln \frac{y+5}{2}$$

(*c*) Let $f(x) = x^5 + x$. Since $f'(x) = 5x^4 + 1 > 0$, f is an increasing function (Theorem 17.3), and therefore one-one. But if we write $y = x^5 + x$, we have no obvious way of solving the equation for x in terms of y.

37.2 INVERSES OF RESTRICTED TRIGONOMETRIC FUNCTIONS

For a periodic function to become one-one—and so to have an inverse—its domain has to be restricted to some subset of one period.

Inverse Sine

The domain of $f(x) = \sin x$ is restricted to $[-\pi/2, \pi/2]$, on which the function is one-one (in fact, increasing; see Fig. 37-2(*a*)). The inverse function, $f^{-1}(x) = \sin^{-1} x$, is called the *inverse sine* of x (or,

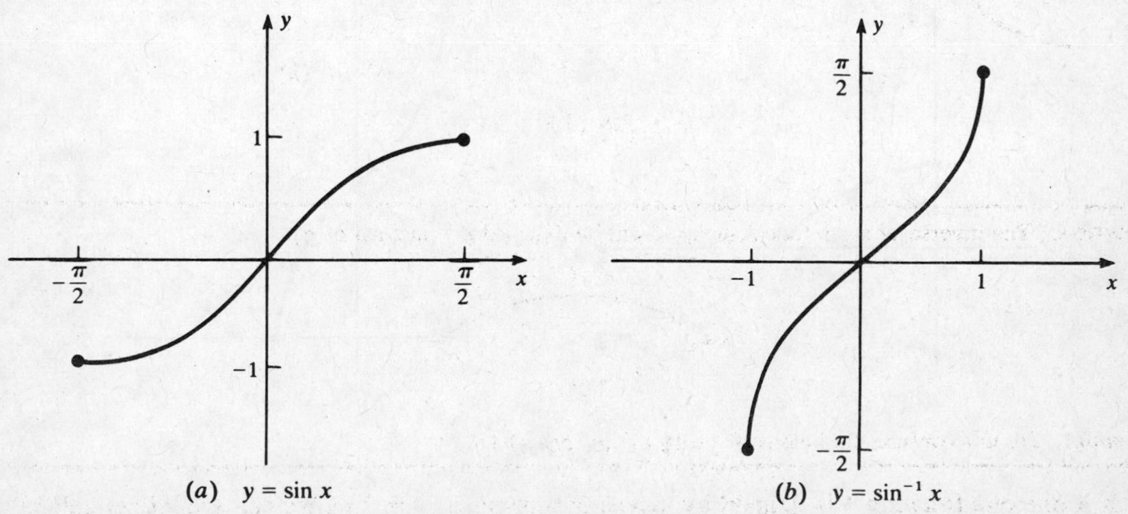

Fig. 37-2

sometimes, the *arc sine* of x, written arc sin x or arcsin x). Its domain is $[-1, 1]$. Thus:

$$y = \sin x \quad \left(-\frac{\pi}{2} \le x \le \frac{\pi}{2}\right) \quad \Leftrightarrow \quad x = \sin^{-1} y \quad (-1 \le y \le 1)$$

The graph of $\sin^{-1} x$ is given in Fig. 37-2(*b*). By a general theorem, the inverse of an increasing function is itself an increasing function. You will find it helpful to think of $\sin^{-1} x$ as the angle (between $-\pi/2$ and $\pi/2$) whose sine is x.

EXAMPLES

(*a*) $\sin^{-1} 1 = \frac{\pi}{2}$ (*b*) $\sin^{-1} \frac{\sqrt{3}}{2} = \frac{\pi}{3}$ (*c*) $\sin^{-1} \frac{\sqrt{2}}{2} = \frac{\pi}{4}$ (*d*) $\sin^{-1} \frac{1}{2} = \frac{\pi}{6}$

(*e*) $\sin^{-1} 0 = 0$ (*f*) $\sin^{-1} -\frac{1}{2} = -\frac{\pi}{6}$ (*g*) $\sin^{-1} -1 = -\frac{\pi}{2}$

The derivative of $\sin^{-1} x$ may be found by the general method of Problem 35.9. In this case, a shortcut evaluation is possible, using implicit differentiation. Let $y = \sin^{-1} x$; then

$$\sin y = x$$
$$\frac{d}{dx}(\sin y) = \frac{d}{dx}(x)$$
$$(\cos y)\frac{dy}{dx} = 1 \quad \text{[by the Chain Rule]}$$
$$\frac{dy}{dx} = \frac{1}{\cos y} = \frac{1}{+\sqrt{1-\sin^2 y}} = \frac{1}{+\sqrt{1-x^2}}$$

where the positive square root is implied because $\cos y > 0$ for $-\pi/2 < y < \pi/2$. Thus we have found:

$$D_x(\sin^{-1} x) = \frac{1}{\sqrt{1-x^2}} \quad \text{or} \quad \int \frac{1}{\sqrt{1-x^2}}\,dx = \sin^{-1} x + C \tag{37.1}$$

The importance of the inverse sine function in calculus lies mainly in the integration formula (*37.1*).

The procedure for the other trigonometric functions is exactly analogous to that for the sine function. In every case, a restriction of the domain is chosen that will lead to a simple formula for the derivative of the inverse function.

Inverse Cosine (Arc Cosine)

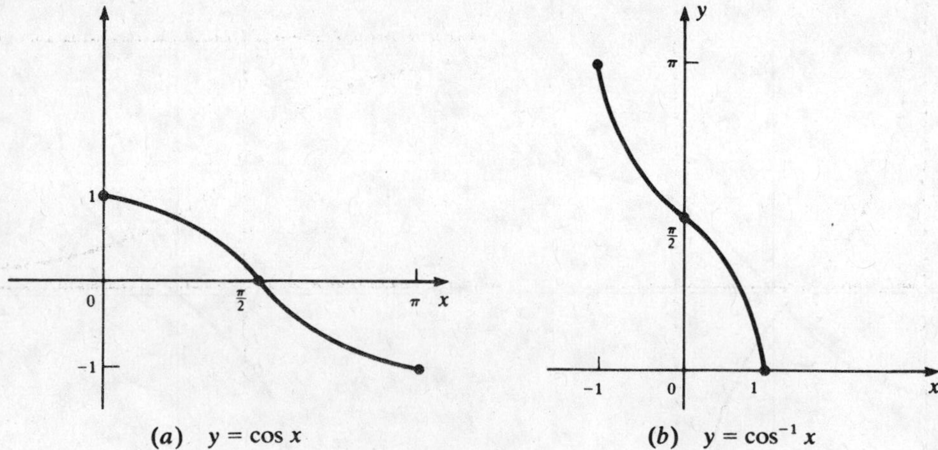

Fig. 37-3

$$y = \cos x \quad (0 \le x \le \pi) \quad \Leftrightarrow \quad x = \cos^{-1} y \quad (-1 \le y \le 1)$$

Interpret $\cos^{-1} x$ as the angle (between 0 and π) whose cosine is x.

$$D_x(\cos^{-1} x) = -\frac{1}{\sqrt{1-x^2}} \qquad \text{or} \qquad \int \frac{1}{\sqrt{1-x^2}}\,dx = -\cos^{-1} x + C \qquad (37.2)$$

Observe that $-\cos^{-1} x$ and $\sin^{-1} x$ have the same derivative; see Problem 37.17.

Inverse Tangent (Arc Tangent)

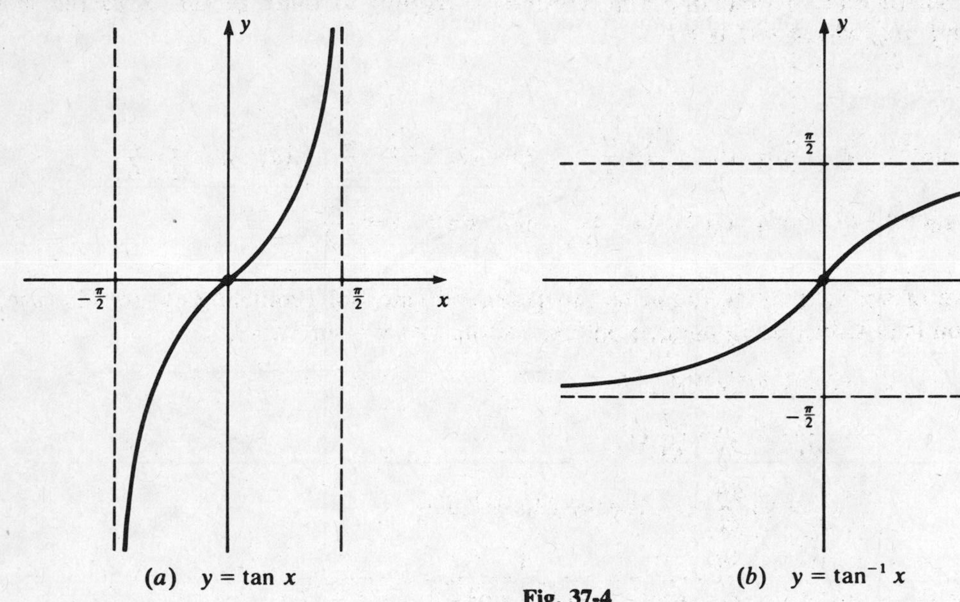

(a) $y = \tan x$ (b) $y = \tan^{-1} x$

Fig. 37-4

$$y = \tan x \quad \left(-\frac{\pi}{2} < x < \frac{\pi}{2}\right) \quad \Leftrightarrow \quad x = \tan^{-1} y \quad (\text{all } y)$$

Interpret $\tan^{-1} x$ as the angle (between $-\pi/2$ and $\pi/2$) whose tangent is x.

$$D_x(\tan^{-1} x) = \frac{1}{1+x^2} \qquad \text{or} \qquad \int \frac{1}{1+x^2}\,dx = \tan^{-1} x + C \qquad (37.3)$$

Inverse Cotangent (Arc Cotangent)

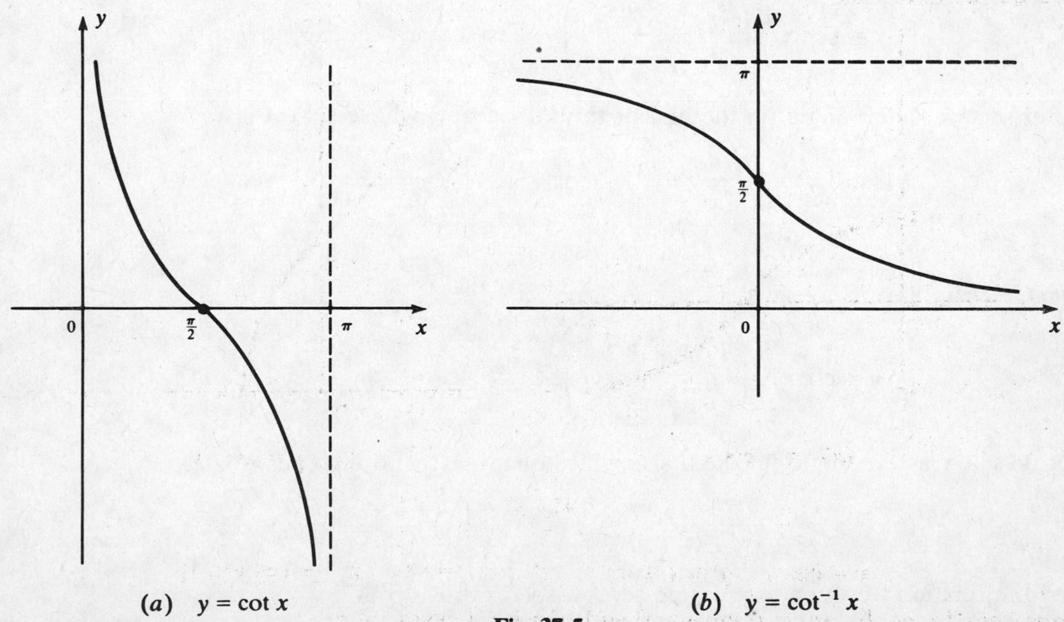

(a) $y = \cot x$ (b) $y = \cot^{-1} x$

Fig. 37-5

$$y = \cot x \quad (0 < x < \pi) \qquad \Leftrightarrow \qquad x = \cot^{-1} y \quad (\text{all } y)$$

Interpret $\cot^{-1} x$ as the angle (between 0 and π) whose cotangent is x.

$$D_x(\cot^{-1} x) = -\frac{1}{1+x^2} \qquad \text{or} \qquad \int \frac{1}{1+x^2}\,dx = -\cot^{-1} x + C \tag{37.4}$$

For the relation between $\cot^{-1} x$ and $\tan^{-1} x$, see Problem 37.17.

Inverse Secant (Arc Secant)

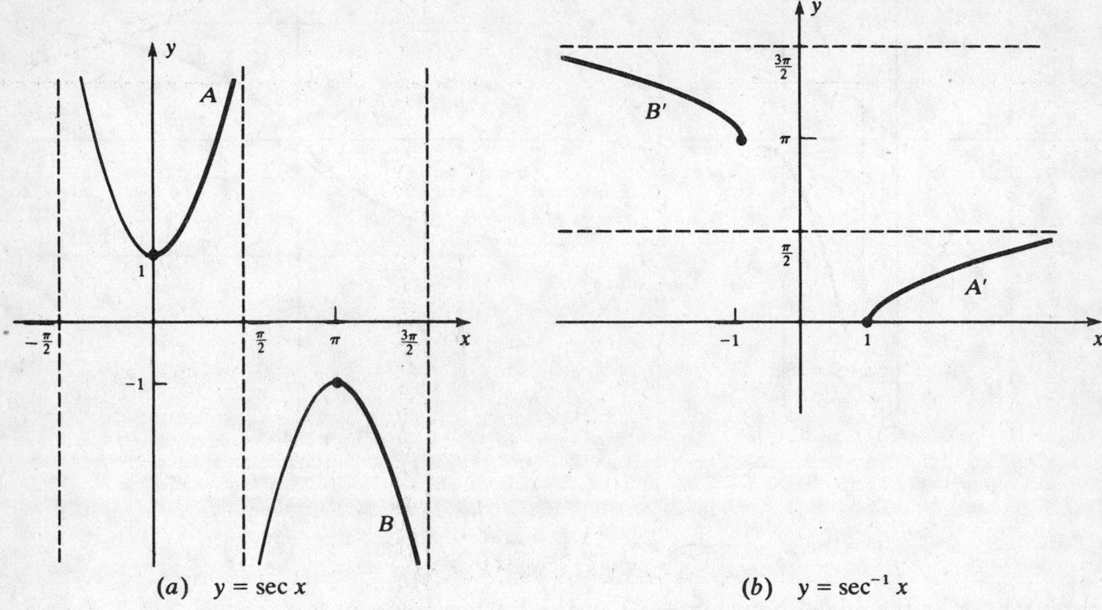

(*a*) $y = \sec x$ (*b*) $y = \sec^{-1} x$

Fig. 37-6

The secant function is restricted to two separate subintervals of the fundamental period $(-\pi/2, 3\pi/2)$:

$$y = \sec x \quad \begin{pmatrix} 0 \le x < \pi/2 \\ \text{or} \\ \pi \le x < 3\pi/2 \end{pmatrix} \qquad \Leftrightarrow \qquad x = \sec^{-1} y \quad (|y| \ge 1)$$

Interpret $\sec^{-1} x$ as the angle (in the first or third quadrant) whose secant is x.

$$D_x(\sec^{-1} x) = \frac{1}{x\sqrt{x^2-1}} \qquad \text{or} \qquad \int \frac{1}{x\sqrt{x^2-1}}\,dx = \sec^{-1} x + C \tag{37.5}$$

Inverse Cosecant (Arc Cosecant)

$$y = \csc x \quad \begin{pmatrix} 0 < x \le \pi/2 \\ \text{or} \\ \pi < x \le 3\pi/2 \end{pmatrix} \qquad \Leftrightarrow \qquad x = \csc^{-1} y \quad (|y| \ge 1)$$

Interpret $\csc^{-1} x$ as the angle (in the first or third quadrant) whose cosecant is x.

$$D_x(\csc^{-1} x) = -\frac{1}{x\sqrt{x^2-1}} \qquad \text{or} \qquad \int \frac{1}{x\sqrt{x^2-1}}\,dx = -\csc^{-1} x + C \tag{37.6}$$

For the relation between $\csc^{-1} x$ and $\sec^{-1} x$, see Problem 37.17.

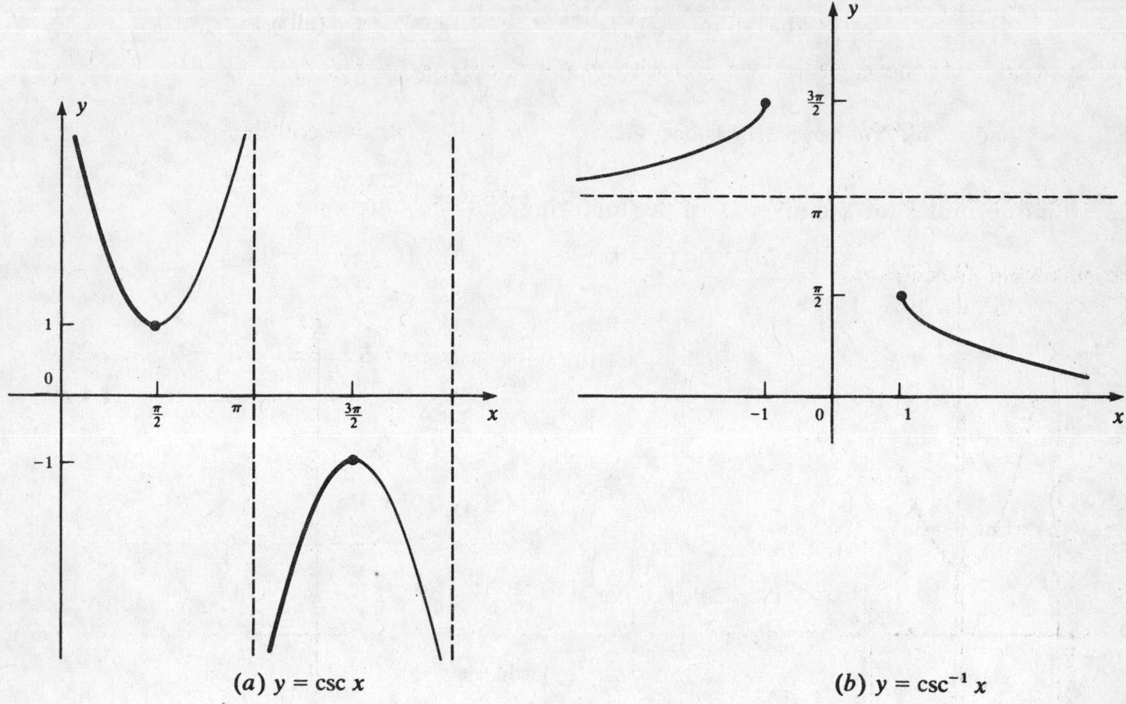

(a) $y = \csc x$

(b) $y = \csc^{-1} x$

Fig. 37-7

Solved Problems

37.1 A one-one function has an inverse; prove, conversely, that a function with an inverse is one-one.

If the domain of f consists of a single point, there is nothing to prove. Let f have the inverse g, but suppose, contrary to hypothesis, that the domain of f contains two distinct numbers, u and v, such that $f(u) = f(v)$. Then,

$$u = g(f(u)) = g(f(v)) = v$$

a contradiction. We must then admit that $f(u) \neq f(v)$; i.e., that f is one-one.

37.2 Determine whether each of the following functions f is one-one. If it is one-one, find a formula for its inverse f^{-1}.

(a) $f(x) = x^2 + 6x - 7$ (b) $f(x) = \ln(x^2 + 1)$ (c) $f(x) = 5x^3 - 2$

(a)
$$f(x) = x^2 + 6x - 7 = (x + 7)(x - 1)$$

Because $f(-7) = f(1)\ (= 0)$, f is not one-one.

(b) Obviously, $f(+1) = f(-1)$; f is not one-one.

(c) If $y = 5x^3 - 2$,

$$y + 2 = 5x^3$$

$$\frac{y+2}{5} = x^3$$

$$\sqrt[3]{\frac{y+2}{5}} = x$$

Thus, $f(x) = 5x^3 - 2$ has an inverse:

$$f^{-1}(y) = \sqrt[3]{\frac{y+2}{5}}$$

and so, by Problem 37.1, f is one-one.

37.3 Find formulas for the inverses of the following one-one functions:

$$(a)\quad f(x) = 10x - 4 \qquad (b)\quad f(x) = 3e^x + 1$$

(a) Let $y = 10x - 4$. Then,

$$y + 4 = 10x \qquad \text{or} \qquad \frac{y+4}{10} = x$$

Hence $$f^{-1}(y) = \frac{y+4}{10}$$

(b) Let $y = 3e^x + 1$. Then,

$$y - 1 = 3e^x \qquad \text{or} \qquad \frac{y-1}{3} = e^x \qquad \text{or} \qquad \ln\frac{y-1}{3} = x$$

Hence $$f^{-1}(y) = \ln\frac{y-1}{3}$$

37.4 Find the following values:

$$(a)\quad \sin^{-1}\left(-\frac{1}{2}\right) \qquad (b)\quad \cos^{-1}\left(-\frac{1}{2}\right) \qquad (c)\quad \tan^{-1}(-1)$$

(a) $$\sin^{-1}\left(-\frac{1}{2}\right) = \text{the angle in } \left[-\frac{\pi}{2}, \frac{\pi}{2}\right] \text{ whose sine is } -\frac{1}{2} = -\frac{\pi}{6}$$

(b) $$\cos^{-1}\left(-\frac{1}{2}\right) = \text{the angle in } [0, \pi] \text{ whose cosine is } -\frac{1}{2} = \pi - \frac{\pi}{3} = \frac{2\pi}{3}$$

(c) $$\tan^{-1}(-1) = \text{the angle in } \left(-\frac{\pi}{2}, \frac{\pi}{2}\right) \text{ whose tangent is } -1 = -\frac{\pi}{4}$$

37.5 Evaluate (a) $\cot^{-1}\sqrt{3}$, (b) $\sec^{-1} 2$, (c) $\csc^{-1}(2/\sqrt{3})$.

(a) $\cot^{-1}\sqrt{3}$ is the angle θ between 0 and π such that $\cot\theta = \sqrt{3}$. This angle is $\pi/6$ (see Fig. 37-8). Note that an inverse trigonometric function must be represented by an angle *in radians*.

(b) The angle θ whose secant is +2 must lie in the first quadrant (the secant function is negative in the third quadrant). From Fig. 37-8, $\theta = \pi/3$.

(c) The angle θ whose cosecant is $+2/\sqrt{3}$ must lie in the first quadrant (the cosecant function is negative in the third quadrant). From Fig. 37-8, $\theta = \pi/3$.

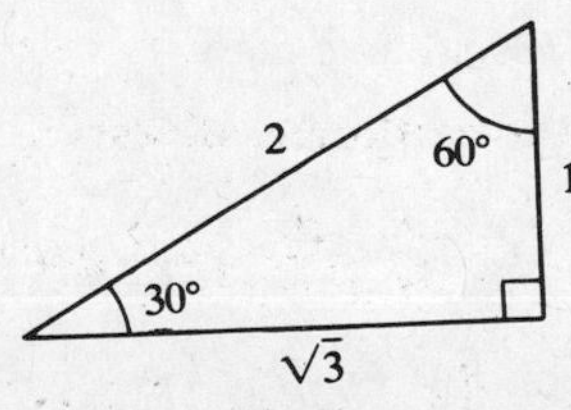

Fig. 37-8

37.6 Find the derivatives of

(*a*) $\sin^{-1} 2x$ (*b*) $\cos^{-1} x^2$ (*c*) $\tan^{-1}\frac{x}{2}$ (*d*) $\sec^{-1} x^3$

(*a*)
$$D_x(\sin^{-1} 2x) = \frac{1}{\sqrt{1-(2x)^2}} D_x(2x) \quad \text{[Chain Rule]}$$
$$= \frac{1}{\sqrt{1-4x^2}}(2) = \frac{2}{\sqrt{1-4x^2}}$$

(*b*)
$$D_x(\cos^{-1} x^2) = \frac{-1}{\sqrt{1-(x^2)^2}} D_x(x^2) \quad \text{[Chain Rule]}$$
$$= \frac{-1}{\sqrt{1-x^4}}(2x) = \frac{-2x}{\sqrt{1-x^4}}$$

(*c*)
$$D_x\left(\tan^{-1}\frac{x}{2}\right) = \frac{1}{1+\left(\frac{x}{2}\right)^2} D_x\left(\frac{x}{2}\right) \quad \text{[Chain Rule]}$$
$$= \frac{1}{1+\frac{x^2}{4}}\left(\frac{1}{2}\right) = \left(\frac{4}{4}\,\frac{1}{1+\frac{x^2}{4}}\right)\left(\frac{1}{2}\right)$$
$$= \left(\frac{4}{4+x^2}\right)\left(\frac{1}{2}\right) = \frac{2}{4+x^2}$$

(*d*)
$$D_x(\sec^{-1} x^3) = \frac{1}{x^3\sqrt{(x^3)^2-1}} D_x(x^3) \quad \text{[Chain Rule]}$$
$$= \frac{1}{x^3\sqrt{x^6-1}}(3x^2) = \frac{3}{x\sqrt{x^6-1}}$$

37.7 Assuming that $\sec^{-1} x$ is differentiable, verify that

$$D_x(\sec^{-1} x) = \frac{1}{x\sqrt{x^2-1}}$$

Let $y = \sec^{-1} x$; then

$$\sec y = x$$
$$D_x(\sec y) = D_x(x)$$
$$(\sec y \tan y)\frac{dy}{dx} = 1 \quad \text{[Chain Rule]}$$
$$\frac{dy}{dx} = \frac{1}{\sec y \tan y} = \frac{1}{x \tan y}$$

But the identity $1 + \tan^2 y = \sec^2 y$ gives:

$$1 + \tan^2 y = x^2 \qquad \text{or} \qquad \tan^2 y = x^2 - 1 \qquad \text{or} \qquad \tan y = \pm\sqrt{x^2-1}$$

By definition of $\sec^{-1} x$, angle y lies in the first or third quadrant, where the tangent is positive. Hence,

$$\tan y = +\sqrt{x^2-1} \qquad \text{and} \qquad \frac{dy}{dx} = \frac{1}{x\sqrt{x^2-1}}$$

37.8 Prove the following formulas for antiderivatives:

(*a*)
$$\int \frac{dx}{a^2+x^2} = \frac{1}{a}\tan^{-1}\frac{x}{a} + C$$

(*b*)
$$\int \frac{dx}{\sqrt{a^2-x^2}} = \sin^{-1}\frac{x}{a} + C \qquad \text{for } a > 0$$

(c) $$\int \frac{dx}{x\sqrt{x^2-a^2}} = \frac{1}{a}\sec^{-1}\frac{x}{a} + C \qquad \text{for } a > 0$$

NOTATION It is common to write

$$\int \frac{dx}{f(x)} \qquad \text{for} \qquad \int \frac{1}{f(x)}\,dx$$

In each case, we employ the Chain Rule to show that the derivative of the function on the right is the integrand on the left.

(a) $$D_x\left(\frac{1}{a}\tan^{-1}\frac{x}{a}\right) = \frac{1}{a}D_x\left(\tan^{-1}\frac{x}{a}\right) = \frac{1}{a}\frac{1}{1+\left(\frac{x}{a}\right)^2}\left(\frac{1}{a}\right)$$

$$= \frac{1}{a^2}\frac{1}{1+\frac{x^2}{a^2}} = \frac{1}{a^2+x^2}$$

(b) $$D_x\left(\sin^{-1}\frac{x}{a}\right) = \frac{1}{\sqrt{1-\left(\frac{x}{a}\right)^2}}\left(\frac{1}{a}\right) = \frac{1}{\sqrt{1-\frac{x^2}{a^2}}}\left(\frac{1}{\sqrt{a^2}}\right) \qquad \left[\begin{matrix} a>0 \text{ implies} \\ a=\sqrt{a^2}\end{matrix}\right]$$

$$= \frac{1}{\sqrt{a^2-x^2}} \qquad [\sqrt{u}\,\sqrt{v} = \sqrt{uv}]$$

(c) $$D_x\left(\frac{1}{a}\sec^{-1}\frac{x}{a}\right) = \frac{1}{a}D_x\left(\sec^{-1}\frac{x}{a}\right) = \frac{1}{a}\frac{1}{\frac{x}{a}\sqrt{\left(\frac{x}{a}\right)^2-1}}\left(\frac{1}{a}\right)$$

$$= \frac{1}{x\sqrt{\frac{x^2}{a^2}-1}}\left(\frac{1}{\sqrt{a^2}}\right) = \frac{1}{x\sqrt{x^2-a^2}}$$

37.9 Find the following antiderivatives:

(a) $\int \frac{dx}{x^2+4}$ (b) $\int \frac{dx}{\sqrt{9-x^2}}$ (c) $\int \frac{dx}{x^2+4x+5}$ (d) $\int \frac{2x\,dx}{x^2-2x+7}$

(a) By Problem 37.8(a), with $a = 2$,

$$\int \frac{dx}{x^2+4} = \frac{1}{2}\tan\frac{x}{2} + C$$

(b) By Problem 37.8(b), with $a = 3$,

$$\int \frac{dx}{\sqrt{9-x^2}} = \sin^{-1}\frac{x}{3} + C$$

(c) Complete the square: $x^2+4x+5 = (x+2)^2+1$. Thus,

$$\int \frac{dx}{x^2+4x+5} = \int \frac{dx}{(x+2)^2+1}$$

Let $u = x+2$; then $du = dx$, and

$$\int \frac{dx}{(x+2)^2+1} = \int \frac{du}{u^2+1} = \tan^{-1}u + C = \tan^{-1}(x+2) + C$$

(d) Let $u = x^2-2x+7$. Then $du = (2x-2)\,dx$; so, rewrite the antiderivative as

$$I_1 + I_2 \equiv \int \frac{(2x-2)\,dx}{x^2-2x+7} + \int \frac{2\,dx}{x^2-2x+7}$$

Now $$I_1 = \int \frac{du}{u} = \ln|u| + C = \ln|x^2-2x+7| + C$$

To compute I_2, complete the square: $x^2 - 2x + 7 = (x-1)^2 + 6$. Now let $v = x - 1$, $dv = dx$, to obtain:

$$I_2 = \int \frac{2\,dv}{v^2+6} = 2\left(\frac{1}{\sqrt{6}}\right)\tan^{-1}\frac{v}{\sqrt{6}} + C \quad \text{[by Problem 37.8(}a\text{)]}$$
$$= \frac{2}{\sqrt{6}}\tan^{-1}\frac{x-1}{\sqrt{6}} + C$$

By completing the square we showed incidentally that $x^2 - 2x + 7 > 0$ for all x. Hence, the final answer may be written as

$$I_1 + I_2 = \ln(x^2 - 2x + 7) + \frac{2}{\sqrt{6}}\tan^{-1}\frac{x-1}{\sqrt{6}} + C$$

37.10 Evaluate $\sin(2\cos^{-1}(-\frac{1}{3}))$.

Let $\theta = \cos^{-1}(-\frac{1}{3})$; then, by Theorems 26.8 and 26.1,

$$\sin 2\theta = 2\sin\theta\cos\theta = 2(\pm\sqrt{1-\cos^2\theta})(\cos\theta)$$

By definition of the function $\cos^{-1}$, angle θ is in the second quadrant (its cosine being negative), and so its sine is positive. Therefore, the plus sign must be taken in the above formula:

$$\sin 2\theta = 2(\sqrt{1-\cos^2\theta})(\cos\theta) = 2\left(\sqrt{1-\left(-\frac{1}{3}\right)^2}\right)\left(-\frac{1}{3}\right)$$
$$= -\frac{2}{3}\sqrt{1-\frac{1}{9}} = -\frac{2}{3}\sqrt{\frac{8}{9}} = -\frac{2\sqrt{8}}{3\sqrt{9}}$$
$$= -\frac{2}{3}\frac{2\sqrt{2}}{3} = -\frac{4\sqrt{2}}{9}$$

37.11 Show that $\sin^{-1} x$ and $\tan^{-1} x$ are odd functions.

More generally, we can prove that if a one-one function is odd (as are the restricted $\sin x$ and the restricted $\tan x$, but not the *restricted* $\cot x$), then its inverse function is odd. In fact, if f is an odd one-one function and g is its inverse, then

$$\begin{aligned} g(-f(x)) &= g(f(-x)) && [\text{since } f(x) = -f(-x)] \\ &= -x && [\text{since } g \text{ is inverse of } f] \\ &= -g(f(x)) && [\text{since } g \text{ is inverse of } f] \end{aligned}$$

which shows that g is odd.

(Could a one-one function be *even*?)

Supplementary Problems

37.12 For each of the following functions f, determine whether it is one-one; and, if it is, find a formula for the inverse function f^{-1}.

(*a*) $f(x) = \frac{x}{2} + 3$ (*b*) $f(x) = |x|$ (*c*) $f(x) = \frac{3x-5}{x+2}$

(*d*) $f(x) = (x-1)^4$ (*e*) $f(x) = (x-1)^5$ (*f*) $f(x) = \frac{1}{x}$

(*g*) $f(x) = \frac{3x+5}{x}$ (*h*) $f(x) = \frac{1}{x^3+4}$ (*i*) $f(x) = \frac{x+2}{x-1}$

37.13 Evaluate:

(a) $\cos^{-1}\left(-\frac{\sqrt{3}}{2}\right)$ (b) $\sin^{-1}\left(-\frac{\sqrt{2}}{2}\right)$ (c) $\tan^{-1} 1$ (d) $\tan^{-1}\frac{\sqrt{3}}{3}$

(e) $\sec^{-1}\sqrt{2}$ (f) $\sec^{-1}\left(-\frac{2\sqrt{3}}{3}\right)$ (g) $\csc^{-1}(-\sqrt{2})$ (h) $\cot^{-1}(-1)$

37.14 (a) Let $\theta = \cos^{-1}(\frac{1}{3})$. Find the values of $\sin\theta$, $\tan\theta$, $\cot\theta$, $\sec\theta$, and $\csc\theta$. (b) Let $\theta = \sin^{-1}(-\frac{1}{4})$. Find $\cos\theta$, $\tan\theta$, $\cot\theta$, $\sec\theta$, and $\csc\theta$.

37.15 Compute the following values:

(a) $\sin\left(\cos^{-1}\frac{4}{5}\right)$ (b) $\tan\left(\sec^{-1}\frac{13}{5}\right)$ (c) $\cos\left(\sin^{-1}\frac{3}{5} + \sec^{-1} 3\right)$

(d) $\sin\left(\cos^{-1}\frac{1}{5} - \tan^{-1} 2\right)$ (e) $\sin^{-1}(\sin\pi)$

(*Hints*: (b) $5^2 + 12^2 = 13^2$; (c) $\cos(u+v) = \cos u\cos v - \sin u\sin v$; (e) is a trick question.)

37.16 Find the domain and range of the function $f(x) = \cos(\tan^{-1} x)$.

37.17 Differentiate

(a) $\sin^{-1} x + \cos^{-1} x$ (b) $\tan^{-1} x + \cot^{-1} x$ (c) $\sec^{-1} x + \csc^{-1} x$ (d) $\tan^{-1} x + \tan^{-1}\frac{1}{x}$

(e) Explain the significance of your answers.

37.18 Differentiate

(a) $x\tan^{-1} x$ (b) $\sin^{-1}\sqrt{x}$ (c) $\tan^{-1}(\cos x)$ (d) $\ln(\cot^{-1} 3x)$

(e) $e^x\cos^{-1} x$ (f) $\ln(\tan^{-1} x)$ (g) $\csc^{-1}\frac{1}{x}$ (h) $x\sqrt{a^2 - x^2} + a^2\sin^{-1}\frac{x}{a}$

(i) $\tan^{-1}\left(\frac{a+x}{1-ax}\right)$ (j) $\sin^{-1}\frac{1}{x} + \sec^{-1} x$ (k) $\tan^{-1}\frac{2}{x}$

37.19 What identity is implied by the result of Problem 37.18(*i*)?

37.20 Find the following antiderivatives:

(a) $\int \frac{dx}{4+x^2}$ (b) $\int \frac{dx}{4+9x^2}$ (c) $\int \frac{dx}{\sqrt{25-x^2}}$ (d) $\int \frac{dx}{\sqrt{25-16x^2}}$

(e) $\int \frac{dx}{(x-3)\sqrt{x^2-6x+8}}$ (f) $\int \frac{dx}{\sqrt{3-2x^2}}$ (g) $\int \frac{dx}{2+7x^2}$ (h) $\int \frac{dx}{x\sqrt{x^2-4}}$

(i) $\int \frac{dx}{x\sqrt{9x^2-16}}$ (j) $\int \frac{dx}{x\sqrt{3x^2-2}}$ (k) $\int \frac{dx}{(1+x)\sqrt{x}}$ (l) $\int \frac{x\,dx}{x^4+9}$

(m) $\int \frac{dx}{\sqrt{6x-x^2}}$ (n) $\int \frac{2x\,dx}{\sqrt{6x-x^2}}$ (o) $\int \frac{x\,dx}{x^2+8x+20}$ (p) $\int \frac{x^3\,dx}{x^2-2x+4}$

(q) $\int \frac{x\,dx}{\sqrt{4-x^4}}$ (r) $\int \frac{e^x\,dx}{4+e^{2x}}$ (s) $\int \frac{\cos x\,dx}{5+\sin^2 x}$

(*Hints*: (b) let $u = 3x$; (d) let $u = 4x$; (e) let $u = x - 3$; (l) let $u^2 = x^4 + 9$; (m) complete the square in $x^2 - 6x$; (n) $D_x(6x - x^2) = 6 - 2x$; (p) divide x^3 by $x^2 - 2x + 4$.)

37.21 Find an equation of the tangent line to the graph of $y = \sin^{-1}(x/3)$ at the origin.

37.22 A ladder which is 13 feet long leans against a wall. The bottom of the ladder is sliding away from the base of the wall at the rate of 5 ft/sec. How fast is the radian measure of the angle between the ladder and the ground changing at the moment when the bottom of the ladder is 12 feet from the base of the wall?

37.23 The beam from a lighthouse 3 miles from a straight coastline turns at the rate of 5 revolutions per minute. How fast is the point P at which the beam hits the shore moving when that point is 4 miles from the point A on the shore directly opposite the lighthouse? (See Fig. 37-9.)

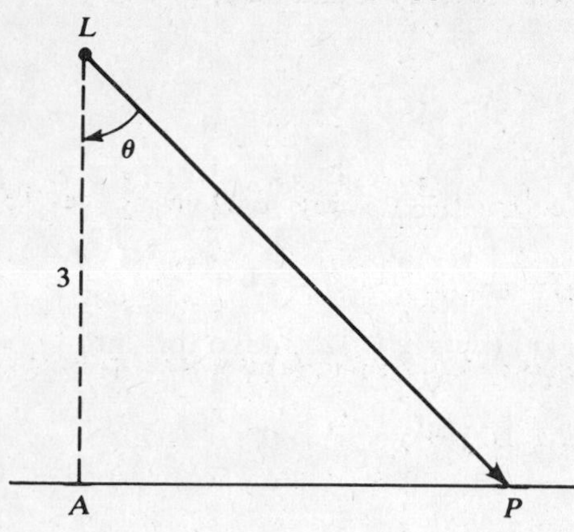

Fig. 37-9

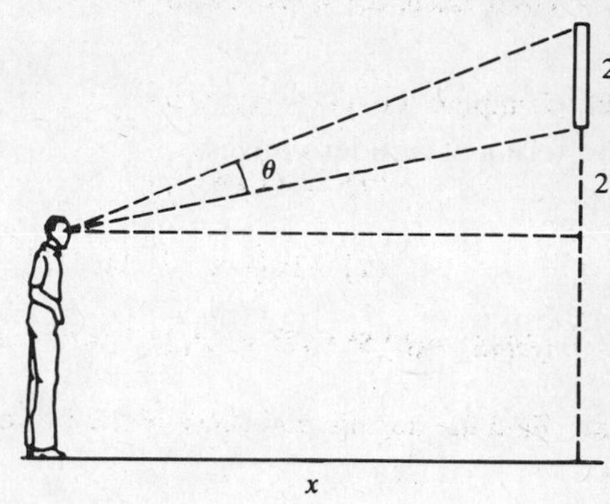

Fig. 37-10

37.24 Find the area under the curve
$$y = \frac{1}{1+x^2}$$
above the x-axis, and between the lines $x = 0$ and $x = 1$.

37.25 Find the area under the curve
$$y = \frac{1}{\sqrt{1-x^2}}$$
above the x-axis, and between the lines $x = 0$ and $x = 1/2$.

37.26 The region $\mathscr{R}$ under the curve
$$y = \frac{1}{x^2\sqrt{x^2-1}}$$
above the x-axis, and between the lines $x = 2/\sqrt{3}$ and $x = 2$, is revolved around the y-axis. Find the volume of the resulting solid.

37.27 Use the formula for arc length, (*32.2*), to find the circumference of a circle of radius r. (*Hint*: Find the arc length of the part of the circle $x^2 + y^2 = r^2$ in the first quadrant, and multiply by 4.)

37.28 A person is viewing a painting hung high on a wall. The vertical dimension of the painting is 2 feet and the bottom of the painting is 2 feet above the eye level of the viewer. Find the distance x that the viewer should stand from the wall in order to maximize the angle θ subtended by the painting. (*Hint*: Express θ as the difference of two inverse cotangents; see Fig. 37-10.)

37.29 For which values of x is each of the following equations true?

(*a*) $\sin^{-1}(\sin x) = x$ (*b*) $\cos^{-1}(\cos x) = x$ (*c*) $\sin^{-1}(-x) = -\sin^{-1} x$

37.30 Find y' by implicit differentiation:

(*a*) $x^2 - x\tan^{-1} y = \ln y$ (*b*) $\cos^{-1} xy = e^{2y}$

37.31 Sketch the graph of $y = \tan^{-1} x - \ln\sqrt{1+x^2}$.

37.32 Assuming that $\cot^{-1} x$ and $\csc^{-1} x$ are differentiable, use implicit differentiation to derive the formulas for their derivatives.

Chapter 38

Integration by Parts

In this chapter, we shall learn one of the most useful techniques for finding antiderivatives. Let f and g be differentiable functions; the Product Rule tells us that

$$\frac{d}{dx}(f(x)g(x)) = f(x)g'(x) + g(x)f'(x)$$

or, in terms of antiderivatives,

$$f(x)g(x) = \int [f(x)g'(x) + g(x)f'(x)]\,dx = \int f(x)g'(x)\,dx + \int g(x)f'(x)\,dx$$

The substitutions $u = f(x)$ $(du = f'(x)\,dx)$ and $v = g(x)$ $(dv = g'(x)\,dx)$ take this into

$$uv = \int u\,dv + \int v\,du$$

or

$$\int u\,dv = uv - \int v\,du \quad \textbf{integration by parts}$$

The idea behind integration by parts is to replace the "difficult" integration $\int u\,dv$ by the "easy" integration $\int v\,du$.

EXAMPLES

(*a*) Find $\int xe^x\,dx$. This will have the form $\int u\,dv$ if we choose

$$u = x \qquad \text{and} \qquad dv = e^x\,dx$$

Now, $dv = v'(x)\,dx$; so if $v'(x)$ is to equal e^x,

$$v = \int e^x\,dx = e^x + C$$

and we take the simplest case, $C = 0$. Then, with $du = dx$,

$$\int u\,dv = uv - \int v\,du$$

$$\int xe^x\,dx = xe^x - \int e^x\,dx$$

$$= xe^x - e^x + C = e^x(x-1) + C$$

Notice that everything depends on a wise choice of u and v. If we had instead picked $u = e^x$ and

$$v = \int x\,dx = \frac{x^2}{2}$$

we would have obtained

$$\int e^x x\,dx = e^x\frac{x^2}{2} - \int \frac{x^2}{2}e^x\,dx$$

which is true enough, but of little use in evaluating $\int xe^x\,dx$.

(*b*) Find $\int x \ln x\,dx$. Choose $u = \ln x$ and

$$dv = x\,dx \qquad \text{or} \qquad v = \int x\,dx = \frac{x^2}{2}$$

(This choice is pretty much forced, unless the antiderivative of ln x is known; see Example (*c*) below.)

$$\int u\,dv = uv - \int v\,du$$

$$\int x \ln x\,dx = (\ln x)\frac{x^2}{2} - \int \frac{x^2}{2}\left(\frac{1}{x}\right)dx = \frac{x^2}{2}\ln x - \frac{1}{2}\int x\,dx$$

$$= \frac{x^2}{2}\ln x - \frac{1}{2}\left(\frac{x^2}{2}\right) + C = \frac{x^2}{4}(2\ln x - 1) + C$$

(*c*) Find $\int \ln x\,dx$. Let $u = \ln x$ and $dv = dx$:

$$u = \ln x \qquad dv = dx$$

$$du = \frac{1}{x}dx \qquad v = x$$

Then

$$\int \ln x\,dx = x\ln x - \int x\left(\frac{1}{x}\right)dx = x\ln x - \int 1\,dx$$

$$= x\ln x - x + C = x(\ln x - 1) + C$$

(*d*) Sometimes, two integrations by parts are necessary. Consider $\int e^x \cos x\,dx$. Let $u = e^x$ and $dv = \cos x\,dx$. Thus,

$$u = e^x \qquad dv = \cos x\,dx$$

$$du = e^x\,dx \qquad v = \sin x$$

and

$$\int e^x \cos x\,dx = e^x \sin x - \int e^x \sin x\,dx \tag{1}$$

Let us try to find $\int e^x \sin x\,dx$ by another integration by parts. Let $u = e^x$ and $dv = \sin x\,dx$:

$$u = e^x \qquad dv = \sin x\,dx$$

$$du = e^x\,dx \qquad v = -\cos x$$

Thus

$$\int e^x \sin x\,dx = -e^x \cos x - \int (-e^x \cos x)\,dx$$

$$= -e^x \cos x + \int e^x \cos x\,dx$$

Substitute this expression for $\int e^x \sin x\,dx$ in (*1*) and solve the resulting equation for the desired antiderivative:

$$\int e^x \cos x\,dx = e^x \sin x - \left(-e^x \cos x + \int e^x \cos x\,dx\right)$$

$$\int e^x \cos x\,dx = e^x \sin x + e^x \cos x - \int e^x \cos x\,dx$$

$$2\int e^x \cos x\,dx = e^x \sin x + e^x \cos x = e^x(\sin x + \cos x)$$

$$\int e^x \cos x\,dx = \frac{e^x(\sin x + \cos x)}{2} + C$$

Solved Problems

38.1 Find $\int xe^{-x}\,dx$.

Let
$$u = x \qquad dv = e^{-x}\,dx$$
$$du = dx \qquad v = \int e^{-x}\,dx = -e^{-x}$$

Integration by parts gives

$$\int xe^{-x}\,dx = -xe^{-x} - \int (-e^{-x})\,dx = -xe^{-x} + \int e^{-x}\,dx$$
$$= -xe^{-x} - e^{-x} + C = -e^{-x}(x+1) + C$$

Another method would consist in making the change of variable $x = -t$ and using Example (*a*) of this chapter.

38.2 (*a*) Establish the *reduction formula*

$$\int x^n e^x\,dx = x^n e^x - n\int x^{n-1}e^x\,dx \tag{1}$$

for $\int x^n e^x\,dx$ $(n = 1, 2, 3, \ldots)$. (*b*) Compute $\int x^2 e^x\,dx$.

(*a*) Let

$$u = x^n \qquad dv = e^x\,dx$$
$$du = nx^{n-1}\,dx \qquad v = e^x$$

and integrate by parts:

$$\int x^n e^x\,dx = x^n e^x - \int e^x(nx^{n-1})\,dx = x^n e^x - n\int x^{n-1}e^x\,dx$$

(*b*) For $n = 1$, (*1*) gives:

$$\int xe^x\,dx = xe^x - \int e^x\,dx = xe^x - e^x = (x-1)e^x$$

as in Example (*a*). We neglect the arbitrary constant C until the end of the calculation. Now let $n = 2$:

$$\int x^2e^x\,dx = x^2e^x - 2\int xe^x\,dx = x^2e^x - 2[(x-1)e^x]$$
$$= [x^2 - 2(x-1)]e^x = (x^2 - 2x + 2)e^x + C$$

38.3 Find $\int \tan^{-1} x\,dx$.

Let
$$u = \tan^{-1} x \qquad dv = dx$$
$$du = \frac{1}{1+x^2}\,dx \qquad v = x$$

Hence
$$\int \tan^{-1} x\,dx = x\tan^{-1} x - \int \frac{x}{1+x^2}\,dx$$

In the integral on the right, make the change of variable $w = 1 + x^2$, $dw = 2x\,dx$:

$$\int \frac{x}{1+x^2}\,dx = \frac{1}{2}\int \frac{dw}{w} = \frac{1}{2}\ln|w| = \frac{1}{2}\ln|1+x^2|$$
$$= \frac{1}{2}\ln(1+x^2) \quad [\text{since } 1+x^2 > 0]$$

Thus
$$\int \tan^{-1} x\,dx = x\tan^{-1} x - \frac{1}{2}\ln(1+x^2) + C$$

38.4 Find $\int \cos^2 x\,dx$.

Let
$$u = \cos x \qquad dv = \cos x\,dx$$
$$du = -\sin x\,dx \qquad v = \sin x$$

Then
$$\int \cos^2 x\,dx = \cos x \sin x - \int (\sin x)(-\sin x)\,dx$$
$$= \cos x \sin x + \int \sin^2 x\,dx$$
$$= \cos x \sin x + \int (1-\cos^2 x)\,dx$$
$$= \cos x \sin x + \int 1\,dx - \int \cos^2 x\,dx$$

Solving this equation for $\int \cos^2 x\,dx$,
$$2\int \cos^2 x\,dx = \cos x \sin x + \int 1\,dx = \cos x \sin x + x$$
$$\int \cos^2 x\,dx = \frac{1}{2}(\cos x \sin x + x) + C$$

This result is more easily obtained by use of Theorem 26.8:
$$\int \cos^2 x\,dx = \int \left(\frac{1}{2}\cos 2x + \frac{1}{2}\right)dx = \frac{1}{2}\int \cos 2x\,dx + \frac{1}{2}\int 1\,dx$$

38.5 Find $\int x \tan^{-1} x\,dx$.

Let
$$u = \tan^{-1} x \qquad dv = x\,dx$$
$$du = \frac{1}{1+x^2}\,dx \qquad v = \frac{x^2}{2}$$

Then
$$\int x \tan^{-1} x\,dx = \frac{x^2}{2}\tan^{-1} x - \frac{1}{2}\int \frac{x^2}{1+x^2}\,dx$$

But
$$\frac{x^2}{1+x^2} = \frac{(1+x^2)-1}{1+x^2} = \frac{1+x^2}{1+x^2} - \frac{1}{1+x^2} = 1 - \frac{1}{1+x^2}$$

and so
$$\int \frac{x^2}{1+x^2}\,dx = \int 1\,dx - \int \frac{1}{1+x^2}\,dx = x - \tan^{-1} x$$

Hence
$$\int x \tan^{-1} x\,dx = \frac{x^2}{2}\tan^{-1} x - \frac{1}{2}(x - \tan^{-1} x) = \frac{1}{2}(x^2 \tan^{-1} x - (x - \tan^{-1} x))$$
$$= \frac{1}{2}(x^2 \tan^{-1} x - x + \tan^{-1} x) = \frac{1}{2}[(\tan^{-1} x)(x^2+1) - x] + C$$

Supplementary Problems

38.6 Compute:

(a) $\int x^2 e^{-x}\,dx$ (b) $\int e^x \sin x\,dx$ (c) $\int x^3 e^x\,dx$ (d) $\int \sin^{-1} x\,dx$

(e) $\int x \cos x\,dx$ (f) $\int x^2 \sin x\,dx$ (g) $\int \cos(\ln x)\,dx$ (h) $\int x \cos(5x-1)\,dx$

(*i*) $\int e^{ax} \cos bx\, dx$	(*j*) $\int \sin^2 x\, dx$	(*k*) $\int \cos^3 x\, dx$	(*l*) $\int \cos^4 x\, dx$
(*m*) $\int xe^{3x}\, dx$	(*n*) $\int x \sec^2 x\, dx$	(*o*) $\int x \cos^2 x\, dx$	(*p*) $\int (\ln x)^2\, dx$
(*q*) $\int x \sin 2x\, dx$	(*r*) $\int x \sin x^2\, dx$	(*s*) $\int \frac{\ln x}{x^2}\, dx$	(*t*) $\int x^2 e^{3x}\, dx$
(*u*) $\int x^2 \tan^{-1} x\, dx$	(*v*) $\int \ln (x^2+1)\, dx$	(*w*) $\int \frac{x^3}{\sqrt{1+x^2}}\, dx$	(*x*) $\int x^2 \ln x\, dx$

(*Hint*: Integration by parts is not a good method for (*r*).)

38.7 Let $\mathscr{R}$ be the region bounded by the curve $y = \ln x$, the x-axis, and the line $x = e$. (*a*) Find the area of $\mathscr{R}$. (*b*) Find the volume generated by revolving $\mathscr{R}$ about (i) the x-axis, (ii) the y-axis.

38.8 Let $\mathscr{R}$ be the region bounded by the curve $y = x^{-1} \ln x$, the x-axis, and the line $x = e$. Find (*a*) the area of $\mathscr{R}$, (*b*) the volume of the solid generated by revolving $\mathscr{R}$ about the y-axis, (*c*) the volume of the solid generated by revolving $\mathscr{R}$ about the x-axis. (*Hint*: (*c*) In the volume integral, let $u = (\ln x)^2$, $v = -1/x$, and use Problem 38.6(*s*).)

38.9 Derive from Problem 38.8(*c*) the (good) bounds: $2.5 \le e \le 2.823$. (*Hint*: By Problem 30.3(*c*),

$$0 \le \int_1^e \left(\frac{\ln x}{x}\right)^2 dx \le \frac{e-1}{e^2}$$

or

$$0 \le 2 - \frac{5}{e} \le \frac{e-1}{e^2}$$

The left-hand inequality gives $e \ge 5/2$; the right-hand inequality gives $e \le (3+\sqrt{7})/2$.)

38.10 Let $\mathscr{R}$ be the region under one arch of the curve $y = \sin x$, above the x-axis, and between $x = 0$ and $x = \pi$. Find the volume of the solid generated by revolving $\mathscr{R}$ about (*a*) the x-axis, (*b*) the y-axis.

38.11 If n is a positive integer, find:

$$(a)\quad \int_0^{2\pi} x \cos nx\, dx \qquad (b)\quad \int_0^{2\pi} x \sin nx\, dx$$

38.12 For $n = 2, 3, 4, \ldots$, find reduction formulas for

$$(a)\quad \int \cos^n x\, dx \qquad (b)\quad \int \sin^n x\, dx$$

(*c*) Use these formulas to check the answers to Problems 38.4, 38.6(*k*), 38.6(*l*), and 38.6(*j*).

38.13 (*a*) Find a reduction formula for $\int \sec^n x\, dx$ $(n = 2, 3, 4, \ldots)$. (*b*) Use this formula, together with Problem 34.5(*b*), to compute

$$\text{(i)}\quad \int \sec^3 x\, dx \qquad \text{(ii)}\quad \int \sec^4 x\, dx$$

Chapter 39

Trigonometric Integrands; Trigonometric Substitutions

39.1 INTEGRATION OF TRIGONOMETRIC FUNCTIONS

We already know the antiderivatives of some simple combinations of the basic trigonometric functions; in particular, we have derived all the formulas given in the second column of Appendix B. Let us now look at more complicated cases.

EXAMPLES

(*a*) Consider $\int \sin^k x \cos^n x\, dx$, where the nonnegative integers k and n are *not both even.* If, say, k is odd ($k = 2j+1$), rewrite the integral as

$$\int \sin^{2j} x \cos^n x \sin x\, dx = \int (\sin^2 x)^j \cos^n x \sin x\, dx$$
$$= \int (1-\cos^2 x)^j \cos^n x \sin x\, dx$$

Now the change of variable $u = \cos x$, $du = -\sin x\, dx$, produces a polynomial integrand. (For n odd, the substitution $u = \sin x$ would be made instead.) For instance:

$$\int \cos^2 x \sin^5 x\, dx = \int \cos^2 x \sin^4 x \sin x\, dx$$
$$= \int \cos^2 x\,(1-\cos^2 x)^2 \sin x\, dx$$
$$= \int u^2(1-u^2)^2(-1)\, du = -\int u^2(1-2u^2+u^4)\, du$$
$$= -\int (u^2 - 2u^4 + u^6)\, du = -\left(\frac{u^3}{3} - 2\frac{u^5}{5} + \frac{u^7}{7}\right) + C$$
$$= -\frac{1}{3}\cos^3 x + \frac{2}{5}\cos^5 x - \frac{1}{7}\cos^7 x + C$$

(*b*) The same antiderivative as in (*a*), but with k and n *both even*; say, $k = 2p$ and $n = 2q$. Then, in view of the half-angle identities

$$\cos^2 x = \frac{1+\cos 2x}{2} \qquad \sin^2 x = \frac{1-\cos 2x}{2}$$

we can write

$$\int \sin^k x \cos^n x\, dx = \int (\sin^2 x)^p (\cos^2 x)^q\, dx$$
$$= \int \left(\frac{1-\cos 2x}{2}\right)^p \left(\frac{1+\cos 2x}{2}\right)^q dx$$
$$= \frac{1}{2^{p+q}} \int (1-\cos 2x)^p (1+\cos 2x)^q\, dx$$

When the binomials are multiplied out, the integrand will appear as a polynomial in $\cos 2x$:

$$1 + (q-p)(\cos 2x) + \cdots \pm (\cos 2x)^{p+q}$$

and so

$$\int \sin^k x \cos^n x\, dx = \frac{1}{2^{p+q}}\left[\int 1\, dx + (q-p)\int (\cos 2x)\, dx + \cdots \pm \int (\cos 2x)^{p+q}\, dx\right] \tag{1}$$

On the right-hand side of (*1*) are antiderivatives of *odd powers* of $\cos 2x$, which may be evaluated by the method of Example (*a*), and antiderivatives of *even powers* of $\cos 2x$, to which the half-angle formula may be applied again. Thus, if the sixth power were present, we would write

$$\int (\cos 2x)^6\,dx = \int (\cos^2 2x)^3\,dx = \int \left(\frac{1+\cos 4x}{2}\right)^3 dx$$

would expand the polynomial in $\cos 4x$, and so forth. Eventually the process must end in a final answer, as is shown in the following specific case.

$$\begin{aligned}
\int \cos^2 x \sin^4 x\,dx &= \int (\cos^2 x)(\sin^2 x)^2\,dx \\
&= \int \left(\frac{1+\cos 2x}{2}\right)\left(\frac{1-\cos 2x}{2}\right)^2 dx \\
&= \int \left(\frac{1+\cos 2x}{2}\right)\left(\frac{1-2\cos 2x+\cos^2 2x}{4}\right) dx \\
&= \frac{1}{8}\int [1(1-2\cos 2x+\cos^2 2x)+(\cos 2x)(1-2\cos 2x+\cos^2 2x)]\,dx \\
&= \frac{1}{8}\int (1-2\cos 2x+\cos^2 2x+\cos 2x-2\cos^2 2x+\cos^3 2x)\,dx \\
&= \frac{1}{8}\int (1-\cos 2x-\cos^2 2x+\cos^3 2x)\,dx \\
&= \frac{1}{8}\left(\int 1\,dx-\int \cos 2x\,dx-\int \cos^2 2x\,dx+\int \cos^3 2x\,dx\right) \\
&= \frac{1}{8}\left(x-\frac{\sin 2x}{2}-\int \frac{1+\cos 4x}{2}\,dx+\int (\cos 2x)(1-\sin^2 2x)\,dx\right) \\
&= \frac{1}{8}\left(x-\frac{\sin 2x}{2}-\frac{1}{2}\left(x+\frac{\sin 4x}{4}\right)+\int \cos 2x\,dx-\frac{1}{2}\int u^2\,du\right) \quad [\text{letting } u = \sin 2x] \\
&= \frac{1}{8}\left(x-\frac{\sin 2x}{2}-\frac{x}{2}-\frac{\sin 4x}{8}+\frac{\sin 2x}{2}-\frac{1}{2}\frac{\sin^3 2x}{3}\right)+C \\
&= \frac{1}{8}\left(\frac{x}{2}-\frac{\sin 4x}{8}-\frac{\sin^3 2x}{6}\right)+C \\
&= \frac{x}{16}-\frac{\sin 4x}{64}-\frac{\sin^3 2x}{48}+C
\end{aligned}$$

(*c*) From Problem 34.5(*a*), we know how to integrate the first power of $\tan x$:

$$\int \tan x\,dx = \ln|\sec x| + C$$

Higher powers are handled by means of a reduction formula: we have, for $n = 2, 3, \ldots,$

$$\begin{aligned}
\int \tan^n x\,dx &= \int \tan^{n-2} x\,(\tan^2 x)\,dx = \int \tan^{n-2} x\,(\sec^2 x - 1)\,dx \\
&= \int \tan^{n-2} x \sec^2 x\,dx - \int \tan^{n-2} x\,dx \\
&= \int u^{n-2}\,du - \int \tan^{n-2} x\,dx \quad [\text{let } u = \tan x] \\
&= \frac{\tan^{n-1} x}{n-1} - \int \tan^{n-2} x\,dx \qquad (39.1)
\end{aligned}$$

Similarly, from

$$\int \sec x\,dx = \ln|\sec x + \tan x| + C$$

and the reduction formula of Problem 38.13(*a*), we can integrate all powers of $\sec x$.

(*d*) Antiderivatives of the forms $\int \sin Ax \cos Bx\,dx$, $\int \sin Ax \sin Bx\,dx$, $\int \cos Ax \cos Bx\,dx$, can be computed by using the identities

$$\sin Ax \cos Bx = \frac{1}{2}(\sin (A+B)x + \sin (A-B)x)$$

$$\sin Ax \sin Bx = \frac{1}{2}(\cos (A-B)x - \cos (A+B)x)$$

$$\cos Ax \cos Bx = \frac{1}{2}(\cos (A-B)x + \cos (A+B)x)$$

For instance:

$$\int \sin 8x \sin 3x\,dx = \int \frac{1}{2}(\cos 5x - \cos 11x)\,dx = \frac{1}{2}\left(\frac{\sin 5x}{5} - \frac{\sin 11x}{11}\right) + C$$

39.2 TRIGONOMETRIC SUBSTITUTIONS

To find the antiderivative of a function involving such expressions as

$$\sqrt{a^2+x^2} \qquad \text{or} \qquad \sqrt{a^2-x^2} \qquad \text{or} \qquad \sqrt{x^2-a^2}$$

it is often helpful to substitute a trigonometric function for x.

EXAMPLES

(*a*) Evaluate
$$\int \sqrt{x^2+2}\,dx$$

None of the methods already available is of any use here. Let us make the substitution $x = \sqrt{2}\tan\theta$, where $-\pi/2 < \theta < \pi/2$; equivalently,

$$\theta = \tan^{-1}\frac{x}{\sqrt{2}}$$

Figure 39-1 illustrates the relationship between x and θ, with θ interpreted as an angle. We have $dx = \sqrt{2}\sec^2\theta\,d\theta$, and, from Fig. 39-1,

$$\frac{\sqrt{x^2+2}}{\sqrt{2}} = \sec\theta \qquad \text{or} \qquad \sqrt{x^2+2} = \sqrt{2}\sec\theta$$

where $\sec\theta > 0$ (as it must be). Thus,

$$\begin{aligned}
\int \sqrt{x^2+2}\,dx &= \int (\sqrt{2}\sec\theta)(\sqrt{2}\sec^2\theta)\,d\theta = 2\int \sec^3\theta\,d\theta \\
&= \sec\theta\tan\theta + \ln|\sec\theta + \tan\theta| \qquad \text{[by Problem 38.13(}b\text{)]} \\
&= \frac{\sqrt{x^2+2}}{\sqrt{2}}\,\frac{x}{\sqrt{2}} + \ln\left|\frac{\sqrt{x^2+2}}{\sqrt{2}} + \frac{x}{\sqrt{2}}\right| \\
&= \frac{x\sqrt{x^2+2}}{2} + \ln\frac{|\sqrt{x^2+2}+x|}{\sqrt{2}} \qquad \left[\left|\frac{a}{b}\right| = \frac{|a|}{|b|}\right] \\
&= \frac{x\sqrt{x^2+2}}{2} + \ln|\sqrt{x^2+2}+x| - \ln\sqrt{2} \\
&= \frac{x\sqrt{x^2+2}}{2} + \ln|\sqrt{x^2+2}+x| + C
\end{aligned}$$

Note how the constant $-\ln\sqrt{2}$ was absorbed in the general constant C in the last step. The absolute value signs in the logarithm may be dropped, since, for all x,

$$\sqrt{x^2+2} + x > 0$$

(You should verify this.)

In general, if $\sqrt{x^2+a^2}$ occurs in an integrand, try the substitution $x = a\tan\theta$, with $-\pi/2 < \theta < \pi/2$.

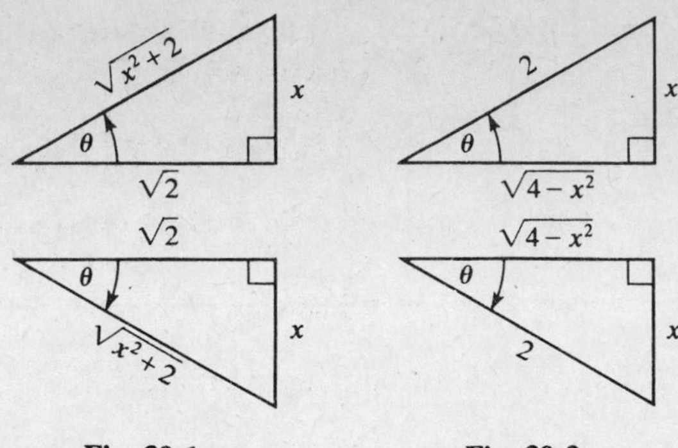

Fig. 39-1 **Fig. 39-2**

(*b*) Evaluate

$$\int \frac{\sqrt{4-x^2}}{x}\,dx$$

Make the substitution $x = 2\sin\theta$, where $-\pi/2 \le \theta \le \pi/2$; equivalently,

$$\theta = \sin^{-1}\frac{x}{2}$$

The angle-interpretation of θ is given in Fig. 39-2. We have:

$$dx = 2\cos\theta\,d\theta \qquad \frac{\sqrt{4-x^2}}{x} = \cot\theta$$

(note that $\cot\theta$ and x are necessarily of the same sign). Thus,

$$\begin{aligned}\int \frac{\sqrt{4-x^2}}{x}\,dx &= \int (\cot\theta)(2\cos\theta)\,d\theta = 2\int\left(\frac{\cos\theta}{\sin\theta}\right)(\cos\theta)\,d\theta\\ &= 2\int\frac{\cos^2\theta}{\sin\theta}\,d\theta = 2\int\frac{1-\sin^2\theta}{\sin\theta}\,d\theta\\ &= 2\left(\int\csc\theta\,d\theta - \int\sin\theta\,d\theta\right) = 2(\ln|\csc\theta-\cot\theta| + \cos\theta)\\ &= 2\left(\ln\left|\frac{2}{x}-\frac{\sqrt{4-x^2}}{x}\right| + \frac{\sqrt{4-x^2}}{2}\right)\\ &= 2\ln\left|\frac{2-\sqrt{4-x^2}}{x}\right| + \sqrt{4-x^2}\\ &= \ln\left(\frac{2-\sqrt{4-x^2}}{x}\right)^2 + \sqrt{4-x^2} + C\end{aligned}$$

where, in the last step, we used $\ln u^2 = \ln|u|^2 = 2\ln|u|$.

In general, occurrence of $\sqrt{a^2-x^2}$ in an integrand suggests the substitution $x = a\sin\theta$, with $-\pi/2 \le \theta \le \pi/2$.

(*c*) Evaluate

$$\int \frac{\sqrt{x^2-4}}{x^3}\,dx$$

Let $x = 2\sec\theta$, where $0 \le \theta < \pi/2$ or $\pi \le \theta < 3\pi/2$; equivalently,

$$\theta = \sec^{-1}\frac{x}{2}$$

We have:

$$dx = 2\sec\theta\tan\theta\,d\theta \qquad \sqrt{x^2-4} = 2\tan\theta$$

(Note that $\tan\theta \ge 0$ for θ in the first or third quadrant.) Thus,

$$\int \frac{\sqrt{x^2-4}}{x^3}\,dx = \int \frac{2\tan\theta}{8\sec^3\theta}(2\sec\theta\tan\theta)\,d\theta$$
$$= \frac{1}{2}\int \frac{\tan^2\theta}{\sec^2\theta}\,d\theta = \frac{1}{2}\int \frac{\sin^2\theta}{\cos^2\theta\sec^2\theta}\,d\theta$$
$$= \frac{1}{2}\int \sin^2\theta\,d\theta = \frac{1}{4}(\theta - \sin\theta\cos\theta)$$
$$= \frac{1}{4}\left(\sec^{-1}\frac{x}{2} - \frac{\sqrt{x^2-4}}{x}\frac{2}{x}\right) + C$$
$$= \frac{1}{4}\left(\sec^{-1}\frac{x}{2} - \frac{2\sqrt{x^2-4}}{x^2}\right) + C$$

In general, whenever the expression $\sqrt{x^2-a^2}$ occurs, try the substitution $x = a\sec\theta$.

Solved Problems

39.1 Find $\int \sin^3 x \cos^2 x\,dx$.

The exponent of $\sin x$ is odd.

$$\int \sin^3 x\cos^2 x\,dx = \int \sin^2 x\cos^2 x\,(\sin x)\,dx = \int (1-\cos^2 x)\cos^2 x\,(\sin x)\,dx$$
$$= -\int (\cos^2 x - \cos^4 x)\,d(\cos x) = -\frac{\cos^3 x}{3} + \frac{\cos^5 x}{5} + C$$

39.2 Find $\int \cos^4 x \sin^4 x\,dx$.

The exponents are both even; in addition, they are equal. This allows an improvement on the method of Section 39.1, Example (*b*).

$$\int \cos^4 x\sin^4 x\,dx = \int \left(\frac{\sin 2x}{2}\right)^4 dx = \frac{1}{16}\int \sin^4 2x\,dx$$
$$= \frac{1}{16}\int \left(\frac{1-\cos 4x}{2}\right)^2 dx$$
$$= \frac{1}{64}\int (1 - 2\cos 4x + \cos^2 4x)\,dx$$
$$= \frac{1}{64}\left(\int 1\,dx - \frac{1}{2}\int \cos u\,du + \frac{1}{4}\int \cos^2 u\,du\right) \quad [\text{letting } u = 4x]$$
$$= \frac{1}{64}\left[x - \frac{1}{2}\sin u + \frac{1}{8}(u + \sin u\cos u)\right]$$
$$= \frac{1}{64}\left[x - \frac{1}{2}\sin 4x + \frac{x}{2} + \frac{\sin 4x\cos 4x}{8}\right] + C$$
$$= \frac{1}{64}\left[\frac{3x}{2} - \frac{\sin 4x}{2} + \frac{\sin 8x}{16}\right] + C$$
$$= \frac{1}{128}\left[3x - \sin 4x + \frac{\sin 8x}{8}\right] + C$$

39.3 Find (*a*) $\int \cos^5 x\,dx$, (*b*) $\int \sin^4 x\,dx$.

(*a*)
$$\int \cos^5 x\,dx = \int \cos^4 x\,(\cos x)\,dx = \int (\cos^2 x)^2(\cos x)\,dx$$
$$= \int (1-\sin^2 x)^2(\cos x)\,dx = \int (1 - 2\sin^2 x + \sin^4 x)(\cos x)\,dx$$

Let $u = \sin x$. Then $du = \cos x\,dx$, and

$$\int \cos^5 x\,dx = \int (1 - 2u^2 + u^4)\,du = u - \frac{2u^3}{3} + \frac{u^5}{5}$$

$$= \sin x - \frac{2\sin^3 x}{3} + \frac{\sin^5 x}{5} + C$$

(*b*) This antiderivative was essentially obtained in Problem 39.2:

$$\int \sin^4 x\,dx = 2\int \sin^4 2u\,du \quad [\text{let } u = x/2]$$

$$= 2 \cdot \frac{16}{128}\left(3u - \sin 4u + \frac{\sin 8u}{8}\right)$$

$$= \frac{1}{4}\left(\frac{3x}{2} - \sin 2x + \frac{\sin 4x}{8}\right) + C$$

39.4 Find $\int \tan^5 x\,dx$.

From the reduction formula (*39.1*),

$$\int \tan^3 x\,dx = \frac{\tan^2 x}{2} - \int \tan x\,dx$$

$$= \frac{\tan^2 x}{2} - \ln|\sec x|$$

$$\int \tan^5 x\,dx = \frac{\tan^4 x}{4} - \int \tan^3 x\,dx$$

$$= \frac{\tan^4 x}{4} - \frac{\tan^2 x}{2} + \ln|\sec x| + C$$

39.5 Show how to find $\int \tan^p x \sec^q x\,dx$ (*a*) when q is even, (*b*) when p is odd. (*c*) Illustrate both techniques with $\int \tan^3 x \sec^4 x\,dx$, and show that the two answers are equivalent.

(*a*) Let $q = 2r$ $(r = 1, 2, 3, \ldots)$; then,

$$\int \tan^p x \sec^{2r} x\,dx = \int \tan^p x \sec^{2(r-1)} x (\sec^2 x)\,dx$$

$$= \int \tan^p x\,(1 + \tan^2 x)^{r-1}(\sec^2 x)\,dx$$

since $1 + \tan^2 x = \sec^2 x$. Now the substitution $u = \tan x$, $du = \sec^2 x\,dx$, produces a polynomial integrand.

(*b*) Let $p = 2s + 1$ $(s = 0, 1, 2, \ldots)$; then,

$$\int \tan^{2s+1} x \sec^q x\,dx = \int \tan^{2s} x \sec^{q-1} x\,(\sec x \tan x)\,dx$$

$$= \int (\sec^2 x - 1)^s \sec^{q-1} x\,(\sec x \tan x)\,dx$$

since $\tan^2 x = \sec^2 x - 1$. Now let $v = \sec x$, $dv = \sec x \tan x\,dx$, to obtain a polynomial integrand.

(*c*) By (*a*),

$$\int \tan^3 x \sec^4 x\,dx = \int \tan^3 x\,(1 + \tan^2 x)(\sec^2 x)\,dx$$

$$= \int u^3(1 + u^2)\,du = \int (u^3 + u^5)\,du$$

$$= \frac{u^4}{4} + \frac{u^6}{6} = \frac{\tan^4 x}{4} + \frac{\tan^6 x}{6} + C$$

By (*b*),

$$\int \tan^3 x \sec^4 x\,dx = \int (\sec^2 x - 1)\sec^3 x\,(\sec x \tan x)\,dx$$

$$= \int (v^2 - 1)v^3\,dv = \int (v^5 - v^3)\,dv$$

$$= \frac{v^6}{6} - \frac{v^4}{4} = \frac{\sec^6 x}{6} - \frac{\sec^4 x}{4} + C$$

Since $1 + u^2 = v^2$,

$$\frac{v^6}{6} - \frac{v^4}{4} = \frac{4v^6 - 6v^4}{24} = \frac{4(1+u^2)^3 - 6(1+u^2)^2}{24}$$

ALGEBRA $\quad (1+t)^3 = 1 + 3t + 3t^2 + t^3$

$$= \frac{4(1 + 3u^2 + 3u^4 + u^6) - 6(1 + 2u^2 + u^4)}{24}$$

$$= \frac{4 + 12u^2 + 12u^4 + 4u^6 - 6 - 12u^2 - 6u^4}{24}$$

$$= \frac{6u^4 + 4u^6 - 2}{24} = \frac{u^4}{4} + \frac{u^6}{6} - \frac{1}{12}$$

and so the two expressions for $\int \tan^3 x \sec^4 x\,dx$ are equivalent (the $-1/12$ is soaked up by the arbitrary constant C).

39.6 Find $\int \tan^2 x \sec x\,dx$.

Problem 39.5 is of no help here.

$$\int \tan^2 x \sec x\,dx = \int (\sec^2 x - 1)\sec x\,dx = \int (\sec^3 x - \sec x)\,dx$$

$$= \frac{1}{2}(\sec x \tan x + \ln|\sec x + \tan x|) - \ln|\sec x + \tan x| + C$$

[by Problem 38.13(*b*)]

$$= \frac{1}{2}\sec x \tan x - \frac{1}{2}\ln|\sec x + \tan x| + C$$

39.7 Prove the trigonometric identity

$$\sin Ax \cos Bx = \frac{1}{2}[\sin(A+B)x + \sin(A-B)x]$$

The sum and difference formulas of Theorem 26.6 give

$$\sin(A+B)x = \sin(Ax + Bx) = \sin Ax \cos Bx + \cos Ax \sin Bx$$
$$\sin(A-B)x = \sin(Ax - Bx) = \sin Ax \cos Bx - \cos Ax \sin Bx$$

and so, by addition, $\sin(A+B)x + \sin(A-B)x = 2 \sin Ax \cos Bx$.

39.8 Compute the value of

$$\int_0^{2\pi} \sin nx \cos kx\,dx$$

for positive integers n and k.

Case 1: $n \neq k$. By Problem 39.7, with $A = n$ and $B = k$,

$$\int_0^{2\pi} \sin nx \cos kx\, dx = \frac{1}{2}\int_0^{2\pi} [\sin (n+k)x + \sin (n-k)x]\, dx$$
$$= -\frac{1}{2}\left(\frac{\cos (n+k)x}{n+k} + \frac{\cos (n-k)x}{n-k}\right)\Bigg]_0^{2\pi} = 0$$

because $\cos px$ is, for p an integer, a periodic function, of period 2π.

Case 2: $n = k$. Then, by the double-angle formula for the sine function,

$$\int_0^{2\pi} \sin nx \cos nx\, dx = \frac{1}{2}\int_0^{2\pi} \sin 2nx\, dx = -\frac{1}{2}\left(\frac{\cos 2nx}{2n}\right)\Bigg]_0^{2\pi} = 0$$

39.9 Find

$$\frac{dx}{\sqrt{x^2+9}}$$

The presence of $\sqrt{x^2+9}$ suggests letting $x = 3\tan\theta$. Then $dx = 3\sec^2\theta\, d\theta$, and

$$\sqrt{x^2+9} = \sqrt{9\tan^2\theta + 9} = \sqrt{9(\tan^2\theta+1)} = 3\sqrt{\sec^2\theta} = 3\sec\theta$$

So,

$$\int \frac{dx}{\sqrt{x^2+9}} = \int \frac{3\sec^2\theta}{3\sec\theta}\, d\theta = \int \sec\theta\, d\theta = \ln|\sec\theta + \tan\theta|$$
$$= \ln\left|\frac{\sqrt{x^2+9}}{3} + \frac{x}{3}\right| + C = \ln\left|\frac{\sqrt{x^2+9}+x}{3}\right| + C$$
$$= \ln|\sqrt{x^2+9}+x| + K = \ln(\sqrt{x^2+9}+x) + K$$

NOTE

$$\ln\left|\frac{\sqrt{x^2+9}+x}{3}\right| = \ln|\sqrt{x^2+9}+x| - \ln 3$$

and the constant $-\ln 3$ can be absorbed in the arbitrary constant. Furthermore,

$$\sqrt{x^2+9}+x > 0$$

39.10 Find

$$\int \frac{dx}{x^2\sqrt{3-x^2}}$$

The presence of $\sqrt{3-x^2}$ suggests the substitution $x = \sqrt{3}\sin\theta$. Then,

$$dx = \sqrt{3}\cos\theta\, d\theta$$
$$\sqrt{3-x^2} = \sqrt{3-3\sin^2\theta} = \sqrt{3(1-\sin^2\theta)} = \sqrt{3}\sqrt{\cos^2\theta} = \sqrt{3}\cos\theta$$

and

$$\int \frac{dx}{x^2\sqrt{3-x^2}} = \int \frac{\sqrt{3}\cos\theta\, d\theta}{(3\sin^2\theta)(\sqrt{3}\cos\theta)} = \frac{1}{3}\int \frac{d\theta}{\sin^2\theta} = \frac{1}{3}\int \csc^2\theta\, d\theta$$
$$= -\frac{1}{3}\cot\theta + C$$

But

$$\cot\theta = \frac{\cos\theta}{\sin\theta} = \frac{\sqrt{3-x^2}}{\sqrt{3}}\bigg/\frac{x}{\sqrt{3}} = \frac{\sqrt{3-x^2}}{x}$$

Hence

$$\int \frac{dx}{x^2\sqrt{3-x^2}} = -\frac{\sqrt{3-x^2}}{3x} + C$$

39.11 Find
$$\int \frac{x^2}{\sqrt{x^2-4}}\,dx$$

The occurrence of $\sqrt{x^2-4}$ suggests the substitution $x = 2\sec\theta$. Then,

$$dx = 2\sec\theta\tan\theta\,d\theta$$

$$\sqrt{x^2-4} = \sqrt{4\sec^2\theta - 4} = \sqrt{4(\sec^2\theta - 1)} = 2\sqrt{\tan^2\theta} = 2\tan\theta$$

and
$$\int \frac{x^2\,dx}{\sqrt{x^2-4}} = \int \frac{(4\sec^2\theta)(2\sec\theta\tan\theta)\,d\theta}{2\tan\theta} = 4\int \sec^3\theta\,d\theta$$

$$= 2(\sec\theta\tan\theta + \ln|\sec\theta + \tan\theta|) \quad \text{[by Problem 38.13(}b\text{)]}$$

$$= 2\left(\frac{x}{2}\frac{\sqrt{x^2-4}}{2} + \ln\left|\frac{x}{2} + \frac{\sqrt{x^2-4}}{2}\right|\right)$$

$$= \frac{x\sqrt{x^2-4}}{2} + 2\ln\left|\frac{x+\sqrt{x^2-4}}{2}\right| + C$$

$$= \frac{x\sqrt{x^2-4}}{2} + 2\ln|x+\sqrt{x^2-4}| + K$$

where $K = C - 2\ln 2$ (compare Problem 39.9).

Supplementary Problems

39.12 Find the following antiderivatives:

(*a*) $\int \sin x\cos^2 x\,dx$ (*b*) $\int \cos^2 3x\,dx$ (*c*) $\int \sin^4 x\cos^5 x\,dx$ (*d*) $\int \cos^6 x\,dx$

(*e*) $\int \cos^6 x\sin^2 x\,dx$ (*f*) $\int \tan^2\frac{x}{2}\,dx$ (*g*) $\int \tan^6 x\,dx$ (*h*) $\int \sec^5 x\,dx$

(*i*) $\int \tan^2 x\sec^4 x\,dx$ (*j*) $\int \tan^3 x\sec^3 x\,dx$ (*k*) $\int \tan^4 x\sec x\,dx$ (*l*) $\int \sin 2x\cos 2x\,dx$

(*m*) $\int \sin\pi x\cos 3\pi x\,dx$ (*n*) $\int \sin 5x\sin 7x\,dx$ (*o*) $\int \cos 4x\cos 9x\,dx$

39.13 Prove the following identities:

(*a*) $\sin Ax\sin Bx = \frac{1}{2}[\cos(A-B)x - \cos(A+B)x]$

(*b*) $\cos Ax\cos Bx = \frac{1}{2}[\cos(A-B)x + \cos(A+B)x]$

39.14 Calculate the following definite integrals, where the positive integers n and k are distinct:

(*a*) $\int_0^{2\pi} \sin nx\sin kx\,dx$ (*b*) $\int_0^{2\pi} \sin^2 nx\,dx$

39.15 Evaluate

(*a*) $\int \frac{\sqrt{x^2-1}}{x}\,dx$ (*b*) $\int \frac{x^2}{\sqrt{4-x^2}}\,dx$ (*c*) $\int \frac{\sqrt{1+x^2}}{x}\,dx$ (*d*) $\int \frac{x}{\sqrt{2-x^2}}\,dx$

(*e*) $\int \frac{dx}{x^2\sqrt{x^2-9}}$ (*f*) $\int \frac{dx}{(4-x^2)^{3/2}}$ (*g*) $\int \frac{dx}{(x^2+9)^2}$ (*h*) $\int \frac{dx}{x\sqrt{16-9x^2}}$

(*i*) $\int x^2\sqrt{1-x^2}\,dx$ (*j*) $\int e^{3x}\sqrt{1-e^{2x}}\,dx$ (*k*) $\int \frac{dx}{(x^2-6x+13)^2}$

(*Hint*: In (*k*), complete the square.)

39.16 Find the arc length of the parabola $y = x^2$ from $(0, 0)$ to $(2, 4)$.

39.17 Find the arc length of the curve $y = \ln x$ from $(1, 0)$ to $(e, 1)$.

39.18 Find the arc length of the curve $y = e^x$ from $(0, 1)$ to $(1, e)$.

39.19 Find the arc length of the curve $y = \ln \cos x$ from $(0, 0)$ to $(\pi/3, -\ln 2)$.

39.20 Find the area enclosed by the ellipse

$$\frac{x^2}{9} + \frac{y^2}{4} = 1$$

Chapter 40

Integration of Rational Functions: The Method of Partial Fractions

This chapter will give a general method for evaluating indefinite integrals of the type

$$\int \frac{N(x)}{D(x)}\,dx$$

where $N(x)$ and $D(x)$ are polynomials. That is to say, we shall show how to find the antiderivative of any rational function $f(x) = N(x)/D(x)$ (see Section 9.3). Two assumptions will be made, neither of which is really restrictive: (i) the leading coefficient (the coefficient of the highest power of x) in $D(x)$ is $+1$; (ii) $N(x)$ is of lower degree than $D(x)$ (i.e., $f(x)$ is a *proper* rational function).

EXAMPLES

(i)

$$\frac{8x^4}{-\frac{1}{7}x^{10} + 3x - 11} = \left(\frac{-7}{-7}\right)\frac{8x^4}{-\frac{1}{7}x^{10} + 3x - 11} = \frac{-56x^5}{+1x^{10} - 21x + 77}$$

(ii) Consider the improper rational function

$$f(x) = \frac{x^4 + 7x}{x^2 - 1}$$

Long division (see Fig. 40-1) yields

$$f(x) = x^2 + 1 + \frac{7x + 1}{x^2 - 1}$$

Consequently,

$$\int f(x)\,dx = \int (x^2 + 1)\,dx + \int \frac{7x+1}{x^2-1}\,dx = \frac{x^3}{3} + x + \int \frac{7x+1}{x^2-1}\,dx$$

and the problem reduces to finding the antiderivative of a proper rational function.

$$\begin{array}{r@{}l} & x^2 + 1 \\ x^2 - 1 \,\big)\, & x^4 + 7x \\ & \underline{x^4 - x^2} \\ & \quad x^2 + 7x \\ & \quad \underline{x^2 - 1} \\ & \qquad 7x + 1 \end{array}$$

Fig. 40-1

The theorems that follow hold for polynomials with arbitrary real coefficients. However, for simplicity, we shall illustrate them only with polynomials whose coefficients are integers.

Theorem 40.1: Any polynomial $D(x)$ with leading coefficient $+1$ can be expressed as the product of *linear factors*, of the form $x - a$, and *irreducible quadratic factors* (that cannot be factored further), of the form $x^2 + bx + c$, repetition of factors being allowed.

As explained in Section 7.4, the real roots of $D(x)$ determine its linear factors.

EXAMPLES

(*a*)
$$x^2 - 1 = (x-1)(x+1)$$

Here, the polynomial has two real roots (± 1) and therefore falls into two linear factors.

(*b*)
$$x^3 + 2x^2 - 8x - 21 = (x-3)(x^2+5x+7)$$

The root $x = 3$, which generates the linear factor $x - 3$, was found by testing the divisors of 21. Division of $D(x)$ by $x - 3$ yielded the polynomial $x^2 + 5x + 7$. This polynomial is irreducible, since, by the quadratic formula, its roots are

$$x = \frac{-b \pm \sqrt{b^2 - 4c}}{2} = \frac{-5 \pm \sqrt{-3}}{2}$$

which are not real numbers.

Theorem 40.2 (*Partial Fractions Representation*): Any (proper) rational function $f(x) = N(x)/D(x)$ may be written as a sum of simpler, proper rational functions. Each summand has as denominator *one* of the linear or quadratic factors of $D(x)$, raised to some power.

By Theorem 40.2, $\int f(x)\,dx$ is given as a sum of simpler antiderivatives—antiderivatives which, in fact, can be found by the techniques already known to us.

It will now be shown how to construct the partial fractions representation and to integrate it term by term.

Case 1: D(x) falls into nonrepeated linear factors.

The partial fractions representation of $f(x)$ is

$$\frac{N(x)}{(x-a_1)(x-a_2)\cdots(x-a_n)} = \frac{A_1}{x-a_1} + \frac{A_2}{x-a_2} + \cdots + \frac{A_n}{x-a_n}$$

The constant numerators $A_1, \ldots, A_n$ are evaluated as in the following

EXAMPLE
$$\frac{2x+1}{(x+1)(x-1)} = \frac{A_1}{x+1} + \frac{A_2}{x-1}$$

Clear the denominators by multiplying both sides by $(x+1)(x-1)$:

$$(x+1)(x-1)\frac{2x+1}{(x+1)(x-1)} = (x+1)(x-1)\frac{A_1}{x+1} + (x+1)(x-1)\frac{A_2}{x-1}$$

$$2x + 1 = A_1(x-1) + A_2(x+1) \qquad (1)$$

In (*1*), substitute individually the roots of $D(x)$. With $x = -1$,

$$-1 = A_1(-2) + 0 \qquad \text{or} \qquad A_1 = \frac{1}{2}$$

and with $x = 1$,

$$3 = 0 + A_2(2) \qquad \text{or} \qquad A_2 = \frac{3}{2}$$

With all constants known, the antiderivative of $f(x)$ will be a sum of terms of the form

$$\int \frac{A}{x-a}\,dx = A \ln|x-a|$$

Case 2: D(x) falls into linear factors, at least one of which is repeated.

This is treated in the same manner as in Case 1, except that a repeated factor $(x-a)^k$ gives rise to a sum of the form

$$\frac{A_1}{x-a} + \frac{A_2}{(x-a)^2} + \cdots + \frac{A_k}{(x-a)^k}$$

EXAMPLE $$\frac{3x+1}{(x-1)^2(x-2)}=\frac{A_1}{x-1}+\frac{A_2}{(x-1)^2}+\frac{A_3}{x-2}$$

Multiply by $(x-1)^2(x-2)$:

$$3x+1=A_1(x-1)(x-2)+A_2(x-2)+A_3(x-1)^2 \tag{2}$$

Letting $x=1$,

$$4=0+A_2(-1)+0 \qquad \text{or} \qquad A_2=-4$$

Letting $x=2$,

$$7=0+0+A_3(1) \qquad \text{or} \qquad A_3=7$$

The remaining numerator, A_1, is determined by the condition that the coefficient of x^2 on the right side of (2) be zero (since it is zero on the left side). Thus,

$$A_1+A_3=0 \qquad \text{or} \qquad A_1=-A_3=-7$$

(More generally, we use all the roots of $D(x)$ to determine the "highest" A's, and then compare coefficients—of as many powers of x as necessary—to find the remaining A's.)

Now the antiderivative of $f(x)$ will consist of terms in $\ln|x-a|$, plus at least one term of the form

$$\int \frac{A}{(x-a)^j}\,dx=\frac{B}{(x-a)^{j-1}} \quad (j\geq 2)$$

Case 3: $D(x)$ has quadratic factors, but none is repeated.

In this case, each quadratic factor x^2+bx+c contributes a term

$$\frac{Ax+B}{x^2+bx+c}$$

to the partial fractions representation.

EXAMPLE $$\frac{x^2-1}{(x^2+1)(x+2)}=\frac{A_1}{x+2}+\frac{A_2x+A_3}{x^2+1}$$

Multiply by $(x^2+1)(x+2)$:

$$x^2-1=A_1(x^2+1)+(A_2x+A_3)(x+2) \tag{3}$$

Let $x=-2$:

$$3=A_1(5)+0 \qquad \text{or} \qquad A_1=\frac{3}{5}$$

Comparing coefficients of x^0 (the constant terms):

$$-1=A_1+2A_3 \qquad \text{or} \qquad A_3=-\frac{1}{2}(1+A_1)=-\frac{1}{2}\left(\frac{8}{5}\right)=-\frac{4}{5}$$

Comparing coefficients of x^2:

$$1=A_1+A_2 \qquad \text{or} \qquad A_2=1-A_1=\frac{2}{5}$$

The sum for $\int f(x)\,dx$ will now include, besides terms arising from any linear factors, at least one term of the form

$$\int \frac{Ax+B}{x^2+bx+c}\,dx=\int \frac{A\left(x+\frac{b}{2}\right)+C}{\left(x+\frac{b}{2}\right)^2+\delta^2}\,dx \qquad \left[\begin{array}{l} C\equiv B-\dfrac{Ab}{2} \\ \delta^2\equiv c-\dfrac{b^2}{4}>0 \end{array}\right]$$

$$= \int \frac{Au + C}{u^2 + \delta^2}\, du \quad \left[\text{let } u = x + \frac{b}{2}\right]$$

$$= A \int \frac{u\, du}{u^2 + \delta^2} + C \int \frac{du}{u^2 + \delta^2}$$

$$= \frac{A}{2} \ln (u^2 + \delta^2) + \frac{C}{\delta} \tan^{-1} \frac{u}{\delta}$$

(For a guarantee that δ is a *real* number, see Problem 40.7.)

Case 4: $D(x)$ has at least one repeated quadratic factor.

A repeated quadratic factor $(x^2 + bx + c)^k$ contributes to the partial fractions representation the expression

$$\frac{A_1x + A_2}{ax^2 + bx + c} + \frac{A_3x + A_4}{(ax^2 + bx + c)^2} + \cdots + \frac{A_{2k-1}x + A_{2k}}{(ax^2 + bx + c)^k}$$

The computations in this case may be long and tedious.

EXAMPLE
$$\frac{x^3 + 1}{(x^2 + 1)^2} = \frac{A_1x + A_2}{x^2 + 1} + \frac{A_3x + A_4}{(x^2 + 1)^2}$$

(the simplest possible $D(x)$). Multiply by $(x^2 + 1)^2$:

$$x^3 + 1 = (A_1x + A_2)(x^2 + 1) + A_3x + A_4 \tag{4}$$

Compare coefficients of x^3:

$$1 = A_1$$

Compare coefficients of x^2:

$$0 = A_2$$

Compare coefficients of x:

$$0 = A_1 + A_3 \qquad \text{or} \qquad A_3 = -A_1 = -1$$

Compare coefficients of x^0:

$$1 = A_2 + A_4 \qquad \text{or} \qquad A_4 = 1 - A_2 = 1$$

The new contribution to $\int f(x)\, dx$ will consist of one or more terms of the form

$$\int \frac{Ax + B}{(x^2 + bx + c)^j}\, dx = A \int \frac{u\, du}{(u^2 + \delta^2)^j} + C \int \frac{du}{(u^2 + \delta^2)^j} \quad \text{[as in Case 3]}$$

$$= \frac{E}{(u^2 + \delta^2)^{j-1}} + F \int \cos^{2(j-1)} \theta\, d\theta \quad [\text{let } u = \delta \tan \theta]$$

and we know how to evaluate the trigonometric integral (see Problem 38.12(a) or Example (b) of Section 39.1).

Solved Problems

40.1 Evaluate

$$\int \frac{2x^3 + x^2 - 6x + 7}{x^2 + x - 6}\,dx$$

The numerator has greater degree than the denominator. Therefore, divide the numerator by the denominator:

$$\begin{array}{r|l} & 2x \;\; - 1 \\ x^2 + x - 6 & 2x^3 + x^2 - 6x + 7 \\ & 2x^3 + 2x^2 - 12x \\ & \quad - x^2 + 6x + 7 \\ & \quad - x^2 - x + 6 \\ & \qquad 7x + 1 \end{array}$$

Thus
$$\frac{2x^3 + x^2 - 6x + 7}{x^2 + x - 6} = 2x - 1 + \frac{7x + 1}{x^2 + x - 6}$$

Next, factor the denominator: $x^2 + x - 6 = (x + 3)(x - 2)$. The partial fractions decomposition has the form (Case 1):

$$\frac{7x + 1}{(x + 3)(x - 2)} = \frac{A_1}{x + 3} + \frac{A_2}{x - 2}$$

Multiply by the denominator $(x + 3)(x - 2)$:

$$7x + 1 = A_1(x - 2) + A_2(x + 3)$$

Let $x = 2$: $15 = 0 + 5A_2$, or $A_2 = 3$.
Let $x = -3$: $-20 = -5A_1 + 0$, or $A_1 = 4$.

Thus
$$\frac{7x + 1}{(x + 3)(x - 2)} = \frac{4}{x + 3} + \frac{3}{x - 2}$$

and
$$\int \frac{2x^3 + x^2 - 6x + 7}{x^2 + x - 6}\,dx = \int (2x - 1)\,dx + \int \frac{4}{x + 3}\,dx + \int \frac{3}{x - 2}\,dx$$
$$= x^2 - x + 4 \ln |x + 3| + 3 \ln |x - 2| + C$$

40.2 Find

$$\int \frac{x^2\,dx}{x^3 - 3x^2 - 9x + 27}$$

Testing the factors of 27, we find that 3 is a root of $D(x)$. Dividing $D(x)$ by $x - 3$ yields

$$x^3 - 3x^2 - 9x + 27 = (x - 3)(x^2 - 9) = (x - 3)(x - 3)(x + 3) = (x - 3)^2(x + 3)$$

and so the partial fractions representation is (Case 2):

$$\frac{x^2}{(x - 3)^2(x + 3)} = \frac{A_1}{x - 3} + \frac{A_2}{(x - 3)^2} + \frac{A_3}{x + 3}$$

Multiply by $(x - 3)^2(x + 3)$:

$$x^2 = A_1(x - 3)(x + 3) + A_2(x + 3) + A_3(x - 3)^2$$

Let $x = 3$: $9 = 0 + 6A_2 + 0$, or $A_2 = 3/2$.
Let $x = -3$: $9 = 0 + 0 + A_3(-6)^2$, or $A_3 = 1/4$.

Compare coefficients of x^2: $1 = A_1 + A_3$, or $A_1 = 1 - A_3 = 3/4$.

Thus
$$\frac{x^2}{x^3-3x^2-9x+27} = \frac{3}{4}\frac{1}{x-3} + \frac{3}{2}\frac{1}{(x-3)^2} + \frac{1}{4}\frac{1}{x+3}$$

and
$$\int \frac{x^2\,dx}{x^3-3x^2-9x+27} = \frac{3}{4}\ln|x-3| - \frac{3}{2}\frac{1}{x-3} + \frac{1}{4}\ln|x+3| + C$$

40.3 Find

$$\int \frac{x+1}{x(x^2+2)}\,dx$$

This is Case 3:

$$\frac{x+1}{x(x^2+2)} = \frac{A_1}{x} + \frac{A_2x + A_3}{x^2+2}$$

Multiply by $x(x^2+2)$:

$$x + 1 = A_1(x^2+2) + x(A_2x + A_3)$$

Let $x = 0$: $1 = 2A_1 + 0$, or $A_1 = 1/2$.
Compare coefficients of x^2: $0 = A_1 + A_2$, or $A_2 = -A_1 = -1/2$.
Compare coefficients of x: $1 = A_3$.

Thus
$$\frac{x+1}{x(x^2+2)} = \frac{1}{2}\left(\frac{1}{x}\right) + \frac{(-1/2)x+1}{x^2+2}$$

and
$$\int \frac{x+1}{x(x^2+2)}\,dx = \frac{1}{2}\int \frac{1}{x}\,dx - \frac{1}{2}\int \frac{x\,dx}{x^2+2} + \int \frac{dx}{x^2+2}$$

Because the quadratic factor, x^2+2, is a complete square, we can perform the integrations on the right without a change of variable:

$$\int \frac{x+1}{x(x^2+2)}\,dx = \frac{1}{2}\ln|x| - \frac{1}{4}\ln(x^2+2) + \frac{1}{\sqrt{2}}\tan^{-1}\frac{x}{\sqrt{2}} + C$$

40.4 Evaluate

$$\int \frac{1}{1 - \sin x + \cos x}\,dx$$

Observe that the integrand is *a rational function of* $\sin x$ *and* $\cos x$. Any rational function of the six trigonometric functions reduces to a function of this type, and the method we shall use to solve this particular problem will work for any such function.

Make the change of variable $z = \tan(x/2)$; that is, $x = 2\tan^{-1} z$. Then:

$$dx = \frac{2}{1+z^2}\,dz$$

and, by Theorem 26.8,

$$\sin x = 2\sin\frac{x}{2}\cos\frac{x}{2} = 2\frac{\tan x/2}{\sec^2 x/2}$$
$$= 2\frac{\tan x/2}{1+\tan^2 x/2} = \frac{2z}{1+z^2}$$

$$\cos x = 1 - 2\sin^2\frac{x}{2} = 1 - 2\frac{\tan^2 x/2}{\sec^2 x/2}$$
$$= 1 - 2\frac{\tan^2 x/2}{1+\tan^2 x/2} = 1 - \frac{2z^2}{1+z^2} = \frac{1-z^2}{1+z^2}$$

When these substitutions are made, the resulting integrand will be a rational function of z (because compositions and products of rational functions are rational functions). The method of partial fractions can then be applied.

$$\int (1-\sin x+\cos x)^{-1}\,dx = \int \left(1-\frac{2z}{1+z^2}+\frac{1-z^2}{1+z^2}\right)^{-1}\frac{2}{1+z^2}\,dz$$

$$= \int \left(\frac{(1+z^2)-2z+(1-z^2)}{1+z^2}\right)^{-1}\frac{2}{1+z^2}\,dz$$

$$= \int \left(\frac{2-2z}{1+z^2}\right)^{-1}\frac{2}{1+z^2}\,dz = \int \frac{1+z^2}{2-2z}\,\frac{2}{1+z^2}\,dz$$

$$= \int \frac{1}{1-z}\,dz = -\ln|1-z|$$

$$= -\ln\left|1-\tan\frac{x}{2}\right| + C$$

40.5 Find

$$\int \frac{x\,dx}{(x+1)(x^2+2x+2)^2}$$

This is Case 4 for $D(x)$, and so

$$\frac{x}{(x+1)(x^2+2x+2)^2} = \frac{A_1}{x+1}+\frac{A_2x+A_3}{x^2+2x+2}+\frac{A_4x+A_5}{(x^2+2x+2)^2}$$

Multiply by $(x+1)(x^2+2x+2)^2$:

$$x = A_1(x^2+2x+2)^2+(A_2x+A_3)(x+1)(x^2+2x+2)+(A_4x+A_5)(x+1)$$

or, partially expanding the right-hand side,

$$x = A_1(x^4+4x^3+8x^2+8x+4)+(A_2x+A_3)(x^3+3x^2+4x+2)+(A_4x+A_5)(x+1) \qquad (1)$$

In (*1*), let $x=-1$: $-1 = A_1(1)^2 = A_1$.
Compare coefficients of x^4: $0 = A_1+A_2$, or $A_2 = -A_1 = 1$.
Compare coefficients of x^3: $0 = 4A_1+3A_2+A_3$, or $A_3 = -4A_1-3A_2 = 1$.
Compare coefficients of x^2: $0 = 8A_1+4A_2+3A_3+A_4$, or $A_4 = -8A_1-4A_2-3A_3 = 1$.
Compare coefficients of x^0: $0 = 4A_1+2A_3+A_5$, or $A_5 = -4A_1-2A_3 = 2$.
Therefore,

$$\int \frac{x\,dx}{(x+1)(x^2+2x+2)^2} = -1\int\frac{dx}{x+1}+\int\frac{(x+1)\,dx}{x^2+2x+2}+\int\frac{(x+2)\,dx}{(x^2+2x+2)^2}$$

$$= -\ln|x+1| + \int\frac{u\,du}{u^2+1}+\int\frac{u\,du}{(u^2+1)^2}+\int\frac{du}{(u^2+1)^2}$$

$$\left[\text{Cases 3 and 4: let } u = x+\frac{2}{2}\right]$$

$$= -\ln|x+1|+\frac{1}{2}\ln(u^2+1)-\frac{1}{2}\left(\frac{1}{u^2+1}\right)$$

$$+\int \cos^2\theta\,d\theta \quad [\text{Case 4: let } u = 1\tan\theta]$$

$$= -\ln|x+1|+\frac{1}{2}\ln(x^2+2x+2)-\frac{1}{2}\left(\frac{1}{x^2+2x+2}\right)$$

$$+\frac{\theta}{2}+\frac{\sin 2\theta}{4}$$

Now,

$$\theta = \tan^{-1}u = \tan^{-1}(x+1)$$

and (see Problem 40.4)

$$\sin 2\theta = \frac{2\tan\theta}{1+\tan^2\theta} = \frac{2(x+1)}{x^2+2x+2}$$

so that we have, finally,

$$\int \frac{x\,dx}{(x+1)(x^2+2x+2)^2} = -\ln|x+1| + \frac{1}{2}\ln(x^2+2x+2) - \frac{1}{2}\left(\frac{1}{x^2+2x+2}\right)$$
$$+\frac{1}{2}\tan^{-1}(x+1) + \frac{1}{2}\left(\frac{x+1}{x^2+2x+2}\right) + C$$
$$= \frac{1}{2}\left[\ln(x^2+2x+2) + \frac{x}{x^2+2x+2} + \tan^{-1}(x+1)\right] - \ln|x+1| + C$$

Supplementary Problems

40.6 Find the following antiderivatives:

(*a*) $\int \frac{dx}{x^2-9}$ (*b*) $\int \frac{x\,dx}{(x+2)(x+3)}$ (*c*) $\int \frac{x^4-4x^2+x+1}{x^2-4}\,dx$

(*d*) $\int \frac{2x^2+1}{(x-1)(x-2)(x-3)}\,dx$ (*e*) $\int \frac{x^2-4}{x^3-3x^2-x+3}\,dx$ (*f*) $\int \frac{x^3+1}{x(x+3)(x+2)(x-1)}\,dx$

(*g*) $\int \frac{x\,dx}{x^4-13x^2+36}$ (*h*) $\int \frac{x-5}{x^2(x+1)}\,dx$ (*i*) $\int \frac{2x\,dx}{(x-2)^2(x+2)}$

(*j*) $\int \frac{x+4}{x^3+6x^2+9x}\,dx$ (*k*) $\int \frac{x^4\,dx}{x^3-2x^2-7x-4}$ (*l*) $\int \frac{dx}{x(x^2+5)}$

(*m*) $\int \frac{x^2\,dx}{(x-1)(x^2+4x+5)}$ (*n*) $\int \frac{dx}{(x^2+1)(x^2+4)}$ (*o*) $\int \frac{x^4+1}{x^3+9x}\,dx$

(*p*) $\int \frac{dx}{x(x^2+1)^2}$ (*q*) $\int \frac{x^2\,dx}{(x-1)(x^2+4)^2}$ (*r*) $\int \frac{x^3+1}{x(x^2+x+1)^2}\,dx$

(*s*) $\int \frac{x-1}{x^3+2x^2-x-2}\,dx$ (*t*) $\int \frac{x^2+2}{x(x^2+5x+6)}\,dx$ (*u*) $\int \frac{dx}{1+e^x}$

40.7 Show that $p(x) = x^2 + bx + c$ is irreducible if and only if

$$c - \frac{b^2}{4} > 0$$

(*Hint*: A quadratic polynomial is irreducible when and only when it has no linear factor; that is (by Theorem 7.2), when and only when it has no real root.)

40.8 (*a*) Find the area of the region in the first quadrant under the curve

$$y = \frac{1}{x^3+27}$$

and to the left of the line $x = 3$. (*b*) Find the volume of the solid generated by revolving the region of (*a*) around the y-axis.

40.9 Find $$\int \frac{dx}{1-\sin x}.$$

(*Hint*: See Problem 40.4.)

40.10 Find

$$\int \frac{\cos x\, dx}{\sin x - 1}$$

(*a*) by the method of Problem 40.4, (*b*) by substituting $u = \sin x$. (*c*) Verify that your answers are equivalent.

40.11 Evaluate the following integrals involving fractional powers:

(*a*) $\int \frac{dx}{\sqrt[3]{x} - x}$ (*b*) $\int \frac{dx}{1 + \sqrt[4]{x - 1}}$ (*c*) $\int \frac{dx}{x\sqrt{1 + 3x}}$ (*d*) $\int \frac{dx}{\sqrt[3]{x} + \sqrt{x}}$

(*e*) $\int \frac{dx}{\sqrt{\sqrt{x} + 1}}$ (*f*) $\int \sqrt{1 + e^x}\, dx$ (*g*) $\int \frac{x^{2/3}\, dx}{x + 1}$

(*Hints*: (*a*) Let $x = z^3$; (*b*) let $x - 1 = z^4$; (*c*) let $1 + 3x = z^2$; (*d*) let $x = z^6$.)

Answers to Supplementary Problems

CHAPTER 1

1.8 (*a*) $u \le 0$; (*b*) $x \ge 3$; (*c*) $x \le 3$.

1.9 If $u \ge 0$, $u = |u|$; if $u < 0$, $u = -|u|$.

1.10 Use (*1.3*): (*a*) $x = \frac{1}{2}$ or $x = -\frac{7}{2}$; (*b*) $x = \frac{8}{5}$ or $x = \frac{6}{5}$.

1.11 (*a*) $0 < x < 2$; (*b*) $-3 \le x \le -\frac{1}{3}$; (*c*) $x < -6$ or $x > -2$; (*d*) $x \le 1$ or $x \ge 4$; (*e*) $2 \le x \le 4$ or $-4 \le x \le -2$; (*f*) $-8 < x < -4$.

1.12 (*a*) $x > -5$; (*b*) $-13 < x < -3$; (*c*) $x < -\frac{1}{2}$ or $x > \frac{1}{6}$; (*d*) $-1 < x < 3$; (*e*) $-1 < x < 1$; (*f*) $1 \le x < \frac{3}{2}$.

1.13 (*a*) $x > 0$ or $x < -2$; (*b*) $-4 < x < 1$; (*c*) $x < 1$ or $x > 5$; (*d*) $-8 < x < 1$; (*e*) $-1 < x < 4$; (*f*) $-1 < x < 0$ or $x > 1$; (*g*) $x < -7$ or $-\frac{1}{2} < x < 3$.

1.14 $|b| \cdot \left|\dfrac{a}{b}\right| = \left|b \cdot \dfrac{a}{b}\right| = |a|$.

1.15 (*a*) By (*1.5*). (*b*) By (*1.5*) and (*a*): $|a^3| = |a^2a| = |a^2|\,|a| = |a|^2\,|a| = |a|^3$.
(*c*) $|a^n| = |a|^n$ for all positive integers *n*.

1.16 Use (*1.3*): (*a*) $x = 5$ or $x = \frac{1}{3}$; (*b*) $x = \frac{9}{4}$ or $x = \frac{1}{10}$; (*c*) $x = 8$.

1.17 (*a*) $\frac{1}{3} < x < 5$; (*b*) $\frac{1}{2} \le x \le \frac{3}{4}$.

1.19 Yes: $(a^2)^2 = a^4$ and $a^2 \ge 0$.

1.20 No: $\sqrt{a^2} < \sqrt{b^2}$ implies $|a| < |b|$, but $a < b$ does not hold when, for example, $a = 1$ and $b = -2$.

1.21

−1 0 1 2 3 4

B D O I C A

$\overline{IA} = 3$, $\overline{AI} = 3$, $\overline{OC} = \frac{3}{2}$, $\overline{BC} = \frac{5}{2}$, $\overline{IB} + \overline{BD} = 2 + \frac{2}{3} = \frac{8}{3}$, $\overline{ID} = \frac{4}{3}$, $\overline{IB} + \overline{BC} = 2 + \frac{5}{2} = \frac{9}{2}$, $\overline{IC} = \frac{1}{2}$.

1.22 (*a*) $b = 10$; (*b*) $b = -5$; (*c*) $b = -5$.

CHAPTER 2

2.4 $A(0, 4)$, $B(2, 2)$, $C(4, 0)$, $D(-3, 1)$, $E(0, -4)$, $F(2, -3)$.

2.6 (*a*) 5; (*b*) 5; (*c*) 2; (*d*) 8.

2.7 area (right triangle) $= \dfrac{1}{2}(\overline{AC})(\overline{AB}) = \dfrac{1}{2}(5)(10) = 25$.

2.8 $(3, 4)$.

2.9 $(-1, 1)$ and $(3, 0)$.

2.10 $(0, 2)$, $(6, 2)$, $(4, -4)$.

2.11 $(2, y)$ for some real number *y*.

2.12 (*a*) $\sqrt{34}$; (*b*) $3\sqrt{2}$; (*c*) $\frac{\sqrt{389}}{4}$.

2.13 (*a*) isosceles only; (*b*) right only, area = 10; (*c*) isosceles right, area = 17.

2.14 $k = 5$.

2.15 (*a*) no; (*b*) yes.

2.16 (*a*) (4, 2); (*b*) $\left(\frac{5}{4}, 2\right)$; (*c*) $\left(\frac{5+\sqrt{2}}{2}, 2\right)$.

2.17 (5, 8).

CHAPTER 3

3.6 See Fig. A-1.

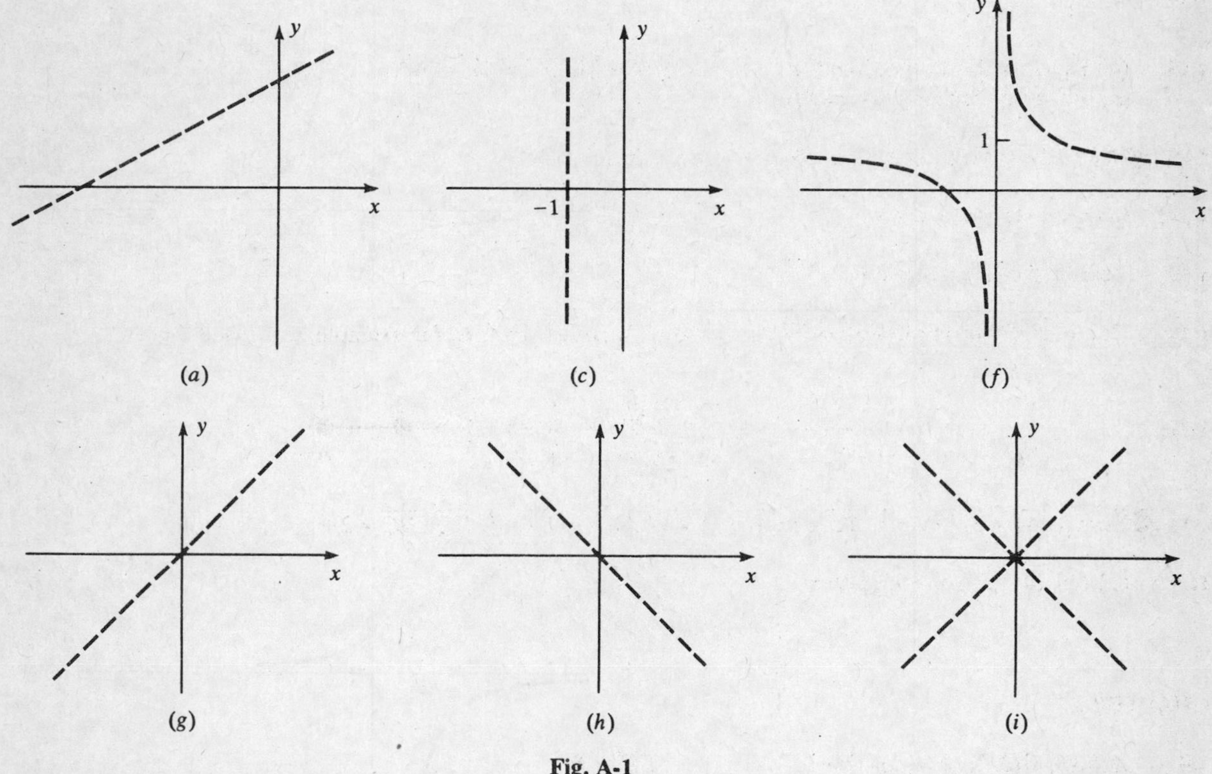

Fig. A-1

3.7 See Fig. A-2, page 306.

3.8 See Fig. A-3, page 306.

3.9 See Fig. A-4, page 307.

3.10 $x^2 = 4py$ (parabola).

3.11 (*a*) $(x-4)^2+(y-3)^2=1$; (*b*) $(x+1)^2+(y-5)^2=2$; (*c*) $x^2+(y-2)^2=16$; (*d*) $(x-3)^2+(y-3)^2=18$; (*e*) $(x-4)^2+(y+1)^2=20$; (*f*) $(x-1)^2+(y-2)^2=5$.

3.12 (*a*) circle (center (6, −10), radius 11); (*b*) circle (center (0, −15), radius 14); (*c*) null set; (*d*) circle (center $(\frac{1}{4}, 0)$, radius $\frac{1}{4}$); (*e*) point (−1, 1); (*f*) circle (center (−3, −2), radius 7).

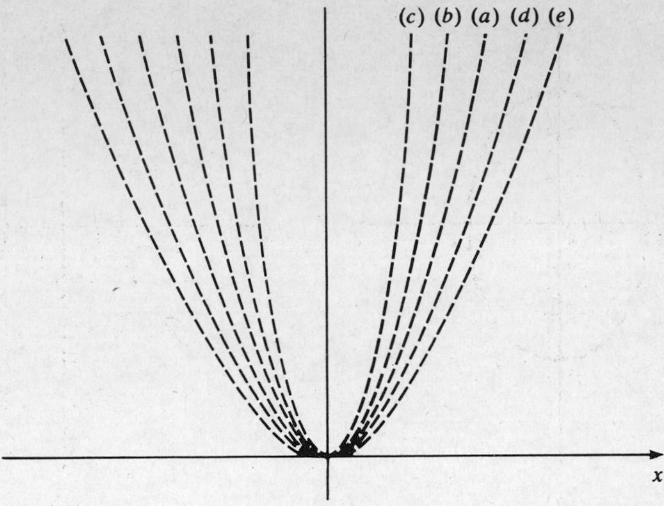

Fig. A-2

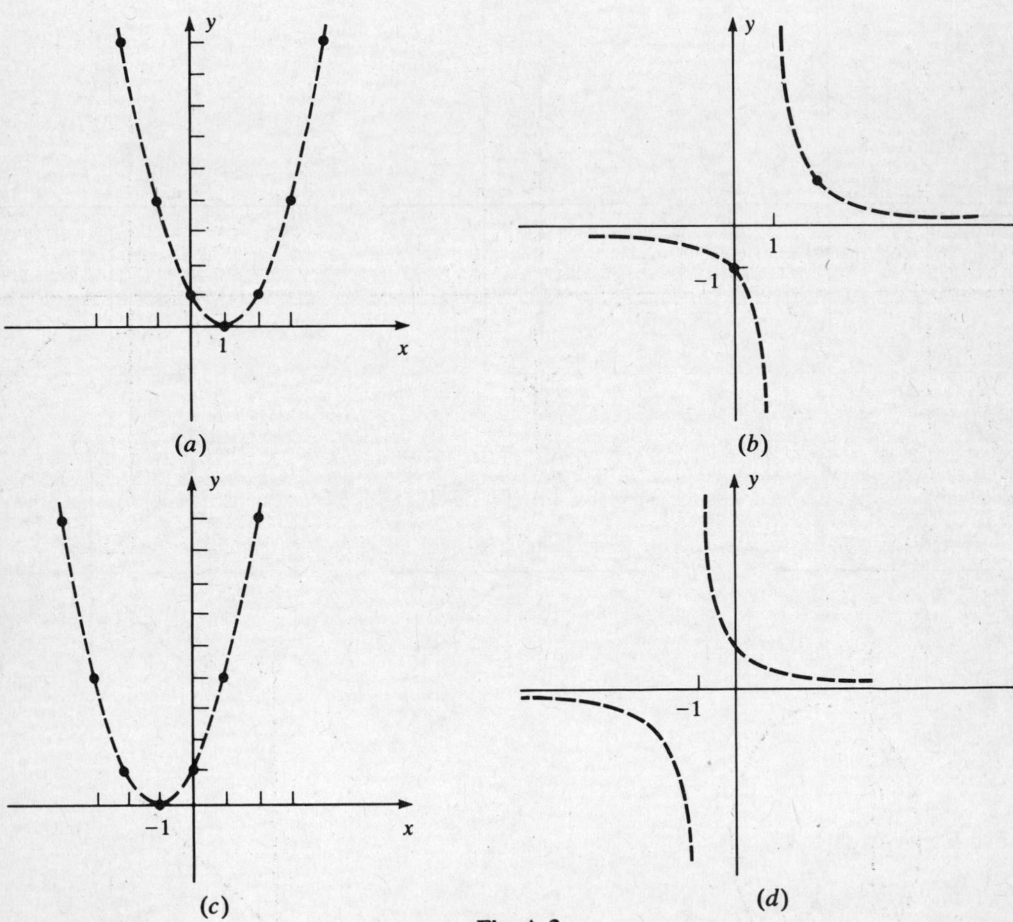

Fig. A-3

3.13 (b) $4F < D^2 + E^2$.

3.14 $(x-3)^2 + (y+2)^2 = 100$.

3.15 $k = 2$ and $k = -\frac{4}{5}$.

3.16 $(x-1)^2 + (y-3)^2 = 9$, $(x-7)^2 + (y-3)^2 = 9$, $(x-1)^2 + (y-9)^2 = 9$, $(x-7)^2 + (y-9)^2 = 9$.

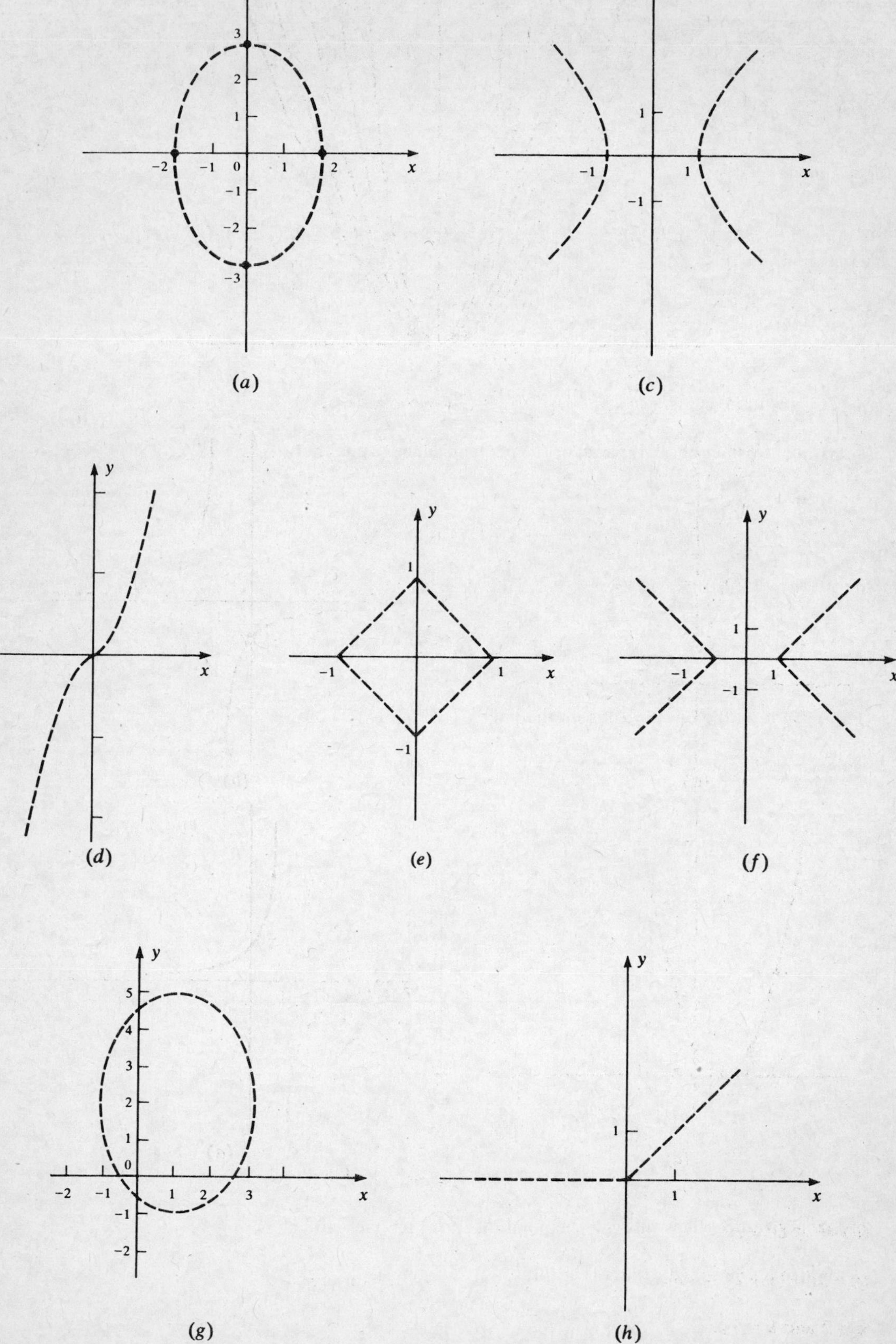

Fig. A-4

CHAPTER 4

4.7 (*a*) $\frac{y-5}{x-2}=\frac{1}{3}$ or $\frac{y-4}{x+1}=\frac{1}{3}$; (*b*) $\frac{y}{x}=4$ or $\frac{y-4}{x-1}=4$; (*c*) $\frac{y+1}{x-7}=-1$ or $\frac{y-7}{x+1}=-1$.

4.8 (*a*) $y=\frac{5}{6}x+\frac{14}{3}$; (*b*) $y=2x-1$; (*c*) $y=-\frac{2}{3}x+2$; (*d*) $y=4$; (*e*) $y=5x-1$; (*f*) $y=-3x+16$; (*g*) $y=-\frac{3}{4}x+\frac{13}{4}$; (*h*) $y=0$; (*i*) $y=-3x+7$; (*j*) $y=\frac{2}{5}x$; (*k*) $y=3$; (*l*) $y=x$.

4.9 (*a*) $m=5$, $b=4$, $(1,9)$; (*b*) $m=\frac{7}{4}$, $b=-2$, $(4,5)$; (*c*) $m=-4$, $b=2$, $(1,-2)$; (*d*) $m=0$, $b=2$, $(1,2)$; (*e*) $m=-\frac{4}{3}$, $b=4$, $(3,0)$.

4.10 $k=9$.

4.11 No.

4.12 (*a*) Yes. (*c*) In all cases. (*d*) $k=-\frac{1}{4}$.

4.13 (*a*) parallel; (*b*) neither; (*c*) parallel; (*d*) perpendicular; (*e*) neither.

4.14 (*a*) $y=\frac{9}{5}x+32$; (*b*) $-40°$.

4.15 (*a*) 10; (*b*) -15; (*c*) $-\frac{5}{4}$; (*d*) $\frac{15}{2}$.

4.16 (*a*) $y=-\frac{1}{11}x+\frac{23}{11}$; (*b*) $y=-4x+32$; (*c*) $y=\frac{7}{2}x-\frac{59}{4}$.

4.17 $(12,9)$ is not on the line; $(6,3)$ is on the line.

4.21 $4x+3y-9>0$ and $x>1$; see Fig. A-5.

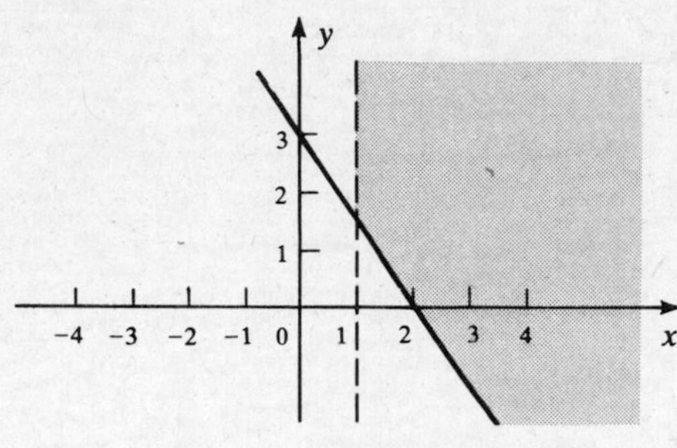

Fig. A-5

4.22 $x<200/3$.

4.24 (*a*) all nonvertical lines through the point $(0,2)$; (*b*) all lines with slope 3.

4.25 (*a*) horizontal lines; (*b*) (i) 2, (ii) 4, (iii) $\frac{1}{5}$, (iv) 3, (v) none.

4.26 (*a*) $y=-\frac{2}{3}x+3$; (*b*) $y=-x+9$; (*c*) $y=\frac{1}{6}x+\frac{43}{12}$.

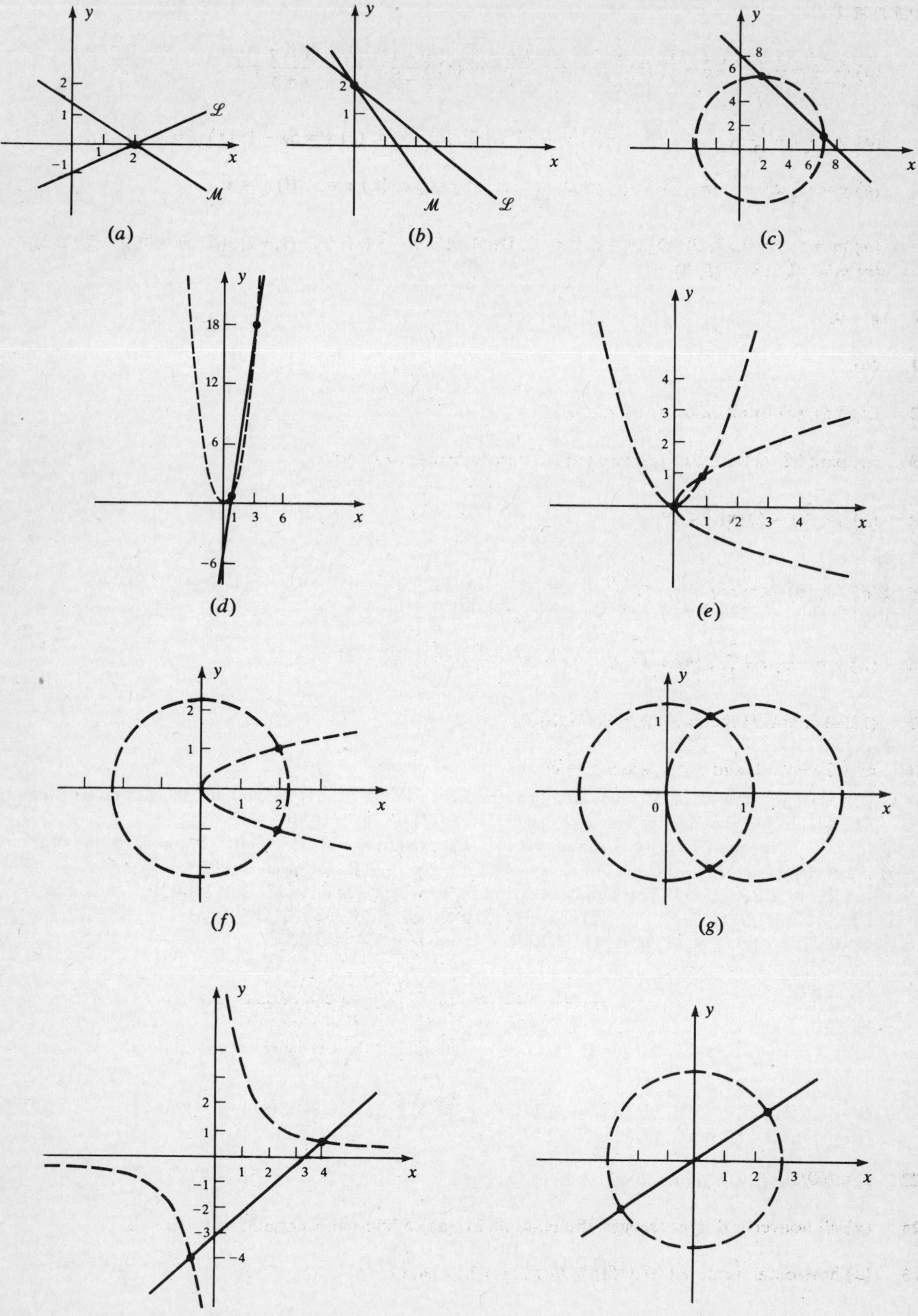

Fig. A-6

CHAPTER 5

5.4 (*a*) (2, 0); (*b*) (0, 2); (*c*) (7, 1) and (2, 6); (*d*) (1, 2) and (3, 18); (*e*) (0, 0) and (1, 1); (*f*) $(2, \sqrt{2})$ and $(2, -\sqrt{2})$; (*g*) $\left(\frac{1}{2}, \frac{\sqrt{3}}{2}\right)$ and $\left(\frac{1}{2}, -\frac{\sqrt{3}}{2}\right)$; (*h*) (4, 1) and (−1, −4); (*i*) $\left(\frac{9}{13}\sqrt{13}, \frac{6}{13}\sqrt{13}\right)$ and $\left(-\frac{9}{13}\sqrt{13}, -\frac{6}{13}\sqrt{13}\right)$. See Fig. A-6, page 309.

5.5 $\frac{8}{5}\sqrt{5} \approx 3.58$.

5.7 $x = 50$.

5.8 center $\left(\frac{9}{2}, \frac{9}{2}\right)$, radius $\frac{3}{2}\sqrt{10}$.

5.9 $x = 0$ and $y = -\frac{4}{3}x$.

5.10 (33/2, 34).

CHAPTER 6

6.5 (*a*) y-axis; (*b*) origin; (*c*) x-axis, y-axis, origin; (*d*) x-axis, y-axis, origin; (*e*) x-axis; (*f*) none; (*g*) origin; (*h*) none; (*i*) y-axis; (*j*) y-axis; (*k*) origin; (*l*) none.

6.6 (*a*) $x^2 + xy + y^2 = 1$; (*b*) $y^3 + xy^2 - x^3 = 8$; (*c*) $x^2 + 12x - 3y = 1$.

CHAPTER 7

7.8 (Let **R** denote the set of real numbers. In each answer, the first set is the domain, the second set the range. The graphs are sketched in Fig. A-7, page 311.) (*a*) **R**, $(-\infty, 4]$; (*b*) $[0, \infty)$, $(-\infty, 0]$; (*c*) $[-2, 2]$, range of H is $[0, 2]$, range of J is $[-2, 0]$; (*d*) $(-\infty, -2] \cup [2, \infty)$ (the union or "sum" of the two intervals), $[0, \infty)$; (*e*) **R**, $[0, \infty)$; (*f*) **R**, set of all integers; (*g*) **R**, set of all integers; (*h*) $\mathbf{R} - \{0\}$ (the set of all real numbers except 0), $\mathbf{R} - \{0\}$; (*i*) $\mathbf{R} - \{1\}$, $\mathbf{R} - \{0\}$; (*j*) **R**, **R**; (*k*) **R**, **R**; (*l*) **R**, $[2, \infty)$; (*m*) $\{1, 2, 4\}$, $\{-1, 3\}$; (*n*) $\mathbf{R} - \{-2\}$, $\mathbf{R} - \{-4\}$; (*o*) **R**, $(-\infty, 2] \cup \{4\}$; (*p*) $\mathbf{R} - \{0\}$, $\{-1, 1\}$; (*q*) **R**, $[2, \infty)$; (*r*) **R**, **R**; (*s*) **R**, $[0, 1)$; (*t*) **R**, **R**.

7.9 (*c*) and (*d*).

7.10 (*a*) $f(x) = \frac{2}{x^3}$, domain is $\mathbf{R} - \{0\}$; (*b*) $f(x) = \frac{x-1}{x+1}$, domain is $\mathbf{R} - \{-1\}$; (*c*) $f(x) = x$, domain is **R**.

7.11 (*a*) domain is $\mathbf{R} - \{2, 3\}$, range is $(0, \infty) \cup (-\infty, -4]$; (*b*) $(-1, 1)$, $(1, \infty)$; (*c*) $(-1, \infty)$, $(0, 2]$; (*d*) $[0, 4)$, $[-1, 2]$; (*e*) **R**, $[0, \infty)$.

7.12 (*a*) $k = -8$; (*b*) f is not defined when $x = 0$, but g is.

7.13 Of the infinitely many correct answers, some examples are: (*a*) $f(x) = 2x$ for $0 < x < 1$; (*b*) $f(x) = 5x - 1$ for $0 \le x < 1$; (*c*) $f(x) = \begin{cases} 0 & \text{if } x = 0 \\ 1 & \text{if } x > 0 \end{cases}$; (*d*) $f(x) = (x-1)^2 + 1$ for $x < 1$ or $1 < x < 2$.

7.14 (*a*) y-axis; (*b*) none; (*c*) y-axis for both; (*d*) y-axis; (*e*) none; (*f*) none; (*g*) none; (*h*) origin; (*i*) none; (*j*) origin; (*k*) origin; (*l*) none; (*m*) none; (*n*) none; (*o*) none; (*p*) origin; (*q*) none; (*r*) none; (*s*) none; (*t*) origin.

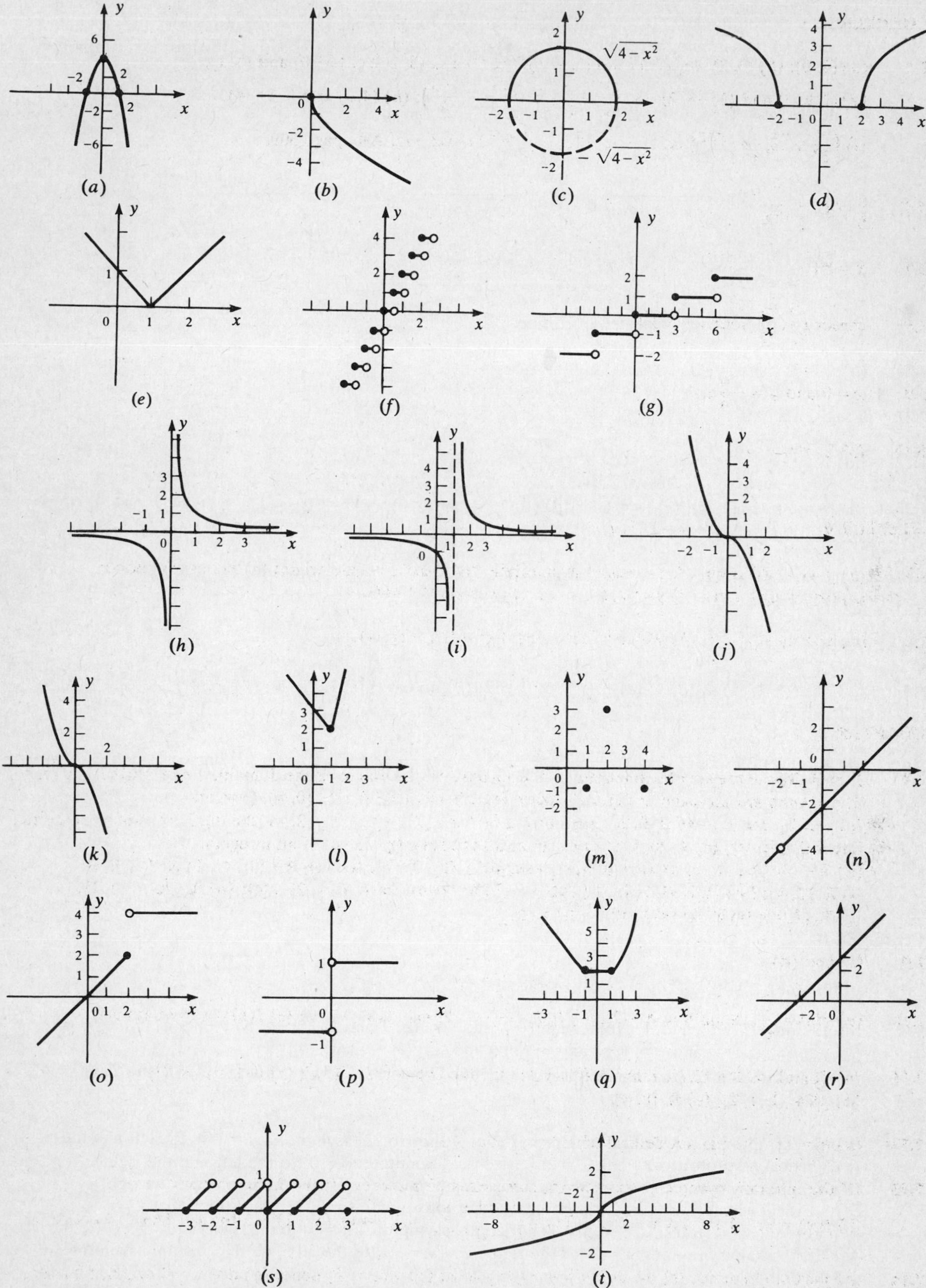

Fig. A-7

7.15 (*a*) even; (*b*) neither; (*c*) both even; (*d*) even; (*e*) neither; (*f*) neither; (*g*) neither; (*h*) odd; (*i*) neither; (*j*) odd; (*k*) odd; (*l*) neither; (*m*) neither; (*n*) neither; (*o*) neither; (*p*) odd; (*q*) neither; (*r*) neither; (*s*) neither; (*t*) odd.

7.16 (*a*) no; (*b*) yes; (*c*) $k = 0$; (*d*) $k = 2$; (*e*) yes; $f(x) = 0$ for all x.

7.17 (*a*) $2x + h - 2$ (*b*) 1 (*c*) $3x^2 + 3hx + h^2$

(*d*) $\dfrac{1}{\sqrt{x+h}+\sqrt{x}}$ (*e*) $\dfrac{1}{\sqrt{x+h}+\sqrt{x}} - 5$ (*f*) $\dfrac{|x+h|-|x|}{h}$

7.18 (*a*) $1, -1, 3, -3$; (*b*) $-2, 4, -4$; (*c*) $2, -2, 3$; (*d*) 2; (*e*) $-2, -3, -4$; (*f*) $-2, 1+\sqrt{2}, 1-\sqrt{2}$; (*g*) $4, \sqrt{2}, -\sqrt{2}$.

7.19 One, two (one of them repeated), or three.

7.20 (*a*) $k = 3$; (*b*) $k = -2$.

7.21 9 and -12.

7.22 (iii).

7.23 (*a*) $(-\infty, 3)$; (*b*) $[1, \infty)$; (*c*) $(-\infty, 1/3]$; (*d*) $(10/7, \infty)$; (*e*) $(2, 3)$; (*f*) $[-3, 2)$; (*g*) $(-3, 1)$; (*h*) $[-1/3, 3]$; (*i*) $(2, 3)$; (*j*) $[-\sqrt{6}, \sqrt{6}]$; (*k*) $(-1, 3)$.

7.24 (*a*) $x < -2$ or $x > 1$; (*b*) $-2 < x < 0$ or $x > 1$.

CHAPTER 8

8.5 (*a*) 7; (*b*) -4; (*c*) 4; (*d*) 1; (*e*) 0; (*f*) 36; (*g*) 12; (*h*) no limit; (*i*) 7; (*j*) 20; (*k*) 37; (*l*) -8; (*m*) $1/2\sqrt{3}$; (*n*) 1/4; (*o*) 1/4.

8.6 (*a*) $\lim_{h\to 0} (6x + 3h) = 6x$; (*b*) $\lim_{h\to 0} \dfrac{-1}{(x+h+1)(x+1)} = -\dfrac{1}{(x+1)^2}$; (*c*) $\lim_{h\to 0} 7 = 7$; (*d*) $\lim_{h\to 0} (3x^2 + 3xh + h^2) = 3x^2$; (*e*) $\lim_{h\to 0} \dfrac{1}{\sqrt{x+h}+\sqrt{x}} = \dfrac{1}{2\sqrt{x}}$; (*f*) $\lim_{h\to 0} (10x + 5h - 2) = 10x - 2$.

8.9 (*d*).

CHAPTER 9

9.6 (*a*) $+\infty$; (*b*) $-\infty$; (*c*) $+\infty$; (*d*) $-\infty$; (*e*) $-\infty$; (*f*) $+\infty$.

9.7 (*a*) 12 and 11; (*b*) 1 and -1; (*c*) $-\infty$; (*d*) $+\infty$ and $-\infty$; (*e*) 1 and 1; (*f*) $-\infty$ and $+\infty$; (*g*) 0 and 0; (*h*) $\frac{3}{5}$ and $\frac{3}{5}$; (*i*) $+\infty$ and $-\infty$; (*j*) 0 and 0; (*k*) 4 and -4; (*l*) $+\infty$ and $-\infty$; (*m*) 0 and undefined (the denominator is undefined when $x^3 < -5$); (*n*) 0 and 0; (*o*) 2 and 2; (*p*) $+\infty$ and $-\infty$; (*q*) $-\frac{1}{2}$; (*r*) $+\infty$; (*s*) $+\infty$.

9.8 (*a*) No asymptotes. (*b*) Vertical asymptote: $x = -\frac{2}{3}$; horizontal asymptote: $y = \frac{2}{3}$ (to the left and the right). (*c*) Vertical asymptotes: $x = -3$, $x = 2$; horizontal asymptote: $y = 0$ (to the left and the right). (*d*) Vertical asymptotes: $x = -\frac{5}{2}$, $x = 1$; horizontal asymptote: $y = 0$ (to the left and the right). (*e*) Vertical asymptote: $x = -2$; horizontal asymptote: $y = 0$ (to the left and the right). (*f*) Vertical asymptotes: $x = -4$, $x = 2$; horizontal asymptotes: $y = 1$ (to the right) and $y = -1$ (to the left). (*g*) Horizontal asymptotes: $y = 2$ (to the right) and $y = -2$ (to the left). (*h*) Horizontal asymptote: $y = 0$ (to the right).

9.9 (*c*).

CHAPTER 10

10.5 (*a*) continuous everywhere; (*b*) continuous for $x \neq 0$; (*c*) continuous everywhere; (*d*) continuous for $x \neq -2$; (*e*) continuous everywhere; (*f*) continuous except at $x = 1$ and $x = 2$.

10.6 (*a*) continuous on the right but not on the left at $x = 0$; (*b*) discontinuous on both the left and the right at any nonnegative integer x; (*c*) no points of discontinuity; (*d*) continuous on the left but not on the right at $x = -3$ and at $x = 2$.

10.7 (*a*) $f(x) = \begin{cases} 0 & \text{if } -2 \leq x < -1 \\ 1 & \text{if } -1 \leq x \leq 1 \\ 0 & \text{if } \ 1 < x \leq 2 \end{cases}$

(*b*) $g(x) = [x]$

(*c*) $h(x) = \begin{cases} 1 & \text{if } x \geq 0 \\ 0 & \text{if } x < 0 \end{cases}$

10.8 (*a*) $x = 4$, $x = -1$; (*b*) $\lim_{x \to 4} f(x)$ does not exist, $\lim_{x \to -1} f(x) = -3/5$; (*c*) $x = 4$ (vertical asymptote), $y = 0$ (horizontal asymptote).

10.9 (*a*) $x = 0$; (*b*) $x = 0$ (vertical asymptote), no horizontal asymptote.

10.10 (*a*) no; (*b*) no; (*c*) yes; (*d*) no.

10.11 $c = 8$.

10.12 (*a*) yes; (*b*) yes; (*c*) no.

10.13 (*b*) 1/8.

10.14 discontinuous for all x.

CHAPTER 11

11.6 (*a*) (i) slope = $4x + 1$, (ii) $y = 2x - \frac{1}{8}$, (iii) see Fig. A-8(*a*); (*b*) (i) slope = x^2, (ii) $y = 4x - \frac{13}{3}$, (iii) see Fig. A-8(*b*); (*c*) (i) slope = $2x - 2$, (ii) $y = -1$, (iii) see Fig. A-8(*c*); (*d*) (i) slope = $8x$, (ii) $y = 4x + 2$, (iii) see Fig. A-8(*d*).

11.7 (3, 9).

11.8 (1, 1) and $(-1, -1)$.

11.9 $y = -3x + \frac{28}{27}$.

11.10 (3, 5) and $\left(-\frac{1}{3}, \frac{55}{9}\right)$.

11.12 (6, 36) and $(-2, 4)$.

11.13 $y = \frac{1}{4}x + 1$.

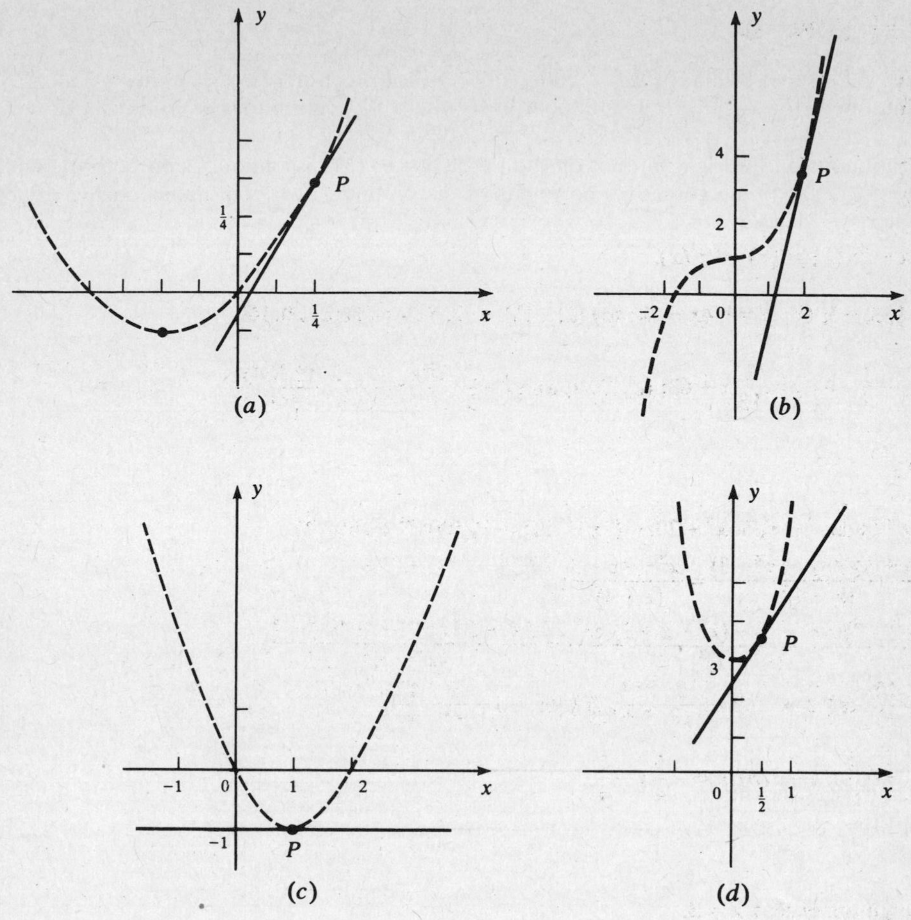

Fig. A-8

CHAPTER 12

12.6 (*a*) 2; (*b*) $\frac{2}{3}x-7$; (*c*) $6x+3$; (*d*) $4x^3$.

12.7 (*a*) $9x^2-8x+5$; (*b*) $-40x^4+3\sqrt{3}x^2+4\pi x$; (*c*) $39x^{12}-50x^9+20x$; (*d*) $102x^{50}+36x^{11}-28x+\sqrt[3]{7}$.

12.8 (*a*) $21x^6-x^4$; (*b*) $6x-5$; (*c*) $2x^3+5$; (*d*) $21t^6-24t$; (*e*) $5\sqrt{2}x^4-3x^2$.

12.9 (*a*) $y=-7x+1$; (*b*) $y=66x-153$; (*c*) $y=3$.

12.10 (*a*) $y=20x+2$ and $y=-44x+2$; (*b*) $y=x+4$ and $y=\frac{85}{4}x-\frac{65}{4}$.

12.11 $y=-x+1$.

12.12 (2, 2).

12.13 (*a*) $D_x(x^5)|_{x=3}=405$; (*b*) $D_x(5x^4)|_{x=1/3}=\frac{20}{27}$.

12.14 $f'(x)=8x$ ((iii) gives $f'(u)=ku$; then choose $u=v=1$ in (iii)).

12.15 (a) $x = \frac{1}{2}$; (b) $(0, -\frac{1}{2})$.

12.16 $c = \dfrac{16}{81}$.

12.17 $b = \dfrac{17}{2}$.

12.18 (a) $4x - 2$; (b) $f^{\#}(x) = 2f'(x)$.

12.19 (a) $-1, 3, -3$; (b) $y = -4x - 12$; (c) $(2, -15)$, $(-2, 5)$, or $(11/2, 1105/8)$.

12.20 (a) $3, -\dfrac{7}{3}$; (b) $y = \dfrac{1}{15}x + 63$; (c) $(3, 0)$ and $\left(-\dfrac{5}{9}, \dfrac{2^{14}}{3^5}\right) = \left(-\dfrac{5}{9}, \dfrac{16\,384}{243}\right)$

CHAPTER 13

13.6 (a) $(x^{100} + 2x^{50} - 3)(56x^7 + 20) + (7x^8 + 20x + 5)(100x^{99} + 100x^{49})$;
(b) $\dfrac{(x+4)(2x) - (x^2 - 3)}{(x+4)^2} = \dfrac{x^2 + 8x + 3}{(x+4)^2}$;
(c) $\dfrac{(x^3+7)(5x^4 - 1) - (x^5 - x + 2)(3x^2)}{(x^3+7)^2} = \dfrac{2x^7 + 35x^4 + 3x^3 - 6x^2 - 7}{(x^3+7)^2}$;
(d) $-\dfrac{15}{x^6}$; (e) $24x^2 - 2x + \dfrac{2}{x^2} - \dfrac{12}{x^4}$; ($f$) $9x^2 + 1 - \dfrac{3}{x^4} + \dfrac{12}{x^5}$.

13.7 (a) $y = -\dfrac{1}{4}x + \dfrac{3}{4}$; ($b$) $y = -\dfrac{5}{4}x - \dfrac{7}{4}$.

13.8 $-1/4$.

13.9 At all points except $x = 3$.

13.10 (b) At $x = 0$ and $x = 4$.

CHAPTER 14

14.7 (a) max = 13 (at $x = -2$), min = -7 (at $x = 3$); (b) max = -1 (at $x = -1$), min = $-129/8$ (at $x = 7/4$);
(c) max = 3 (at $x = 1$), min = $-31/27$ (at $x = -1/3$); (d) max = 1 (at $x = 0$), min = -11 (at $x = -1$);
(e) max = 99 (at $x = 4$), min = -9 (at $x = 2$); (f) max = $-1/5$ (at $x = -3$), min = $-1/4$ (at $x = -4$);
(g) max = 5/4 (at $x = 4$), min = 3/4 (at $x = 2$); (h) max = 1/3 (at $x = 1$), min = -1 (at $x = -1$);
(i) max = 14/3 (at $x = 2$), min = $-2/27$ (at $x = 1/3$).

14.8 75 yards east-west by 50 yards north-south.

14.9 60 yards parallel to stream, 30 yards perpendicular to stream.

14.10 50 miles per hour.

14.11 $200.

14.12 $x = 350$, at $65 per radio.

14.13 (a) side of base = 5 ft, height = 5 ft; (b) side of base = $5\sqrt{2}$ ft, height = $5/\sqrt{2}$ ft.

14.14 length = 105 ft, width (parallel to divider) = 60 ft.

14.15 175.

14.16 (*a*) $x = 50, y = 50$; (*b*) $x = 100, y = 0$ or $x = 0, y = 100$; (*c*) $x = 50, y = 50$.

14.17 $\ell = 314$ meters, $w = 628/\pi \approx 200$ meters.

14.18 (*a*) the entire wire for the circle; (*b*) $\dfrac{\pi L}{\pi + 4}$ for the circle, $\dfrac{4L}{\pi + 4}$ for the square.

14.19 The whole wire for the square.

14.20 1000 sets.

14.21 $r = 2$ ft, $h = \dfrac{5}{3}$ ft.

14.22 $h = \dfrac{2a}{\sqrt{3}}, r = \dfrac{\sqrt{2}a}{\sqrt{3}}$.

13.23 the equilateral triangle of side $p/3$.

14.24 (*a*) $\left(\dfrac{25}{16}, \pm\dfrac{3}{16}\sqrt{231}\right)$; (*b*) $(-5, 0)$.

14.25 east-west dimension = 80 ft, north-south dimension = 48 ft.

14.26 (*a*) $21 \le x \le 100$ m; (*b*) 20 400 m^2 (when $x = 100$ m).

14.27 15 tons

CHAPTER 15

15.7 (*a*) $(f \circ g)(x) = \dfrac{2}{3x+1}$, $(g \circ f)(x) = \dfrac{6}{x+1}$; (*b*) $x^6 + 2x^3 - 5$, $(x^2 + 2x - 5)^3$; (*c*) 49, 3; (*d*) x^6, x^6; (*e*) x, x; (*f*) $x^2 - 4$, $x^2 - 4$.

15.8 (*a*) all x; (*b*) $x = -1/4$; (*c*) $x = 1/3$; (*d*) $x = 0$; (*e*) $x = \pm\sqrt{2}$.

15.9 (*a*) $f(x) = x^3 - x^2 + 2$, $g(x) = x^7$; (*b*) $f(x) = 8 - x$, $g(x) = x^4$; (*c*) $f(x) = 1 + x^2$, $g(x) = \sqrt{x}$; (*d*) $f(x) = x^2 - 4$, $g(x) = x^{-1}$.

15.10 (*a*) $4(x^3 - 2x^2 + 7x - 3)^3(3x^2 - 4x + 7)$ (*b*) $15(7 + 3x)^4$ (*c*) $-4(2x-3)^{-3}$
(*d*) $-18x(3x^2 + 5)^{-4}$ (*e*) $(4x^2 - 3)(x + 5)^2(28x^2 + 80x - 9)$ (*f*) $-15\dfrac{(x+2)^2}{(x-3)^4}$
(*g*) $\dfrac{20x(x^2 - 2)}{(2x^2 + 1)^3}$ (*h*) $\dfrac{4(1 - 6x)}{(3x^2 - x + 5)^2}$

15.11 (*a*) $\dfrac{3}{2}x^{-1/4}$ (*b*) $\dfrac{x(2 - 9x^3)}{(1 - 3x^3)^{2/3}}$ (*c*) $\dfrac{1}{(x^2+1)^{3/2}}$
(*d*) $\dfrac{x(21x - 8)}{4(7x^3 - 4x^2 + 2)^{3/4}}$ (*e*) $-\dfrac{3}{2}\dfrac{1}{\sqrt{x + 2(\sqrt{x} - 1)^3}}$ (*f*) $6x^{-1/4} + x^{-3/4} + \dfrac{1}{3}x^{-4/3}$
(*g*) $\dfrac{16x}{3\sqrt[3]{4x^2 + 3}}$ (*h*) $-\dfrac{2 + \sqrt{3x}}{2\sqrt{x^3}}$ (*i*) $-\dfrac{1}{4\sqrt{4 - \sqrt{4 + x}}\sqrt{4 + x}}$

15.12 $y = -\dfrac{3}{50}x + \dfrac{8}{25}$.

15.13 $y = -\frac{5}{3}x + 10$.

15.14 (*a*) $(g \circ f)(x) = \left(\frac{x+2}{x-2}\right)^2 - 4 = \frac{-3x^2 + 20x - 12}{(x-2)^2}$, $(g \circ f)'(x) = -\frac{8(x+2)}{(x-2)^3}$.

15.15 (*a*) max $= \sqrt{2}/2$ at $x = 1$, min $= -\sqrt{2}/2$ at $x = -1$; (*b*) max $= 216$ at $x = 3$, min $= -36$ at $x = -4$; (*c*) max $= 3$ at $x = -1$, min $= 1$ at $x = 1$.

15.16 R is 8/5 miles from A.

15.17 $H'(x) = 0$.

15.18 x^2.

15.19 $3x^2\, G(x^3)$.

15.20 all real numbers except 0 and 1.

15.21 $f'(-x) = f'(x)$ (the derivative is an even function).

15.22 12.

15.23 (*a*) domain $\left[-\frac{1}{3}, \infty\right)$, range $[0, \infty)$; (*b*) $y = \frac{3}{8}x + \frac{17}{8}$; (*c*) (1, 2).

15.24 base $= \sqrt{2}$, height $= \sqrt{2}/2$.

CHAPTER 16

16.4 (*a*) $\frac{y-2x}{2y-x}$; (*b*) (2, 4), (−2, −4); (*c*) (4, 2), (−4, −2).

16.5 (*a*) $\frac{5x}{2y}$; (*b*) $k = \pm 21$.

16.6 (*a*) $-\frac{x}{y}$; (*b*) $\frac{8x^3 - 3x^2y - 2}{1 + x^3}$ or $\frac{3x^2(2x-y)^2 + 4y}{4x}$; (*c*) $-\frac{y^2}{x^2}$ or $-(y-1)^2$; (*d*) $-\frac{\sqrt{y}}{\sqrt{x}} = 1 - \frac{1}{\sqrt{x}}$; (*e*) $\frac{3x^2 - 2y}{2x + 3y^2}$; (*f*) $\frac{21(7x-1)^2}{8y^3}$; (*g*) $-\frac{4x}{9y}$.

16.7 (*a*) $y = \frac{2}{9}x + \frac{4}{3}$; (*b*) $y = -\frac{\sqrt{3}}{12}x + \frac{2\sqrt{3}}{3}$; (*c*) $y = \frac{4}{13}x + \frac{22}{13}$; (*d*) $y = \frac{17}{20}x - \frac{7}{5}$; (*e*) $y = \frac{5}{2}x - \frac{11}{2}$.

16.8 (*a*) $y = -\frac{8}{3}x + \frac{19}{3}$; (*b*) $y = -3x + 7$; (*c*) $y = -\frac{45}{32}x + \frac{917}{32}$; (*d*) $y = \frac{4}{3}x$.

16.9 2/3.

CHAPTER 17

17.10 (*a*) $f'(1) = 0$; (*b*) $f'(\sqrt{3}/3) = 0$; (*c*) $f'(2\sqrt{3}/9) = 0$; (*d*) $f'\left(1 + \frac{\sqrt{6}}{3}\right) = 0$; (*e*) f not continuous at $x = 1$; (*f*) $f'(1/2) = 0$; (*g*) $f'(1) = 0$; (*h*) f not differentiable at $x = 1$.

17.11 (*a*) $1 < c < 4$; (*b*) $c = 7/2$; (*c*) $c = 81/16$; (*d*) $c = 4 - \sqrt{3}$; (*e*) $c = \pm\sqrt{2}/2$; (*f*) $c = 4 - 2\sqrt{2}$.

17.12 The function is increasing in the first-listed region, decreasing in the second: (*a*) everywhere, nowhere; (*b*) nowhere, everywhere; (*c*) $x>2$, $x<2$; (*d*) $x<-2$, $x>-2$; (*e*) $-1<x<0$, $0<x<1$; (*f*) $-3<x<0$, $0<x<3$; (*g*) $x<1$ or $x>5$, $1<x<5$; (*h*) $x<-1$ or $x>1$, $-1<x<0$ or $0<x<1$; (*i*) $x<-2$ or $x>2$, $-2<x<2$.

17.14 (*d*).

17.17 $\dfrac{343}{3\sqrt{3}}=\dfrac{343\sqrt{3}}{9}$.

17.19 (*a*) the points (1, 1), $\left(\dfrac{3+\sqrt{5}}{2}, 1\right)$, and $\left(\dfrac{3-\sqrt{5}}{2}, 1\right)$; (*b*) 0, 2 (double root); (*c*) [0, 3].

CHAPTER 18

18.4 (*a*) $s_0 - s = 16t^2$; (*b*) (i) 0.25, (ii) 1, (iii) 2, (iv) 2.5.

18.5 4 sec later.

18.6 108.8 mph.

18.7 (*a*) 112 ft, 192 ft; (*b*) 4 sec; (*c*) 256 ft; (*d*) 8 sec; (*e*) 128 ft/sec $= v_0$.

18.8 (*a*) 5 sec; (*b*) 176 ft/sec; (*c*) 3 sec; (*d*) 1.5 sec.

18.9 (*a*) It is always moving to the right. (*b*) 3 miles.

18.10 (*a*) $t>4$; (*b*) $t<4$; (*c*) $t=4$; (*d*) never; (*e*) 27 units.

18.11 (*a*) $t<\frac{1}{2}$ hr and $t>1$ hr; (*b*) $\frac{1}{2}<t<1$ hr; (*c*) $t=\frac{1}{2}$ hr and $t=1$ hr; (*d*) 0.75 mph.

18.12 13 units. **18.13** 320 ft/sec. **18.14** (*a*) $t=0$ and $t=5$; (*b*) $t=2.5$; (*c*) no.

CHAPTER 19

19.4 $30 per set. **19.5** $7 per unit. **19.6** 24 square feet per foot.

19.7 40 km/s (at $t=1000$ s). **19.8** (*a*) 18 gal/sec (i.e., $G'=-18$); (*b*) 54 gal/sec (i.e., $\Delta G/\Delta t=-54$).

19.9 (*a*) 9; (*b*) 6. **19.10** $y=2$.

CHAPTER 20

20.5 24/7 ft/min. **20.6** $3.14/\pi \approx 1$ ft/min. **20.7** 8 ft/sec. **20.8** 64 ft/sec.

20.9 $12/\pi \approx 3.82$ mm/s. **20.10** $4/\pi \approx 1.27$ in/sec. **20.11** $400\pi \approx 1256$ m^2/min.

20.12 (*a*) 10 mm; (*b*) increasing at $6/\sqrt{5} \approx 2.69$ mm/s. **20.13** 36 units per second.

20.14 decreasing at 3/2 units per second. **20.15** 30 mm^3/s. **20.16** $r=1/\pi$.

20.17 increasing at 6 units per second. **20.18** 500 km/h. **20.19** $4\sqrt{2} \approx 5.64$ mph.

20.20 $\dfrac{16}{25}\left(\dfrac{3.14}{\pi}\right) \approx 0.64$ m/min. **20.21** $(-1, -\frac{3}{2})$. **20.22** 5 ft/sec. **20.23** 6 ft/sec.

CHAPTER 21

21.4 $(a)\ \frac{50}{7} \approx 7.1429$; $(b)\ \frac{53}{9} \approx 8.8333$; $(c)\ \frac{373}{75} \approx 4.9733$; $(d)\ \frac{12.35}{3} \approx 4.1167$; $(e)\ \frac{159}{320} \approx 0.4969$; $(f)\ \frac{65}{96} \approx 0.6771$; $(g)\ \frac{193}{480} \approx 0.4021$.

21.5 0.994 cm^3. **21.6** $\frac{1920\pi}{77} \approx 78.3$ gal. **21.7** $(a)\ 5\pi \approx 15.7\ cm^3$; $(b)\ 5.1\pi \approx 16.0\ cm^3$.

21.8 9%. **21.9** $\frac{1}{8\pi}$ mile ≈ 210 feet. **21.10** (*a*) 6; (*b*) 6. **21.11** $\max(1.8725, 1.8475) = 1.8725\ ft^2$.

CHAPTER 22

22.5 $(a)\ -\frac{2}{x^3}$; $(b)\ 6\pi x$; $(c)\ -\frac{1}{4}(x+5)^{-3/2} = -\frac{1}{4(\sqrt{x+5})^3}$; $(d)\ -\frac{2}{9}(x-1)^{-5/3} = -\frac{2}{9(\sqrt[3]{x-1})^5}$; $(e)\ \frac{1}{4}(2-x^2)(x^2+1)^{-7/4}$; $(f)\ 4(x-3)(5x^2-6x-3)$; $(g)\ \frac{4x+5}{(1-x)^4}$.

22.6 $(a)\ -\frac{1}{y^3}$; $(b)\ -\frac{1}{y^3}$; $(c)\ -\frac{2x}{y^5}$; $(d)\ \frac{2}{(x+2y)^3}$.

22.7 (*a*) and (*b*) $y'' = \frac{1}{2}x^{-3/2}$; (*c*) (*a*).

22.8 $(a)\ y' = 16x^3 - 4x,\ y'' = 48x^2 - 4,\ y''' = 96x,\ y^{(4)} = 96,\ y^{(n)} = 0$ for $n > 4$;
$(b)\ y' = 4x + 1 - x^{-2},\ y'' = 4 + 2x^{-3},\ y''' = -(2\cdot 3)x^{-4}, \ldots, y^{(n)} = (-1)^n n!\, x^{-(n+1)}$;
$(c)\ y' = \frac{1}{2}x^{-1/2},\ y'' = -\frac{1}{4}x^{-3/2},\ y''' = \frac{3}{8}x^{-5/2},\ y^{(4)} = -\frac{15}{16}x^{-7/2}, \ldots,$
$y^{(n)} = (-1)^{n-1}\frac{(1)(3)(5)\cdots(2n-3)}{2^n}x^{-(2n-1)/2}$;
$(d)\ y' = -2(x-1)^{-2},\ y'' = 4(x-1)^{-3},\ y''' = -12(x-1)^{-4}, \ldots, y^{(n)} = (-1)^n 2(n!)(x-1)^{-(n+1)}$;
$(e)\ y' = -(3+x)^{-2},\ y'' = 2(3+x)^{-3},\ y''' = -6(3+x)^{-4}, \ldots, y^{(n)} = (-1)^n n!(3+x)^{-(n+1)}$.

22.9 (*a*) $a = 2 \neq 0$; (*b*) -3; (*c*) 0.

22.10 $y'' = -\frac{2}{3}$. **22.11** $\frac{dy}{dx} = 0,\ \frac{d^2y}{dx^2} = -\frac{1}{2}$. **22.12** (*a*) $K = \frac{1}{7},\ L = \frac{3}{7}$; (*b*) no.

CHAPTER 23

23.4 (*a*) Concave upward for all *x*. (*b*) Upward for $x > -5$, downward for $x < -5$; $I(-5, 221)$. (*c*) Upward for $x < -5$ or $x > -4$, downward for $-5 < x < -4$; $I(-5, 1371)$, $I(-4, 1021)$. (*d*) Upward for $x > 1/2$, downward for $x < 1/2$; no inflection points. (*e*) Upward for $x < 3$, downward for $x > 3$; $I(3, 162)$.

23.5 (*a*) 1.5 (min); (*b*) 0 (max), 3 (min), -3 (min); (*c*) 4 (min), $-\frac{2}{3}$ (max); (*d*) 0 (max), 2 (min); (*e*) 0 (min).

23.6 See Fig. A-9, page 320. **23.7** (*d*). **23.8** *B* and *E*.

23.9 (*a*) one; (*b*) none; (*c*) parabola. **23.10** (*f*). **23.11** $-4 < k < 0$.

23.12 See Fig. A-10, page 321.

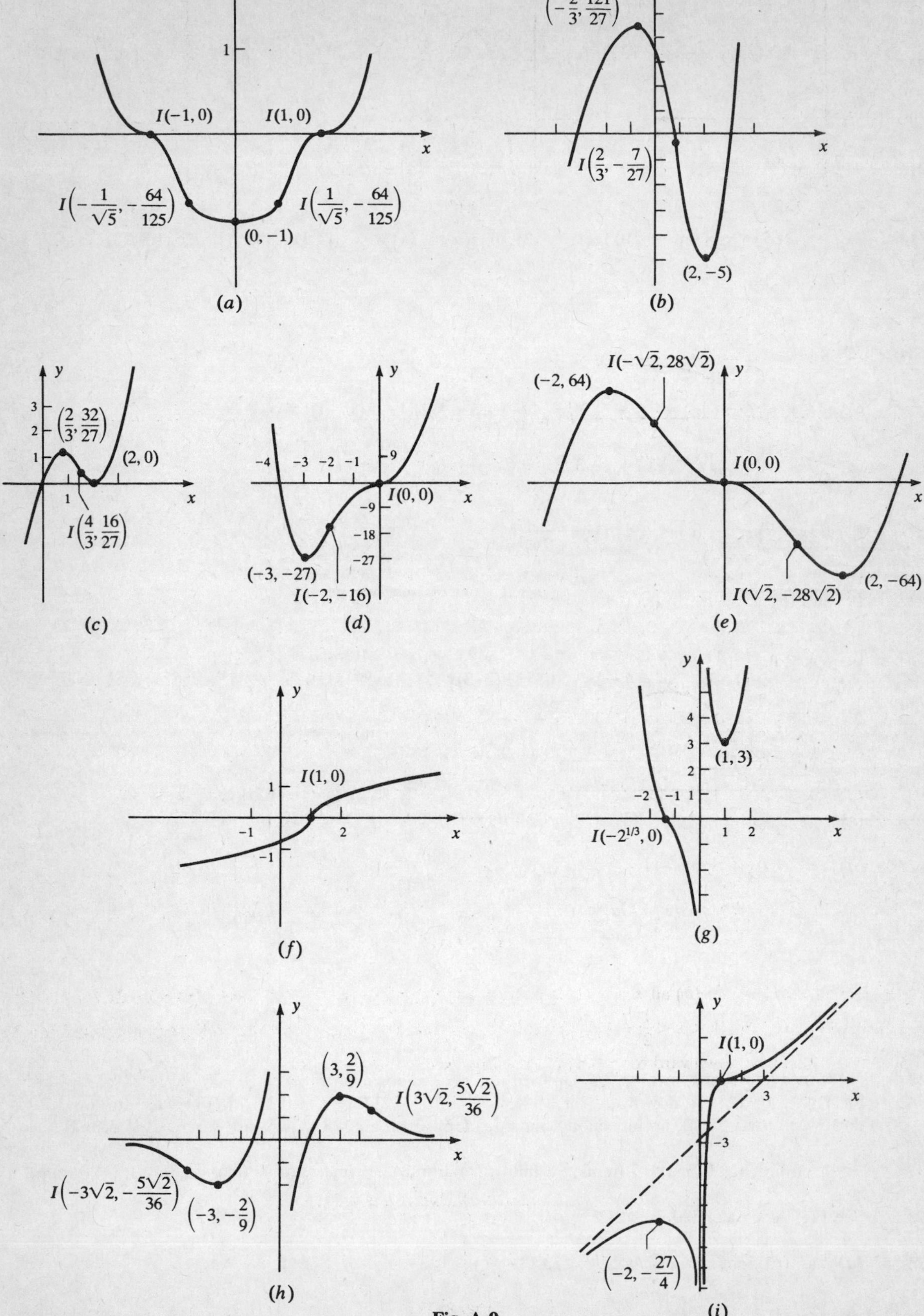

$I(-1, 0)$
$I(1, 0)$
$I\left(-\frac{1}{\sqrt{5}}, -\frac{64}{125}\right)$
$I\left(\frac{1}{\sqrt{5}}, -\frac{64}{125}\right)$
$(0, -1)$
(a)
$\left(-\frac{2}{3}, \frac{121}{27}\right)$
$I\left(\frac{2}{3}, -\frac{7}{27}\right)$
$(2, -5)$
(b)
$\left(\frac{2}{3}, \frac{32}{27}\right)$
$(2, 0)$
$I\left(\frac{4}{3}, \frac{16}{27}\right)$
(c)
$(-3, -27)$
$I(-2, -16)$
$I(0, 0)$
(d)
$I(-\sqrt{2}, 28\sqrt{2})$
$(-2, 64)$
$I(0, 0)$
$I(\sqrt{2}, -28\sqrt{2})$
$(2, -64)$
(e)
$I(1, 0)$
(f)
$(1, 3)$
$I(-2^{1/3}, 0)$
(g)
$\left(3, \frac{2}{9}\right)$
$I\left(3\sqrt{2}, \frac{5\sqrt{2}}{36}\right)$
$I\left(-3\sqrt{2}, -\frac{5\sqrt{2}}{36}\right)$
$\left(-3, -\frac{2}{9}\right)$
(h)
$I(1, 0)$
$\left(-2, -\frac{27}{4}\right)$
(i)

Fig. A-9

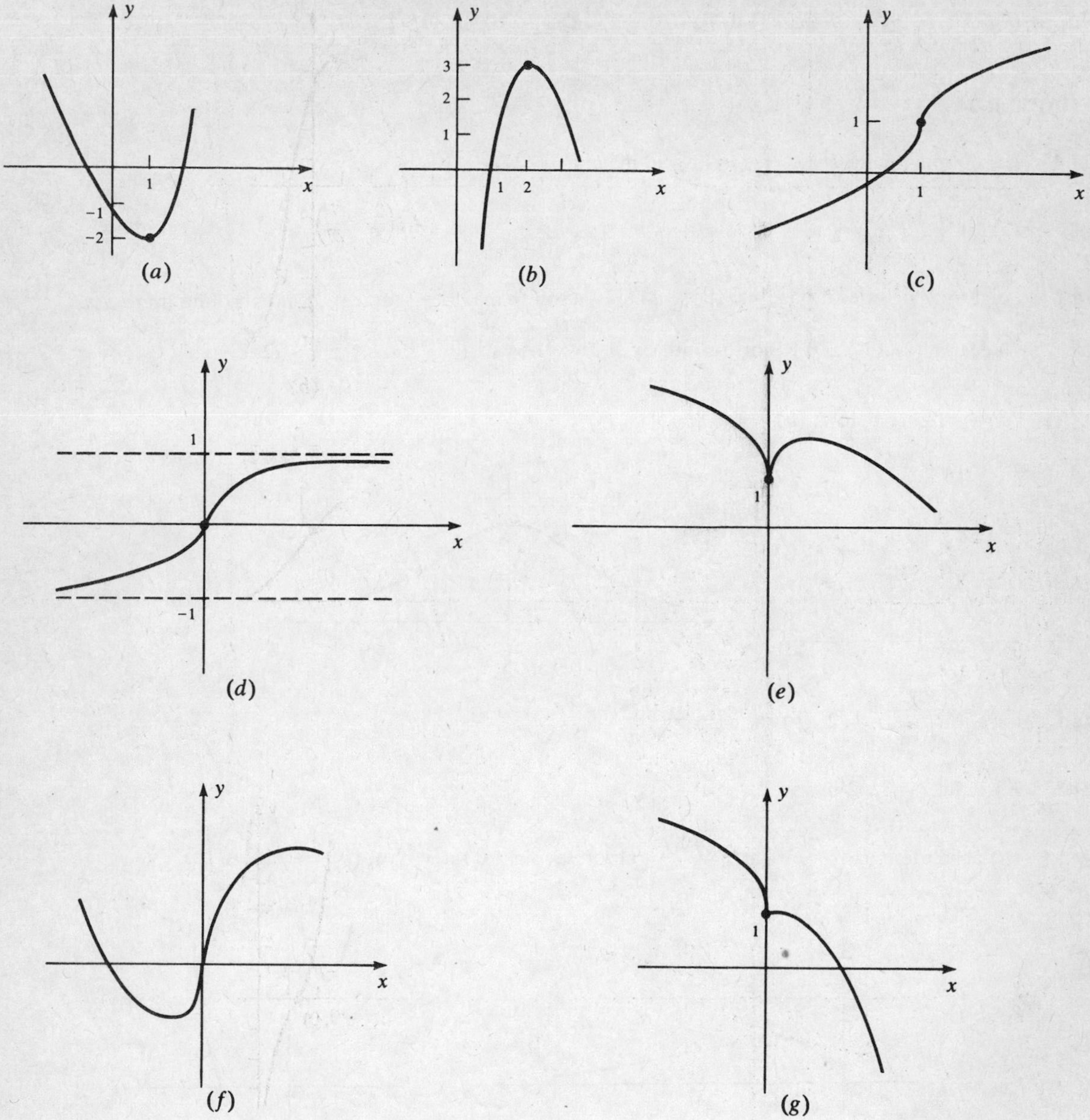

(*a*) (*b*) (*c*)

(*d*) (*e*)

(*f*) (*g*)

Fig. A-10

23.13 (*a*) all x in $[-1, 2]$; (*b*) all x in $[-1, 2]$ but $x = 1$ ($f'(x) = 2x - 1$ for $x > 1$, and $f'(x) = 1 - 2x$ for $x < 1$); (*c*) $1 < x < 2$ or $-1 < x < \frac{1}{2}$; (*d*) $f''(x) = 2$ for $x > 1$, and $f''(x) = -2$ for $x < 1$; (*e*) concave upward for $x > 1$, concave downward for $x < 1$; (*f*) see Fig. A-11.

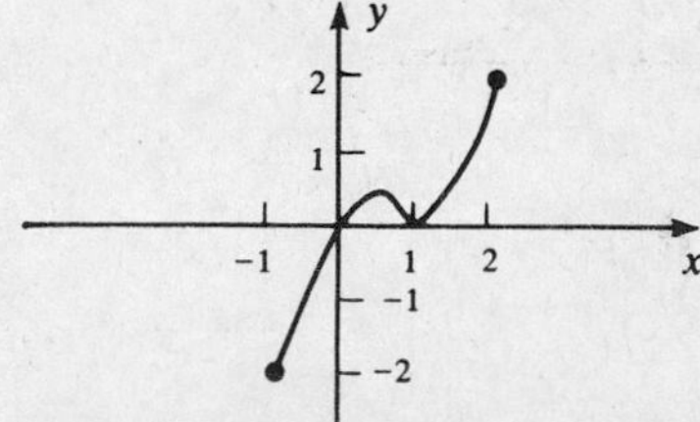

Fig. A-11

23.15 $k = -4$. **23.16** $A = C = 0, D = 1, B = -4, k = \pm\sqrt{2}$.

CHAPTER 24

24.4 (*a*) no maximum; (*b*) 10 meters by 10 meters. **24.5** (0, 0).

24.6 $\left(\frac{3}{2}, \frac{1}{2}\right), \left(-\frac{3}{2}, \frac{1}{2}\right)$.

24.7 height = 7 ft, side of base = 6 ft. **24.8** absolute max = −2 (at $x = 0$), no absolute min.

24.9 10 cm. **24.10** 200 ft north-south by 50 ft east-west.

24.11 (*a*) see figure below; (*b*) $\left(-1, \frac{1}{2}\right)$.

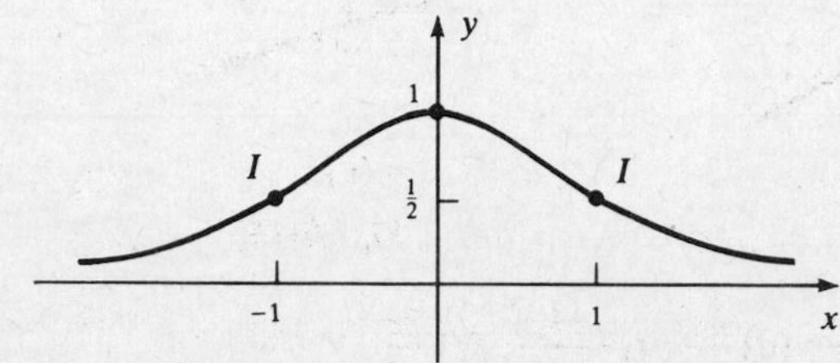

24.12 (*a*) $r = \sqrt[3]{\frac{k}{2\pi}}, \frac{h}{r} = 2$; (*b*) $r = \sqrt[3]{\frac{k}{4}}, \frac{h}{r} = \frac{4}{\pi}$.

24.13 $h = 6$ in, $r = 3\sqrt{2}$ in.

24.14 (*a*) absolute max $= \frac{2\sqrt{3}}{9}\left(\text{at } x = \frac{\sqrt{2}}{2}\right)$, absolute min = 0 (at $x = 0$); (*b*) see Fig. A-12.

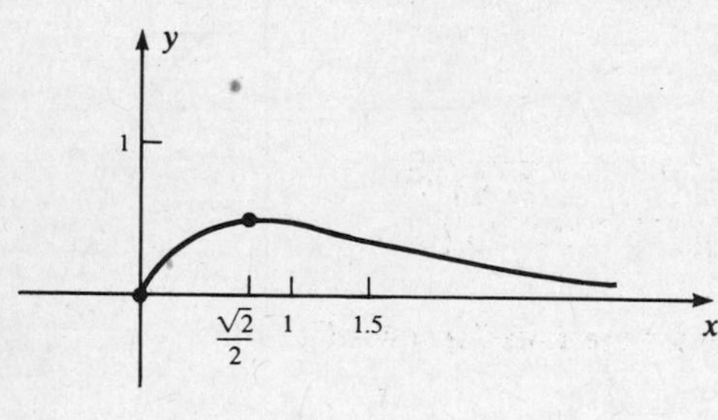

Fig. A-12

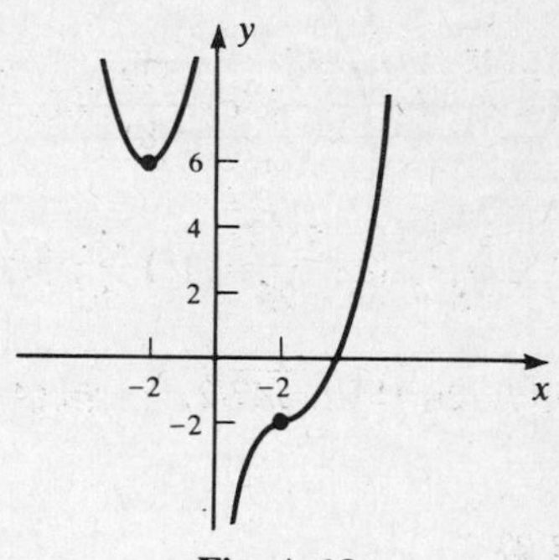

Fig. A-13

24.15 height = 8 m, side of base = 4 m. **24.16** 1500 per day. **24.17** $x = 2, y = 6$.

24.18 (*a*) $k = -8$; (*b*) see Fig. A-13; (*c*) $k = 0$. **24.19** $\left(\frac{3}{4}, \frac{3}{4}\right), \left(-\frac{3}{4}, -\frac{3}{4}\right)$.

CHAPTER 25

25.6 (*a*) $\frac{\pi}{5}$; (*b*) $\frac{\pi}{12}$; (*c*) $\frac{\pi}{90}$; (*d*) $\frac{1}{2}$; (*e*) $\frac{4\pi}{5}$.

25.7 (*a*) $\frac{360}{\pi}$; (*b*) 36; (*c*) 105; (*d*) 225; (*e*) 210. **25.8** 12 cm.

25.9 (*a*) $s = 2\pi$; (*b*) $r = \dfrac{11}{7\pi}$; (*c*) $\theta = \dfrac{\pi}{4}$ (*d*) $\theta = \dfrac{3}{2}$; (*e*) $s = \dfrac{3\pi}{2}$; (*f*) $r = \dfrac{6.28318}{\pi} \approx 2$; (*g*) $s = \dfrac{20\pi}{3}$.

25.10 $A = \dfrac{r^2\theta}{2}$. **25.11** See Fig. A-14.

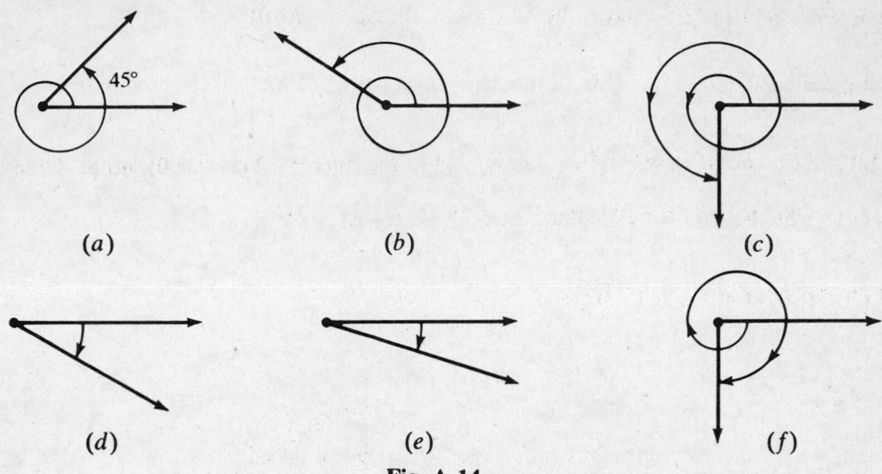

Fig. A-14

25.12 (*a*) $\dfrac{\pi}{4}$; (*b*) $\dfrac{3\pi}{4}$; (*c*) $\dfrac{3\pi}{2}$; (*d*) $\dfrac{5\pi}{3}$; (*e*) $\dfrac{11\pi}{6}$; (*f*) $\dfrac{3\pi}{2}$.

CHAPTER 26

26.7 (*a*) $-\dfrac{\sqrt{3}}{2}$; (*b*) $\dfrac{\sqrt{3}}{2}$; (*c*) $-\dfrac{\sqrt{3}}{2}$; (*d*) $\dfrac{\sqrt{2}}{2}$; (*e*) $\dfrac{\sqrt{2}}{4}(1+\sqrt{3})$; (*f*) $\dfrac{\sqrt{2}}{4}(\sqrt{3}-1)$; (*g*) $\dfrac{\sqrt{2}}{4}(\sqrt{3}-1)$; (*h*) 0.3256 (to four decimal places); (*i*) 0.2079 (to four decimal places).

26.8 (*a*) $\dfrac{4\sqrt{5}}{9}$; (*b*) $\dfrac{8\sqrt{5}}{81}$; (*c*) $-\dfrac{79}{81}$; (*d*) $\dfrac{\sqrt{5}}{3}$; (*e*) $\dfrac{2}{3}$.

26.9 (*a*) $-\dfrac{\sqrt{21}}{5}$; (*b*) $\dfrac{4\sqrt{21}}{25}$; (*c*) $\dfrac{17}{25}$; (*d*) $-\sqrt{\dfrac{5-\sqrt{21}}{10}}$; (*e*) $\sqrt{\dfrac{5+\sqrt{21}}{10}}$.

26.10 1/5. **26.11** $\sqrt{247}/2$. **26.13** (*a*) and (*c*) are false for $\theta = \pi/4$.

CHAPTER 27

27.9 (*a*) $p = \dfrac{3\pi}{2}$, $f = \dfrac{4}{3}$, $A = 1$, see Fig. A-15(*a*); (*b*) $p = \dfrac{2\pi}{3}$, $f = 3$, $A = \dfrac{1}{2}$, see Fig. A-15(*b*).

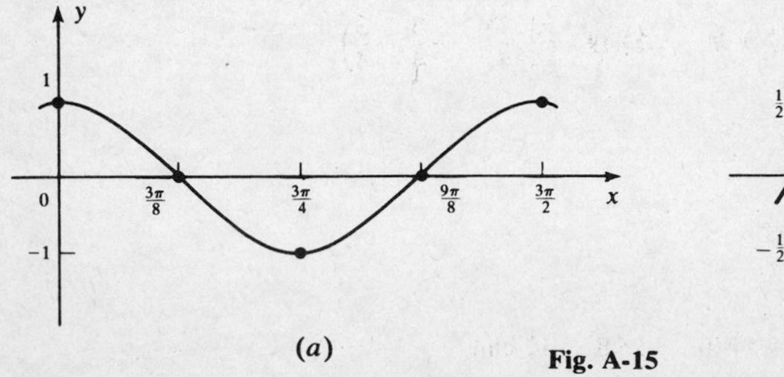

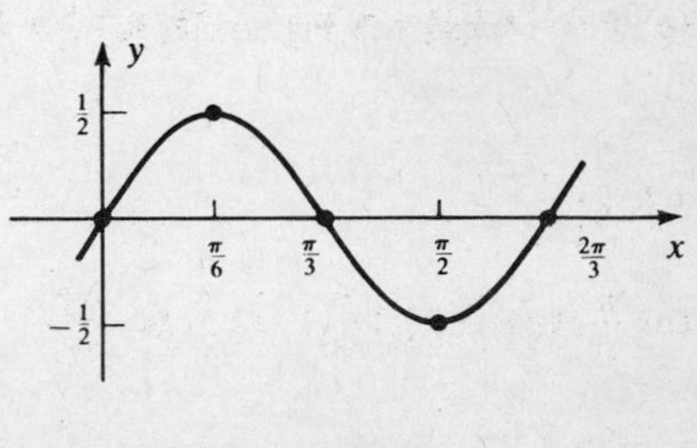

Fig. A-15

27.10 (*a*) $p = 6\pi$, $f = \frac{1}{3}$, $A = 2$; (*b*) $p = \pi$, $f = 2$, $A = 1$; (*c*) $p = \frac{4\pi}{5}$, $f = \frac{5}{2}$, $A = 5$; (*d*) $p = \frac{\pi}{2}$, $f = +4$, $A = 1$.

27.11 (*a*) $n\pi$ (for all integers n); (*b*) $2\pi n$ (for all integers n); (*c*) $\frac{\pi}{2} + 2\pi n$ (for all integers n); (*d*) $k\pi$ (for all odd integers k); (*e*) $\frac{3\pi}{2} + 2\pi n$ (for all integers n); (*f*) $\pm\frac{\pi}{3} + 2\pi n$ (for all integers n); (*g*) $\frac{\pi}{4} + 2\pi n$ and $\frac{3\pi}{4} + 2\pi n$ (for all integers n); (*h*) $\pm\frac{\pi}{6} + 2\pi n$ (for all integers n).

27.12 (*a*) $12 \sin^2 x \cos x$; (*b*) $\cos x - 2 \sin x$; (*c*) $x \cos x + \sin x$; (*d*) $2x(\cos 2x - x \sin 2x)$; (*e*) $\frac{x \cos x - \sin x}{x^2}$; (*f*) $\frac{x \sin x + 2 \cos x - 2}{x^3}$; (*g*) $5(3 \cos x \cos 3x - \sin 3x \sin x)$; (*h*) $-4 \cos 2x \sin 2x = -2 \sin 4x$; (*i*) $-4x \sin (2x^2 - 3)$; (*j*) $15 \sin^2 (5x + 4) \cos (5x + 4)$; (*k*) $-\frac{\sin 2x}{\sqrt{\cos 2x}}$.

27.13 (*a*) $\frac{1}{3}$; (*b*) 0; (*c*) $\frac{2}{3}$; (*d*) 1; (*e*) $\frac{4}{5}$; (*f*) $\frac{1}{4}$; (*g*) 0; (*h*) -3.

27.14 Figure A-16 (page 325) shows the graph for one period of the function, except in the case of the aperiodic function (*f*).

27.15 (*a*) $\max = 2\pi$ (at 2π), $\min = 0$ (at 0);
(*b*) $\max = \sqrt{2}$ $\left(\text{at } \frac{3\pi}{4}\right)$, $\min = -1$ (at 0);
(*c*) $\max = \frac{5}{4}$ $\left(\text{at } \frac{\pi}{6} \text{ and } \frac{5\pi}{6}\right)$, $\min = 1$ $\left(\text{at } 0, \frac{\pi}{2}, \text{ and } \pi\right)$;
(*d*) $\max = \frac{3}{2}\sqrt{3}$ $\left(\text{at } \frac{\pi}{3}\right)$, $\min = -\frac{3}{2}\sqrt{3}$ $\left(\text{at } \frac{5\pi}{3}\right)$;
(*e*) $\max = \frac{5}{4}$ (at π), $\min = 0$ $\left(\text{at the two roots of } \cos x = \frac{1}{4}\right)$;
(*f*) $\max = \frac{5\pi + 3\sqrt{3}}{6}$ $\left(\text{at } \frac{5\pi}{3}\right)$, $\min = \frac{\pi - 3\sqrt{3}}{6}$ $\left(\text{at } \frac{\pi}{3}\right)$;
(*g*) $\max = 5$ $\left(\text{at } x_0, \text{ where } \sin x_0 = \frac{3}{5} \text{ and } \frac{\pi}{2} < x_0 < \pi\right)$, $\min = -5$ (at $x_0 + \pi$).

27.16 (*b*) (i) amplitude = 5, period = 2π; (ii) amplitude = 13, period = π; (iii) amplitude = $\sqrt{6}$, period = 2π.

27.17 (*a*) $\frac{\sqrt{2}}{2}$; (*b*) $-\frac{\sqrt{3}}{2}$. **27.18** $A = \pi$. **27.19** $y = \frac{\sqrt{3}}{2}x + \frac{9 - 2\pi\sqrt{3}}{12}$.

27.20 $y = \frac{2\sqrt{3}}{3}x + \frac{27 - 4\pi\sqrt{3}}{18}$. **27.21** $n = 4$. **27.22** (*a*) see Fig. A-17; (*b*) yes; (*c*) no.

27.23 (*a*) 1; (*b*) 0.

27.24 (*a*) $y' = -\frac{1}{\sin y} = \pm\frac{1}{\sqrt{1 - x^2}}$; (*b*) $y' = -\frac{y \cos (xy)}{x \cos (xy) - 2y}$.

27.25 (*a*) $-\frac{1}{8}$ rad/sec; (*b*) $200\sqrt{3}$ km/h.

27.26 (*a*) $\theta = \frac{\pi}{2}$; (*b*) $\theta = 0$.

27.27 (*a*) $\max = \frac{\pi}{2} - 1$ $\left(\text{at } \frac{\pi}{2}\right)$, $\min = 0$ (at 0).

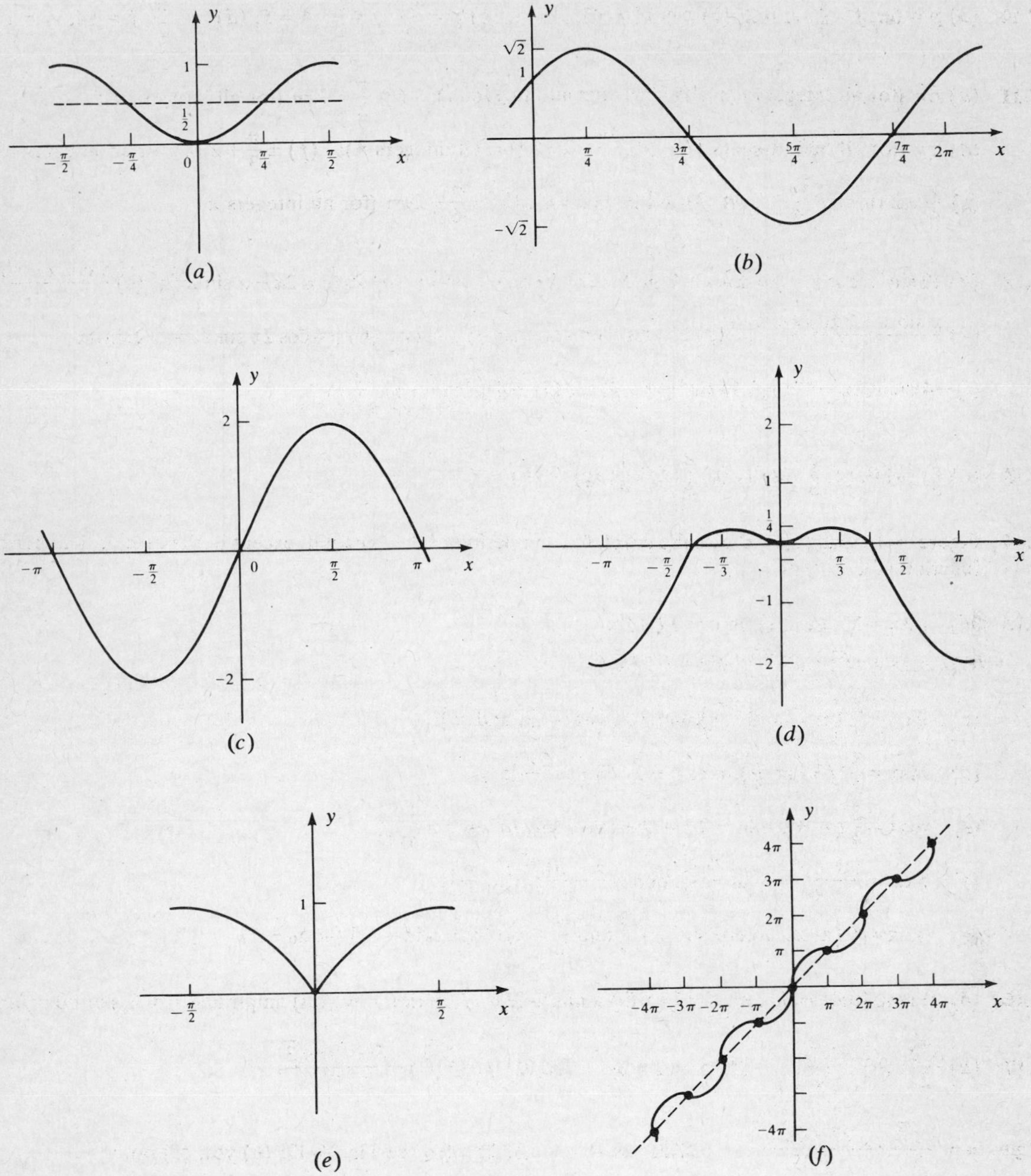

(a)

(b)

(c)

(d)

(e)

(f)

Fig. A-16

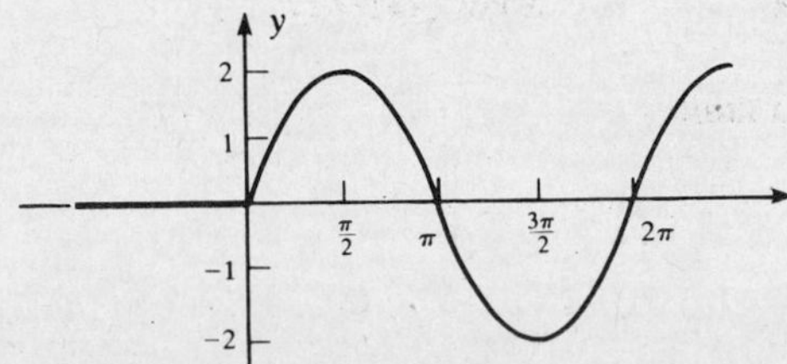

Fig. A-17

CHAPTER 28

28.8 See Fig. A-18.

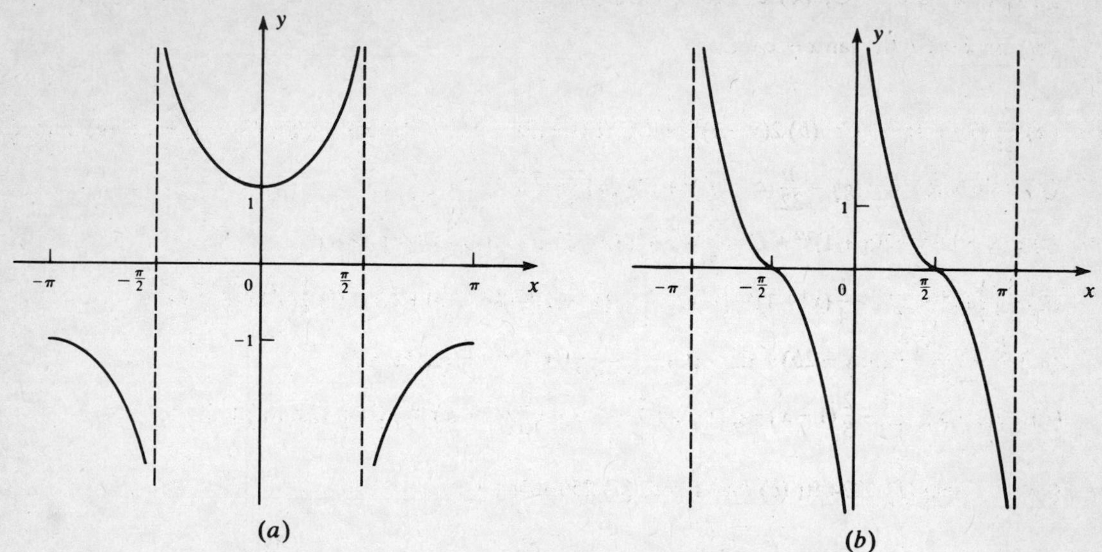

Fig. A-18

28.10 (a) $\sec^2 \frac{x}{2}$; (b) $(\sec x)(\sec x - \tan x)$; (c) $-2 \cot x \csc^2 x$; (d) $3 - 4 \csc^2 4x$; (e) $6 \sec^2 3x \tan 3x$; (f) $-3 \cot (3x-5) \csc (3x-5)$; (g) $-\frac{\cot \sqrt{x} \csc \sqrt{x}}{2\sqrt{x}}$.

28.11 (a) $y' = -\frac{y \sec^2 (xy)}{x \sec^2 (xy) - 1} = -\frac{y(y^2+1)}{x(y^2+1)-1}$; (b) $y' = \frac{\csc^2 x \cot x}{\sec^2 y \tan y}$; (c) $y' = \frac{3}{2} \frac{\cos x}{\tan (y+1) \sec^2 (y+1)}$;

(d) $y' = \frac{2 \tan (x+y) \sec^2 (x+y)}{1 - 2 \tan (x+y) \sec^2 (x+y)} = \frac{2(1+y) \tan (x+y)}{1 - 2(1+y) \tan (x+y)}$.

28.12 $y = 4\left(x - \frac{\pi}{3}\right) + \sqrt{3}$. **28.13** $y = -\frac{\sqrt{3}}{8}\left(x - \frac{\pi}{6}\right) + 4$. **28.14** (a) 1; (b) 8.

28.15 320 rad/h $= \frac{4}{45}$ rad/s $\approx 5°$ per second. **28.16** $\tan (\alpha_2 - \alpha_1) = 1.5$, $\alpha_2 - \alpha_1 \approx 56°$.

28.17 $\tan (\alpha_2 - \alpha_1) = 2$, $\alpha_2 - \alpha_1 \approx 63°$. **28.18** $\tan (\alpha_2 - \alpha_1) = 3$, $\alpha_2 - \alpha_1 \approx 71°$. **28.19** $y = \frac{4x - \pi}{8(\pi^2 + 1)}$.

28.20 (a) rel. max. at $\pi/4$, rel. min. at $3\pi/4$, vertical asymptote at $\pi/2$, inflection points at 0 and π; (b) rel. max. at $2\pi/3$, rel. min. at $\pi/3$, vertical asymptote at $\pi/2$, inflection points at 0 and π.

28.21 On all intervals where it is defined: $\left(-\frac{\pi}{2}, \frac{\pi}{2}\right)$, $\left(\frac{\pi}{2}, \frac{3\pi}{2}\right)$, etc.

CHAPTER 29

29.6 (a) $\frac{x^4}{2} - \frac{5}{3}x^3 + \frac{3}{2}x^2 + x + C$; (b) $5x - 2\sqrt{x} + C$; (c) $\frac{8}{5}x^{5/4} + C$; (d) $3x^{5/3} + C$; (e) $-\frac{1}{x^3} + C$; (f) $\frac{2}{7}x^{7/2} - \frac{2}{3}x^{3/2} + C = \frac{2}{21}x^{3/2}(3x^2 - 7) + C$; (g) $-\frac{1}{2x^2} + \frac{1}{4x^4} + C = \frac{1}{4x^4}(1 - 2x^2) + C$;

(*h*) $\frac{6}{5}x^{5/2} - \frac{4}{3}x^{3/2} + 2x^{1/2} + C = \frac{2}{15}\sqrt{x}(9x^2 - 10x + 15) + C$; (*i*) $-3\cos x + 5\sin x + C$;

(*j*) $7\tan x - \sec x + C$; (*k*) $x^3 - \cot x + C$; (*l*) $\frac{2\sqrt{3}}{5}x^{5/2} + C = \frac{2x^2\sqrt{3x}}{5} + C$;

(*m*) $\sin x + C$; (*n*) $\tan x - x + C$.

29.7 (*a*) $\frac{2}{21}(7x+4)^{3/2} + C$; (*b*) $2(x-1)^{1/2} + C$; (*c*) $\frac{1}{39}(3x-5)^{13} + C$; (*d*) $-\frac{1}{3}\cos(3x-1) + C$; (*e*) $2\tan\frac{x}{2} + C$;

(*f*) $2\sin\sqrt{x} + C$; (*g*) $-\frac{1}{32}(4-2t^2)^8 + C$; (*h*) $\frac{1}{4}(x^3+5)^{4/3} + C$;

(*i*) $\frac{2}{3}(x+1)^{3/2} - 2(x+1)^{1/2} + C = \frac{2}{3}\sqrt{x+1}\,(x-2) + C$; (*j*) $\frac{3}{5}(x-1)^{5/3} + C$;

(*k*) $\frac{3}{4}\left[\frac{1}{7}(x^4+1)^{7/3} - \frac{1}{4}(x^4+1)^{4/3}\right] + C = \frac{3}{112}(x^4+1)^{4/3}(4x^4-3) + C$; (*l*) $\frac{1}{5}\sqrt{1+5x^2} + C$;

(*m*) $\frac{2(ax+b)^{3/2}}{15a^2}(3ax-2b) + C$; (*n*) $-\frac{1}{3\sin 3x} + C = -\frac{1}{3}\csc 3x + C$;

(*o*) $-2(1-x)^{3/2}\left[\frac{1}{3} - \frac{2}{5}(1-x) + \frac{1}{7}(1-x)^2\right] + C = -\frac{2}{105}(1-x)^{3/2}(15x^2+12x+8) + C$.

29.9 (*a*) $t = 7$ sec; (*b*) 1024 ft; (*c*) $t = 15$ sec; (*d*) 256 ft/sec.

29.10 (*a*) $v = t^2 - 3t + 4$; (*b*) $x = \frac{t^3}{3} - \frac{3t^2}{2} + 4t$; (*c*) at $x = \frac{40}{3}$ when $t = 4$, at $x = -\frac{35}{6}$ when $t = -1$; (*d*) $-1 < t < 4$.

29.11 (*a*) $v = \frac{t^3}{3} - \frac{13}{3}t + 4$; (*b*) $x = \frac{t^4}{12} - \frac{13t^2}{6} + 4t$;

(*c*) at $x = \frac{23}{12}$ when $t = 1$, at $x = -\frac{3}{4}$ when $t = 3$, and at $x = -\frac{88}{3}$ when $t = -4$; (*d*) $t < -4$ and $1 < t < 3$.

29.12 (*a*) 160 ft/sec; (*b*) 400 ft. **29.13** (*a*) 3 seconds; (*b*) 99 feet. **29.14** 7 units.

29.15 (*a*) $y = \frac{2x^3}{3} - 5x - 1$; (*b*) $y = 6x^2 + x + 1$.

CHAPTER 30

30.4 (*a*) 24; (*b*) $\frac{5}{3}$; (*c*) $\frac{13}{3}$. **30.5** $A_2 - A_1 - A_3$. **30.6** (*c*) 20. **30.9** (*a*) π.

30.10 (*a*) 15; (*b*) 110; (*c*) $\frac{3+\sqrt{3}}{2}$; (*d*) 15.

CHAPTER 31

31.10 (*a*) 24; (*b*) $\frac{\sqrt{2}}{2}$; (*c*) $\sqrt{3}$; (*d*) $\frac{2046}{5}$; (*e*) $\frac{8\sqrt{5}-25}{2}$; (*f*) $\frac{5}{2}$.

31.11 (*a*) $\frac{\sqrt{3}-1}{2}$; (*b*) 27; (*c*) $\frac{9}{2}$; (*d*) $\frac{13}{3}$; (*e*) $\frac{26}{3}$; (*f*) $\frac{1}{3}$; (*g*) $\frac{35}{6}$.

31.12 (*a*) $\frac{1}{2}$; (*b*) $\frac{1}{2}$; (*c*) $\frac{8}{3}(1-2\sqrt{2})$; (*d*) $\frac{2}{3}(2\sqrt{2}-1)$; (*e*) $\frac{254}{7} - \frac{248}{5} + \frac{56}{3}$; (*f*) $\frac{2}{45}[(11)^3(383) - 256] = \frac{113\,226}{5}$;

(*g*) $\frac{349\,522}{7}$; (*h*) $\frac{1}{9}$; (*i*) $\frac{8}{3}$; (*j*) $\frac{5}{2}$; (*k*) $\frac{223}{160}$. **31.13** (*a*) $\frac{3}{4}$; (*b*) $\frac{4}{\pi}$; (*c*) $\frac{1}{3}$.

31.14 (*a*) $c = \frac{3}{2}$; (*b*) $c = \frac{1}{\sqrt[3]{4}}$; (*c*) $c = \sqrt{3}$. **31.15** (*a*) $\frac{857}{35}$; (*b*) $\frac{1}{6}$. **31.16** $\frac{\pi}{4}$.

31.17 $\frac{x_2 - x_1}{T}$. **31.18** (*a*) $\sqrt{5 + 7x^2}$; (*b*) $-\sin^3 x$; (*c*) $2\sqrt[3]{x^4 + 1}$. **31.19** 0.

31.20 (*a*) $6x\sqrt{243x^{10} + 1}$; (*b*) $f(h(x)) \cdot h'(x)$. **31.21** $b = \sqrt[n]{3}$. **31.22** 1. **31.23** 0.

31.24 (*a*) $4x$; (*b*) ± 2. **31.25** (*a*) 0; (*b*) $\frac{1}{2}$. **31.26** $b = 4$. **31.27** $\sqrt[3]{6}$.

31.28 All three are equal. **31.29** $c = \frac{\pi}{3}$.

31.30 All values such that x^k is integrable on $[0, 2]$; these include all positive values.

31.31 (*a*) $x = \frac{\sin 3t}{3}$; (*b*) $\frac{2}{3\pi}$; (*c*) $0 < t < \frac{\pi}{6}$; (*d*) $\frac{1}{3}$ and $-\frac{1}{3}$. **31.32** $\frac{2}{\pi}$.

31.33 4 m. **31.34** approximation $= \frac{101}{400} = 0.2525$, exact value $= 0.25$.

CHAPTER 32

32.7 (*a*) $\frac{52}{3}$ (Fig. A-19(*a*)); (*b*) $\frac{16}{3}$ (Fig. A-19(*b*)); (*c*) $\frac{5}{4}$ (Fig. A-19(*c*)).

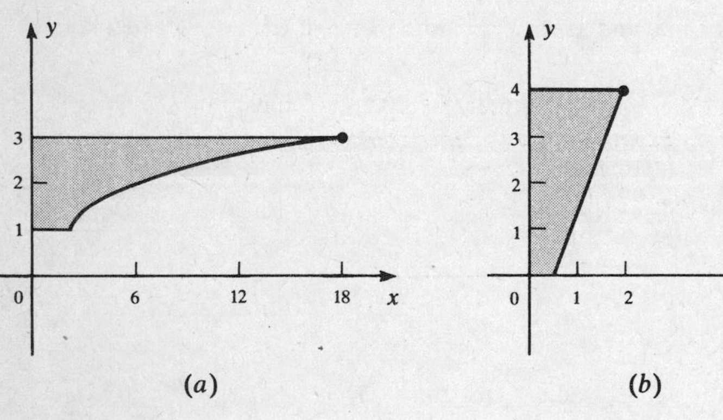

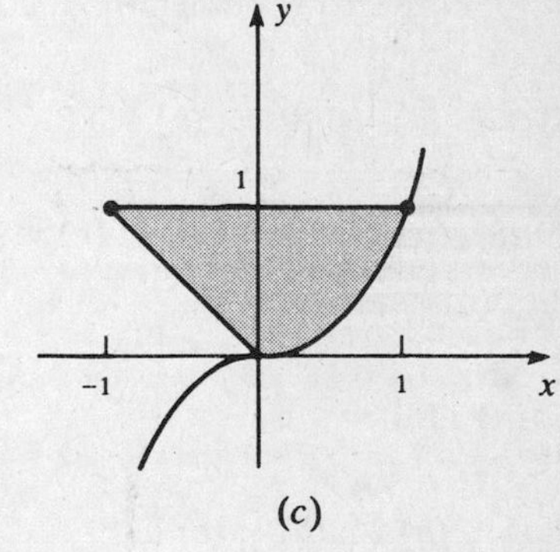

Fig. A-19

32.8 (*a*) $\frac{1}{12}$; (*b*) $\frac{1}{12}$ (*c*) $\frac{5}{3}$; (*d*) $\frac{1}{6}$; (*e*) $\sqrt{2} - 1$; (*f*) $\frac{125}{6}$; (*g*) $\frac{9}{2}$; (*h*) $\frac{5}{12}$; (*i*) $\frac{1}{8}$; (*j*) 9; (*k*) $\frac{125}{6}$; (*l*) $\frac{1}{3}$; (*m*) $\frac{2}{15}$.

32.9 (*a*) $\frac{33}{16}$; (*b*) $\sqrt{10}$; (*c*) $\frac{1}{27}(80\sqrt{10} - 13\sqrt{13})$; (*d*) 9; (*e*) $\frac{1097}{480}$.

CHAPTER 33

33.7 (*a*) $\frac{6\pi}{7}$; (*b*) $\frac{3\pi}{5}$; (*c*) $\frac{4\pi}{3}$; (*d*) $\frac{3\pi}{10}$; (*e*) $\frac{4\pi}{3}[r^3 - (r^2 - a^2)^{3/2}]$; (*f*) $\frac{2\pi}{3}(r^3 - a^3) - \pi a(r^2 - a^2)$; (*g*) $\frac{17}{15}(64\pi)$;
(*h*) $2\pi^2 ba^2$; (*i*) $\frac{2\pi}{3}$; (*j*) $\frac{27\pi}{5}$; (*k*) $\frac{27\pi}{2}$; (*l*) $\frac{2\pi}{3}$; (*m*) 4π.

33.8 (*a*) $\frac{4}{3}r^3$; (*b*) $\frac{2\sqrt{3}}{9}r^3$; (*c*) $\frac{1}{3}r^2 h$; (*d*) $\frac{1}{3}\left(\frac{1}{2}ab\right)c = \frac{1}{6}abc$.

CHAPTER 34

34.9 (*a*) $\frac{4}{4x-1}$; (*b*) $\frac{3(\ln x)^2}{x}$; (*c*) $\frac{1}{2x\sqrt{\ln x}}$; (*d*) $\frac{1}{x \ln x}$; (*e*) $x(1+2\ln x)$; (*f*) $\frac{2}{(x-1)(x+1)}$; (*g*) $\frac{5}{5x-2}$.

34.10 (*a*) $\frac{1}{3}\ln|x| + C$; (*b*) $\frac{1}{7}\ln|7x-2| + C$; (*c*) $\frac{1}{4}\ln|x^4-1| + C$; (*d*) $\ln|\sin x| + C$; (*e*) $\ln(\ln x) + C$;
(*f*) $-\frac{1}{2}\ln|1-\sin 2x| + C$; (*g*) $x^3 + 2\ln|x| + \frac{3}{2x^2} + C$; (*h*) $\ln|\tan x| + C$;
(*i*) $-2\ln|1-\sqrt{x}| + C = \ln\frac{1}{(1-\sqrt{x})^2} + C$; (*j*) $\frac{1}{2}(\ln x)^2 + C$.

34.11 (*a*) $y' = \frac{4x^2(3-x^2)}{\sqrt{4-x^2}}$; (*b*) $y' = \frac{(x-2)^4\sqrt[3]{x+5}}{\sqrt{x^2+4}}\left(\frac{4}{x-2} + \frac{1}{3(x+5)} - \frac{x}{x^2+4}\right)$;
(*c*) $y' = \frac{\sqrt{x^2-1}\sin x}{(2x+3)^4}\left(\frac{x}{x^2-1} + \cot x - \frac{8}{2x+3}\right)$; (*d*) $y' = -\frac{4}{3}\sqrt[3]{\frac{x+2}{x-2}}\,\frac{1}{(x+2)(x-2)}$.

34.13 (*a*) $\ln 2 + \ln 5$; (*b*) $-\ln 2$; (*c*) $-\ln 5$; (*d*) $2\ln 5$; (*e*) $\frac{1}{2}\ln 2$; (*f*) $\frac{1}{3}\ln 5$; (*g*) $-(2\ln 2 + \ln 5)$; (*h*) $12\ln 2$.

34.14 $y = x - 1$. **34.15** $\ln 2 - \frac{7}{24}$. **34.16** $\frac{2\ln 2}{3}$. **34.17** $2\pi \ln 2$. **34.18** See Fig. A-20.

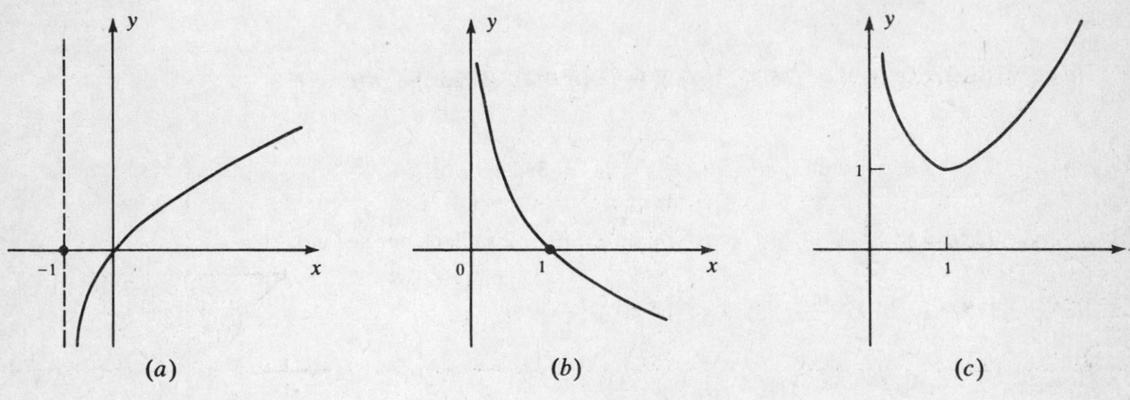

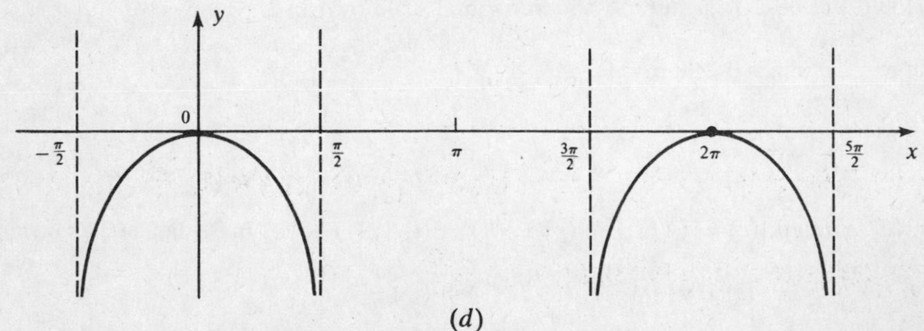

Fig. A-20

34.19 (*a*) $v = \frac{1}{2}t^2 - t + 6\ln|t| + 2$; (*b*) $\frac{67}{2} + 12\ln 3$.

34.20 (*a*) $y' = \frac{x}{y(x^2+y^2-1)}$; (*b*) $y' = \frac{y(2x+1)}{x(y-1)}$; (*c*) $y' = \frac{1}{y(3xy+3y^3-2)}$. **34.21** $\frac{1}{3}$.

34.23 0.6938 (to four decimals, $\ln 2 = 0.6931$).

CHAPTER 35

35.10 (*a*) $\frac{1}{x}$; (*b*) $-x$; (*c*) x^2; (*d*) $x^{1+\ln 3}$; (*e*) $x-1$; (*f*) $x-\ln x$.

35.11 (*a*) $-e^{-x}$; (*b*) $-\frac{e^{1/x}}{x^2}$; (*c*) $(-\sin x)e^{\cos x}$; (*d*) $e^x \sec^2 e^x$; (*e*) $\frac{e^x(x-1)}{x^2}$; (*f*) $e^x\left(\frac{1}{x}+\ln x\right)$; (*g*) $\pi x^{\pi-1}$; (*h*) $(\ln \pi)\pi^x$; (*i*) 2; (*j*) e^x+e^{-x}.

35.12 (*a*) $\frac{1}{3}e^{3x}+C$; (*b*) $-e^{-x}+C$; (*c*) $\frac{2}{3}(\sqrt{e^x-2})^3+C$; (*d*) $-e^{\cos x}+C$; (*e*) $\frac{1}{2\ln 3}3^{2x}+C$; (*f*) $2e^{x/2}+C$; (*g*) $\frac{x^{\pi+1}}{\pi+1}+C$; (*h*) $\frac{1}{3}e^{3x}+C$; (*i*) $e-\ln(e^x+1)+C$; (*j*) $\frac{2^{x^3}}{3\ln 2}+C$; (*k*) $-\frac{1}{4}e^{-x^4}+C$.

35.13 (*a*) $\frac{1}{x(e^y-1)}$; (*b*) $1+\frac{2x}{e^{y-x}\sec^2 e^{y-x}}=1+\frac{2x}{e^{y-x}(1+x^4)}$; (*c*) $\frac{2y^2}{y^2e^y-e^{1/y}}$; (*d*) $-\frac{2x+ye^{xy}}{2y+xe^{xy}}$; (*e*) $\cot x$.

35.14 (*a*) $(\ln 3)(\cos x)3^{\sin x}$; (*b*) $\frac{\ln 2}{2}e^x 2^{0.5e^x}$; (*c*) $(2\ln x)x^{(\ln x)-1}$; (*d*) $\frac{1}{x}[1+\ln(\ln x)](\ln x)^{\ln x}$; (*e*) $\frac{y}{2}\left(\frac{1}{x+1}+\frac{1}{x+2}\right)=\frac{2x+3}{2\sqrt{(x+1)(x+2)}}$.

35.15 (*a*) $\frac{\ln 2}{3}$; (*b*) $\frac{1}{\sqrt[3]{e}}$; (*c*) $\ln 2$; (*d*) e^e; (*e*) 2.

35.16 (*a*) $e-1$; (*b*) $\frac{\pi}{2}(e^2-1)$. **35.17** (*a*) 2; (*b*) $\pi(e^2+1)$. **35.18** $\pi(e-1)$.

35.19 $\max = e \quad \left(\text{at } x=\frac{\pi}{2}\right)$, $\min=\frac{1}{e} \quad \left(\text{at } x=-\frac{\pi}{2}\right)$. **35.20** $n^n e^{nx}$.

35.21 (*a*) $y'=(2\cos x)e^{\sin x}$, $y''=2e^{\sin x}(\cos^2 x-\sin x)$; (*b*) -2 radians per second.

35.22 (*a*) $v=3e^{3t}+1$; (*b*) $2+\frac{\ln 3}{3}$; (*c*) $x=e^{3t}+t-1$.

35.23 $y=2x+2$. **35.24** See Fig. A-21, page 331.

35.25 In Fig. 35-1(*a*) and Fig. A-21(*d*), multiply the horizontal scale by $1/\ln 2$.

35.28 $\max=e^{-1}$ (at $x=e$), $\min=0$ (at $x=1$).

CHAPTER 36

36.9 (*a*) $+\infty$; (*b*) 1; (*c*) 0; (*d*) 0; (*e*) -1; (*f*) $+\infty$; (*g*) 1; (*h*) $\frac{1}{2}$; (*i*) 1; (*j*) 1; (*k*) $\ln 3-\ln 2$; (*l*) $+\infty$; (*m*) 0; (*n*) 1; (*o*) 0; (*p*) e^3; (*q*) $+\infty$; (*r*) 0; (*s*) $1/\pi$; (*t*) 3/7; (*u*) 3; (*v*) 4; (*w*) $\frac{1}{2}$.

36.10 See Fig. A-22, page 332. **36.11** 8 times. **36.12** $80.

36.13 $1690\frac{\ln 10}{\ln 2}\approx 1690(3.32)\approx 5611$ years. **36.14** $e^{-(12\ln 2)/23}\approx 0.7$ g.

36.15 $T=\frac{\ln 2}{\ln 10-3\ln 2}\approx\frac{0.6931}{2.3026-2.0793}\approx 3.1$ years.

36.16 $\ln\left(1+\frac{r}{100}\right)$. **36.17** 69.31 days. **36.18** 3.125 g.

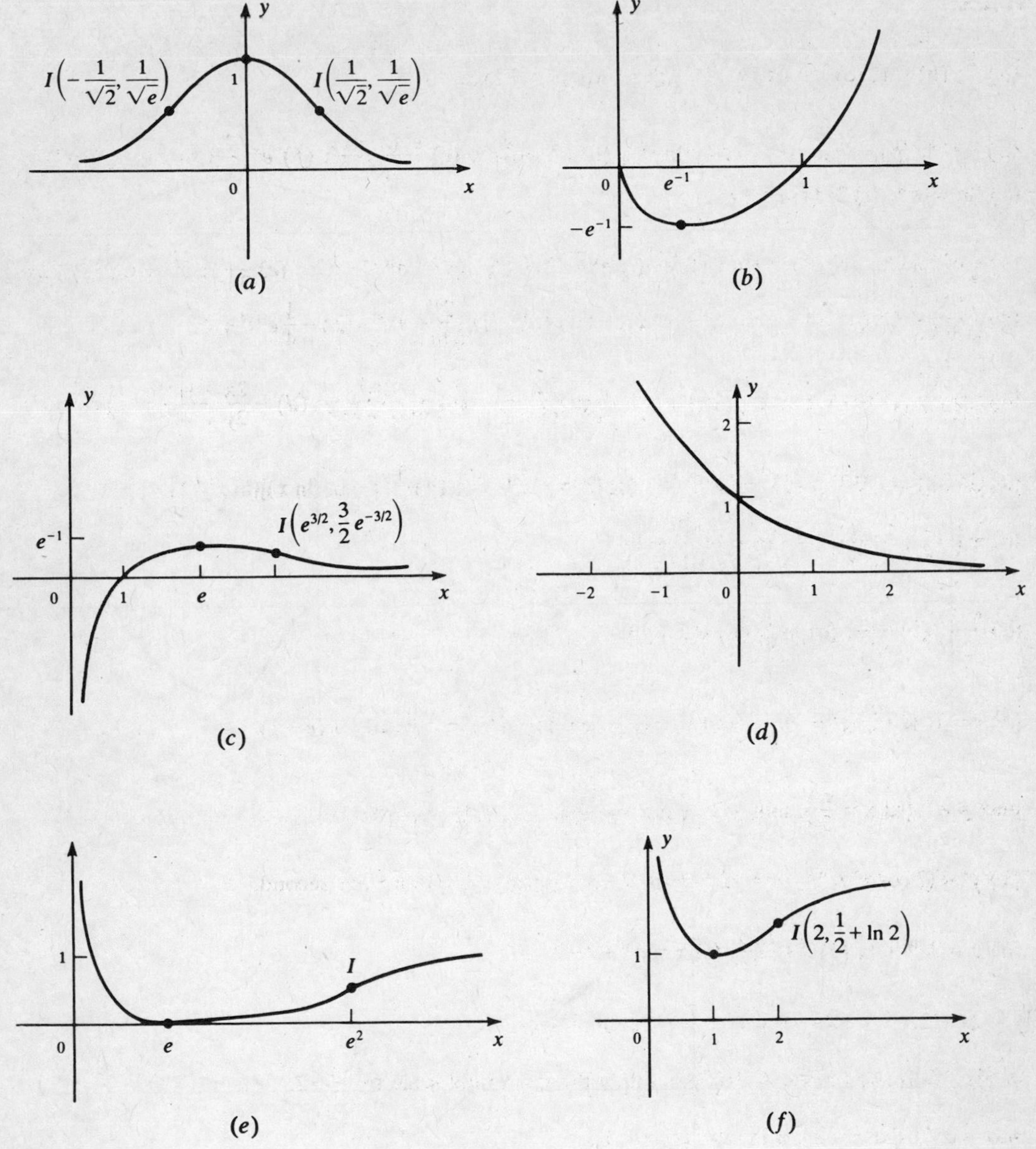

Fig. A-21

CHAPTER 37

37.12 (*a*) $f^{-1}(y) = 2y - 6$; (*c*) $f^{-1}(y) = \frac{5+2y}{3-y}$; (*e*) $f^{-1}(y) = \sqrt[5]{y} + 1$; (*f*) $f^{-1}(y) = \frac{1}{y} = f(y)$; (*g*) $f^{-1}(y) = \frac{5}{y-3}$;

(*h*) $f^{-1}(y) \quad \sqrt[3]{\frac{1}{y} - 4}$; (*i*) $f^{-1}(y) = \frac{y+2}{y-1}$.

37.13 (*a*) $\frac{5\pi}{6}$; (*b*) $-\frac{\pi}{4}$; (*c*) $\frac{\pi}{4}$; (*d*) $\frac{\pi}{6}$; (*e*) $\frac{\pi}{4}$; (*f*) $\frac{7\pi}{6}$; (*g*) $\frac{5\pi}{4}$; (*h*) $\frac{3\pi}{4}$.

37.14 (*a*) $\sin\theta = \frac{2\sqrt{2}}{3}$, $\tan\theta = 2\sqrt{2}$, $\cot\theta = \frac{\sqrt{2}}{4}$, $\sec\theta = 3$, $\csc\theta = \frac{3\sqrt{2}}{4}$;

(*b*) $\cos\theta = \frac{\sqrt{15}}{4}$, $\tan\theta = -\frac{\sqrt{15}}{15}$, $\cot\theta = -\sqrt{15}$, $\sec\theta = \frac{4\sqrt{15}}{15}$, $\csc\theta = -4$.

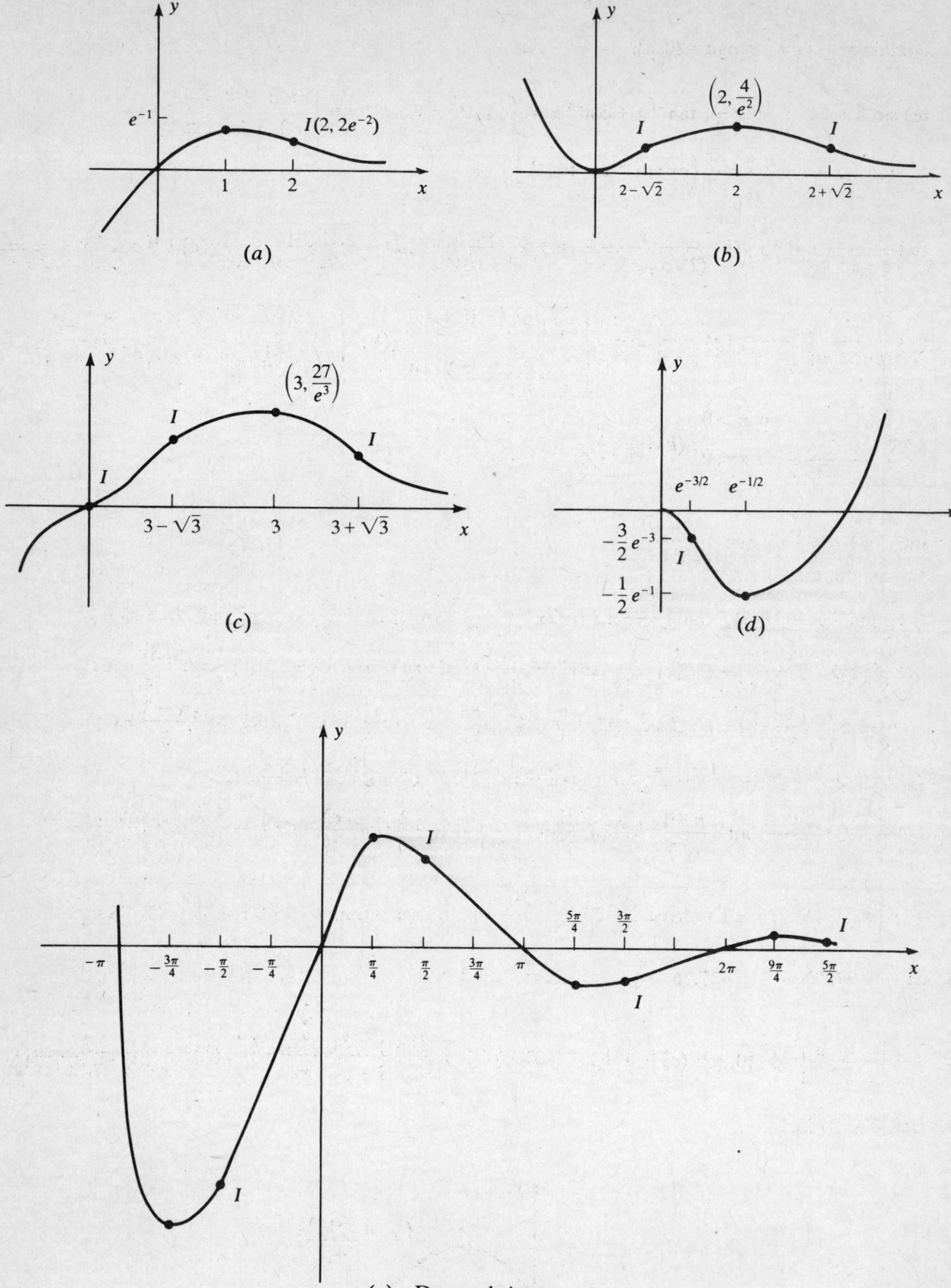

(e) Damped sine wave

Fig. A-22

37.15 $(a)\ \frac{3}{5}$; $(b)\ \frac{12}{5}$; $(c)\ \frac{2}{15}(2-3\sqrt{2})$; $(d)\ \frac{2\sqrt{5}}{25}(\sqrt{6}-1)$; (e) 0.

37.16 domain $= (-\infty, \infty)$, range $= (0, 1]$.

37.17 $(e)\ \sin^{-1}x + \cos^{-1}x = \frac{\pi}{2}$, $\tan^{-1}x + \cot^{-1}x = \frac{\pi}{2}$, $\sec^{-1}x + \csc^{-1}x = \begin{cases} 5\pi/2 & \text{for } x \le -1 \\ \pi/2 & \text{for } x \ge 1 \end{cases}$,

$\tan^{-1}x + \tan^{-1}\frac{1}{x} = \frac{\pi}{2}\left(\text{or } \cot^{-1}x = \tan^{-1}\frac{1}{x}\right)$.

37.18 $(a)\ \frac{x}{1+x^2} + \tan^{-1}x$; $(b)\ \frac{1}{(2\sqrt{x})\sqrt{1-x}}$; $(c)\ -\frac{\sin x}{1+\cos^2 x}$; $(d)\ \frac{-3}{(1+9x^2)\cot^{-1}3x}$; $(e)\ e^x\left(\cos^{-1}x - \frac{1}{\sqrt{1-x^2}}\right)$;

$(f)\ \frac{1}{(\tan^{-1}x)(1+x^2)}$; $(g)\ \frac{\sqrt{x^2}}{x\sqrt{1-x^2}} = \begin{cases} \frac{1}{\sqrt{1-x^2}} & \text{if } x > 0 \\ \frac{-1}{\sqrt{1-x^2}} & \text{if } x < 0 \end{cases}$; $(h)\ \begin{cases} 2\sqrt{a^2-x^2} & \text{if } a > 0 \\ -\frac{2x^2}{\sqrt{a^2-x^2}} & \text{if } a < 0 \end{cases}$; $(i)\ \frac{1}{1+x^2}$;

$(j)\ \begin{cases} 0 & \text{if } x > 0 \\ \frac{2}{x\sqrt{x^2-1}} & \text{if } x < 0 \end{cases}$; $(k)\ -\frac{2}{x^2+4}$.

37.19 $\tan^{-1}u + \tan^{-1}v = \tan^{-1}\frac{u+v}{1-uv}$.

37.20 $(a)\ \frac{1}{2}\tan^{-1}\frac{x}{2} + C$; $(b)\ \frac{1}{6}\tan^{-1}\frac{3x}{2} + C$; $(c)\ \sin^{-1}\frac{x}{5} + C$; $(d)\ \frac{1}{4}\sin^{-1}\frac{4x}{5} + C$; $(e)\ \sec^{-1}(x-3) + C$;

$(f)\ \frac{1}{\sqrt{2}}\sin^{-1}\left(\frac{\sqrt{6}\,x}{3}\right) + C$; $(g)\ \frac{1}{\sqrt{14}}\tan^{-1}\left(\frac{\sqrt{14}\,x}{2}\right) + C$; $(h)\ \frac{1}{2}\sec^{-1}\frac{x}{2} + C$; $(i)\ \frac{1}{4}\sec^{-1}\frac{3x}{4} + C$;

$(j)\ \frac{1}{\sqrt{2}}\sec^{-1}\left(\frac{\sqrt{6}\,x}{2}\right) + C$; $(k)\ 2\tan^{-1}\sqrt{x} + C$; $(l)\ \frac{1}{6}\sec^{-1}\frac{\sqrt{x^4+9}}{3} + C$; $(m)\ \sin^{-1}\frac{x-3}{3} + C$;

$(n)\ 2\left(3\sin^{-1}\frac{x-3}{3} - \sqrt{6x-x^2}\right) + C$; $(o)\ \frac{1}{2}\ln(x^2+8x+20) - 2\tan^{-1}\frac{x+4}{2} + C$;

$(p)\ \frac{x^2}{2} + 2x - \frac{8\sqrt{3}}{3}\tan^{-1}\frac{x-1}{\sqrt{3}} + C$; $(q)\ \frac{1}{2}\sin^{-1}\frac{x^2}{2} + C$; $(r)\ \frac{1}{2}\tan^{-1}\frac{e^x}{2} + C$; $(s)\ \frac{1}{\sqrt{5}}\tan^{-1}\left(\frac{\sin x}{\sqrt{5}}\right) + C$.

37.21 $y = \frac{1}{3}x$. **37.22** -1 rad/sec. **37.23** $\frac{250\pi}{3}$ miles per minute $= 5000\pi$ mph $\approx 15\,708$ mph.

37.24 $\frac{\pi}{4}$. **37.25** $\frac{\pi}{6}$. **37.26** $\frac{\pi^2}{3}$. **37.28** $2\sqrt{2}$ ft.

37.29 $(a)\ \left[-\frac{\pi}{2}, \frac{\pi}{2}\right]$; $(b)\ [0, \pi]$; $(c)\ [-1, 1]$. **37.30** $(a)\ \frac{y(1+y^2)(2x - \tan^{-1}y)}{1+y^2+xy}$; $(b)\ -\frac{y}{x+2e^{2y}\sqrt{1-x^2y^2}}$.

37.31 See Fig. A-23.

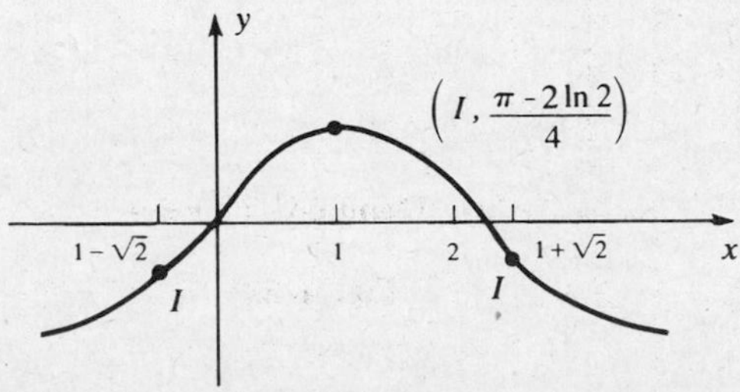

Fig. A-23

CHAPTER 38

38.6 (*a*) $-e^{-x}(x^2+2x+2)+C$; (*b*) $\frac{e^x(\sin x-\cos x)}{2}+C$; (*c*) $e^x(x^3-3x^2+6x-6)+C$;
(*d*) $x\sin^{-1}x+\sqrt{1-x^2}+C$; (*e*) $x\sin x+\cos x+C$; (*f*) $2x\sin x+(\cos x)(2-x^2)+C$;
(*g*) $\frac{x}{2}[\cos(\ln x)+\sin(\ln x)]+C$; (*h*) $\frac{1}{25}[5x\sin(5x-1)+\cos(5x-1)]+C$;
(*i*) $\frac{e^{ax}}{a^2+b^2}(b\sin bx+a\cos bx)+C$; (*j*) $\frac{1}{2}(x-\sin x\cos x)+C$;
(*k*) $\sin x-\frac{\sin^3 x}{3}+C=\frac{\sin x}{3}(2+\cos^2 x)+C$; (*l*) $\frac{1}{4}\left(\frac{3}{2}x+\sin 2x+\frac{1}{8}\sin 4x\right)+C$; (*m*) $\frac{e^{3x}}{9}(3x-1)+C$;
(*n*) $x\tan x-\ln|\sec x|+C$; (*o*) $\frac{1}{8}(2x^2+2x\sin 2x-\cos 2x)+C$; (*p*) $x(\ln x)^2-2x(\ln x-1)+C$;
(*q*) $\frac{1}{4}(\sin 2x-2x\cos 2x)+C$; (*r*) $-\frac{1}{2}\cos x^2+C$; (*s*) $-\frac{1}{x}(1+\ln x)+C$; (*t*) $\frac{e^{3x}}{27}(9x^2-6x+2)+C$;
(*u*) $\frac{1}{6}(2x^3\tan^{-1}x-x^2+\ln(1+x^2))+C$; (*v*) $x\ln(x^2+1)-2(x-\tan^{-1}x)+C$; (*w*) $\frac{\sqrt{1+x^2}}{3}(x^2-2)+C$;
(*x*) $\frac{x^3}{9}(3\ln x-1)+C$.

38.7 (*a*) 1; (*b*) (i) $\pi(e-2)$, (ii) $\frac{\pi}{2}(1+e^2)$.

38.8 (*a*) $\frac{1}{2}$; (*b*) 2π; (*c*) $\pi\left(2-\frac{5}{e}\right)$. **38.10** (*a*) $\frac{\pi^2}{2}$; (*b*) $2\pi^2$. **38.11** (*a*) 0; (*b*) $-\frac{2\pi}{n}$.

38.12 (*a*) $\int\cos^n x\,dx=\frac{\cos^{n-1}x\sin x}{n}+\frac{n-1}{n}\int\cos^{n-2}x\,dx$;
(*b*) $\int\sin^n x\,dx=-\frac{\sin^{n-1}x\cos x}{n}+\frac{n-1}{n}\int\sin^{n-2}x\,dx$.

38.13 (*a*) $\int\sec^n x\,dx=\frac{\sec^{n-2}x\tan x}{n-1}+\frac{n-2}{n-1}\int\sec^{n-2}x\,dx$;
(*b*) (i) $\frac{1}{2}(\sec x\tan x+\ln|\sec x+\tan x|)+C$, (ii) $\frac{\sec^2 x\tan x}{3}+\frac{2}{3}\tan x+C=(\tan x)\left(1+\frac{\tan^2 x}{3}\right)+C$.

CHAPTER 39

39.12 (*a*) $-\frac{1}{3}\cos^3 x+C$; (*b*) $\frac{1}{2}x+\frac{\sin 6x}{12}+C$; (*c*) $\frac{1}{5}\sin^5 x-\frac{2}{7}\sin^7 x+\frac{1}{9}\sin^9 x+C$;
(*d*) $\frac{1}{8}\left(\frac{5x}{2}+2\sin 2x+\frac{3}{8}\sin 4x-\frac{1}{6}\sin^3 2x\right)+C$; (*e*) $\frac{1}{16}\left(\frac{5}{8}x-\frac{\sin 4x}{8}+\frac{\sin^3 2x}{3}-\frac{\sin 8x}{64}\right)+C$;
(*f*) $2\tan\frac{x}{2}-x+C$; (*g*) $\frac{1}{5}\tan^5 x-\frac{1}{3}\tan^3 x+\tan x-x+C$;
(*h*) $\frac{1}{4}\sec^3 x\tan x+\frac{3}{8}\sec x\tan x+\frac{3}{8}\ln|\sec x+\tan x|+C$; (*i*) $\frac{1}{5}\tan^5 x+\frac{1}{3}\tan^3 x+C$;
(*j*) $\frac{1}{5}\sec^5 x-\frac{1}{3}\sec^3 x+C$; (*k*) $\frac{1}{4}\sec^3 x\tan x-\frac{5}{8}\sec x\tan x+\frac{3}{8}\ln|\sec x+\tan x|+C$; (*l*) $-\frac{1}{8}\cos 4x+C$;
(*m*) $\frac{1}{8\pi}(2\cos 2\pi x-\cos 4\pi x)+C$; (*n*) $\frac{1}{24}(6\sin 2x-\sin 12x)+C$; (*o*) $\frac{1}{2}\left(\frac{\sin 5x}{5}+\frac{\sin 13x}{13}\right)+C$.

39.14 (*a*) 0; (*b*) π.

39.15 (*a*) $\sqrt{x^2-1}-\sec^{-1}x+C$; (*b*) $2\sin^{-1}\frac{x}{2}-\frac{x}{2}\sqrt{4-x^2}+C$; (*c*) $\sqrt{1+x^2}+\ln\frac{\sqrt{1+x^2}-1}{|x|}+C$;
(*d*) $-\sqrt{2-x^2}+C$; (*e*) $\frac{1}{9}\frac{\sqrt{x^2-9}}{x}+C$; (*f*) $\frac{x}{4\sqrt{4-x^2}}+C$; (*g*) $\frac{1}{54}\left(\tan^{-1}\frac{x}{3}+\frac{3x}{x^2+9}\right)+C$;

$(h)\ \frac{1}{4}\ln\frac{4-\sqrt{16-9x^2}}{|3x|}+C=\frac{1}{4}\ln\frac{4-\sqrt{16-9x^2}}{|x|}+K$; $(i)\ \frac{1}{8}(\sin^{-1}x-x(1-2x^2)\sqrt{1-x^2})+C$;
$(j)\ \frac{1}{8}(\sin^{-1}e^x-e^x(1-2e^{2x})\sqrt{1-e^{2x}})+C$; $(k)\ \frac{1}{16}\left(\tan^{-1}\frac{x-3}{2}-\frac{2(x-3)}{x^2-6x+13}\right)+C$.

39.16 $\sqrt{17}+\frac{1}{4}\ln(\sqrt{17}+4)$. **39.17** $\sqrt{e^2+1}-(1+\sqrt{2})+\ln\frac{\sqrt{e^2+1}-1}{\sqrt{2}-1}$.

39.18 Same answer as to Problem 39.17 (because the two arcs are mirror images in the line $y=x$).

39.19 $\ln(2+\sqrt{3})$.

39.20 $\pi\sqrt{9}\sqrt{4}=6\pi$.

CHAPTER 40

40.6 $(a)\ \frac{1}{6}\ln\left|\frac{x-3}{x+3}\right|+C$; $(b)\ 3\ln|x+3|-2\ln|x+2|+C$; $(c)\ \frac{x^3}{3}+\frac{1}{4}\ln|x+2|+\frac{3}{4}\ln|x-2|+C$;
$(d)\ \frac{3}{2}\ln|x-1|-9\ln|x-2|+\frac{19}{2}\ln|x-3|+C$; $(e)\ \frac{3}{4}\ln|x-1|-\frac{3}{8}\ln|x+1|+\frac{5}{8}\ln|x-3|+C$;
$(f)\ -\frac{1}{6}\ln|x|+\frac{13}{6}\ln|x+3|-\frac{7}{6}\ln|x+2|+\frac{1}{6}\ln|x-1|+C$; $(g)\ \frac{1}{10}\ln\left|\frac{x^2-9}{x^2-4}\right|+C$;
$(h)\ 6\ln\left|\frac{x}{x+1}\right|+\frac{5}{x}+C$; $(i)\ \frac{1}{4}\ln\left|\frac{x-2}{x+2}\right|-\frac{1}{x-2}+C$; $(j)\ \frac{4}{9}\ln\left|\frac{x}{x+3}\right|+\frac{1}{3}\frac{1}{x+3}+C$;
$(k)\ \frac{x^2}{2}+2x+\frac{1}{25}\left(266\ln|x-4|+9\ln|x+1|+\frac{55}{x+1}\right)+C$; $(l)\ \frac{1}{10}\ln\left(\frac{x^2}{x^2+5}\right)+C$;
$(m)\ \frac{1}{10}\ln|x-1|+\frac{9}{20}\ln(x^2+4x+5)-\frac{13}{10}\tan^{-1}(x+2)+C$; $(n)\ \frac{1}{3}\tan^{-1}x-\frac{1}{6}\tan^{-1}\frac{x}{2}+C$;
$(o)\ \frac{x^2}{2}+\frac{1}{9}(\ln|x|-41\ln(x^2+9))+C$; $(p)\ \ln|x|-\frac{1}{2}\ln(x^2+1)+\frac{1}{2}\frac{1}{x^2+1}+C$;
$(q)\ \frac{1}{25}\ln|x-1|-\frac{1}{50}\ln(x^2+4)+\frac{3}{100}\tan^{-1}\frac{x}{2}+\frac{1}{10}\frac{x-4}{x^2+4}+C$;
$(r)\ \ln|x|-\frac{1}{2}\ln(x^2+x+1)-\frac{\sqrt{3}}{9}\tan^{-1}\left(\frac{2x+1}{\sqrt{3}}\right)-\frac{2}{3}\frac{x-1}{x^2+x+1}+C$; $(s)\ \ln\left|\frac{x+1}{x+2}\right|+C$;
$(t)\ \frac{1}{3}\ln|x|+\frac{11}{3}\ln|x+3|-3\ln|x+2|+C$; $(u)\ x-\ln(1+e^x)+C$.

40.8 $(a)\ \frac{1}{81}(3\ln 2+\pi\sqrt{3})$; $(b)\ \frac{1}{27}(\pi\sqrt{3}-3\ln 2)$.

40.9 $\frac{2}{1-\tan\frac{x}{2}}+C$.

40.10 $(a)\ \ln(1-\sin x)+C$; $(b)\ \ln|\sin x-1|+C$; $(c)\ \sin x\le 1$.

40.11 $(a)\ -\frac{3}{2}\ln|1-x^{2/3}|+C$; $(b)\ 4\left[\frac{(\sqrt[4]{x-1})^3}{3}-\frac{\sqrt{x-1}}{2}+\sqrt[4]{x-1}-\ln(1+\sqrt[4]{x-1})\right]+C$;
$(c)\ \ln\left|\frac{\sqrt{1+3x}-1}{\sqrt{1+3x}+1}\right|+C$; $(d)\ 2\sqrt{x}-3\sqrt[3]{x}+6\sqrt[6]{x}-\ln(1+\sqrt[6]{x})+C$; $(e)\ \frac{4}{3}(\sqrt{x}-2)\sqrt{\sqrt{x}+1}+C$;
$(f)\ 2\sqrt{1+e^x}+\ln\left(\frac{\sqrt{1+e^x}-1}{\sqrt{1+e^x}+1}\right)+C=2\sqrt{1+e^x}+2\ln(\sqrt{1+e^x}-1)-x+C$;
$(g)\ \frac{3}{2}x^{2/3}+\ln|x^{1/3}+1|-\frac{1}{2}\ln\left|\frac{x+1}{x^{1/3}+1}\right|-\sqrt{3}\tan^{-1}\left(\frac{2x^{1/3}-1}{\sqrt{3}}\right)+C$.

Appendix A

Trigonometric Formulas

$\cos^2\theta + \sin^2\theta = 1$

$\cos(\theta + 2\pi) = \cos\theta$

$\sin(\theta + 2\pi) = \sin\theta$

$\cos(-\theta) = \cos\theta$

$\sin(-\theta) = -\sin\theta$

$\cos(u + v) = \cos u \cos v - \sin u \sin v$

$\cos(u - v) = \cos u \cos v + \sin u \sin v$

$\sin(u + v) = \sin u \cos v + \cos u \sin v$

$\sin(u - v) = \sin u \cos v - \cos u \sin v$

$\sin 2\theta = 2\sin\theta\cos\theta$

$$\cos 2\theta = \cos^2\theta - \sin^2\theta$$
$$= 2\cos^2\theta - 1 = 1 - 2\sin^2\theta$$

$$\cos^2\frac{\theta}{2} = \frac{1+\cos\theta}{2}$$

$$\sin^2\frac{\theta}{2} = \frac{1-\cos\theta}{2}$$

$$\tan x = \frac{\sin x}{\cos x} = \frac{1}{\cot x}$$

$$\cot x = \frac{\cos x}{\sin x} = \frac{1}{\tan x}$$

$$\sec x = \frac{1}{\cos x}$$

$$\csc x = \frac{1}{\sin x}$$

$\tan(-x) = -\tan x$

$\tan(x + \pi) = \tan x$

$1 + \tan^2 x = \sec^2 x$

$1 + \cot^2 x = \csc^2 x$

$$\tan(u + v) = \frac{\tan u + \tan v}{1 - \tan u \tan v}$$

$$\tan(u - v) = \frac{\tan u - \tan v}{1 + \tan u \tan v}$$

$$\cos\left(\frac{\pi}{2} - \theta\right) = \sin\theta;\ \sin(\pi - \theta) = \sin\theta;\ \sin(\theta + \pi) = -\sin\theta$$

$$\sin\left(\frac{\pi}{2} - \theta\right) = \cos\theta;\ \cos(\pi - \theta) = -\cos\theta;\ \cos(\theta + \pi) = -\cos\theta$$

Law of Cosines: $c^2 = a^2 + b^2 - 2ab\cos\theta$

Law of Sines: $\dfrac{\sin A}{a} = \dfrac{\sin B}{b} = \dfrac{\sin C}{c}$

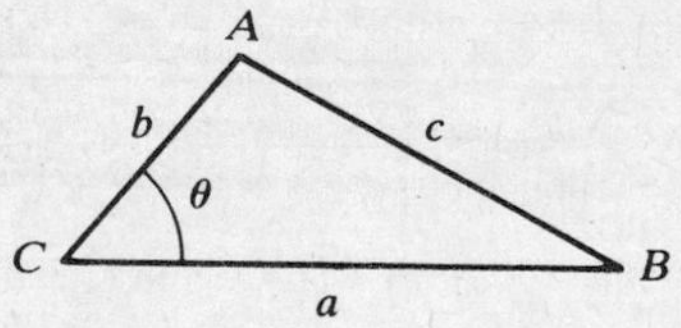

A
b
c
θ
C
B
a

Appendix B

Basic Integration Formulas

$$\int a\,dx = ax + C$$

$$\int x^r\,dx = \frac{x^{r+1}}{r+1} + C \qquad (r \neq -1)$$

$$\int \frac{1}{x}\,dx = \ln|x| + C$$

$$\int \frac{dx}{a^2+x^2} = \frac{1}{a}\tan^{-1}\frac{x}{a} + C$$

$$\int \frac{dx}{\sqrt{a^2-x^2}} = \sin^{-1}\frac{x}{a} + C \qquad (a>0)$$

$$\int \frac{dx}{x\sqrt{x^2-a^2}} = \frac{1}{a}\sec^{-1}\frac{x}{a} + C \qquad (a>0)$$

$$\int e^x\,dx = e^x + C$$

$$\int a^x\,dx = \frac{a^x}{\ln a} + C \qquad (a>0)$$

$$\int xe^x\,dx = e^x(x-1) + C$$

$$\int \ln x\,dx = x\ln x - x + C$$

$$\int \sin x\,dx = -\cos x + C$$

$$\int \cos x\,dx = \sin x + C$$

$$\int \tan x\,dx = \ln|\sec x| + C$$

$$\int \sec x\,dx = \ln|\sec x + \tan x| + C$$

$$\int \cot x\,dx = \ln|\sin x| + C$$

$$\int \csc x\,dx = \ln|\csc x - \cot x| + C$$

$$\int \sec^2 x\,dx = \tan x + C$$

$$\int \csc^2 x\,dx = -\cot x + C$$

$$\int \sec x \tan x\,dx = \sec x + C$$

$$\int \csc x \cot x\,dx = -\csc x + C$$

$$\int \sin^2 x\,dx = \frac{x}{2} - \frac{\sin 2x}{4} + C$$

$$\int \cos^2 x\,dx = \frac{x}{2} + \frac{\sin 2x}{4} + C$$

$$\int \tan^2 x\,dx = \tan x - x + C$$

Appendix C

Geometric Formulas

(A = area, C = circumference, V = volume, S = lateral surface area)

Triangle

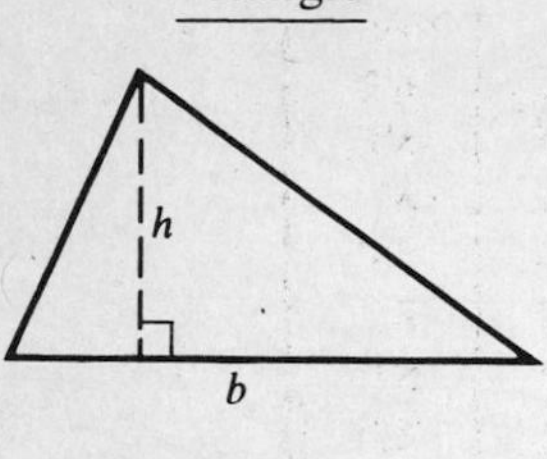

$$A = \frac{1}{2} bh$$

Trapezoid

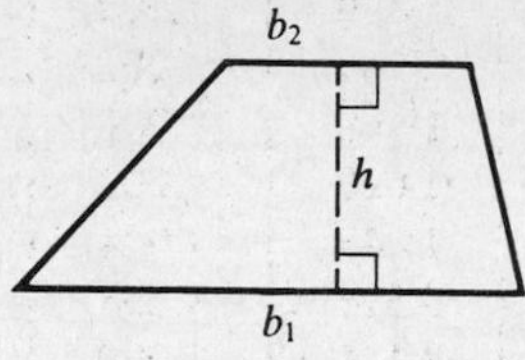

$$A = \frac{1}{2}(b_1 + b_2)h$$

Parallelogram

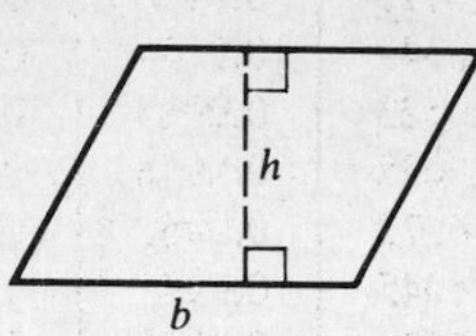

$$A = bh$$

Circle

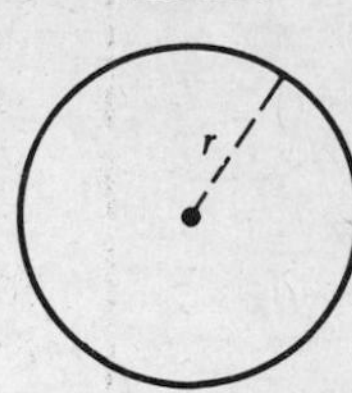

$$A = \pi r^2$$
$$C = 2\pi r$$

Sphere

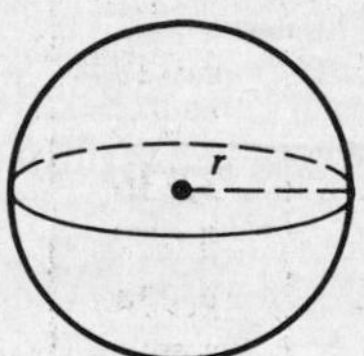

$$V = \frac{4}{3}\pi r^3$$
$$S = 4\pi r^2$$

Cylinder

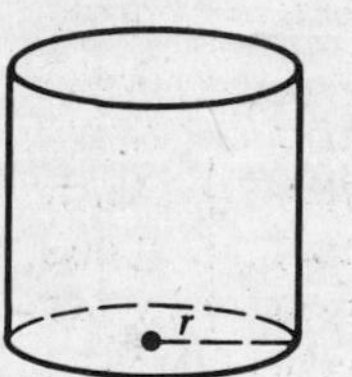

$$V = \pi r^2 h$$
$$S = 2\pi rh$$

Cone

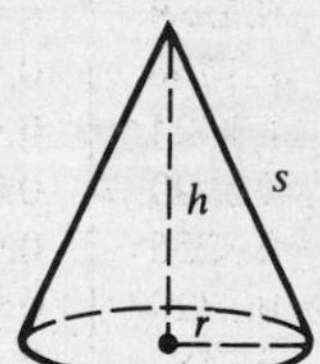

$$V = \frac{1}{3}\pi r^2 h$$
$$S = \pi rs = \pi r\sqrt{r^2 + h^2}$$

Appendix D

Trigonometric Functions

	sin	cos	tan	cot	sec	csc	
0°	0.0000	1.0000	0.0000	....	1.000	....	90°
1°	0.0175	0.9998	0.0175	57.29	1.000	57.30	89°
2°	0.0349	0.9994	0.0349	28.64	1.001	28.65	88°
3°	0.0523	0.9986	0.0524	19.08	1.001	19.11	87°
4°	0.0698	0.9976	0.0699	14.30	1.002	14.34	86°
5°	0.0872	0.9962	0.0875	11.43	1.004	11.47	85°
6°	0.1045	0.9945	0.1051	9.514	1.006	9.567	84°
7°	0.1219	0.9925	0.1228	8.144	1.008	8.206	83°
8°	0.1392	0.9903	0.1405	7.115	1.010	7.185	82°
9°	0.1564	0.9877	0.1584	6.314	1.012	6.392	81°
10°	0.1736	0.9848	0.1763	5.671	1.015	5.759	80°
11°	0.1908	0.9816	0.1944	5.145	1.019	5.241	79°
12°	0.2079	0.9781	0.2126	4.705	1.022	4.810	78°
13°	0.2250	0.9744	0.2309	4.331	1.026	4.445	77°
14°	0.2419	0.9703	0.2493	4.011	1.031	4.134	76°
15°	0.2588	0.9659	0.2679	3.732	1.035	3.864	75°
16°	0.2756	0.9613	0.2867	3.487	1.040	3.628	74°
17°	0.2924	0.9563	0.3057	3.271	1.046	3.420	73°
18°	0.3090	0.9511	0.3249	3.078	1.051	3.236	72°
19°	0.3256	0.9455	0.3443	2.904	1.058	3.072	71°
20°	0.3420	0.9397	0.3640	2.747	1.064	2.924	70°
21°	0.3584	0.9336	0.3839	2.605	1.071	2.790	69°
22°	0.3746	0.9272	0.4040	2.475	1.079	2.669	68°
23°	0.3907	0.9205	0.4245	2.356	1.086	2.559	67°
24°	0.4067	0.9135	0.4452	2.246	1.095	2.459	66°
25°	0.4226	0.9063	0.4663	2.145	1.103	2.366	65°
26°	0.4384	0.8988	0.4877	2.050	1.113	2.281	64°
27°	0.4540	0.8910	0.5095	1.963	1.122	2.203	63°
28°	0.4695	0.8829	0.5317	1.881	1.133	2.130	62°
29°	0.4848	0.8746	0.5543	1.804	1.143	2.063	61°
30°	0.5000	0.8660	0.5774	1.732	1.155	2.000	60°
31°	0.5150	0.8572	0.6009	1.664	1.167	1.942	59°
32°	0.5299	0.8480	0.6249	1.600	1.179	1.887	58°
33°	0.5446	0.8387	0.6494	1.540	1.192	1.836	57°
34°	0.5592	0.8290	0.6745	1.483	1.206	1.788	56°
35°	0.5736	0.8192	0.7002	1.428	1.221	1.743	55°
36°	0.5878	0.8090	0.7265	1.376	1.236	1.701	54°
37°	0.6018	0.7986	0.7536	1.327	1.252	1.662	53°
38°	0.6157	0.7880	0.7813	1.280	1.269	1.624	52°
39°	0.6293	0.7771	0.8098	1.235	1.287	1.589	51°
40°	0.6428	0.7660	0.8391	1.192	1.305	1.556	50°
41°	0.6561	0.7547	0.8693	1.150	1.325	1.524	49°
42°	0.6691	0.7431	0.9004	1.111	1.346	1.494	48°
43°	0.6820	0.7314	0.9325	1.072	1.367	1.466	47°
44°	0.6947	0.7193	0.9657	1.036	1.390	1.440	46°
45°	0.7071	0.7071	1.000	1.000	1.414	1.414	45°
	cos	sin	cot	tan	csc	sec	

Appendix E

Natural Logarithms

n	$\ln n$	n	$\ln n$	n	$\ln n$	n	$\ln n$
0.0	—	3.4	1.2238	6.8	1.9169	11	2.3979
0.1	−2.3026	3.5	1.2528	6.9	1.9315	12	2.4849
0.2	−1.6094	3.6	1.2809	7.0	1.9459	13	2.5649
0.3	−1.2040	3.7	1.3083	7.1	1.9601	14	2.6391
0.4	−0.9163	3.8	1.3350	7.2	1.9741	15	2.7081
0.5	−0.6931	3.9	1.3610	7.3	1.9879	16	2.7726
0.6	−0.5108	4.0	1.3863	7.4	2.0015	17	2.8332
0.7	−0.3567	4.1	1.4110	7.5	2.0149	18	2.8904
0.8	−0.2231	4.2	1.4351	7.6	2.0281	19	2.9444
0.9	−0.1054	4.3	1.4586	7.7	2.0142	20	2.9957
1.0	0.0000	4.4	1.4816	7.8	2.0541	25	3.2189
1.1	0.0953	4.5	1.5041	7.9	2.0669	30	3.4012
1.2	0.1823	4.6	1.5261	8.0	2.0794	35	3.5553
1.3	0.2624	4.7	1.5476	8.1	2.0919	40	3.6889
1.4	0.3365	4.8	1.5686	8.2	2.1041	45	3.8067
1.5	0.4055	4.9	1.5892	8.3	2.1163	50	3.9120
1.6	0.4700	5.0	1.6094	8.4	2.1282	55	4.0073
1.7	0.5306	5.1	1.6292	8.5	2.1401	60	4.0943
1.8	0.5878	5.2	1.6487	8.6	2.1518	65	4.1744
1.9	0.6419	5.3	1.6677	8.7	2.1633	70	4.2485
2.0	0.6931	5.4	1.6864	8.8	2.1748	75	4.3175
2.1	0.7419	5.5	1.7047	8.9	2.1861	80	4.3820
2.2	0.7885	5.6	1.7228	9.0	2.1972	85	4.4427
2.3	0.8329	5.7	1.7405	9.1	2.2083	90	4.4998
2.4	0.8755	5.8	1.7579	9.2	2.2192	95	4.5539
2.5	0.9163	5.9	1.7750	9.3	2.2300	100	4.6052
2.6	0.9555	6.0	1.7918	9.4	2.2407	200	5.2983
2.7	0.9933	6.1	1.8083	9.5	2.2513	300	5.7038
2.8	1.0296	6.2	1.8245	9.6	2.2618	400	5.9915
2.9	1.0647	6.3	1.8405	9.7	2.2721	500	6.2146
3.0	1.0986	6.4	1.8563	9.8	2.2824	600	6.3069
3.1	1.1314	6.5	1.8718	9.9	2.2925	700	6.5511
3.2	1.1632	6.6	1.8871	10	2.3026	800	6.6846
3.3	1.1939	6.7	1.9021			900	6.8024

Appendix F

Exponential Functions

x	e^x	e^{-x}	x	e^x	e^{-x}
0.00	1.0000	1.0000	2.5	12.182	0.0821
0.05	1.0513	0.9512	2.6	13.464	0.0743
0.10	1.1052	0.9048	2.7	14.880	0.0672
0.15	1.1618	0.8607	2.8	16.445	0.0608
0.20	1.2214	0.8187	2.9	18.174	0.0550
0.25	1.2840	0.7788	3.0	20.086	0.0498
0.30	1.3499	0.7408	3.1	22.198	0.0450
0.35	1.4191	0.7047	3.2	24.533	0.0408
0.40	1.4918	0.6703	3.3	27.113	0.0369
0.45	1.5683	0.6376	3.4	29.964	0.0334
0.50	1.6487	0.6065	3.5	33.115	0.0302
0.55	1.7333	0.5769	3.6	36.598	0.0273
0.60	1.8221	0.5488	3.7	40.447	0.0247
0.65	1.9155	0.5220	3.8	44.701	0.0224
0.70	2.0138	0.4966	3.9	49.402	0.0202
0.75	2.1170	0.4724	4.0	54.598	0.0183
0.80	2.2255	0.4493	4.1	60.340	0.0166
0.85	2.3396	0.4274	4.2	66.686	0.0150
0.90	2.4596	0.4066	4.3	73.700	0.0136
0.95	2.5857	0.3867	4.4	81.451	0.0123
1.0	2.7183	0.3679	4.5	90.017	0.0111
1.1	3.0042	0.3329	4.6	99.484	0.0101
1.2	3.3201	0.3012	4.7	109.95	0.0091
1.3	3.6693	0.2725	4.8	121.51	0.0082
1.4	4.0552	0.2466	4.9	134.29	0.0074
1.5	4.4817	0.2231	5	148.41	0.0067
1.6	4.9530	0.2019	6	403.43	0.0025
1.7	5.4739	0.1827	7	1096.6	0.0009
1.8	6.0496	0.1653	8	2981.0	0.0003
1.9	6.6859	0.1496	9	8103.1	0.0001
2.0	7.3891	0.1353	10	22026	0.00005
2.1	8.1662	0.1225			
2.2	9.0250	0.1108			
2.3	9.9742	0.1003			
2.4	11.023	0.0907			

Index